Birkhäuser

Mathematics and the Built Environment
Volume 1

Series editor

Kim Williams, Kim Williams Books, Torino, Italy

More information about this series at http://www.springer.com/series/15181

Michael J. Ostwald · Josephine Vaughan

The Fractal Dimension of Architecture

Birkhäuser

Michael J. Ostwald
School of Architecture and Built
Environment
The University of Newcastle
Newcastle, NSW
Australia

Josephine Vaughan
School of Architecture and Built
Environment
The University of Newcastle
Newcastle, NSW
Australia

Mathematics and the Built Environment
ISBN 978-3-319-32424-1 ISBN 978-3-319-32426-5 (eBook)
DOI 10.1007/978-3-319-32426-5

Library of Congress Control Number: 2016942907

Printed on acid-free paper

This book is published under the trade name Birkhäuser.
The registered company is Springer International Publishing AG Switzerland
(www.birkhauser-science.com)

Preface

This book describes a unique way of measuring, analysing and comparing buildings using fractal dimensions. A fractal dimension is a mathematical determination of the typical or characteristic level of complexity in an image or object. Thus, fractal dimensions provide a rigorous measure of the extent to which an object, say a building, is relatively simple, plain or smooth at one extreme, or complex, jagged and rough at the other.

After introducing the method for calculating fractal dimensions in Part I of the book, Part II presents the results of a major study of the plans and elevations of eighty-five canonical houses designed or constructed between 1901 and 2007. The houses include works by Le Corbusier, Eileen Gray, Mies van der Rohe, Frank Lloyd Wright, Robert Venturi, Denise Scott Brown, Frank Gehry, Peter Eisenman, John Hejduk, Richard Meier, Kazuyo Sejima, Ryue Nishizawa, Yoshiharu Tsukamoto, Momoyo Kajima, Glenn Murcutt and Peter Stutchbury. The eighty-five houses are measured to examine trends in individual designer's works, across different stylistic movements and over more than a century of shifting social patterns and aesthetic tastes. These trends are encapsulated in a series of three hypotheses which are proposed in the introduction and examined in the book's conclusion.

In addition to the results of this overarching study, five specific arguments about architecture are also tested using mathematical evidence. The first of these is concerned with the way the formal expression of modernist architecture is allegedly shaped in response to its orientation and address. The second examines claims about the changing visual experience of walking through one of Frank Lloyd Wright's houses and the third is about the extent to which façade permeability (the presence of windows and doors) shapes the formal expression of a building. The fourth of these studies examines arguments about frontality and rotation in the early domestic architecture of Eisenman, Hejduk and Meier. The fifth and final study investigates the degree to which Murcutt's architecture is shaped by either literal or phenomenal transparency. These secondary studies all use variants of the fractal analysis method that are attuned to testing specific architectural properties.

As a result of this combined approach—a primary overarching study and five secondary studies—this book does not possess a neat, singular conclusion about architecture that can be summarised in a paragraph. Instead, the newly developed measures are used to illuminate a large number of beliefs about design, including arguments pertaining to changing trends in planning and expression and the extent to which different stylistic movements are visually differentiable from each other. Furthermore, the data are used to distinguish between diverse approaches to spatial planning, form-making and architectural expression. Thus, the majority of the results of this research are presented sequentially, at the end of specific sections and chapters.

To give some context to the research, this book has been written for people with backgrounds in architecture, urban design, interior design and design computing. It has also been written and framed in such a way that it is accessible to postgraduate students, as well as to professionals and academics. For this reason, the level of mathematics used is relatively general and only basic statistical methods are employed. This descriptive approach has been taken to the data because, with no detailed inferences being drawn from it about the relationship between the designs studied here and the larger body of domestic architecture produced in the same period, there is no need for more complex statistical analysis. For the same reason, the mathematical results are typically analysed using the critical-interpretative techniques of design theorists and historians. Thus, a common approach in this book is to use numbers, charts and simple statistical measures (average, median, standard deviation) in parallel with scholarly arguments, to reach a reasoned conclusion about an issue. More mathematically inclined readers are invited to undertake their own analysis of the data or follow links to our other publications which contain more detailed results. Similarly, design theorists and historians are free to interpret the results in their own terms or read our papers, cited in the text, which offer a more intricate interpretation of the philosophical basis for some of this material.

At this juncture, it is also useful to provide a note about authorship and how we will refer to our past research. For much of the last ten years, we—Michael Ostwald and Josephine Vaughan—have jointly published our research into applications of computational fractal analysis in design. Across twenty-five co-authored papers and chapters, we have gradually developed and refined the theory and practice of fractal analysis for architectural and urban applications. The intellectual content of the present book is shaped by these publications, a few of which have been substantially revised and expanded for inclusion here. However, prior to this time Michael Ostwald separately published a large body of research on philosophical, theoretical and historical connections between architecture, non-linear mathematics and fractal geometry. Furthermore, he also worked closely with several other co-authors on this early research. For this reason, throughout the present book we will refer to past research published by Michael alone, or in partnership with other colleagues, in the third person. In contrast, we will tend to refer to our joint research in the first person, and in this way hope to remain clear about authorship.

The software used for the majority of the calculations in this book is called *ArchImage*. We developed and refined this software with the support of colleagues from computer science and software engineering at the University of Newcastle (see the Acknowledgments section for full details). *ArchImage's* basic properties are described in Chap. 6 and it is available for download through the authors' websites.

In this book, we present the results of our mathematical analysis of more than 625 reconstructed architectural plans and elevations and over 200 specially prepared views of famous buildings. Using software that has been specially authored for this project, over five million separate pieces of data were extracted from these images and subjected to over 9000 mathematical operations to measure the dimensional properties of eighty-five designs. To the best of our knowledge, this is the largest mathematical study ever undertaken into architectural design and the largest single application of fractal analysis in any field. We hope that through this research the reader will be inspired to think about architecture—its history, theory and analysis—in a new way.

Newcastle, Australia
2016

Michael J. Ostwald
Josephine Vaughan

Acknowledgements

Several past and present colleagues have contributed to the development of ideas contained in this book. In particular, we wish to thank Stephan Chalup, Steven Nicklin and Chris Tucker who worked with us on stages of this research and made valuable contributions to it. We are also indebted to the ideas of Carl Bovill who published important early research in this field. Special thanks also to Anna Mätzener and Sarah Goob (Birkhäuser, Basel), Thomas Hempfling (Springer, Basel) and to series editor for *Mathematics for the Built Environment*, Kim Williams.

ArchImage software was used for the majority of the calculations in this book. Naomi Henderson authored the prototype version of this software with Michael Ostwald and Stephan Chalup. Steven Nicklin wrote the final version of *ArchImage* with Stephan Chalup and ourselves. In addition, our research has also been ably assisted by the efforts of Michael Dawes, Maria Roberts and Ian Owen, along with Romi McPherson, Lachlan Seegers, Jasmine Richardson, Raeana Henderson and Kelly Campbell. The Australian Research Council (ARC) supported this project through the award of a Discovery Grant (DP1094154) and a Future Fellowship (FT0991309).

Some sections of this book are derived from material that was previously published in journals and chapters and has been substantially revised, expanded or updated for the present work. Specifically, in Chap. 3, the worked examples were initially developed by Michael J. Ostwald and Michael Dawes, and the first of these was previously presented as part of: Özgür Ediz and Michael J. Ostwald, 2012. 'The Süleymaniye Mosque', *ARQ*, 16(2). Chapter 4 is a revised and expanded version of: Michael J. Ostwald and Josephine Vaughan, 2013. 'Representing Architecture for Fractal Analysis', *Architectural Science Review*, 56(3). Chapter 5 includes revised sections and results from two previously published papers: Michael J. Ostwald, 2013. 'The Fractal Analysis of Architecture', *Environment and Planning B*, 40; and Michael J. Ostwald and Josephine Vaughan, 2013. 'Limits and Errors', *ArS: Architectural Science Research,* 7. In Chap. 7, the background section and part of the additional application were adapted from, respectively: Michael J. Ostwald and

Josephine Vaughan, 2011. 'The Mathematics of Domestic Modernism (1922–1934)', *Design Principles and Practices*, 4(6); and Josephine Vaughan and Michael J. Ostwald, 2009. 'A Quantitative Comparison between the Formal Complexity of Le Corbusier's Pre-Modern (1905–1912) and Early Modern (1922–1928) Architecture', *Design Principles and Practices*, 3(4). Chapter 8 includes cases presented in preliminary form in the following: Josephine Vaughan and Michael J. Ostwald, 2011. 'The Relationship Between the Fractal Dimension of Plans and Elevations in the Architecture of Frank Lloyd Wright', *ArS: Architectural Science Research,* 4; and Michael J. Ostwald and Josephine Vaughan, 2010. 'The Mathematics of Style in the Architecture of Frank Lloyd Wright', in Paul S. Geller (ed). *Built Environment: Design, Management and Applications*, Nova: New York. In Chap. 10, some of the project descriptions were adapted from the following: Michael J. Ostwald and Josephine Vaughan, 2013. 'Differentiating the Whites', *Empirical Studies in the Arts*, 31(1). Finally, the additional methodological application in Chap. 11 was developed from the following: Josephine Vaughan and Michael J. Ostwald, 2015. 'Measuring the Significance of Façade Transparency in Australian Regionalist Architecture', *Architectural Science Review.* Full details of these publications are contained in the references. We gratefully acknowledge the advice and support of referees and editors involved in the production of these works.

Contents

About the Authors

Michael J. Ostwald is a Professor and Dean of architecture at the University of Newcastle (Australia) and a visiting professor at RMIT University. He has previously been a professorial research fellow at Victoria University Wellington, an Australian Research Council (ARC) Future Fellow at Newcastle and a visiting fellow at ANU, MIT and UCLA, amongst other institutions. Michael has a Ph.D. in architectural history and theory and a D.Sc. in design mathematics and computing. Under the auspices of the Byera Hadley scholarship, he completed postdoctoral research on baroque geometry at the CCA (Montreal) and at the Loeb Archives (Harvard). He is a co-editor-in-chief of the *Nexus Network Journal: Architecture and Mathematics* and on the editorial boards of *ARQ* and *Architectural Theory Review*. He has authored more than 400 scholarly publications and his architectural designs have been published and exhibited internationally. Michael J. Ostwald is a co-editor with Kim Williams of the landmark two-volume reference work, *Architecture and Mathematics from Antiquity to the Future* (Birkhäuser, 2015).

Josephine Vaughan is a research academic at the University of Newcastle (Australia). Josephine has undergraduate and graduate qualifications in architecture and design and after running her own practice for almost a decade, she is currently completing a Ph.D. Josephine has authored twenty-six research papers with Michael Ostwald on the topic of computational fractal and dimensional analysis. She has been on the organising committees for international conferences on architectural science and design computing and has chaired conference sessions on these topics and regularly reviews research in this field for international journals.

Chapter 1
Introduction

This book presents the results of an investigation of eighty-five houses that have been designed by some of the world's most respected architects and have since become enshrined in the history of twentieth-century architecture. These houses include, amongst many others, Le Corbusier's *Villa Savoye*, Wright's *Robie House*, Gray's *E.1027*, Mies's *Edith Farnsworth House*, Venturi's *Vanna Venturi House*, Meier's *Douglas House* and Murcutt's *Marie Short House*. The majority of these eighty-five designs have been repeatedly published and analysed by scholars; they have been studied by students and used as precedents by practitioners. These designs serve as benchmarks against which other works are tested and, as such, they have an enduring presence in the historiographical landscape of architecture. However, given that they are already so well known, what is to be gained by revisiting them?

Like the majority of designs that have been identified by historians as canonical works, the eighty-five houses analysed in this book are understood almost exclusively in *qualitative* terms. That is, the properties that make them special or significant are documented and communicated using textual descriptions, supplemented by photographic or graphic media. The value of the visual media is assumed to be self-evident in such cases, leaving the descriptive text with the burden of providing the reader with an understanding of these designs. Such texts are invariably presented using a combination of comparative and denotative terms. Thus, these designs are characterised by historians and scholars as having 'less ornamentation', 'bigger windows', 'denser planning structures', 'industrial detailing', 'richer iconography', 'stronger horizontal lines' and 'more articulated social structures'. They are 'more richly textured', 'starkly geometrical', 'tectonically conservative' and 'phenomenally enlivened'. These examples are typical of the qualitative descriptions used to explain the characteristics of architecture and the significance of these buildings in a larger historical context. Variations of these phrases are repeated in almost every major architectural reference work. They represent a combination of professional judgment, informed personal opinion and received wisdom. There is nothing intrinsically wrong with this way of constructing

M.J. Ostwald and J. Vaughan, *The Fractal Dimension of Architecture*,
Mathematics and the Built Environment 1, DOI 10.1007/978-3-319-32426-5_1

the history and theory of architecture, but there are valuable alternative approaches that can be used to question the traditional classification of these buildings and promote a new way of understanding them. One such alternative method has been chosen for the present book, giving it a distinctive starting point from which to selectively rethink the properties of some of the world's most famous buildings.

1.1 Rationale and Aims

This book uses a *quantitative*, mathematical and computational approach to understand the properties of the eighty-five designs and how they relate to each other. It commences by viewing architecture as a special type of dimensional data, which can be measured in individual designs, and then compared with other works of the same era, by the same architect and even across more than a century of architectural production. However, while this starting point deliberately avoids adopting the normative interpretations of these famous works, it is not productive to examine architecture solely using numbers. Buildings serve human functions, they enable critical social structures and they embody cultural values. Architecture is not just space and form divorced from purpose, geography or human aspiration. Therefore, this book provides a social and historical context for all of the designs it examines, prior to undertaking a new mathematical analysis of each. Once the analysis is complete, the resultant data is then interpreted in terms of both simple statistical patterns and accepted historical and theoretical readings. In this way the book shifts between qualitative and quantitative approaches, using the former to ground or frame the research and the latter to give it a unique lens through which to study architecture.

While the designs we have chosen to analyse are ones which we believe will be of relevance to the majority of readers, it might be argued that the very act of starting such a project with a large selection of famous buildings—which have already been identified by historians as significant—undermines the purity of the mathematical approach. Certainly, if the purpose of this book was to analyse, for example, housing more generally, then a much larger sample, excluding architect-designed free-standing dwellings, would be more useful. The statistical trends of such a study would then be more critical and confidence in the data more important to demonstrate. The results of such a study would also be more tangible and straightforward. Nevertheless, as previously stated, the buildings analysed in this book are not representative of all housing, but they are representative of the way we understand the stylistic, aesthetic, tectonic and functional expression of the built environment. The landmark works chosen for the present study cast a long shadow over society, its tastes, practices and dreams. Many of these buildings have been catalysts for the design of hundreds, if not thousands of variations on these themes, which have in turn shaped the lives of millions of people around the world. The significance of this study is therefore largely about the way we can illuminate alternative perspectives on these landmark designs.

The mathematical and computational method used throughout this book measures the fractal dimension of an architectural plan, elevation or other representation of a design. A fractal dimension is a rigorous measure of the relative density and diversity of geometric information in an image or object. This property, which is described as either 'characteristic complexity' or 'statistical roughness', is simply a determination of the amount (meaning volume) and distribution (meaning how it is spread over many scales) of geometry in a form. In architectural terms, it could be seen as a mathematical calculation of the extent to which lines, regardless of their purpose, are both present in, and dispersed across, an elevation or plan.

The method for calculating the fractal dimension of an object was first proposed by mathematicians in the 1980s, and was originally known as the 'box-counting' approach. While there are different ways of measuring fractal dimensions, the box-counting variant is the most well known, stable and repeatable, and thus, over time, it has become synonymous with 'fractal analysis'. Many hundreds of scientific and medical studies have been published using variations of the fractal analysis method to measure and compare complex objects, but it is still, despite important past research, poorly understood by architectural scholars and almost completely unknown amongst design students and practitioners. Part of the reason for this situation is that these other fields (engineering, biology, astronomy, geology and medicine) have had a long-term interest in measuring the properties of complex objects and have thus developed stable versions of the method. However, in architecture and design, despite progress in the 1990s, the most accurate and useful variant has only recently been identified. For this reason, Part I of this book contains a clear description of the process of using fractal dimensions to measure architecture, along with a demonstration of its application, a review of its methodological variables and a discussion of its limits. This first part concludes with the presentation of a refined and optimised version of the fractal analysis method for use in architecture. Once the method and its application are explained, it is then used to calculate the fractal dimensions of the plans and elevations of the eighty-five designs. All of these designs are first individually measured, before being compared within sets, then within stylistic movements, and finally across more than a century of architectural design practice. To facilitate the drawing together of these different scales of results, three primary hypotheses have been framed for testing in the conclusion and a further five secondary hypotheses are also tested as an integral part of individual chapters. These primary and secondary hypotheses are described in the following two sections.

1.2 Primary Hypotheses

The method used throughout this book offers a rare opportunity to investigate three different arguments about larger scale issues associated with design. These three issues are concerned with changing social patterns in design, the capacity to distinguish between different stylistic movements in purely aesthetic terms, and the idea that designers possess singular approaches to form-making which can be identified and used to differentiate between them. These three arguments were not the catalyst for this research, but they are contentious issues that can be usefully examined using a large-scale quantitative study of this type. Therefore, three hypotheses have been framed for testing using the complete set of data.

The first hypothesis holds that *as the complexity of social groupings and functions contained within the home has reduced over time, the fractal dimensions of plans and elevations should decrease to reflect this change*. This hypothesis is derived from a growing number of observations about the relationship between social structures and the formal properties of plans and elevations which have been shaped by these structures (Gutman 1972; Hillier and Hanson 1984; Kent 1990; Karlen 2009; Steinfeld and White 2010). Such research suggests that the geometric character of a plan is a reflection of the social relationships that either the plan has been created to accommodate, or modified to support (Hillier 1996). This implies that, for example, houses with larger, more complex plans have been created to accommodate extended families, regular guests or multiple servants (Hanson 1998). In contrast, plans that are less cellular, more open or of reduced scale, serve the needs of smaller families with a less rigid separation of functions (Ostwald 2011a; Ostwald and Dawes 2013a). Similarly, the external expression of a house is at least partially shaped by the degree to which it requires permeability (doors, windows, porticos and the like) and this, in turn, is potentially an indicator of the social structure of its interior (Moore et al. 1974; Blumenson 1979; Ostwald and Vaughan 2011). Therefore, if the modelling of a building's plans and elevations are at least partially shaped by the intricacy of their underlying social structures, then over time the visual complexity of these plans and elevations should reduce in parallel with the gradual simplification of the social structures of Western domesticity. With a time span of over one hundred years, even taking into account the fact that these are all architect-designed houses and therefore tend to be larger and less conventional, a trend reflecting this hypothesis should be observable in the chronological data developed in this book.

The second hypothesis states that in architecture *each stylistic genre or movement possesses a distinct visual character that is measurable using fractal dimensions*. The general thrust of this hypothesis is that the aesthetic expression of a style is, at least in part, formally derived. This is a common argument in architectural theory and critique, one which can be found in the earliest architectural treatises and has been repeated to the present day (Blumenson 1979; Haneman 1984; Calloway and Cromley 1996; Burden 2000; Lewis 2010; Jones 2014). It is also a concept that has been discussed in research that uses fractal analysis to investigate architecture. For example, Wen and Kao (2005) argue that architectural style is visually differentiable using the fractal dimensions of elevations and plans, an idea that was partially rejected by ourselves (Ostwald et al. 2008) but has been the subject of on-going discussion (Bovill 1996; Ostwald and Vaughan 2009b; Ostwald and Vaughan 2010; Lorenz 2011). The reason there is continued speculation about this proposition is that past research has lacked both the sample size and the methodological consistency to usefully test it. The data collected in the present book has both the scale and consistency to finally examine this proposition in an authoritative way. Furthermore, this present book is an ideal setting for testing this hypothesis because the set of buildings considered here are all ones that have been used by historians to classify and differentiate stylistic movements.

The final of the primary hypotheses states that *individual architects will present distinctive patterns of three-dimensional formal and spatial measures across multiple designs*. This hypothesis is not about the visual expression of a larger movement (the topic of the second hypothesis) but of the works of individual designers or partnerships. This view is so pervasive in twentieth-century architectural history and theory that it is rarely explicitly stated in such a way, but it is the basis on which many critical texts have been constructed (Jencks 1977; Baker 1995; Kent 1990; Laseau and Tice 1992; Clark and Pause 2012). Variations of this proposition can also be found in past research using fractal analysis or generation in architecture (Bovill 1996; Eaton 1998; Harris 2007). The suggestion in such research is that architects who have had the opportunity to develop a mature approach to design will tend to produce plans and elevations that might separately have distinctive visual characteristics, but in combination reveal the unique three-dimensional signature of the designer. This too is a question that is well suited to the content of the present book.

1.3 Secondary Hypotheses

In addition to using fractal dimensions to analyse eighty-five buildings, this book also introduces five advanced variations of the method and demonstrates their application to test specific claims about architects, designs or movements. To demonstrate these variations it is useful to frame their applications around the testing of a series of hypotheses. Therefore, five secondary hypotheses have been identified, the first of which is framed in response to the Modernist dictum, 'form follows function'. This hypothesis maintains that *the form of a functional façade is necessarily shaped by a combination of its orientation and address* (being the difference between the public and private elevations of a building). A common argument in architectural design is that in a building where form is shaped by function, the level of detail in a façade should shift in response to changing environmental and programmatic conditions (Feininger 1956; Grillo 1975; Jones 1992; Box 2007; Frederick 2007). The most basic of these environmental conditions is orientation, leading to the expectation that the levels of formal modelling present in each façade should change in response to the direction it is facing (Moore et al. 1974; Leatherbarrow 2000; Ching 2007). A closely related argument is that of address, being founded on the assumption that a functional building will express its public and private façades differently (Venturi 1966; King 2005). If the hypothesis is true, then a particular pattern of fractal dimensions should be visible across a set of functionalist façades, in response to both orientation and address.

The second of these hypotheses is inspired by a series of detailed, phenomenal descriptions of the changing visual experience of movement through one of Frank Lloyd Wright's houses. This hypothesis states that *the degree of visual complexity observed while moving into and through Wright's Robie House will, on average, reduce from beginning to end*. While there are several related arguments about the experience of Wright's architecture, Hildebrand's (1991) proposition, which draws on the work of Jay Appleton (1975, 1988), suggests that the experience of visual complexity and associated feelings of confusion will reduce across the length of the path from the entrance to the living room. One of the major cases Hildebrand uses to support this argument is derived from the experience of Wright's *Robie House*.

The third of these secondary hypotheses is inspired by the arguments of Frampton (1975) and Rowe (1996), who both called for a rejection of elevational 'frontality' in architecture or practice wherein buildings were designed to be appreciated from a single, dominant viewpoint. Instead of frontality, Frampton and Rowe called for architects to adopt a strategy of 'rotation' or '*contrapposto*', an approach wherein the sculptural modelling of a building embraces multiple viewing perspectives. As part of his explanation of the significance of this change, Frampton effectively argues that *the designs of Eisenman, Hejduk and Meier show a more consistent formal expression when viewed from multiple perspectives than from a single perspective*. While Frampton, Rowe and others have offered a more nuanced reading of how this property of rotation works for each of these architects, this general claim provides the fourth hypothesis.

Just as the first of these hypotheses was concerned with the relationship between the formal expression of a façade and its environmental function, so too is the fourth. However, this time the motivation for the hypothesis is drawn from the famous, and some would say facetious, postmodern classification of buildings as either 'ducks' or 'decorated sheds' (Venturi et al. 1972). The former category, the duck, describes a building that has been elaborately shaped or modelled to express its function. The latter, the shed, is a building which uses simple motifs and signs that are often a direct extension of its most banal functional attributes, to communicate its intent. Prior to proposing the theory of ducks and decorated sheds, Venturi (1966) suggested a simpler variation wherein he argued that two related factors—formal modelling and permeability—shape and thereby assist in differentiating, basic design intent. This earlier proposition is the impetus for the fourth hypothesis which states that *a comparison between the number of openings in a façade, and the geometric modelling of that façade, will reveal the extent to which a building is dominated by form or function*.

For the fifth and final secondary hypothesis, the significance of transparency in Australian Regionalist architecture is considered. Façade transparency is allegedly an important part of the character of Australian Regionalist architecture in general and of the architecture of Glenn Murcutt in particular. However, the various accounts of transparency in Murcutt's architecture fail to state whether this property is a result of the limpidity of the materials chosen, or is associated with the sense that a person can pass through and experience Murcutt's architecture. For the purpose of examining this issue, Colin Rowe and Robert Slutzky's (1963) differentiation of two types or architectural transparency is adopted as a starting point. This final hypothesis holds that *transparency plays a critical role in the visual expression of Murcutt's architecture*, and that *this transparency is literal rather than phenomenological*.

1.4 What Is a Fractal?

The word 'fractal' is derived from the Latin word *frangere*, meaning to break or fragment (Mandelbrot 1975). In mathematics the word 'fraction' is derived from the Latin *fractus*, which is the past participle of *frangere*. A fraction is both a value produced by dividing one number into another, and a fragment of a larger whole. The meaning of the word fractal is drawn from both the original Latin and the mathematical variant. In conventional use, the word fractal is used in two contexts, the first to describe a type of irregular dimensionality and the second an infinitely deep geometric set. In order to understand what a fractal dimension is, and the difference between fractal dimensions (the topic of this book) and fractal geometry (the shapes often adopted by architectural designers), it is necessary to briefly delve into the theory of dimensions and the history of fractals.

Architects and designers conventionally talk about and conceptualise shape and form in both two and three dimensions. That is, from the first stages in their education, designers understand that objects (including cities, buildings and furniture) are three-dimensional, although their properties are described using two-dimensional representations (plans, elevations, sections and various perspective and isometric views). While this way of thinking about flat representations as 'two-dimensional' and physical objects as 'three-dimensional' is in common use in society, the theory of dimensionality is actually much more intricate and diverse. As a starting point to understanding this theory, it is first necessary to clarify some of the basic terminology and concepts used.

Mathematicians and scientists sometimes call the world in which we physically exist 'Euclidean space', while philosophers describe it as the 'material world', and architectural theorists define it as 'lived space' or 'experiential space'. This dimension is physically tangible (it can be touched and otherwise sensed) and it has practical material and scale limits, meaning it cannot be infinitely divided or enlarged. To use an architectural example, a building in the material world can be touched, we can move through it physically, and it is made of substances that lose their structure if they are sufficiently weathered or broken down. In contrast, the rigorously theorised or imagined world is described by mathematicians as 'topological space', by philosophers as 'abstract space', and by architectural theorists as 'geometric space'. This imagined space has no direct physicality and no practical limits, but it can still be studied in valuable ways. To use another architectural example, a computer model of a building cannot be touched, it cannot be physically entered and it can be made infinitely small or large without any impact on its geometry. Both the material world (of the building) and the abstract world (of the CAD model) are rigorously defined, dimensional spaces, but as we will see, while architects view them both as three-dimensional, mathematicians and scientists see them differently.

Technically, a dimension is a topological measure of the space-filling properties of an object (Manning 1956). Thus, a dimension is an abstract but still accurate gauge of the extent to which an object occupies space. This space-filling property is also known as the Lebesgue covering dimension (Dieudonne 1994) and while architects talk about only two different dimensions—two-dimensional representations and three-dimensional objects—for a mathematician, a large number of hypothetical dimensions (n) exists in topological space. Mathematically, the relative membership of an object in a dimensional set is determined by calculating the number of coordinates required to define the location of a point on that object. Thus, for example, the corner of a planar surface can be located in space with only a pair of x and y coordinates, while the corner of a cube requires a triad of x, y and z coordinates. For the first of these examples $n = 2$ and for the second $n = 3$; that is, they are respectively in two-dimensional and three-dimensional space (Sommerville 1958). Because a fourth dimension—space-time—has been widely theorised, and further additional dimensions are possible, mathematicians typically talk of space as being n-dimensional (Pierpont 1930; Manning 1956).

Until the early 1970s, mathematicians accepted that n was necessarily a whole number or integer (for example, 1, 2 or 3). Moreover, the Euclidean world was thought of as necessarily a three-dimensional space, with all other dimensions existing only in abstract space. However, in the last few decades the idea that multiple dimensions may exist simultaneously in Euclidean space has become known as the 'theory of general dimensions' (Edgar 2008; Pears 1975). One of the catalysts for this development was the growing realisation that whole number (integer) dimensions are incapable of describing the full complexity of the material world. Probably the most famous of the general dimensions, and the first to methodically develop non-integer values, is the fractal dimension.

In his seminal text *Les Objects Fractals*, Mandelbrot (1975) suggests that Euclidean geometry, the traditional tool used in science to describe natural objects, is fundamentally unable to fulfil this purpose. While science historically considered roughness and irregularity to be an aberration disguising underlying systems with finite values, Mandelbrot suggests that the fragmentation of all naturally occurring phenomena cannot be so easily disregarded. In order to solve this dilemma Mandelbrot (1982) proposed that certain natural structures may be interpreted as lying in the range between traditional integer dimensions. He argues that, for example, if we look at a snowflake under a microscope, it fills more space than a line (n >1.0), yet far less than a surface (n <2.0); its actual dimension is therefore a fraction which is more than one but less than two. Mandelbrot called such fractional, non-integer dimensions 'fractal dimensions'. Mandelbrot's (1975) technical definition of a fractal has been widely paraphrased as 'a set for which one has Hausdorff-Besicovitch dimension greater than topological dimension'; Mandelbrot demonstrated this definition of a type of irregularity using a series of geometric constructions which parametrically repeat themselves to produce evocative and infinitely complex images. In this process Mandelbrot used fractal geometric sets to explain fractal dimensions and vice versa.

It is this early combination of fractal dimensions and fractal geometry that caused much of the later confusion and even forced Mandelbrot to revise his definition. In hindsight, this misunderstanding was almost to be expected, because for a mathematician dimensions are necessarily topological or abstract and should not be confused with the measures used to describe the material world. However, such was the extent to which non-mathematicians assumed that fractal dimensions and fractal geometry were the same things that Mandelbrot was forced to retract his original definition (Feder 1988). Mandelbrot (1986) eventually revised his definition because he was dissatisfied with the way it excluded many mathematical sets from the material world that are visually reminiscent of fractals but which failed to fulfil the precise topological conditions he originally set. Furthermore, some completely chaotic topological sets (those lacking any degree of order) complied with his original definition even though they were not in its spirit. For this reason, by the 1980s Mandelbrot was forced to differentiate between fractal *geometry* and fractal *dimensions*. The former refers to particular geometric sets that only exist in topological space and exhibit high levels of self-similarity, while the latter is a more general term describing the space-filling properties of irregular objects which may exist in either the topological or the material world.

A fractal geometric figure is one that is generated by successively sub-dividing or growing a geometric set using a series of iterative rules. This process produces a figure that has parts, which, under varying levels of magnification, tend to look similar, if not identical, to each other. For example, if the starting geometric set is an equilateral triangle, and the rule says that the middle third of each face of that triangle is replaced by a new equilateral triangular extension, then once the rule is applied the first time, the starting figure transforms from a triangle into a six-pointed star (like an outline of the Star of David). If the same rule is applied to this new shape, it takes each of the twelve faces, again identifies the middle third, and uses it to generate a new triangular extension, which it adds to the face. By now, the new figure has forty-eight faces, and it has begun to resemble a geometric snowflake. Furthermore this operation can be repeated an infinite number of times, generating an endless sequence of geometrically identical, though increasingly smaller-scaled, triangular additions. This special geometric set is known as the Koch Snowflake and regardless of the scale of magnification, the geometry looks the same (Fig. 1.1).

There are many well-known fractal sets like the Koch Snowflake that feature infinitely deep and repetitious shapes. These are often called 'ideal fractals' because they can only exist in the mind, in computer simulations or as algorithmic processes. Nevertheless, these fractal geometric sets have many interesting properties. For example, the Koch Snowflake may have an infinite number of surfaces, but its area will never exceed that of a square drawn around its first iteration (the 'Star of David' stage). Thus, the Koch Snowflake has an infinite boundary length but an area that is forever approaching, but does not reach, a fixed figure. Despite this paradoxical quality, the Koch Snowflake does have one fixed property—its

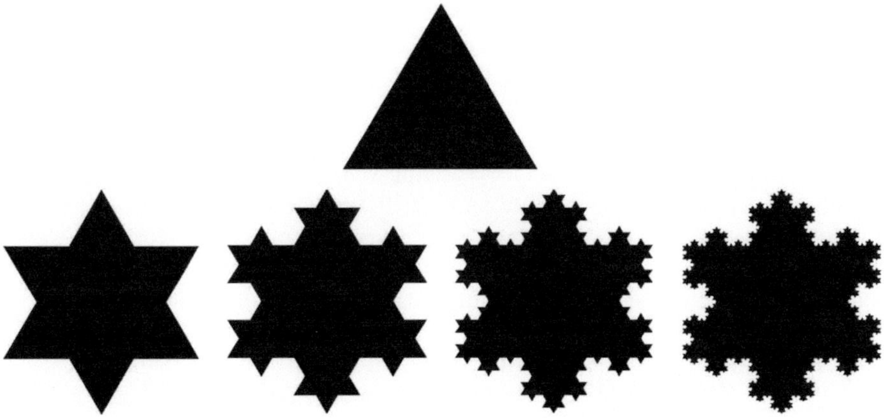

Fig. 1.1 The Koch Snowflake fractal set; starting figure (*above*) and first four iterations (*below*)

characteristic irregularity—and it is possible to use mathematics to accurately calculate how consistently rough or spiky it is. This characteristic irregularity is the fractal dimension of the Koch Snowflake, and using mathematics we can determine that its roughness has a dimension (*D*) of 1.26186. While the method used to measure this irregularity is discussed in greater detail in the course of the following few chapters, the message here is that there is a major difference between fractal geometry (an infinitely deep form, generated by the consistent application of a rule) and fractal dimension (the characteristic roughness of an object).

Mandelbrot (1982) defined fractal geometry as a type of deep geometric phenomena that arises from the application of a system of repetitively applied feedback rules—also known as Iterative Function Systems or IFS (Peitgen and Richter 1986). As a product of the IFS, the resultant geometric figure, when examined at increasingly fine scales, is seen to be self-similar that is, at a variety of ranges, the object in question tends to resemble itself (Kaye 1989). This property is known as 'scaling'. For Mandelbrot, any set may have a fractal dimension, but only sets with a defined scaling pattern can be described as instances of fractal geometry. This distinction is a critical one in architectural analysis, where the two are rarely differentiated and widespread confusion exists about whether or not a building can be fractal (Jencks 1995). From a pure mathematical perspective, buildings may have fractal dimensions, but they are never, in the material world at least, examples of fractal geometry (Ostwald 2001a, 2003). Moreover, buildings are actually part of a general class of objects called multi-fractals, a class that covers most natural and synthetic objects in the material world (Stanley and Meakin 1988). Before returning to the distinction between geometry and dimensionality, the multi-fractal is worthy of a brief diversion.

Ideal mathematical fractals, such as the Koch snowflake or Sierpinski Triangle (Fig. 1.2), possess infinite scalability and singular stable dimensions, and as such they are sometimes called 'uni-fractals'. In contrast, a 'multi-fractal' is an object

Fig. 1.2 The Sierpinski Triangle fractal set; starting figure (*above*) and first four iterations (*below*)

that simultaneously possesses a range of dimensions, each of which is relatively consistent over several scales, but is not continuous (Alber and Peinke 1998). For example, a tree in the material world has several distinct scales at which it exhibits levels of characteristic complexity. Buildings and cities are also multi-fractals; every building has several levels of stable dimensionality, ranging from the cellular, granular, and material to the textural, constructional and formal (Ostwald 2001a, 2003). Despite this, the formal and the textural are the only levels featured in the majority of architectural and urban analyses cited in past research and throughout this book.

Returning to the previous point, the distinction between geometry and dimensionality is important because it differentiates between two separate mathematical processes. The first of these, which includes models like the IFS, is used to identify the structure of a fractal geometric set. The second, made up of a range of related analytical systems, can be used to determine the fractal dimension of a set. While these two methods are independent, a small number of fractal sets effectively have matching structures and dimensions and are therefore ideal for the calibration of analytical methods (Da Silva et al. 2006; Górski et al. 2012). The Koch Snowflake is one such set as it has both a defined IFS and a fixed fractal dimension.

1.5 Measuring Fractal Dimensions

There are multiple ways of mathematically calculating the fractal dimension of an image (where $1.0 < D < 2.0$) or object (were $2.0 < D < 3.0$). For example, Mandelbrot (1982) describes three alternatives, the first of which, the box-counting approach, relies on overlaying different scales of grids and comparing the amount of detail present in each. Often credited to Voss (1986, 1988), technically the box-counting method calculates the Minkowski-Bouligant dimension. In practice though, the method has become so widely accepted that the result is described as either the box-counting dimension or the fractal dimension. Mandelbrot (1982) also presented a second way of calculating the approximate fractal dimension of an image using overlapping circles of different radii and a comparison between the capacities of these circles to cover the outline of an image. The third method described by Mandelbrot was the packing dimension which is based on the capacity of a series of circles to cover an irregular line around an image. This third version imagines that a range of circles, of increasingly reducing size, are iteratively packed inside the borders of that image. A comparison is then constructed between the number of circles, of different scales, needed to 'fill' the object.

Since Mandelbrot first proposed that fractal dimensions were measurable, seven major permutations or approaches have been identified. The first two are the *box-counting* method and the *differential box-counting* method. The other five are the *power spectrum* method, the *power differentiation* method, the *difference statistics* method, the *Kth nearest neighbour* method and the *covering blanket* approach. All of these versions have been evaluated and compared with the outcome that, for most results ($1.2 < D < 1.8$) the box-counting method is the most accurate and useful (Asvestas et al. 2000; Li et al. 2009). Sarker and Chaudhuri (1994) concur, arguing that despite some known issues with higher range results ($D > 1.8$), the box-counting method remains the most reliable approach. This particular issue arises from the fact that for very complex dimensions, the box-counting method begins to lose accuracy at the most complex extreme (Asvestas et al. 2000). This observation is of less concern for architectural analysis than for some other fields, because buildings and most correctly pre-processed urban forms do not fall into the range where $D > 1.8$, and if they do, the level of error does not become substantial until $D > 1.9$ (Ostwald et al. 2009; Ostwald and Vaughan 2013a). However, while the majority of past research into methods of measuring fractal dimensions have confirmed that the box-counting approach is the most accurate and useful (Xie and Xie 1997; Yu et al. 2005) there are reservations about two facets of the method—its repeatability and accuracy. The first of these occurs simply because researchers have failed to publish the methodological settings and variables they have used, which has meant that their results are difficult to replicate. Developing standard or best-practice models and recording them can readily solve this problem. The solution to the second is to undertake detailed experiments to calibrate the particular discipline-specific variation being used, and to measure and quantify its limits.

The box-counting method was first adopted for architectural and urban analysis in the 1990s (Batty and Longley 1994) and since that time has been used for the analysis of a growing number of buildings, ranging from ancient structures to twentieth-century designs (Bovill 1996; Burkle-Elizondo and Valdéz-Cepeda 2001; Rian et al. 2007; Ostwald and Vaughan 2009b, 2010, 2013a). A stable computational version was first presented in 2008 (Ostwald et al. 2008) and the box-counting method is now the accepted version in architectural scholarship as it is 'easy to use and an appropriate method for measuring works of architecture with regard to continuity of roughness over a specific scale-range (coherence of scales)' (Lorenz 2009: 703). However, architectural researchers, like the scientists and mathematicians before them, have also noted that the method has some weaknesses and have identified several specific factors which can dramatically affect the accuracy of the calculation (Bovill 1996; Benguigui et al. 2000; Ostwald et al. 2008; Lorenz 2012). Despite this, it is only in the last few years that solutions to these problems have been identified and their impacts determined (Ostwald 2013; Ostwald and Vaughan 2013b, c).

Using this stable computational version of the method, which is described in detail in the present book, it is now possible to measure the fractal dimensions of the plans and elevations of a wide range of buildings. The data points extracted from these views can then be synthesized into a series of values that are in turn compiled in various ways to produce a series of composite results describing the fractal dimension of a complete building. Once this process is complete the data may be coded with additional information, producing a set of mathematical results that describe the properties of a design, or a set of buildings, or changing formal and spatial patterns over time.

1.6 Book Structure

This book is structured in two parts. Part I provides a background to the topic (Chap. 2), an introduction to the standard method for calculating fractal dimensions (Chap. 3) and a detailed demonstration of how this method has been developed and refined (Chaps 4 and 5). Part II starts with a description of a major research study investigating the fractal dimensions of eighty-five designs by fifteen of the twentieth-century's most respected architects and practices (Chap. 6). The results of this study are reported in Chaps. 7 through 11, grouped into different stylistic movements and presented broadly in chronological order, although there are several overlaps. The complete set of data developed throughout the book is also examined in the conclusion, Chap. 12. The content of each of these chapters is described in more detail hereafter.

This book is about the measurement and classification of architectural form using fractal dimensions. Many readers will be aware that architectural designers and scholars have made a number of statements about fractals over the last four decades. However, there is also widespread confusion about these claims, some of which are of limited validity, while others range in veracity from the naïve to the insightful. To clarify the purpose of this book, and its relationship to past research, Chap. 2 provides a critical overview of attempts to connect architecture with fractals. It commences with a detailed review of the different ways in which scholars and designers have defined or sought to create 'fractal architecture'. Through this process the chapter identifies a series of recurring misunderstandings about the topic, the first of which is the ongoing and deeply problematic merging of the concepts of nature and fractal geometry, as if the two are inherently linked. The second is the tendency to confuse symbolic and actual relationships, along with the associated lack of clarity about the difference between the physical ('geometric space') and phenomenal ('lived space') properties of space. A final theme discussed in Chap. 2 is a lack of differentiation between fractal geometry and fractal dimensions. With the exception of the discussion in Chap. 2, and an explanation previously in this chapter, this book is about fractal dimensions, not fractal geometry.

Chapter 3 provides an introduction to the most common mathematical approach for calculating the fractal dimension of an image or object, the box-counting method. The chapter commences with three examples of applications of the method —a historic window and an elevation from the *Villa Savoye* and the *Robie House*— all of which include a full record of the calculations used. Thereafter, an overview of past applications of the method in architectural and urban analysis is presented along with a summary of results calculated for historic and modern structures. Throughout this section common errors in the architectural application of the method are identified. Such problems, which relate to inconsistent standards in image representation, data processing and methodological application, are responsible for the unfortunate fact that many of the results published previously are so inaccurate as to be unusable. This situation is the catalyst for two chapters that follow, which set out to use a combination of reasoned argument and experimental results to identify an optimal version of the box-counting method for architectural and urban analysis.

The first major problem identified in the review of past research using the box-counting method is that the images being analysed are either inconsistent in their levels of representation or are poorly chosen for their purpose. For this reason, Chap. 4 is entirely focussed on the question of what information in an architectural design should be considered in the measuring of its fractal dimension. There are many ways in which a building may be measured and compared and a similarly large number of reasons why such measures are useful. Chapter 4 adopts a post positivist perspective to these questions, considering both precision and purpose when determining which aspects of a building should be measured and why. Thereafter, a framework is proposed which maps levels of architectural representation against research goals. Five levels of representation are identified in this framework, each aligned to particular questions or issues, and all are illustrated and discussed.

Chapter 5 is concerned with methodological issues associated with optimizing the results of the box-counting method. These issues are divided into three categories: 'pre-processing', which is what occurs before the method commences; 'processing', which is what happens during the application of the method; and 'post processing' of the results to achieve a statistically reliable outcome. Through a series of calibration experiments the chapter identifies the optimal settings or ranges for these three categories, to achieve a reliable and repeatable calculation. Importantly, Chap. 4 also begins to quantify the limits of these factors, and in doing so explains why past attempts to use this method have often failed to produce useful results.

Part II of this book measures and compares 625 architectural plans and elevations, derived from eighty-five individual designs. Chapter 6 commences with an explanation of the rationale for the sample chosen, for both the architects and designs included in the study, and the parameters used to develop the complete list of all of the houses. The source material used for the research is also documented and the way this material was treated in preparation for the study is described. Next, the settings or data processing variables employed are described and tabulated. In the second part of Chap. 6, the stages used throughout the remainder of the research are recorded, including how plans and elevations are treated, measured and coded for both individual designs and sets of designs. The data presentation techniques used in the study are introduced using a hypothetical case, and various approaches to comparing the data are outlined. Many of the definitions of specific measures, including what they mean and how they are produced, are contained in Chap. 6.

Chapter 7 is the first chapter which presents specific results, in this case relating to four sets of Pre-Modern and Modern houses, two by Le Corbusier and one each by Eileen Gray and Mies van der Rohe. The first of these sets, from Le Corbusier, contains his early (1905 to 1912) Arts and Crafts style domestic architecture. Ignored for much of the twentieth-century, and still poorly understood by historians and critics, these designs are now considered important precursors to Modernism. The second set by Le Corbusier includes some of the twentieth-century's most famous houses, the *Villa Savoye*, the *Villa Stein-de Monzie* and the *Maison-Atelier Ozenfant*. Collectively, these houses signalled the rise of functionalism and the concomitant rejection of traditional forms, materials and cellular spatial arrangements. A contemporary of Le Corbusier, Eileen Gray was one of the most influential furniture designers of the early twentieth-century and the third set in this chapter contains five of her architectural designs. Gray was responsible for one of the great masterworks of domestic architecture, the cryptically named *E.1027*. The last set analysed in Chap. 7 is of designs by Mies van der Rohe, including four of his flat-roofed masonry houses in Germany and his iconic *Edith Farnsworth House* in the USA. In the final section of Chap. 7, the adage 'form follows function' is examined using a variant of the fractal analysis method. Specifically, the idea that the external expression of a house can be understood as a function of its orientation (siting) or address (approach), is investigated using five houses each from Le Corbusier and Gray. Such an analysis reveals the degree to which there is any correlation between façade complexity and either climatic conditions or function of internal spaces. This variant involves augmenting the fractal dimension data using absolute (orientation) and relative (address) information derived from the siting of each house.

Frank Lloyd Wright is the focus of Chap. 8, which examines and compares three sets of houses from different stages in his career. Starting with his Prairie Style works, the five houses examined here were completed between 1901 and 1910 and include the *Robie House*, the quintessential example of this style. The second set comprises five of Wright's Textile-block designs, including the renowned *Ennis House*, which were completed between 1923 and 1929. The final set are Wright's Usonian houses, built between 1950 and 1955 and including the *Palmer House*, a design that has been the subject of on-going speculation about its fractal geometric properties. These three styles are all regarded as variants of organic Modernism, an approach to design which, unlike its functionalist European counterpart, embraced a sense of siting, landscape and vegetation as part of its design language. In this chapter the fifteen houses are both examined individually and collectively in order to chart distinct trends across Wright's career in the way he shaped space in both plan and elevation. The last section in Chap. 8 discusses the use of non-orthogonal images (particularly perspectives) for fractal analysis, examining the strengths and weaknesses of such an approach. It then presents a unique application of the fractal analysis method to perspective images to demonstrate how visual perception can be measured and compared, and thereby begin to test a famous theory about the experience of passage through Wright's *Robie House*.

Chapter 9 is about the Late-Modern Avant-Garde, a design approach which was exemplified by the work of the 'New York Five': John Hejduk, Michael Graves, Peter Eisenman, Charles Gwathmey and Richard Meier. One hallmark of their approach was a strong rejection of humanist and symbolic approaches to design in favour of programmed abstraction and an aesthetic expression reminiscent of abstract art. The work of three of these architects is analysed here. Five of Eisenman's numbered (rather than named) designs are featured in the first set. Completed between 1968 and 1976 these works are landmarks in the history of architectural theory, offering a system or 'grammar' for generating a design, regardless of its programmatic intent or function. The second set in this chapter is derived from the work of Hejduk, an educator and theorist who designed a series of rigorously geometric houses as a type of laboratory experiment to test the potential of architecture. Five of his numbered designs, executed on paper between 1954 and 1963, make up the second set analysed in the chapter. The final set of works, which span between 1967 and 1974, is by Richard Meier. Awarded the Pritzker prize in 1984, Meier's highly influential series of early houses feature intricate geometric façades and plans. The secondary analytical approach presented in Chap. 9 is concerned with testing arguments about frontality and rotation which were originally presented as part of the theory explaining the work of the New York Five. To test these ideas, Eisenman's *House I*, Hejduk's *House 7* and Meier's *Hoffman House* are all examined using new, non-cardinal elevations, developed from thirty-six different rotational positions. Then, returning to Frampton's essay about the New York Five, 'Frontality vs Rotation' (1975), the implications of the results are considered.

By the 1950s it was starting to become apparent to some designers and critics that Modern architecture had lost sight of basic social, regional and phenomenal values. One of the first and most academically respected architects to rebel against Modernism's seemingly puritanical fixation on form and function was Robert Venturi. Awarded the Pritzker prize in 1991, and working with Denise Scott Brown for much of his career, Venturi designed the iconic *Vanna Venturi House* in 1964. The set of five houses examined in Chap. 10 includes both Venturi's solo designs and those completed jointly with Scott Brown. The next section in Chap. 10 measures and analyses the fractal dimensions of five houses by Frank Gehry. Awarded the Pritzker Prize in 1989, Gehry, in an intermediate stage of his career, designed a series of experimental houses that have rarely been investigated by scholars or critics. Including both completed and unbuilt works, this set spans between 1978 and 1984 and represents an important shift in Gehry's design style. Chap. 10 concludes with a further analysis of the truism 'form follows function', this time considering the argument that the formal properties of a façade are shaped by the extent to which a façade is permeable. Augmenting the fractal dimension data with information about the frequency of openings in façades, this variant of the method analyses five designs by Venturi and Scott Brown and five by Le Corbusier to see if there is any relationship between formal complexity and functional permeability.

Chapter 11 contains a detailed analysis of twenty-five Minimalist and Regionalist houses. The Minimalist works include two sets of five houses that were completed in Japan between 1994 and 2005. The first set is by Kazuyo Sejima who, along with Ryue Nishizawa her partner in SAANA (Sejima and Nishizawa and Associates), was awarded the Pritzker Prize in 2010. Sejima's designs are notable for their stark simplicity, often featuring blank or frosted glass façades that hide complex, layered interiors. The second Minimalist set comprises designs by Atelier Bow-Wow. Founded by Yoshiharu Tsukamoto and Momoyo Kajima, Atelier Bow-Wow is a Tokyo-based architectural practice whose idiosyncratic, often brightly coloured micro-buildings are typically designed for tiny sites in dense urban neighbourhoods. With simple or inexpensive finishes, their designs are not always minimal in terms of aesthetic expression, but they are in their approach to the relationship between form and inhabitation.

The Regionalist designs analysed in Chap. 11 include a set of five houses by Peter Stutchbury and two sets of five houses by Glenn Murcutt, all of which were built in Australia between 1975 and 2007. Australian Regionalist architecture, while influenced by Modernity and its aesthetic and spatial tropes, is celebrated for the way it promotes both practical and poetic responses to site and climate. Peter Stutchbury's architecture in particular is characterised as being technologically focussed, with tectonic rather than formal properties being central to his Regionalist aesthetic. Murcutt's architecture is more closely aligned to the standard definition of Regionalist architecture. Moreover, Murcutt, who was awarded the Pritzker prize in 2002, is regarded as having produced throughout his career a highly consistent set of rural domestic designs which have similar aesthetic expressions and planning. However, critics have also noted a subtle shift in his approach that occurred in the early 1980s and thus in this chapter two sets of his houses are considered. The first set includes five 'early' designs completed prior to 1983, and the second, five 'later' career works, completed after 1983.

The secondary application of the method in Chap. 11 is to test claims about the importance of transparency in Murcutt's architecture. Many arguments about the aesthetic character of Regionalist Australian architecture identify its relative transparency as a key feature. However, despite such claims, there is on-going confusion about whether this property is literal or phenomenal and indeed whether these designs are, in any way, transparent. Using fractal analysis, Chap. 11 compares the dimensions of the façades of Murcutt's designs when represented in two different ways. The first of these is a conventional opaque presentation of the elevations and the second treats windows as translucent and doors, louvers and screens as open. This method allows for the literal transparency of the façades to be measured. In a further variation, perspective views of one of Murcutt's designs are compared using both translucent and opaque representations, to determine if the phenomenal sense of permeability in the façades is more important than the literal one.

The concluding chapter, Chap. 12, draws together the complete set of primary data developed throughout Part II, combining the results from all of the houses, to test three hypotheses. The first of these hypotheses is about longitudinal trends in the data, the second concerns its usefulness for differentiating architectural styles and the final is about the method's capacity to distinguish the formal and spatial traits of individual designers. The book concludes with a series of classification tables to allow future researchers to categorise additional cases using the data presented here.

1.7 Conclusion

This book has been structured in such a way as to introduce theories and methods before they are applied and thereafter the resultant data is presented in chronological order and sorted within stylistically themed chapters. Despite this structure, the reader can and should approach the material more selectively, in such a way as to meet their specific needs. For example, Chap. 2 is about ideas and propositions that are, for the most part, outside the primary scope of this book, but for many readers this will be an important starting point. Conversely, some readers will be less interested in the finer details of the method and will therefore move straight from Chap. 3 to Chap. 6 in order to start reading the experimental application and results.

The various paths taken through this work are each up to the individual reader. There is no need to read each chapter from beginning to end to benefit from it. However, to interrogate specific results and to understand how many of the mathematical comparisons have been constructed and interpreted, it is necessary to read the methodological chapters. Certain interpretations of the results, when read in isolation, may seem confusing or contradictory. This is why a careful reading of all of the results and how they have been produced can help to explain why such discrepancies exist and what they might imply. This is the nature of research of this type, with its mixed qualitative and quantitative approach, and, in the present case, a body of work which combines mathematics and computing with design history and theory.

Part I
Understanding and Measuring Fractal Dimensions

Chapter 2
Fractals in Architectural Design and Critique

This book is about the analysis of architecture using fractal dimensions. This method and its application are described in detail in the coming chapters, but it must also be acknowledged that the relationship between fractals and architecture has traditionally been both more diverse and more controversial than the scope of this book might imply. For thirty years architectural scholars and designers have opportunistically appropriated images and ideas from fractal geometry along with concepts broadly related to fractal dimensions and non-linear dynamics, and used them for a wide variety of purposes. Some of these appropriations have been motivated by the desire to advance architecture or to offer new ways of understanding design, but many others have a seemingly more superficial or expeditious agenda. In a detailed analysis of the reasons why architects are drawn to adopt ideas from fractal geometry and dimensionality, two of the most common motives identified were 'legitimisation' and 'obfuscation'; respectively, the desire to seek 'authority' from an external body of knowledge and 'appropriation for the purpose of creating mystique' (Ostwald 1998a). This finding is neither unexpected nor innately problematic because philosophers, artists and scientists often have similar motivations for engaging in cross-disciplinary work (Kuhn 1962; Latour 1987; Sokal and Bricmont 1998). But such motivations are a reminder that the relationships between disciplines—like architecture and mathematics—can be based more on convenience than respect.

One of the essential problems when considering such cross-disciplinary connections is that many different types of relationships are possible between seemingly diverse fields. This problem is exacerbated when architecture is considered, because design serves a wide range of functions, from the physical to the social and the symbolic (Ostwald and Williams 2015a). We cannot assume that architecture's purpose can be described simply from a scientific or mathematical perspective; the enduring role of architecture in society is often linked to its material presence, its historic significance or its capacity to represent a set of otherwise intangible values (Giedion 1941; Banham 1960; Pérez-Gómez 1983). Conversely, we cannot suppose that the myriad of other-disciplinary connections evoked or claimed by a design are

M.J. Ostwald and J. Vaughan, *The Fractal Dimension of Architecture*,
Mathematics and the Built Environment 1, DOI 10.1007/978-3-319-32426-5_2

equally valid or meaningful. Thus, this chapter is about the occasionally enlightening but sometimes frustrating and obtuse connections that have been proposed between architecture and fractals. Despite this observation, the purpose of this chapter is not to criticise these proposed connections, but rather to examine a large number of examples where architecture and fractal geometry have been used as a catalyst for discussion of the broader nature of this complex and creative association.

Three common types of relationships between architecture and other fields are those concerned with inspiration, application and accommodation (Ostwald and Williams 2015b). For example, a building design can be inspired by a scientific ideal, it can be designed to take advantage of scientific knowledge, and it can house a scientific function. These are three different types of connection and while it is possible for a building to simultaneously possess all of these properties, it is highly unlikely that all three will actually be related to each other in any coherent way. For example, the shape of a building may be *inspired* by vertebrate biology, the same building may feature an *application* of bio-waste recycling and it may *accommodate* a laboratory for gene analysis. Such a building would fulfil all three of these possible relationships with science, but there is no connection between any of them, and particularly not as they are embodied in the building. Another way of understanding this principle is that there is no essential relationship between how something looks, how it is constructed and what it does. Furthermore, when symbolic, metaphoric or semiotic connections are proposed between architecture and another field, it is especially difficult to convincingly argue that the relationship exists at any deep level. Thus, for example, a building façade may be covered in images of trees, or have leaf-shaped windows, but this does not, in itself, make a building natural, organic or ecological. This is especially the case for buildings that are allegedly inspired by, or designed in accordance with, the principles of fractal geometry.

This chapter is concerned with the way *fractal geometry* and associated imagery and ideas have been used by architectural designers, scholars and critics. In contrast, the remainder of this book is about the way in which architecture can be measured and analysed using *fractal dimensions*. The two approaches offer different ways of considering the relationship between design and geometry. Much like the example of the three types of relationships between science and architecture, there is no explicit connection between fractal measurement and a design that seeks to evoke—through form, texture or tectonics—fractal geometry. Thus, while it is possible to measure the fractal dimension of a building that is inspired by fractal geometry, the two processes—measurement and inspiration—are fundamentally unrelated. Measurement is a universal set of actions, following a strict protocol, which can be repeated for multiple similar objects, while inspiration is an intricate and potentially poetic process, typically unique to an individual. Both of these processes are valid and useful, but they should not be confused with each other.

The purpose of this chapter is therefore to examine the way architects and scholars have incorporated fractals into the design and interpretation of the built environment. We commence with a discussion of the problems of defining fractal architecture and the tension that exists between definitions that are derived from geometric properties and those which are more phenomenological or experiential in their framing. Thereafter we analyse several conscious and subconscious examples of fractals in design. Finally, we consider the use of recursive processes, akin to fractal growth algorithms or Iterative Function Systems (IFS), as a design method. Through this review of past research the chapter provides a conceptual foundation for thinking about fractals in architecture and for positioning the present research in the context of broader architectural debates.

2.1 The Problem of Defining 'Fractal Architecture'

Since the early 1980s, a growing number of scholars and designers have acknowledged the influence of fractals upon architecture (Ostwald 2001a; Joye 2011). Fascinated by its mathematics and imagery, or drawn to possible natural or mystical connections, such architectural writers and designers have promulgated a range of often idiosyncratic interpretations of fractal geometry. Because of the diverse range of motives for adopting fractal geometry, there is neither an agreed upon definition nor a common title for works that use fractals for inspiration, design rationale or form generation. For example, several portmanteau descriptors exist which merge multiple, often dissimilar properties. Probably the best known of these is Charles Jencks's (1995) 'Architecture of the Jumping Universe', an evocative title for an eclectic set of ideas cherry-picked from science, philosophy and art. Similarly, the 'New Baroque' (Kipnis 1993) and the 'Architecture of the Fold' (Eisenman 1993) freely merge concepts from fractal geometry with themes from the writings of Deleuze and Guattari, philosophers who once used fractal geometry as a metaphor for political theory (Ostwald 2000, 2006). The repeated use of other classifications including 'Fractalism', 'Complexitism', 'Complexity Architecture' and 'Non-linear Architecture' have led scholars like Yannick Joye to argue that 'a systematic, encompassing, scholarly treatment of the use and presence of this geometrical language in architecture is missing' (2011: 814).

To further complicate matters, since the 1970s scientists and mathematicians have offered their own definitions of fractal architecture, although these have often been for the purpose of explaining concepts, rather than offering designers a recipe for creating a new architectural style (Ostwald and Moore 1997; Ostwald 2009). The most famous of these definitions, from Benoit Mandelbrot, suggests that certain architectural styles possess formal properties similar to those of various natural fractals. This argument is encapsulated in his statement that 'a high period Beaux Arts building is rich in fractal aspects' (1982: 24), because it possesses 'very many scales of length and favour[s] self-similarity' (1982: 23). Mandelbrot argues that if, for example, the perimeter of a Beaux Arts building like the *Paris Opera* is measured using three different scales—a yardstick divided into feet, a tape measure in inches and a ruler with centimetres—three different lengths will result. While this is true of many buildings, Mandelbrot's identification of the Beaux Arts as being especially fractal is also supported by the way this style actually does feature elements repeated at different scales. Architects have often noted this propensity suggesting that it leads to a particular 'phenomenal complexity' (Poppeliers and Chambers 2003: 93) which is superficially similar to Mandelbrot's reading of this style. Beaux Arts buildings often feature elaborate ornamentation, including decorative relief on walls and window surrounds, grand carved balustrades and repeated motifs in columns, pilasters, archways and tiling or paving. Thus, in a limited sense, Mandelbrot correctly identifies that a Beaux Art building has a rich and complex form and that some of its elements, including columns and archways, are repeated at different scales.

While Mandelbrot's description of fractal architecture is informative, even if it is derived from a lay interpretation of the Beaux Arts, his discussion of architecture which is *not* fractal is equally telling. Mandelbrot states that the 'scalebound' (1982: 24) buildings of Mies van der Rohe are not fractal because, when measured with the three tools mentioned previously (the yardstick, the tape measure and ruler), the dimensions would be the same. Mies, a Modernist architect, was famous for designing ostensibly Phileban geometric structures like the *Edith Farnsworth House* in Illinois. This house, 'with its open plan, glass walls and freestanding partitions, was as pure an exercise in architectural minimalism as Mies could have hoped for' (Friedman 1998: 134). Mandelbrot further identifies Mies's *Seagram Building*—a concrete high-rise structure clad with a curtain wall of bronze and glass with a seemingly simple rectangular form—as being the antithesis of fractal geometry.

What is interesting in the context of Mandelbrot's discussion of fractal architecture is the extent to which his position is phenomenologically defined rather than mathematically determined. Past research has observed that Mandelbrot differentiates the Beaux Arts from Modernism on the basis of obvious and often superficial visual and perceptual differences, rather than on the actual geometric properties of a plan, section or elevation (Gray 1991; Ostwald 2003). The same tension between the mathematical and the philosophical properties of architecture is also present in many discussions about fractal design. For example Carl Bovill (1996), who is clearly aware that 'buildings are not fractals in the same way that mathematical constructs' are, chooses to describe a key 'fractal characteristic' of architecture as the 'progression of interesting detail as one approaches, enters, and uses a building' (1996: 117). Bovill develops this phenomenological reading of architectural form to suggest that in an especially engaging building 'there should always be another smaller-scale' of 'detail that expresses the overall intent of the composition' (1996: 5). But whereas Bovill demonstrates an awareness of the problems of using rigorous mathematical concepts to interrogate architecture, not all scholars make such a clear distinction.

Consider Douglas Boldt's 'Fractalism', an incipient movement which is predicated on the idea that a 'fractal building may be based on a single iteration of a fundamental fractal shape or the shape may reiterate itself in building spaces or details' (2002: 10). While broadly in accordance with the concept of scaling, Boldt's definition is more contentious because, as Mandelbrot observes, a building that is 'rich in fractal aspects' will possess multiple complex, scaled and statistically varied, formal iterations. Furthermore, self-similarity is present in many buildings that would not normally be accepted as having any fractal geometric or phenomenal properties. Thus for Boldt to accommodate both the geometric and phenomenal properties of architecture he expands his definition to include buildings which have curves, look like natural objects or are environmentally friendly. Boldt's definition conflates 'fractal', 'organic' and 'ecological' properties in a way which is common in the rhetoric of architectural designers but which is problematic from both a mathematical and a philosophical perspective.

Descriptions of fractal architecture which draw connections to organic design are relatively common. For example, Derek Thomas defines fractal architecture as a 'contemporary form of organic design' (2012: 185) suggesting that '[e]xpressions of fractal geometry in architecture are essentially organic in character amounting to a continuity or a continuous linking through iterative cues and cognitive association' (2012: 189). He goes on to argue that, '[t]o experience organic form is to appreciate the distinctive interconnections over multiple scales' (2012: 189). In this example, the fractal and the organic are once again seemingly merged when a geometric or formal property is extrapolated to suggest its phenomenal impact. However, neither of these are necessarily true. David Pearson rightly observes that typically in architecture fractal geometry is only 'applied externally' and is 'divorced from the internal functions of the building. The use of geometry and science, alone, does not produce organic design' (2001: 46). It cannot be assumed that 'fractal and organic architecture are essentially the same thing' (Ostwald 2003: 263) or that there is any environmental benefit from shaping a building like a fractal (Ostwald and Wassell 2002).

The counterpoint to this tradition of merging fractal and organic architecture, is the practice of describing Deconstructivist architecture in fractal terms (Jencks 1995; Kelbaugh 2002; Pearson 2001). Ignoring for the moment the philosophical origins of this movement, the formal and visual properties of Deconstructivist architecture include distorted, angled and awkward forms. It is this quality—along with the etymology of the words 'fractured' and 'fractal'—which led architects to often wilfully blur the distinctions between non-linear mathematics, and Deconstructivist architecture. Nikos Salingaros (2004) and Joye (2011) correctly reject any suggestion that Deconstructivist architecture might embody, in any consistent or coherent way, the properties of fractal geometry. In theory, Deconstructivist architecture could possess a limited range of fractal geometric forms, but being fractured and being fractal are very different things. Furthermore, the Derridean and Post-Structuralist foundations of Deconstruction rely on recursive logic structures that have superficial similarities to the lessons of non-linear mathematics but the connection is largely through analogy (Ostwald and Moore 1996a). Nevertheless, from a phenomenological perspective, a Deconstructivist building could possess the same level of experiential appeal as a Beaux Arts building. Thus, Bovill is right to propose that 'Deconstructivist architecture can provide a modern equivalent of the cascade of interesting detail that classical architecture provided' (1996: 185). Despite Salingaros's (2004) criticism of

Deconstructivist architecture for its lack of human scale, there is evidence that it can possess the same level of visual and formal information as a building of any other style or era (Ostwald and Vaughan 2013a).

The merging of phenomenal and geometric properties is also found in Maycon Sedrez and Alice Pereira's proposition that fractals can be present in architecture 'through … recursive patterns, as generative patterns [or] as tools of scale perception' (2012: 99). The first two of these connections are geometric and algorithmic, while the last is more concerned with the senses. In the first instance, recursive architectural features are those that are characterised by both formal repetition and routine geometric construction. For example, Sedrez and Pereira, like Leonard Eaton before them, identify Frank Lloyd Wright's *Palmer House* as an example of fractal design. To support this case, Eaton adopts a narrow definition of fractal geometry as comprising 'a geometrical figure in which an identical motif repeats itself on an ever diminishing scale' (1998: 33). However, James Harris soundly and correctly rejects Eaton's definition stating that it 'points out the misconception that a repetition of a form … constitutes a fractal quality. It is not the repetition of the form or motif but the manner in which it is repeated or its structure and nesting characteristics which are important' (Harris 2007: 98).

Andrew Crompton is similarly critical of proposals like Eaton's noting that, '[f]rom this point of view almost any building can show fractal qualities, one simply has to count the elements of a façade which occur within different ranges of size and see how they increase in number as they get smaller' (2001: 245–246). To demonstrate the fallacy of this position, in a deliberately subversive argument it has been shown that even Mies van der Rohe's *Seagram Building*, the design Mandelbrot chose as the epitome of 'not-fractal', has more than twelve scales of conscious self-similarity (Ostwald and Moore 1996b).

Returning to Sedrez and Pereira's second characteristic of fractal architecture, the generative nature of the formal repetition, Kirti Trivedi cites the Indian temple as an example of a building type that features both recursive and rule-based geometries that conform more closely to the expectations of fractal geometry. Trivedi starts by stressing that fractal geometry is not simply defined by scaling, but also by the systemic and iterative evolution of shapes across multiple scales. Trivedi observes that in certain ancient Indian temples, visually complex shapes are generated through the use of successive 'production rules that are similar to the rules for generating fractals.' Moreover, there appear to be multiple different rule variables which are pertinent to different parts of the temple. In combination these rules, scales and variables operate through 'self-similar iteration in a decreasing scale: repetition, superimposition and juxtaposition' (1989: 249); all of which Trivedi calls 'fractalization'.

Despite such attempts to define 'fractal architecture', the central paradox of the endeavour is that no building can truly possess fractal geometry but every building can possess a fractal dimension (Bovill 1996; Ostwald 2003). Recall that fractal geometry is a system which describes forms that are generated from precise algorithmic rules which are infinitely repeated, whereas a fractal dimension is a measure of how consistently complex or textured an object is. The reason that the idea of fractal architecture is so problematic is that buildings (like trees and mountains) have what is called a 'scaling limit': a point at which any sense of self-similarity either changes (whereby it might be classified as a multi-fractal) or simply breaks down completely. In contrast, fractal geometric forms don't have scaling limits and thus remain similar regardless of their scale. This means that every object, whether natural or synthetic, can have its formal complexity measured or estimated, which is why architecture can have a fractal dimension. Conversely, pure fractal geometry only exists in hypothetical examples, in computer simulations or in philosophical puzzles. This is why architecture in the real world can never be a true example of fractal geometry.

Given this paradox, is it even meaningful to talk about fractal architecture? The proliferation of unsubstantiated quasi-fractal references in architecture and design has tended to undermine the usefulness of the concept and the degree to which it can be taken seriously. In order to overcome this problem, scholars and designers should be careful to describe the way in which they are using fractal geometry: as structure, as form, as ornament, or as inspiration. In certain circumstances it might be meaningful to imagine that '[a]rchitecture could possess a symbolic, metaphoric, metonymic or experiential fractal dimension' (Ostwald 2003) but the limits of the definition must be made clear. Both Mandelbrot's and Bovill's descriptions of the phenomenal properties of buildings which are reminiscent of fractal geometry are reasonable in the limited context in which they are offered. Similarly, within its particular geometric framework, Trivedi's definition is useful. The most important factor is not necessarily whether a geometric or phenomenal view is taken, but rather that each author is clear about their perspective, its purpose and limitations.

2.2 Fractals in Architectural Design

Despite ongoing confusion over definitions, there are many examples of possible connections between fractal geometry and architectural design. More than 200 examples of designs that have been inspired by, or allegedly designed in accordance with, fractal geometry have been identified and analysed (Ostwald 2001a, 2009). There are also other designs which have, purportedly at least, been intuitively led to use fractal geometry, often many hundreds of years before the theory was formulated. Thus, it is helpful in this context to divide the complete set of these works into two broad categories: those completed prior to the formulation and publication of theories of fractal geometry and those completed after. The first category necessarily includes works that demonstrate either intuitive or subconscious evidence of an understanding of the geometric principles underlying fractal geometry. The second category includes works which more explicitly acknowledge a debt to fractal geometry, even though the resultant architecture may not have such a clear relationship. However, within these categories there are also many different possible connections between architecture and fractal geometry, ranging from inspiration to structure, from construction to surface treatment and from applied ornament to algorithmic generator.

2.2.1 Architecture: Pre-formulation of Fractal Theory

A range of historic and traditional buildings have been the subject of ongoing speculation about the extent to which people or cultures have intuitively created geometric constructs which possess seemingly fractal qualities. For example, Ron Eglash (1999) notes the similarities between the geometric patterns found in indigenous African design and the self-similar shapes of fractal geometry. Gerardo Burkle-Elizondo (2001) offers a parallel argument drawing connections between fractal geometry and ancient Mesoamerican pyramids. Several architects and mathematicians have observed that the thirteenth-century plan of Frederick II's *Castel del Monte* possesses self-similarity at two scales, thereby suggesting the start of a sequence of fractal iterations (Schroeder 1991; Götze 1996). Each of these examples is an instance of scaled, geometric repetition which is superficially similar to the geometric scaling found in ideal mathematical fractals. In contrast, researchers have identified fractal properties in the way the classical Greek and Roman orders have been iteratively constructed (Crompton 2002; Capo 2004; Bovill 2009). Bovill (1996) argues that 'fractalesque' design features can be found in Greek and Roman monumental details along with doorway mouldings of English medieval buildings and in the plan of the eighteenth-century Baroque church of the *Madonna di S. Luca* in Bologna. Eilenberger (1986), Schroeder (1991), Crompton (2001), Lorenz (2011) and Samper and Herrera (2014) all suggest that Gothic architecture has fractal properties. Joye even proposes that the Gothic cathedral offers one of 'the most compelling instances of building styles with fractal characteristics' (2011: 820). Through the writings of John Ruskin, several authors have also identified fractal properties in Gothic architectural detailing and craftsmanship (Fuller 1987; Emerson 1991), although their arguments are typically only based on Ruskin's reading of the ethics or logic of geometric construction (Moore and Ostwald 1996, 1997). George Hersey (1999) identifies examples of fractal-like iteration in Renaissance architecture, in eighteenth-century Turkish buildings and in the work of Jean-Nicolas-Louis Durand. In the nineteenth-century, in addition to Mandelbrot's case for the fractalesque features of the *Paris Opera*, he is also one of multiple authors to suggest that the *Eiffel Tower* could be considered structurally fractal, at least for up to four iterations (Mandelbrot 1982; Schroeder 1991; Crompton 2001).

Indian temples provide a more compelling case for an intuitive connection between fractal geometry and architecture, in part because they actually possess, to a limited extent, scaled, self-similar geometric forms that follow a seemingly clear generative process (Lorenz 2011; Sedrez and Pereira 2012). Jinu Louishidha Kitchley (2003) identifies specific fractal qualities in the north Indian temples of the Nagara style as well as south Indian temples of the Dravida style. Trivedi's research analyses Hindu temples in plan, elevation and massing to provide examples of the steps involved in creating the form of these ancient buildings. Trivedi further suggests that the unconscious demonstration of fractal geometry is probably connected to the Hindu belief that religious buildings should depict 'an evolving cosmos of growing complexity, which is self replicating, self-generating, self-similar and dynamic' (1989: 249).

In the early years of the twentieth-century, and in parallel with the rise of interest in organic metaphors for design, several architects, including Frank Lloyd Wright and his mentor Louis Sullivan, began to produce works which were suggestive of fractal geometry in their experiential, planning or ornamental qualities (Kubala 1990). In terms of the first of these three qualities, Bovill proposes that 'Wright's buildings are a good example of this progression of self-similar detail from the large to the small scale' (1996: 116). However, Eaton argues that Wright's architecture only became more perceptually complex after the completion of the Textile-block house *La Miniatura*, a building which Eaton feels has no strong fractal presence or expression. But in terms of the geometry of the plan, as previously noted in this chapter, Eaton suggests that Wright's Usonian work of the 1950s and 1960s features a 'striking anticipation of fractal geometry' (1998: 31). His rationale for this argument is derived from the recurring presence of equilateral triangles, at different scales, in the plan of Wright's *Palmer House*. Forms in this house, ranging from the large triangular slabs of the cast concrete floors down to the triangular shape of the fire-iron rest are noted. Eaton counts 'no less than eleven scales of equilateral triangles ascending and descending from the basic triangle' (1998: 32) leading him to conclude that the *Palmer House* has 'a three-dimensional geometry of bewildering complexity' (1998: 35). While this argument is often repeated (Ferrero et al. 2009), as previously explained in this chapter, it is not especially convincing. Harris suggests that at best the relationship between Wright's plan and fractal geometry is 'analogous' (2007: 98); an appropriate description for a symbolic or metaphoric relationship between fractal geometry and repetitious form in a floor plan.

By the middle of the twentieth-century, Alvar Aalto had begun to produce a series of buildings which featured 'fragmented skylines, voids and irregularity' (Radford and Oksala 2007: 257), properties which promulgated a range of suggestions that Aalto had an intuitive, experiential appreciation of the fractal qualities of nature (Bovill 1996; Radford and Oksala 2007; Suau 2009). Bovill (1996) also describes the *Student Club at Otaniemi*, the work of Finnish architects Reima and Raili Pietilä, as displaying fractal qualities. While this last design was completed after the publication of Mandelbrot's theory of fractal geometry, it represents a continuation of a particular, regionally-nuanced Modernist tradition, rather than the adoption of a new type of geometry. In contrast, the mid-twentieth-century works of Aldo and Hannie van Eyck feature several details and forms which were merely suggestive of fractal geometry (Meuwissen 2008), but after the publication of Mandelbrot's theory they adopted more explicit fractal imagery as part of their conceptual design process for the *European Space and Technology Centre* (Ostwald 2003).

2.2.2 Post-formulation: Architecture Inspired by Fractals

In July of 1978, less than twelve months after the English language publication of Mandelbrot's *Fractals: Form, Chance, and Dimension,* Peter Eisenman exhibited his *House 11a* for the first time. Eisenman described this design as adopting several lessons from complexity theory and fractal geometry including self-similarity and scaling. In the three decades which followed, more than 200 architectural designs or works of architectural theory have been published which have laid claim, in some way, to aspects of chaos theory, nonlinear dynamics or fractal geometry. Some of the architects and firms that have either made explicit reference to complexity science, or have been linked to fractals include: Asymptote, Bolles Wilson, Charles Correa, Carlos Ferrater, Zaha Hadid, Coop Himmelblau, Steven Holl, Arata Isozaki, Kulka and Königs, Fumihiko Maki, Morphosis, Eric Owen Moss, Jean Nouvel, Philippe Samyn, Kazuo Shinohara, Ushida Findlay, Aldo and Hannie van Eyck, and Ben van Berkel and Caroline Bos (UNStudio). In some cases the influence of fractal geometry in a particular architectural project may be obvious, whereas in others it is less clear what the nature of the connection is. For example, one of Charles Correa's designs for a research facility in India features a landscaped courtyard that is tiled in a representation of the fractal Sierpinski triangle. This is an obvious and literal connection that might be appropriate, given the function of the building, but it is potentially little more than an ornamental application (Ostwald and Moore 1997). In contrast, Ushida Findlay produced a three-dimensional map of the design themes they had been investigating at different stages during their joint career. This map, a nested, recursive structure which traces a spiralling path towards a series of design solutions, is visually and structurally similar to a strange attractor; an iconic form in complexity science (Ostwald 1998a). Whereas in Correa's design, fractal geometry is at best a signpost to a larger idea and at worst a prosaic decoration, in the case of Ushida Findlay, an awareness of its structure has offered an insight into the way they design, but this is not always visible in their architecture. Each of these examples is potentially reasonable for their stated purpose, although neither confronts a broad range of themes associated with fractal geometry.

More commonly, architecture that explicitly acknowledges a connection to fractal geometry is inspired by some part of the theory or its imagery even though it does not employ a scientific or mathematical understanding of the concept. Thus, in architecture the fractal tends to serve as a sign, symbol or metaphor representing a connection to something else. For instance, a large number of architectural appropriations of fractal forms are inspired by the desire to suggest a connection to science, nature or ecology, while others use fractals as a means of rejecting the dominant Euclidian geometric tradition in architecture or to suggest some innate authority for a computational approach.

An example of the first of these motivations—to suggest a connection to nature or ecology—is found in the *Botanical Gardens of Medellin*, jointly designed by Plan B Architects and JPRCR Architects, where the architects admit to being 'inspired to attempt a fractal composition' (Martignoni 2008: 55). Haggard and Cooper also use fractal geometry for its 'holistic characteristics and endless scales aiming at the creation of sustainable architecture' (Sedrez and Pereira 2012). In both of these cases, geometric scaling is deployed to evoke a connection to nature; a link which, as previously suggested, might be reasonable in symbolic or phenomenal terms, but does not support any genuine ecological agenda.

The second of the motivations, to reject Euclidean orthodoxies, is found in Kazuo Shinohara's work. Shinohara repeatedly expressed his interest in 'randomness, fuzziness, fractals and chaos' (Graham 2012: 144) as a stimulus for a design approach which is capable of responding to the complexity of contemporary urban environments. Shinohara displays a relatively detailed awareness of the implications of fractal geometry, but his designs tend to comprise simple collisions between Phileban solids (cones, spheres, prisms, and cubes). This is an example of poetic formalism wherein the adoption of an external body of theory (fractal geometry) is used as authority to create a sculptural composition. The resultant building design has no connection, through its form, tectonics or materiality, to fractal geometry.

Zaha Hadid is often described as using fractal geometry (Kelbaugh 2002; Novak 2006; Tiezzi 2006). Bovill (1996) suggests that her *Hong Kong Peak Leisure Club* displays some fractal qualities, while Hadid herself classifies the *Mosque Design* in France 'as a "fractal space", generated by Islamic geometry' (Richardson 2004: 58). However, while Hadid may have been inspired by fractal geometry there is little evidence in her architecture, her design process, sketches or notes to suggest where this inspiration actually found its place in the resultant work. Further examples of mathematical inspiration in architecture include the work of Philippe Samyn who researched 'fractalisation of regular polygons and polyhedrals' (Capron 1993: 90), creating '"harmonic" double curved structures… which are low-cost, lightweight, and easy to erect' (Pearson 2001: 62). Japanese Metabolist Kisho Kurokawa (2000) admits to using fractal geometry as an inspiration in his *Kuala Lumpur International Airport Terminal*. Kurokawa has also used fractal geometry in a more practical way to solve computational modelling and construction challenges in some of his designs (Rawlings 2007).

In all of these examples architects have either accepted, or actively promoted, a connection between their design work and fractal geometry. However, a more contentious category includes works that critics have interpreted as being fractalesque, but without any apparent agreement from the designers involved. For example, Bovill (1996) and Salingaros (1998) separately observe that Lucien Kroll's *The Architecture of Complexity* (1986), contains images and ideas which are suggestive of fractal geometry. This is true, but Kroll does not mention fractals and most of his ideas about complexity relate to the use of modular elements in design and construction. A further contested example is found in the work of Daniel

Libeskind. Derek Thomas, writing about Libeskind's *Jewish Museum*, argues that '[f]ractal geometry can be … discerned in the way the openings on the vast aluminium cladding reflect the form of the plan and section' (2012: 191). Such formalist readings are repeated in a range of scholarly works, often without any apparent awareness of what these claims imply (Jencks 1995). For example, Salingaros (2004) not only rejects such propositions but he is highly critical of Jencks's claim that Gehry's Guggenheim in Bilbao is self-similar and thereby fractal. Salingaros, argues that Jencks 'is misusing the word "fractal" to mean "broken, or jagged" [and]… he has apparently missed the central idea of fractals, which is their recursiveness generating a nested hierarchy of internal connections' (2004: 47).

2.2.3 Fractally-Generated Architecture

The last category of fractal architecture encompasses designs that have been generated using the mathematics, rules or processes of fractal geometry. These examples range from the straightforward proposition to construct a classical fractal set and inhabit it, to more elaborate, computational, algorithmic or scripted approaches to evolving a formal solution to a design problem. Amongst the more literal examples is Bolles Wilson's proposal for the *Forum of Water* in the 1993 Das Schloss Exhibition, a design in the shape of a modified Menger Cube—a classic or ideal fractal object. Menger Cubes have also been proposed as architectural designs in the works of the Russian Paper Architects Turin and Bush, Podyapolsky, and Khomyakov (Ostwald 2010a) and as a façade treatment in the architecture of Steven Holl (2010). Other literal constructions of this type include designs that resemble strange attractors (Tiezzi 2006) and Julia Sets (Dantas 2010).

By adopting fractal geometry as a formal generator, architects have manipulated mathematical fractals to produce shapes, layouts or patterns using both manual and computational techniques. Such methods typically commence with a starting shape and a generating rule that is repeatedly applied to the shape. This process can be used to create a plan, elevation or three-dimensional form. However, fractals generated in this way are potentially problematic as they are rarely suitable for inhabitation. For example, in Eisenman's *House 11a*, an 'L-shaped' form is traced within itself at increasingly smaller scales, until it is 'paradoxically filled with an infinite series of scaled versions of itself rendering it unusable' (Ostwald 2001a: 74–75). While Eisenman's proposal for a house that is uninhabitable, by virtue of its recursive nature, is deliberately provocative, it reflects one of the key practical problems of fractal generation: when to stop the iterative process. Thus, in most circumstances, only a partial generative procedure is used for the building form or surface.

Projects such as *House 11a* led many scholars to posit that Eisenman's architecture has fractal qualities (Jencks 1995; Pearson 2001; Kelbaugh 2002; Tiezzi 2006). Certainly, Eisenman's project for a biological research centre at the

University of Frankfurt offers an example of architectural scaling, rather than fractal scaling, as does his 1985 project *Moving Arrows, Eros and other Errors*. But his motivation in each of these cases is to reject the apparent hegemony of Euclidian geometry rather than to propose a use for fractal geometry, which is why these projects resist more detailed consideration of their formal properties.

Given that fractal geometry was presented to the world through its evocative computer-generated imagery, it is only to be expected that architects would swiftly develop algorithms for creating fractal designs. Unfortunately, the core problem with this endeavour is that the resultant 'organic', 'blob-like' or 'crystalline' forms are often unusable or unfeasible for architectural purposes (Yessios 1987; Akleman et al. 2005; Joye and Van Locke 2007; Wang et al. 2008). Of the large number of these works that have been generated, relatively few have resulted in habitable and constructible forms (Coates et al. 2001). Renato Saleri (2005) developed an interesting alternative to this tradition, which proposed using fractal algorithms to generate building façades from a palette of architectural elements. Thus, rather than producing ambiguous organic shapes, Saleri created elevations and forms which feature windows, doors and other building elements that have been placed in accordance with a set of rules, rather than the needs of inhabitants. In a different way, Harris (2007) uses mathematical transformations and repetitions of simple forms to generate potentially functional architectural designs, mostly skyscrapers. He examines the basic design 'rules' of various styles of architecture or architects and applies these rules to his fractal, generative process. In this manner, he has produced three-dimensional images of buildings which look convincingly similar to real Art Deco towers and designs by Frank Lloyd Wright.

One of the dilemmas with using any generative design algorithm to create architecture is that it is almost always form-based. That is, these methods are used for creating shapes, not for accommodating social structures or functional needs, and lack any sense of the tectonic or material properties of the form (Ostwald 2004; 2010b). Furthermore, such evolved forms typically do not take into account environmental or cultural considerations. As Gert Van Tonder observes, '[f]ractals emulate our natural visual surroundings in terms of structural self-similarity, a fact which unfortunately renders architectural fractals prohibitively expensive to construct, and inefficient as architectural space for human occupation' (2006: 2). This is why, instead of attempting to design a complete building using fractal algorithms, most architectural works in this category only use fractals to generate part of the building. As Holl observes, '[a] real building, of course, cannot be a perfect mathematical figure' (2010: 7).

Several attempts have been made to break away from the form-dominated approach to fractal architecture. In a very early strategy to address the problems inherent in fractally-generated designs, Yessios notes the paradox that, 'if left unrestrained', a fractal process 'will go on forever' (1987: 173), then proceeds to identify the fundamental problem mentioned previously in this section arguing that 'if applied in a 'pure' fashion, [fractal geometry] will create an interesting shape but will never produce a building. A building typically has to respond to a multiplicity of processes, superimposed or interwoven. Therefore, the fractal process needs to be guided, to be constrained and to be filtered' (1987: 173).

In a similar way, almost two decades before architects began to design buildings using fractal algorithms, Schmitt suggested that the resultant designs would be too 'boring' without some additional consideration of the qualities that define architecture. 'The next logical step', Schmitt argues, is towards modes of 'computer creativity' which can develop 'context sensitive associations, preferably in interaction with the user' (1988: 103). Schmitt rejects the simplistic formalism so often associated with fractally-generated design, calling for architecture to be sensitive to external, site-based, and internal, inhabitation-based, qualities. In a more recent reflection of this proposition, Sedrez and Pereira propose a method which commences with a fractal form and then uses functional emergence principles to 'choose the appropriate size for that object, … proposing architectural information through doors/windows, landscaping, furniture and surfaces, finishes, colours' (2012: 100).

The suggestion that a fractal process might be modified in some way to produce a context-sensitive architecture has become a common response to the problems of creating useful (functional, constructible and inhabitable) geometric objects. Frank Gehry's name is often associated with this trend as well as with the wider use of computational fractal models (Jencks 1995; Pearson 2001; Tiezzi 2006; Vyzantiadou 2007). For example, according to Thomas, in Gehry's design process, '[f]ractal geometry is applied through programmed formulae in the software and then manipulated to create the resultant form. … Instead of forcing conventional geometry onto [a] natural landform, the dynamic positioning of architectural form in context with its site using an iterative design syntax of fractal geometry … will present design possibilities in a meaningful way' (2012: 191). While the extent to which Gehry actually uses this technique is debatable (Ostwald 2006), as too is the degree to which this is even possible (Terzidis 2006), the recognition that fractally-generated designs must be modified through the inclusion of a range of site- or context-based measures is a positive development.

A more interesting application of fractal generation to design is associated with 'contextual fit'; that is, the capacity of a new intervention to be 'sympathetic to', or 'in keeping with', the visual character of its surrounding site. Applications of fractal generation have been proposed for a range of urban neighbourhoods and regions (Kobayashi and Battina 2005; Marsault 2005; Saleri 2005). However, much like the architectural examples of generative design, some of these cases are dominated by formalist solutions, which have only limited connection to their sites or cultural contexts. In contrast, Bovill proposes that it is possible to measure the fractal dimension of a site or environment, and then generate a design with the same fractal dimension, to produce a visually coherent addition to a location. For example, 'the fractal dimension of a mountain ridge behind an architectural project could be measured and used to guide the fractal rhythms of the project design. The project design and the site background would then have a similar rhythmic characteristic' (Bovill 1996: 6). Bovill also offers the example of the design of a noise abatement wall with the same fractal dimension as that of the forest behind the wall. Similar concepts have been tested both perceptually and mathematically (Vaughan and Ostwald 2009a; Lorenz 2012). However, cases where generated architectural designs match the dimensions of their context are rare. One possible exception is found in the work of Arthur Stamps (2002) who generated images of high-rise buildings for use in perceptual experiments to test Bovill's theory. Gozubuyuk, Cagdas and Ediz (2006) determined the fractal dimension of the urban layout and typical buildings of historical districts of Turkish cities and then generated a building design with a similar fractal dimension to respond to the existing architectural 'languages' of the districts. Similarly, Wang, Ma and Liu (2008) selected a fractal dimension derived from a geometric 'dust' (a type of fractal set), then used computational algorithms to produce an architectural shape to match that dimension.

Despite the usefulness (or lack thereof) of the generated form, the concept of modelling a building or surface to achieve a distinct fractal dimension is a valid one. For example, Sakai's (2012) team examined the fractal dimension of tree canopies and produced a shelter with a similar dimension, as a means of shedding heat load. The surface was successful for this purpose, although it is uncertain whether this was a by-product of the fractal dimension or was a combined property of the material it was constructed from and its design. Several other geometric surfaces have been applied to major buildings in the past to achieve a type of urban contextual fit. Van Tonder (2006) even observes that using fractals as a surface treatment may allow for a more practical solution to architectural borrowings from non-linearity by using computational layering to create details on a façade. Strategies of this type have already been tested in several major buildings including *Storey Hall* by architects Ashton Raggatt McDougall and LAB Architecture Studio's *Federation Square*, both in Melbourne (Australia). The former building is clad in a bright, tessellated pattern known as Penrose Aperiodic Tiling, and the latter is clad in a more subdued tessellation known as Conway Pinwheel Tiling. Both of these buildings have been described as featuring fractal façades, but neither of these is actually fractal. The Conway tile does scale, but then so too do many other conventional building surfaces that would not be considered fractal, and neither tessellation has a clear structural rule for generational growth. Tessellations are a category of plane-filling topographic structures which are superficially reminiscent of fractals but which actually have a range of innate architectural qualities which have, thus far, largely eluded architects (Ostwald 1998b; Bovill 2012; Ostwald and Williams 2015b).

2.3 Conclusion

Twenty-six years after the publication of his seminal text, Benoit Mandelbrot was asked if he thought that Frank Gehry's work expressed some of the properties of fractal geometry. 'No', Mandelbrot replied, 'I find Gehry repetitive' (Mandebrot qtd. in Obrist 2008). While Mandelbrot then went on to say positive things about the geometric relationships found in Gehry's work, the lack of phenomenal scaling led to his emphatic rejection. In his later life Mandelbrot began to differentiate between two categories of fractals. The first, classical or ideal fractal sets, which he would later call 'uni-fractals' and the second being statistical sets, which he would call 'multi-fractals'. The former category comprises precise, abstract and infinitely scalable geometric sets, which can neither be constructed nor inhabited. The second category, the multi-fractal, includes architecture; a geometric object which cannot have true fractal geometry, but which can have fractal dimensions. While the precise difference between these two properties is described in other chapters in this book, in the present context this distinction is useful for thinking about the idea of fractal architecture. It is difficult, if not impossible, for architecture to provide a consistent, perfect or holistic connection to fractal geometry in any meaningful way. But architecture can, potentially, have multiple different connections to fractal geometry, all of which, within clearly described limits, are informative or useful.

It would be a simple task to list the multitude of inaccurate, incorrect and often bizarre things that architects have said about fractal geometry. But, as Robin Evans notes, 'architects do not produce geometry, they consume it' (1995: xxvi), and we would add, in gluttonous and indefatigable ways. Therefore, the literal appliqué of a fractal image to the side of a building certainly does not make that building more ecologically sustainable, but it might act as a signpost to the concerns or values of its inhabitants. Similarly, a great building could be inspired by fractal geometry, but possess no clear trace in the finished design of the origins of that inspiration. Furthermore, as various scholars have noted when considering the philosophical musings of Deleuze and Guattari (1987), a narrow, superficial use of an idea might be valid, as long as its limitations are clear. The problem arises when, for example, fractal geometry is used as justification for a complex, costly form, because it is allegedly a scientific approach to design, or when architectural critics are drawn to describe a random jumble of forms as fractal and thereby suggest some universal quality is implicit in a design. Conversely, as several of the examples in this chapter demonstrate, it is possible to develop and maintain a phenomenological interpretation of the fractal experience of form. It is also possible to develop a more detailed understanding of both historic and modern buildings, in terms of their repetitive, scaled structures. Thus, to return to the theme developed at the beginning of this chapter, there is a reason why no agreed upon definition of fractal architecture currently exists, but this does not justify abandoning all consideration of fractal geometry in architecture, or as the rest of this book demonstrates, the fractal dimensions of architecture.

We may summarise the three key messages to be found in this chapter as follows. First, the diverse and often controversial definitions of fractal geometry that have previously been developed in architecture need to be framed appropriately if they are to be taken seriously. For example, using experiential descriptors to examine fractalesque qualities in a building may be appropriate, provided that the author does not claim that the reasoning is scientifically based. The most important factor is not necessarily whether a geometric, generative or phenomenal view is taken, but rather that each author is clear about the perspective chosen, its purpose and limitations. Thus, when working with fractal geometry, scholars and designers should be especially careful to ensure that they describe how they are using it: as structure, as form, as ornament or as inspiration. Second, fractal algorithms and other computational methods of generating forms cannot be used to produce a complete, finished design for a building without some input from the designer, either in the decision-making process or in the authoring stage. Fractally-generated designs must be modified through the inclusion of a range of site or context-based measures before they can become designs suitable for habitation. The vast and growing body of examples of computer-evolved buildings all require sensible human input (either through direct intervention or the authoring of parameters to ensure functional and social conditions are met) to create architecture. Finally, architects should remember that there are two completely different approaches to considering fractals in the context of design. The one covered in the majority of this chapter involves *fractal geometry* and its associated imagery, which can provide inspiration for designers. The second approach is about the way in which architecture can be measured and analysed using *fractal dimensions*. As the remainder of this book demonstrates, every object, whether natural or synthetic, can have its formal complexity measured or estimated.

Chapter 3
Introducing the Box-Counting Method

This chapter presents three worked examples of the most basic variation of the box-counting method for calculating the fractal dimension of an image. The first of these uses a small-scale façade element, a window in a historic building, as its subject, and the remaining two examples use elevations of famous houses. Thereafter the chapter provides a background to the application of the box-counting method in architectural and urban analysis and describes the analytical intent and conclusions of this past work. Throughout these sections various weaknesses in the method and its architectural application are identified. In particular, in this chapter we observe that the box-counting method can be highly sensitive to data standards, representational decisions and methodological issues. Thus, even though this chapter provides an explanation of the method, it is only in the two chapters that follow that we develop a reasoned approach to solving two important issues: *which* facets of architecture should be measured, and *how* can we ensure that these measurements are reasonable, repeatable and accurate.

3.1 Introduction

The box-counting method for determining the fractal dimension of an image is probably the best-known approach, in any discipline, for quantifying characteristic visual complexity. This method has been studied extensively and applied in the sciences and mathematics and, over time, several variations of it have been developed for use in different fields. For example, specific versions have been developed for biology (De Vico et al. 2005), neuroscience (Jelinek et al. 2005), mineralogy (Blenkinsop and Sanderson 1999), geology (Grau et al. 2006) and physics (Kruger 1996). The reason these variations exist is that the box-counting method is known to have particular strengths and weaknesses in certain ranges of dimensions and for particular image types. As a result of this, scientists and mathematicians have identified several mathematical refinements, along with a

M.J. Ostwald and J. Vaughan, *The Fractal Dimension of Architecture*,
Mathematics and the Built Environment 1, DOI 10.1007/978-3-319-32426-5_3

range of methodological and data variables that, in combination, can be optimised to meet the needs of different disciplines.

The particular version of the box-counting approach that has typically been used in almost all architectural and urban analysis is known by scientists and mathematicians as the 'basic' or 'naïve' version, because it uses the base mathematical process without any optimisation or refinement (Huang et al. 1994). This particular version commences with, for example, an architectural image, say an elevation of a façade. A grid is then placed over that image and each square in the grid is analysed to see if any of the lines (often called 'information' in the scientific applications) of that elevation drawing are present. The number of boxes with lines in them is then recorded, often by cross-hatching the cell and then counting the number of cells which have been marked in this way. Then a grid of reduced size is overlaid on the same image and the process is repeated, now at a different scale, and the number of boxes with lines in them is also recorded. A mathematical comparison is then made of the number of boxes with detail in the first grid ($N_{(s1)}$) and the number of boxes with detail in the second grid ($N_{(s2)}$). Such a comparison is made by plotting a log-log diagram (log[$N_{(s\#)}$] versus log[1/s#]) for each grid size. The slope of the straight line produced by this comparison is called the box-counting dimension (D_b). This value is calculated for a comparison between two grids (# = 1 and # = 2 in this example) as follows:

$$D_b = \frac{[\log(N_{s2}) - \log(N_{s1})]}{[\log(1/s2) - \log(1/s1)]}$$

where
$N_{(s\#)}$ = the number of boxes in grid number "#" containing some detail
1/s# = the number of boxes in grid number "#" at the base of the grid

When this process is repeated a *sufficient* number of times, for multiple grid overlays on the same image, the average slope can be calculated, producing the fractal dimension (*D*) of the image. The critical, and often forgotten, word in this sentence is *sufficient*; the lower the number of grid comparisons the less accurate the result, the higher the number of comparisons the more accurate the result. In essence, the fractal dimension is the mean result for multiple iterations of this process and an average of only two or three results will necessarily be inaccurate. For example, an average of two figures will likely produce a result with only ±25 % accuracy; or a potential error of 50 %. A comparison of three scales will typically only reduce this to ±22 % accuracy. In order to achieve a useful result at least eight and preferably ten or more comparisons are needed, reducing the error rate to around ±1 % or less. However, this is a somewhat simplistic explanation, because the error rate is also sensitive to other factors, including the quality of the starting image, the configuration and positioning of successive grids and the scaling coefficient (the degree by which each successive grid is reduced in size).

If all of these other factors are optimized, then the error rate will be reduced to such a level that between eight and ten comparisons will be sufficient to achieve a reasonable result. If none of these factors are optimized, then up to one hundred comparisons may be required to achieve a highly accurate result. Keeping this limitation in mind, the mathematics of the method is demonstrated hereafter in three simple examples.

3.1.1 Mosque Window Detail

In this first example the method is demonstrated using a window detail taken from the north-west elevation of the *Süleymaniye Mosque* in Istanbul (Figs. 3.1 and 3.2). Four grid overlays are provided creating three grid comparisons (1–2, 2–3 and 3–4). Each successive grid is half the dimension of the previous one, normally described as using a scaling coefficient of 2:1. This is the most common and practical scaling coefficient used in architectural analysis, but not, as we will see, the most accurate or useful one for generating multiple points for producing a statistically viable result.

i. In the first grid (# = 1), with a 3 × 5 configuration (1/s1 = 3) there are 15 cells ($N_{(s1)}$ = 15) with detail contained in them (Fig. 3.3).
ii. In the second grid (# = 2), with a 6 × 10 configuration (1/s2 = 6) there are 34 cells ($N_{(s2)}$ = 34) with lines contained in them (Fig. 3.4).

Fig. 3.1 Entry façade, north-west elevation of the *Süleymaniye Mosque*, Istanbul

Fig. 3.2 Starting image: window detail

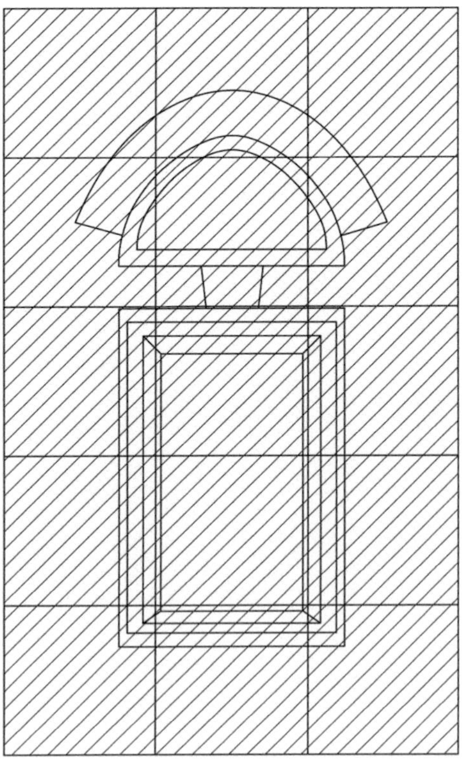

Fig. 3.3 Grid 1: 3 × 5 grid; box count 15 or 1/s1 = 3 and $N_{(s1)}$ = 15

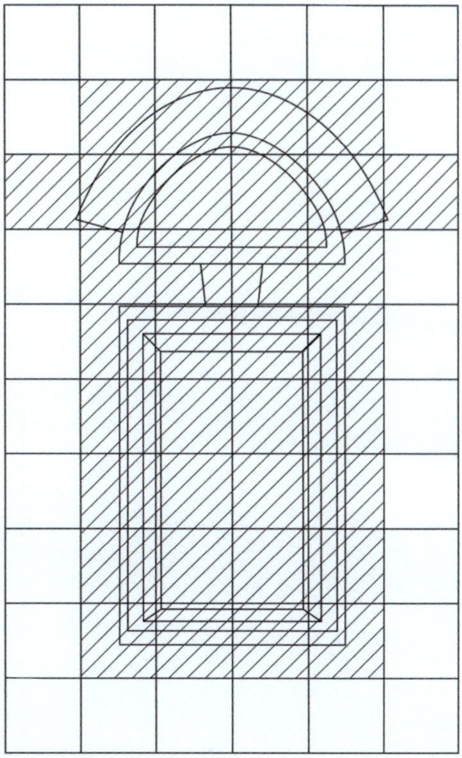

Fig. 3.4 Grid 2: 6 × 10 grid; box count 34 or 1/s2 = 6 and $N_{(s2)} = 34$

iii. In the next grid (# = 3), with a 12 × 20 configuration (1/s3 = 12) there are 88 cells ($N_{(s3)} = 88$) with lines contained in them (Fig. 3.5).
iv. In the final grid in this example (# = 4), with a 24 × 40 configuration (1/s4 = 24) there are 246 cells ($N_{(s4)} = 246$) with lines contained in them (Fig. 3.6).

Before progressing with the calculations, note that in this section figures are rounded to three decimal places and because the scaling coefficient is 2:1 in all cases, the ultimate denominator is always 0.301.

Using the standard formula and the information developed from the review of the grid overlays, the comparison between grid 1 and grid 2 is constructed mathematically as follows:

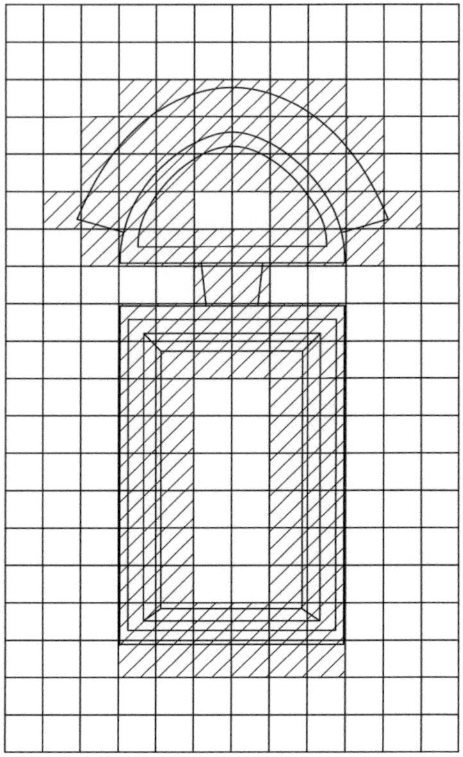

Fig. 3.5 Grid 3: 12 × 20 grid; box count 88 or 1/s3 = 12 and $N_{(s3)}$ = 88

$$D_b = \frac{[\log(N_{s2}) - \log(N_{s1})]}{[\log(1/s2) - \log(1/s1)]}$$
$$D_b = \frac{[\log(34) - \log(15)]}{[\log(6) - \log(3)]}$$
$$D_b = \frac{[1.531 - 1.176]}{[0.778 - 0.477]}$$
$$D_b = \frac{0.355}{0.301}$$
$$D_b = 1.179$$

Thus, the first box-counting dimension of the window is 1.179.
Then grid 2 and grid 3 are compared as follows:

Fig. 3.6 Grid 4: 24 × 40 grid; box count 246 or 1/s4 = 22 and $N_{(s4)}$ = 246

$$D_b = \frac{[\log(88) - \log(34)]}{[\log(12) - \log(6)]}$$
$$D_b = \frac{[1.944 - 1.531]}{[1.079 - 0.778]}$$
$$D_b = \frac{0.413}{0.301}$$
$$D_b = 1.372$$

The second box-counting dimension of the window is 1.372.

Then grid 3 and grid 4 are compared in the same way:

$$D_b = \frac{[\log(246) - \log(88)]}{[\log(24) - \log(12)]}$$
$$D_b = \frac{[2.391 - 1.944]}{[1.380 - 1.079]}$$
$$D_b = \frac{0.446}{0.301}$$
$$D_b = 1.485$$

The last of the three box-counting calculations for the window gives a result of 1.485. The mean for these comparisons—which is an estimate of D, or alternatively a D calculation with a high error rate as a result of such a limited data set—is therefore:

$$D = \frac{1.179 + 1.372 + 1.485}{3}$$
$$D = 1.345$$

The set of results are then graphed in a log-log graph (that is, both scales are logarithmic), with the box-count (y axis) against the box size (x axis). In this first example, the three comparison results appear relatively close to the mean (Fig. 3.7).

3.1.2 *The Robie House*

The second worked example is a partial calculation of the fractal dimension of the west elevation of the *Robie House* (Figs. 3.8 and 3.9). The same comments provided previously about rounding decimal places, scaling coefficient (with ultimate

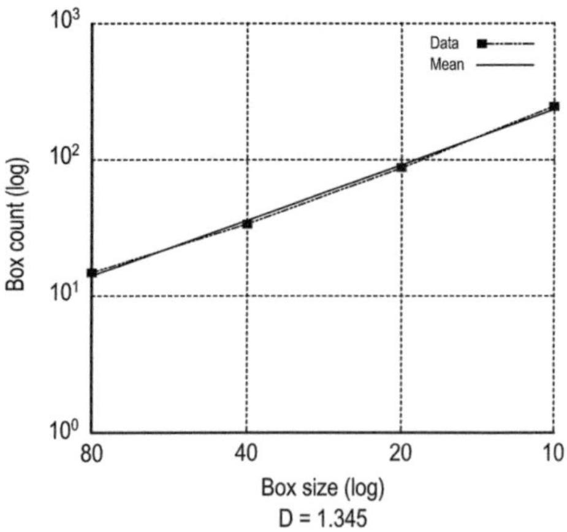

Fig. 3.7 Log-log graph for the first three comparisons of the window detail

Fig. 3.8 The *Robie House*

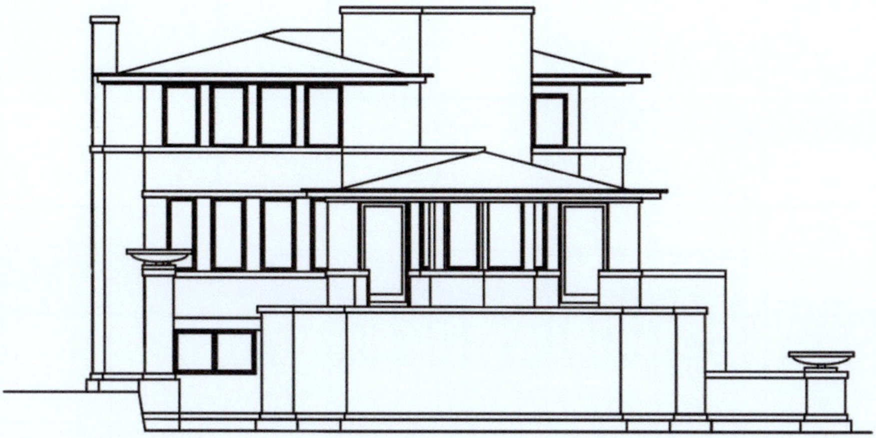

Fig. 3.9 Base image, west elevation

denominator being 0.301) and error rates also apply to this case. In this example, four grids are constructed and three comparison values calculated. The first grid is 5×3 in configuration and has 13 cells with information contained in them and the second grid is 10×6 with 29 cells containing information (Figs. 3.10 and 3.11). The final two grids, respectively three and four, have 20×12 and 40×24 configurations, and 93 and 307 cells with information (Figs. 3.12 and 3.13).

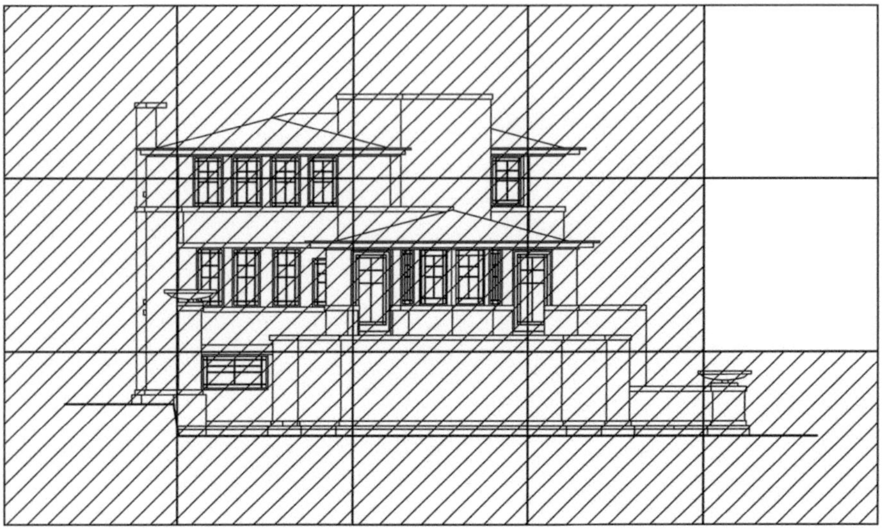

Fig. 3.10 Grid 1: 5 × 3 grid; box count 13 or $1/s1 = 5$ and $N_{(s1)} = 13$

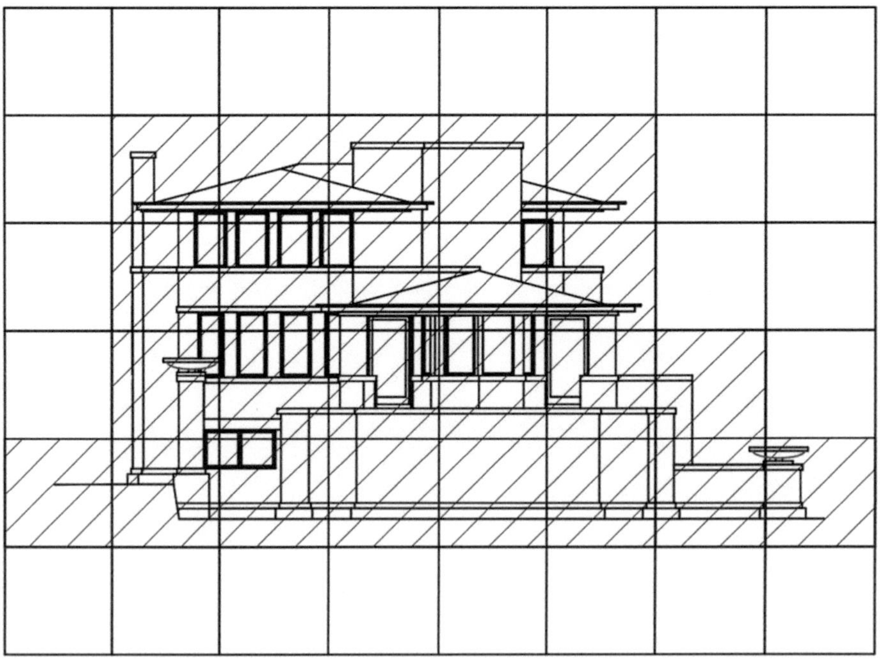

Fig. 3.11 Grid 2: 10 × 6 grid; box count 29 or $1/s2 = 10$ and $N_{(s2)} = 29$

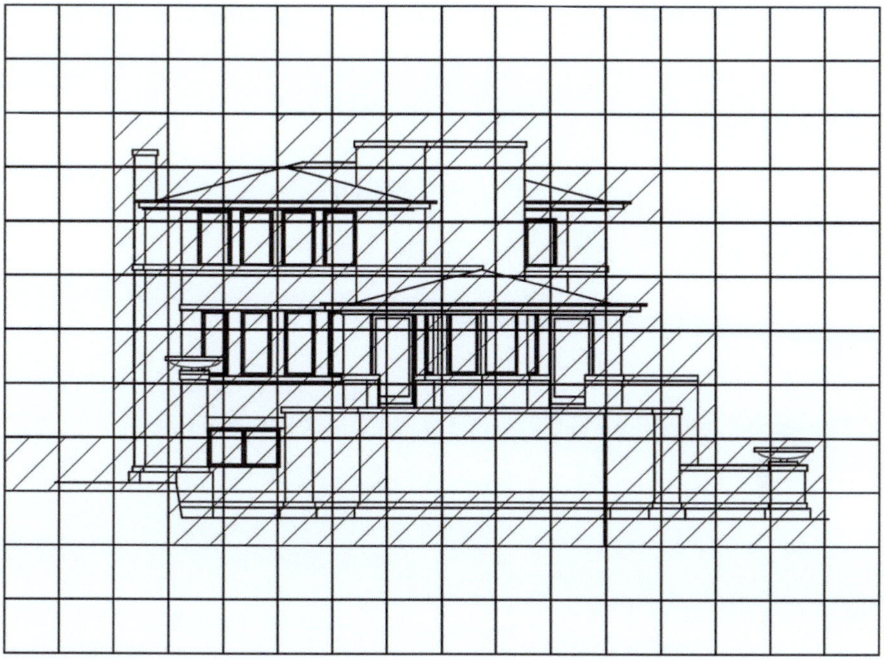

Fig. 3.12 Grid 3: 20 × 12 grid; box count 93 or $1/s3 = 20$ and $N_{(s3)} = 93$

Fig. 3.13 Grid 4: 40 × 24 grid; box count 307 or $1/s4 = 40$ and $N_{(s4)} = 307$

The first comparison between grid 1 and grid 2 is constructed as follows:

$$D_b = \frac{[\log(N_{s2}) - \log(N_{s1})]}{[\log(1/s2) - \log(1/s1)]}$$
$$D_b = \frac{[\log(29) - \log(13)]}{[\log(10) - \log(5)]}$$
$$D_b = \frac{[1.462 - 1.114]}{[1 - 0.699]}$$
$$D_b = \frac{0.348}{0.301}$$
$$D_b = 1.156$$

The second comparison between grid 2 and grid 3 is as follows:

$$D_b = \frac{[\log(93) - \log(29)]}{[\log(20) - \log(10)]}$$
$$D_b = \frac{[1.968 - 1.462]}{[1.301 - 1]}$$
$$D_b = \frac{0.506}{0.301}$$
$$D_b = 1.681$$

The calculation is repeated to compare grids 3 and 4:

$$D_b = \frac{[\log(307) - \log(93)]}{[\log(40) - \log(20)]}$$
$$D_b = \frac{[2.487 - 1.968]}{[1.602 - 1.301]}$$
$$D_b = \frac{0.509}{0.301}$$
$$D_b = 1.724$$

Combining the three box-counting results leads to a fractal dimension estimate of $D = 1.520$. The results are then graphed (Fig. 3.14).

3.1.3 The Villa Savoye

The front elevation, or 'Elevation 1' as Le Corbusier designated it, of the *Villa Savoye* in Poissy (France), is the subject of the third worked example (Figs. 3.15 and 3.16). Once more, four grid overlays are presented leading to three

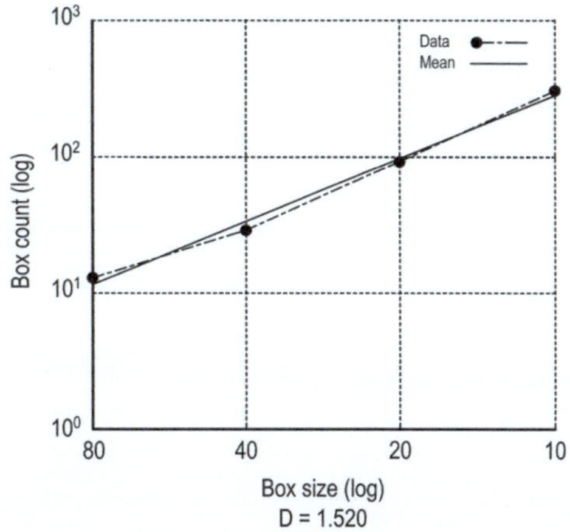

Fig. 3.14 Log-log graph for the first three comparisons of the *Robie House* elevation

Fig. 3.15 Villa Savoye

comparisons. The scaling coefficient is 2:1 and the four configurations are: 5×3, 10×6, 20×12 and 40×24. The box count for the four grids is, respectively, 15, 38, 123 and 331 (Figs. 3.17, 3.18, 3.19 and 3.20).

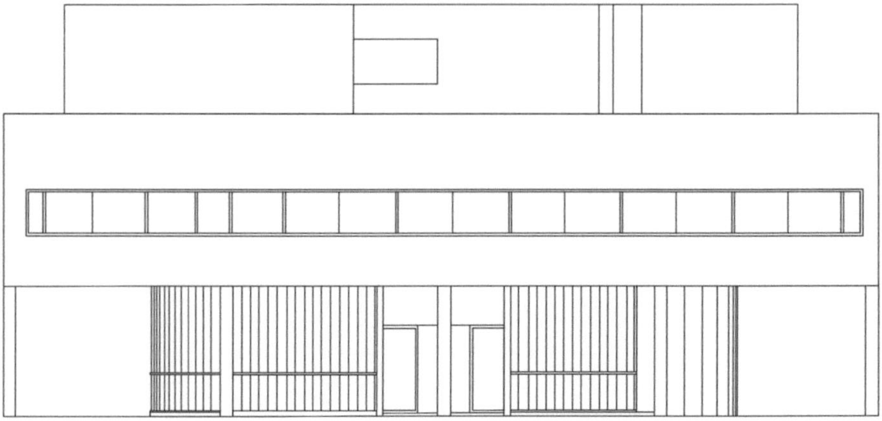

Fig. 3.16 Base image, elevation 1

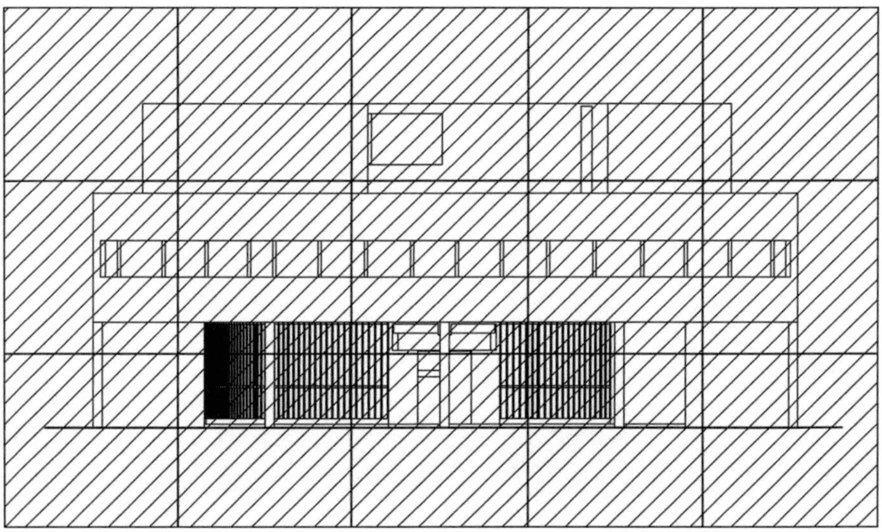

Fig. 3.17 Grid 1: 5 × 3 grid; box count 15 or 1/s1 = 5 and $N_{(s1)} = 15$

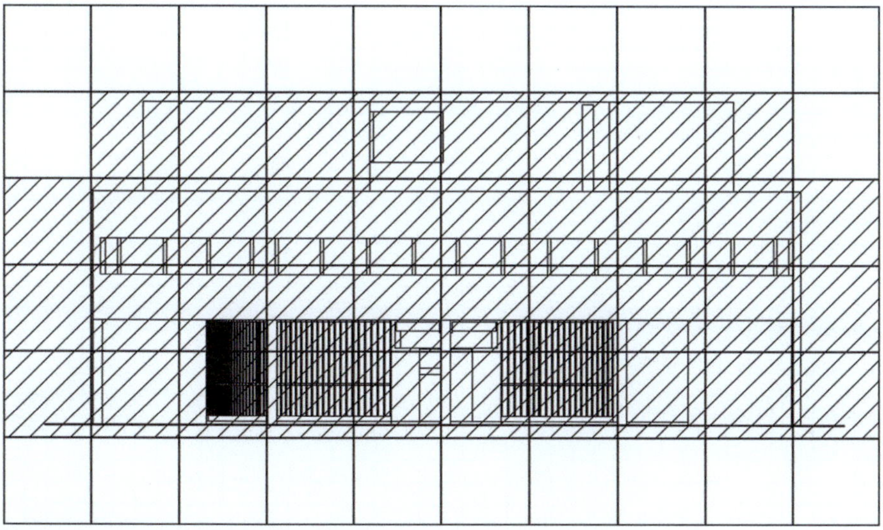

Fig. 3.18 Grid 2: 10 × 6 grid; box count 38 or 1/s2 = 10 and $N_{(s2)} = 38$

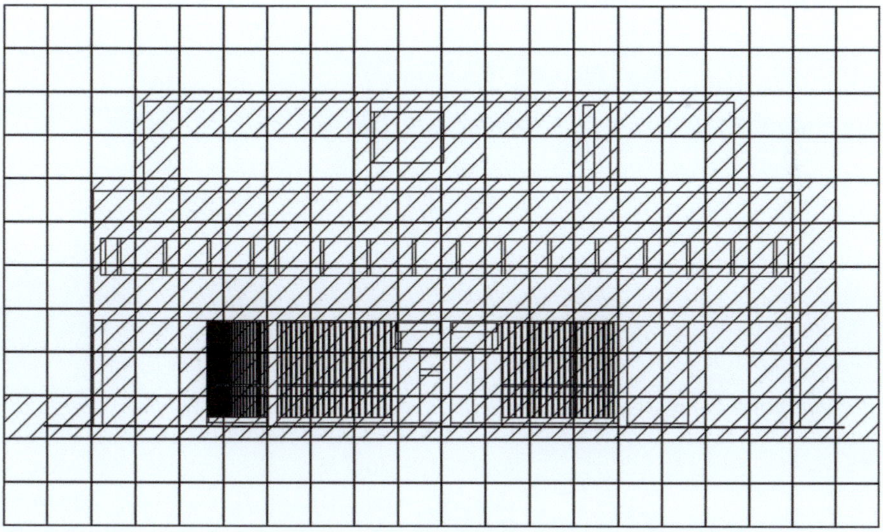

Fig. 3.19 Grid 3: 20 × 12 grid; box count 123 or 1/s3 = 20 and $N_{(s3)} = 123$

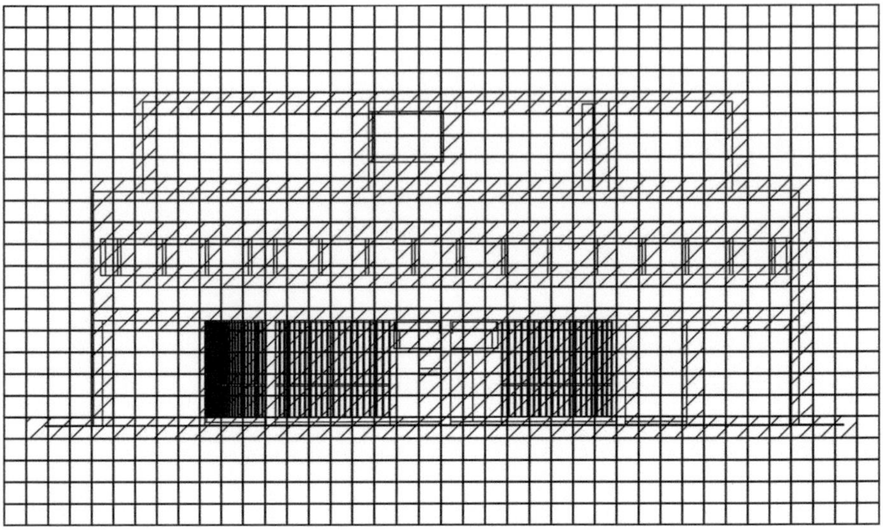

Fig. 3.20 Grid 4: 40 × 24 grid; box count 331 or 1/s4 = 40 and $N_{(s4)} = 331$

The first comparison between grid 1 and grid 2 is constructed as follows:

$$D_b = \frac{[\log(N_{s2}) - \log(N_{s1})]}{[\log(1/s2) - \log(1/s1)]}$$
$$D_b = \frac{[\log(38) - \log(15)]}{[\log(10) - \log(5)]}$$
$$D_b = \frac{[1.580 - 1.176]}{[1 - 0.699]}$$
$$D_b = \frac{0.404}{0.301}$$
$$D_b = 1.134$$

The second comparison between grid 2 and grid 3 follows the same formula:

$$D_b = \frac{[\log(123) - \log(38)]}{[\log(20) - \log(10)]}$$
$$D_b = \frac{[2.090 - 1.580]}{[1.301 - 1]}$$
$$D_b = \frac{0.510}{0.301}$$
$$D_b = 1.694$$

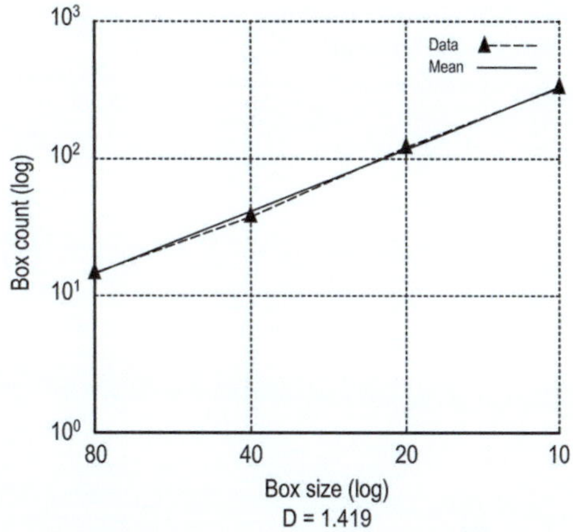

Fig. 3.21 Log-log graph for the first three comparisons of the *Villa Savoye* elevation

The calculation is repeated to compare grids 3 and 4:

$$D_b = \frac{[\log(331) - \log(123)]}{[\log(40) - \log(20)]}$$

$$D_b = \frac{[2.520 - 2.090]}{[1.602 - 1.301]}$$

$$D_b = \frac{0.430}{0.301}$$

$$D_b = 1.429$$

Combining the three box-counting dimensions produces a result of $D = 1.419$. The set of results are then presented in a log-log graph (Fig. 3.21).

3.1.4 Comparison of Results

These three results, while only calculated from a very limited set of data, can also be compared in a reasonably straightforward manner by either charting the data directly (Fig. 3.22) or the trend-lines generated by the three sets of results (Fig. 3.23). Through this comparison it might be possible to suggest that the *Robie House* elevation is the most visually complex of the three ($D = 1.520$), while the *Villa Savoye* elevation is the least complex ($D = 1.345$). While this might broadly reflect our intuitive reading of the elevations, the explanation for the window result, being positioned between the other two, is less readily apparent. Certainly, the orthogonal part of the window frame itself, rather than the arched screen above, is geometrically nested in a way that suggests a scaled and complex form, but the answer to this conundrum is more likely related to the limited data gathered. The full depth of consistent detail in the two elevations is only just beginning to be

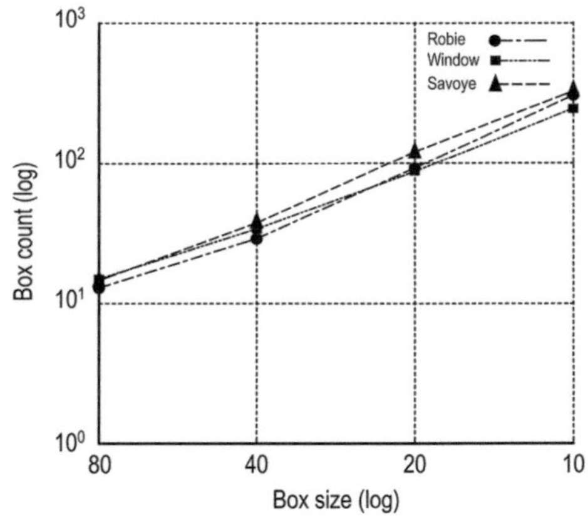

Fig. 3.22 Results for the three worked examples

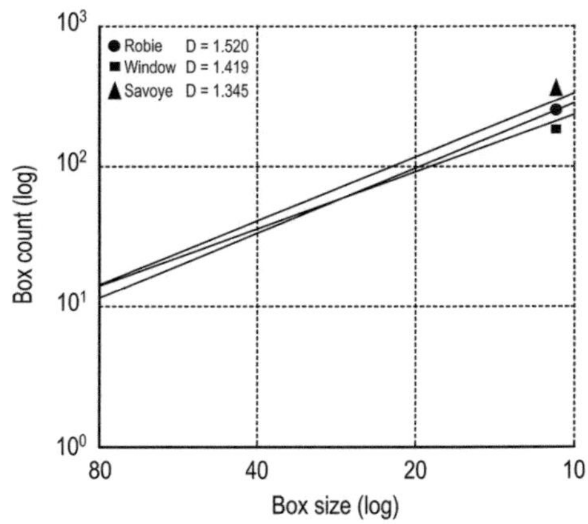

Fig. 3.23 Trendlines for the three worked examples

revealed in three comparisons of grid overlays, but the complete detail of the window is effectively already captured at this scale. Thus, the result for the window is neither statistically supportable in itself (being produced from only three data points) nor very useful for comparative purposes. In contrast, the elevations for the two houses, if appropriately developed over a larger number of grid comparisons, will provide a more refined and accurate measure of their visual character.

The following section provides an overview of past research that has been undertaken using the box-counting method to measure the fractal dimension of the built environment. The earliest example of this type can be traced to 1994 and since that time, such studies have increased in both quantity and breadth of application (Batty and Longley 1994). The scale of these studies varies from the analysis of city plans to measurements of individual buildings and architectural details. Many of these studies were undertaken using a manual version of the method which, much like the worked examples in the present section, rely on a person physically counting the number of details in various grids, then using formulas to calculate the fractal dimension of an image. The more recent examples tend to use purpose-designed or authored software to undertake much larger and more accurate applications of the method. Despite this difference, the basic approach remains the same. However, as several comments in the present section indicate, not all of the results of this basic variation of the method are useful or accurate. Thus, despite an increase in the application of this method, surprisingly few of these past studies include details about the particular variation they employ, or the settings and raw data they use for their calculations. This means that the results of most of these studies are impossible to replicate. Furthermore, there are some serious methodological flaws in a few of the past applications along with more subtle problems with the way authors have interpreted their results. While a small number of these concerns are noted in the next section, its primary purpose is not to be critical of these works, but to describe the breadth of applications of the method. Furthermore, in many of the cases described hereafter—including some of our own early works—the strengths and weakness of the method were so poorly understood at the time they were produced that some mistakes were inevitable. Nevertheless, with more recent advances, including those described in the present book, such flaws should be a thing of the past.

3.2 The Application of Fractal Analysis to the Built Environment

This section divides past research using the box-counting approach broadly by application, starting with research that is focussed on urban forms and then considering those that focus on architecture. While many of the results of these studies are described in the text, most cannot legitimately be compared with each other because they use different starting points (from photographs to sketches and line drawings) and different data extraction and processing procedures (from manual techniques to software supported ones). Thus, as much as some readers might want to delve more deeply into patterns suggested in these disparate results, in the majority of cases no consistent basis is available for constructing such a comparison. Furthermore, for the sake of producing a relatively complete overview, several of our own past publications are included in the discussion. In a few cases the original published results have now been completely revised and refined and are included in later chapters. Thus, in this section we also describe some of our earlier published research, but the more definitive results are contained in the present book.

3.2.1 Urban Analysis

Studies of cities using fractal analysis range from a consideration of urban morphology to measurements of the plans of streets, transport networks and green spaces. Observations about the potential fractal dimension of urban forms began to be published in the late 1980s (Yamagishi et al. 1988) and since then a growing number of different approaches to the fractal analysis of urban plans have been proposed (Oku 1990; Mizuno and Kakei 1990; Rodin and Rodina 2000; Ben-Hamouche 2009). However, the earliest research to specifically use the box-counting method in urban analysis can be traced to Michael Batty and Paul Longley (1994) who employed a variation of the method, which they called 'cell-counting', to examine changes in the growth and form of urban boundaries. Following their work, fractal analysis continued to be used to measure changing urban forms including studies of Tel Aviv (Benguigui et al. 2000) and London (Masucci et al. 2012), along with shifting settlement patterns in Mayan cities (Brown and Witschey 2003). The application of the box-counting method to the analysis of urban form has also been undertaken by Barros-Filho and Sobreira (2005), who examined, amongst other areas, slums in Brazil. The box-counting method has since been used to compare the fractal dimension of street patterns in more than twenty cities (Cardillo et al. 2006) and a worldwide urban classification system using fractal dimensions has been proposed (Encarnação et al. 2012).

In a variation of these urban approaches, the box-counting method has also been used to analyse transportation networks and their impact on settlement patterns, including a comparison between Seoul and Paris (Kim et al. 2003). Lu and Tang (2004) used the method to analyse the connection between city size and transportation networks in Texas, while Thomas and Frankhauser (2013) compared the dimensions of developed spaces and roadways in Belgium. At a smaller scale Eglash (1999) examined plans of part of a Mofou settlement in Cameroon and the urban core of the Turkish city of Amasya, the latter of which has been the subject of several studies about the relationship between the fractal dimension of traditional urban centres and of their surrounding natural context (Bovill 1996; Lorenz 2003; Vaughan and Ostwald 2009a). Green spaces, typically urban parks, have also been measured using box-counting to develop a model for sustainable development (Wang et al. 2011) and to compare the porosity of parks in the USA, China and Argentina (Liang et al. 2013).

All of these examples of the measurement of urban form are focussed on plan views (or aerial photographs, which are treated as a type of plan). An alternative approach is found in a small number of examples that analyse elevations or perspectives of urban forms, in the latter case from the point of view of a pedestrian. In particular, Jon Cooper has led a series of detailed studies of streetscape quality in Oxford (Cooper and Oskrochi 2008; Cooper et al. 2010) and Taipei (Cooper et al. 2013) using the box-counting method. Distant views of city skylines have also been analysed by Stamps (2002) and the visual qualities of city skylines in Amsterdam, Sydney and Suzhou have been measured and compared (Chalup et al. 2008).

In the majority of these examples of urban dimensional analysis, the box-counting method has been used to quantify the characteristic complexity of a city, including its growth patterns, road and rail networks, open spaces and skylines. Several of the studies also display an awareness that fractal dimensions are more informative when used for comparative purposes, or for classifying different types of patterns against a standard value or measure.

3.2.2 Architectural Analysis

The first serious attempt to calculate the fractal dimension of architecture using the box-counting method is found in the work of Carl Bovill, whose *Fractal Geometry in Architecture and Design* (1996) provided the first major exploration of the relationship between fractal geometry and art, music, design and architecture. In that work Bovill not only demonstrated the box-counting method in detail, he also used it to measure the fractal properties of plans and elevations of several canonical buildings, including the south elevation of Wright's *Robie House* and the west elevation of Le Corbusier's *Villa Savoye*. Bovill concluded from this analysis that the *Robie House* elevation is in the order of 10 % more visually complex than the *Villa Savoye* elevation. This comparison seems to confirm the intuitive interpretation that architects have historically offered, that Wright's design, with its elaborate windows, modelling and raked rooflines, has greater and more consistent levels of visual complexity than Le Corbusier's white, geometric façade. This same result is reflected in the worked examples contained in the previous section of the present chapter.

More controversially, Bovill also used the box-counting method to compare architecture and its surrounding context by calculating the fractal dimensions of a row of houses and their mountainous setting. He suggests that the 14 % difference in characteristic visual complexity between these two sets of results demonstrates that 'the indigenous builders somehow applied the rhythms of nature to their housing site layout and elevation design' (1996: 145). While such claims have been examined and criticised (Vaughan and Ostwald 2009a), Bovill's clear and detailed explanation of the method paved the way for many scholars to use this approach for measuring architecture. The remainder of this section reviews the application of the box-counting method to both historic and more contemporary buildings.

Brown et al. argue that the box-counting method is useful for archaeologists because 'it is always important to identify, describe, and quantify variation in material culture' (2005: 54). These concerns are of similar significance for architectural historians who, like archaeologists, are often interested in both the form of a cultural artefact and symbolic meaning. However, applications of the box-counting method to historic buildings also contain a high proportion of arguments which seem to confuse fractal dimensions with fractal geometry, as well as those which try to conflate measured dimensions with mystical or symbolic properties. Within papers which otherwise contain rigorous mathematical analysis, an unexpected range of esoteric and misleading conclusions are recorded, including several which are not supported by the method or its results.

The most common historic buildings that have been the subject of fractal analysis are temples and pyramids. In the latter category, a team led by Klaudia

Oleschko analysed three major Teotihuacan pyramids and six ancient complexes (100 BC–700 AD), as well as four recent buildings in modern day Teotihuacan. The computational analysis was based on digitized black and white aerial photographs of these buildings. The results grouped the images in three fractal dimension ranges, with pyramids $1.8876 < D < 1.8993$, complexes $1.8755 < D < 1.883$ and modern buildings $1.7805 < D < 1.8243$ (Oleschko et al. 2000). Despite the fact that these results only determine fractal dimensions, and not necessarily that the buildings have any fractal geometric qualities, Oleschko's team claims that 'this technique, … confirms the supposition that Teotihuacan was laid out according to a master plan, where each small building may be considered to be a replica of the whole complex' (Oleschko et al. 2000: 1015). Notwithstanding the serious methodological problems inherent in extracting data from aerial photographs (where countless additional features artificially raise the *D* result), a common range of dimensions does not necessarily mean that all of the buildings in a given set were designed in accordance with a similar formal schema; there are other more plausible explanations. For instance, a large number of Gothic church elevations have similar fractal dimensions but this does not mean that the architects responsible for them were all involved in a cult to replicate this form across Europe (Samper and Herrera 2014). Instead, common crafting techniques, materials and details, along with similar technology and iconography, means that a level of consistency would naturally exist.

Two studies of Mesoamerican pyramids and temples (Burkle-Elizondo 2001; Burkle-Elizondo and Valdez-Cepeda 2001) feature interpretations that, like the Teotihuacan case, may be debatable. These studies use the box-counting method to measure the dimension of scanned images of elevations of Mayan, Aztec and Toltec monuments (300 BC–1110 AD). These results superficially suggest that these monuments are ornate, visually complex structures with an average *D* of 1.92. However, before considering Gerardo Burkle-Elizondo's conclusion, it is worth noting that a *D* of 1.92 would be amongst the highest dimensions ever recorded in architecture, being comparable with the dimension of an intricate vascular network or dense tree structure, but it is only for a set of stepped pyramids and some decorative panels. A close review of the images used for the analysis reveals that they are scanned, grey-scale images, which when converted into line drawings, generate a large amount of visual 'noise', including a large number of features which are not actually present in the architecture. Thus, the *D* results are exaggerated by the nature of the starting images. Regardless of the results, Burkle-Elizondo's conclusion, which echoes that of Oleschko, is that, based on the results, 'we think that there undoubtedly existed a mathematical system and a deep geometrical development in Mesoamerican art and architecture, and that they used patterns and "golden units"' (2001: 212). Because the Golden Mean is actually a 'primitive' or 'trivial' fractal, it has a 'known' fractal dimension which is far less than $D = 1.92$. Furthermore, that a culture promulgates a recurring set of geometric patterns is not unexpected, but this is not necessarily a reflection of any deeper level of understanding or significance. These two facts mean that the spirit of

Burkle-Elizondo's conclusion may be correct, but the fractal dimension results are insufficient, in and of themselves, to support this position.

Rian et al. (2007) consider both fractal geometry and fractal dimensions as two distinct and separate aspects of the *Kandariya Mahadev*, an eleventh-century Hindu temple in Northern India. They use the box-counting method to confirm the characteristic complexity of plans, elevations and details of the temple, and a separate diagrammatic analysis provides a breakdown of the monument's fractal-like geometric construction. Their research also reports important information regarding the method used, the results of which identify a close range of high dimensions ($1.7 < D < 1.8$) in the plans, elevations, details and ceiling panels of the ancient temple.

Both Wolfgang Lorenz and Daniele Capo have used the box-counting method to analyse classical Greek and Roman orders. Lorenz (2003) investigated a set of line drawings of the entry elevations of four ancient Grecian temples and found that of the set, the *Treasury of Athens* (c. 490 BC) in Delphi had the lowest fractal dimension ($D = 1.494$) and the *Erechtheion* (c. 400 BC) in Athens had the highest ($D = 1.710$). Lorenz concluded that the dimensions confirmed an intuitive visual reading of the complexity of the different building elements of the temples. Capo (2004) used a modified version of the box-counting method (described as the 'information dimension') to compare the Doric, Corinthian and Composite orders of architecture (600 BC–100 BC). Capo did not publish the resulting dimensions, but concluded that they 'showed a fundamental coherence' (2004: 35).

Architecture of the sixteenth-century Ottoman period in Turkey has been the subject of fractal analysis by several authors. For example, William Bechoefer and Carl Bovill analysed a set of Ottoman houses in the ancient city of Amasya which were an example 'of the most important remaining assemblage of waterfront houses in Anatolia' (1994: 5). They used a limited, manual version of the box-counting method to measure the elevation of the group of five houses, producing a result of $D = 1.717$. This same strip of housing was re-analysed using the manual method by Lorenz in 2003 with a different result ($D = 1.546$) and again, by ourselves (Vaughan and Ostwald 2010a) using *ArchImage* software with a third result ($D = 1.505$). The geometric properties of another group of eight traditional Ottoman houses were measured by Cagdas et al. (2005). Three different facets of these houses, in the Chora district of Istanbul, were considered. First, their combined roof plans ($D = 1.7$) then their building outline ($D = 1.2$) and finally their street elevation ($D = 1.2$).

In a large and technically advanced application of the method, Özgür Ediz and Michael Ostwald analysed the elevations of Mimar Sinan's sixteenth-century *Süleymaniye Mosque* (Ediz and Ostwald 2012) and the *Kılıç Ali Paşa Mosque* (Ostwald and Ediz 2015), both in Istanbul. Ediz and Ostwald used box-counting to provide quantitative data to interpret scholarly arguments about the importance of visual layering in these culturally significant buildings. Consistent and accurate line drawings of elevations of the two mosques were measured with three different levels of detail: the form of the elevations, the form and major ornament of the elevations and the form, plus ornament and with all of the material joints expressed.

For the *Süleymaniye Mosque*, the results for these three different representations were, respectively: $1.598 < D < 1.688$; $1.638 < D < 1.702$; and $1.790 < D < 1.807$. In total, almost 2,000,000 calculations were completed to determine these results for the two buildings. Notably, these mosques are amongst the most richly textured buildings ever constructed, with dense layers of ornament and material joints, and their highest D result is in the order of 1.807. For this reason, any non-integer dimensional measure for architecture that is higher than this should be carefully and critically reviewed before being accepted.

As well as being an important example of an architectural era, the houses analysed from the Ottoman period could also be thought of as examples of vernacular or traditional architecture. Another type of traditional housing that has been analysed using this method is from Poland. Zarnowiecka (2002) found that a traditional Polish cottage had a fractal dimension of $D = 1.514$. When Zarnowiecka expanded her use of the method to determine the effect on the visual complexity of a traditional cottage after being 'modernised', the result changed from $D = 1.386$ to $D = 1.536$. In a similar way Debailleux (2010) analysed thirty-six elevations of vernacular timber-framed structures in rural Belgium. Debailleux extracted line drawings from a set of photographs for the analysis. The complete results were not reported in the paper, but Debailleux concluded that the fractal dimensions were consistent with the different frame types, and the average value for all of the structures was $D = 1.38$. Lorenz (2003), in one of the more extensive studies of traditional architecture using this method, analysed line drawings of sixty-one elevations of vernacular farmhouses in the Italian Dolomite Mountains. He employed a rigorous computational methodology, reporting most of the parameters used, and noted several significant challenges with the process. He concluded that the houses could be grouped into nine characteristic sets with similar fractal dimensions ranging from $1.20 < D < 1.66$.

Perhaps because Bovill demonstrated the box-counting approach to fractal analysis using the work of Frank Lloyd Wright, domestic architecture in general, and Wright's architecture in particular, has remained a common focus of this approach. Bovill claimed that Wright's designs 'provide good examples of a progression of interesting detail from large scale to small scale' (1996: 119). Bovill's initial fractal analysis of the south elevation of Wright's *Robie House* has since generated a detailed response from other scholars and this one façade is probably the most frequently analysed of any example, with at least seven separate box-counting studies published. The results of these studies are discussed in more detail in Chap. 5 but they typically range from $D = 1.520$ (Bovill 1996) to $D = 1.689$ (Vaughan and Ostwald 2010a).

Including the *Robie House*, a total of twenty of Wright's houses have been measured using the box-counting method. Wen and Kao (2005) applied a computational version of the method to plans of five houses by Wright spanning from 1890 to 1937. The results for the houses varied between $D = 1.436$ (*Frank Lloyd Wright Residence*) and $D = 1.626$ (*Harley Brandley House*). The elevations of five of Wright's Prairie Houses (1901–1910) have also been examined using two different computational variations of the box-counting method (Ostwald et al. 2008).

The range of fractal dimensions which were recorded is between $D = 1.505$ (*Zeigler House*) and $D = 1.580$ (*Evans House*). We, the present authors, published preliminary results for an analysis of Wright's Usonian and Textile-block houses (Vaughan and Ostwald 2011). While these results are revised and refined later in the present book, the original range for the Usonian houses was between $D = 1.350$ (*Fawcett House*) and $D = 1.486$ (*Palmer House*) and the average for the set was $D = 1.425$. The fractal dimensions for the Textile-block houses were between $D = 1.506$ (*Freeman House*) and $D = 1.614$ (*La Miniatura*) and the average for the set was $D = 1.538$.

The *Unity Temple* (1905) in Chicago is the only non-domestic building designed by Wright which has been analysed using this method. The fractal dimension of the north elevation of the *Unity Temple* has been the subject of three separate studies. The first two used a manual variation with results of $D = 1.550$ (Bovill 1996) and $D = 1.513$ (Lorenz 2003). The final study on the same image was undertaken using a computational variation, producing a relatively similar measure, $D = 1.574$ (Vaughan and Ostwald 2010a).

Bovill's (1996) other choice for his initial excursion into fractal analysis was the west elevation of Le Corbusier's *Villa Savoye*. Modernist architecture was also the focus of the first recorded application of the box-counting method in architecture: Bechhoefer and Bovill's (1994) analysis of an elevation of a hypothetical two-storey Modernist apartment block ($D = 1.37$). The fractal dimension calculated by Bovill for the *Villa Savoye* ($D = 1.3775$) is lower than his result for the *Robie House*, leading him to suggest that such a variation 'is due to the difference in design approach. Wright's organic architecture called for materials to be used in a way that captured nature's complexity and order. Le Corbusier's purism called for materials to be used in a more industrial way' (1996: 143).

Since Bovill's original assessment of the *Villa Savoye*, it too, like the *Robie House*, has become a regular test subject for attempts to refine the method. As such, this particular case is also discussed in detail in Chap. 5. However, Lorenz (2003) used a manual variation of the method to analyse Bovill's drawing of the north elevation, producing an overall result of $D = 1.306$. This low result led Lorenz to agree with Bovill's claim that Modern architecture lacks 'textural progression' (Bovill 1996: 6). Furthermore, Lorenz suggests that the *Villa Savoye* 'is missing … natural, structural depth' (2003: 41). Our own calculation (Vaughan and Ostwald 2010a) of the same line drawing using a computational method produced a result of $D = 1.544$; higher than Lorenz's and Bovill's, but still lower than the calculated results for the *Robie House*. We also, in collaboration with colleague Chris Tucker, determined a composite result for the entire villa, which averaged the fractal dimension of all of the elevations of the building ($D = 1.480$) (Ostwald et al. 2008). In contrast, Wen and Kao (2005) studied the ground floor plan of the *Villa Savoye* using a computational variation of the method ($D = 1.789$). Most recently, Lorenz returned to measure Bovill's original image using an improved computational method and found difficulties analysing the elevation, observing that if the analysis was of the distant view of the entire elevation, the D value was higher (1.66), compared to an analysis of a specific part of the building where the value was lower

(D = 1.25). This led Lorenz (2012) to conclude that the 'result underlines the tendency of modern architecture towards a clear expression with details on small scales being reduced to a minimum: after higher complexity at the beginning, the data curve quickly flattens, but remains constant' (2012: 511).

Eight other Modernist residential designs by Le Corbusier have also been studied using fractal analysis. Wen and Kao (2005) examined plans of five houses by Le Corbusier spanning five decades (1914–1956). The spread of fractal dimensions for the houses was between D = 1.576 (*Villa Shodan a Ahmedabad*) and D = 1.789 (*Villa Savoye*) and, despite a range of 21 %, the authors concluded that the results were consistent. Ostwald, Vaughan and Tucker measured the fractal dimensions of all elevations of five of Le Corbusier's Modern houses (1922–1928) using two different computational variations of the method (*Benoit* and *ArchImage*) and the results ranged between D = 1.420 (*Weissenhof-Siedlung Villa 13*) and D = 1.515 (*Villa Stein-de Monzie*) and the average for the set was D = 1.481. While these cases are revised in a later chapter of the present book, at the time the original results were published, they challenged Bovill's insistence that Wright's architecture was much more visually complex than Le Corbusier's and concluded that, '[i]f each sequence of five houses, produced over a ten-year period by Wright and by Le Corbusier, is taken in its totality, then there is relatively little difference between the fractal dimension of each architect's works' (2008a: 212). In a further study, we also analysed elevations from a set of five of Le Corbusier's more ornate, Swiss-chalet style homes (1905–1912) from his pre-Modernist period, using two computational methods (Vaughan and Ostwald 2009b). In that analysis we identified a range between D = 1.458 (*Villa Jaquemet*) and D = 1.584 (*Villa Favre-Jacot*).

Possibly due to Bovill's bold statement that 'some modern architecture …is too flat' (1996: 6), several other iconic architectural designs from the Modernist era have been examined using fractal analysis. For example, five houses by Ludwig Mies van der Rohe (1907–1952) were measured by Wen and Kao (2005) with the results ranging between D = 1.4281 (*Alois Riehl House*) and D = 2.561 (*Edith Farnsworth House*). This last result must be considered extremely controversial, and most likely totally incorrect, because, it will be remembered that the D of a two-dimensional image 'must' be within the range between 1.0 and 2.0. Anything outside this range is almost certainly an experimental error. There are possible exceptions, as we will see in a Chap. 5, because the box-counting method can deliver results that are just below 1.0 (say, 0.989) under certain circumstances. However, a result of 2.5 suggests a serious flaw in the method and is most likely a by-product of using a colour or greyscale image that the software has incorrectly processed.

Another major Modernist architect whose work has been examined using this method is Eileen Gray, five of whose designs (1926–1934) were investigated using a computational method (Ostwald and Vaughan 2008). The results for the houses were between D = 1.289 (*House for an Engineer*) and D = 1.464 (*E.1027*). Two additional works of Modernist architecture, Gerrit Rietveld's 1924 *Schröder House* (D = 1.52) and Peter Behrens's 1910 industrial Modernist *Turbine Factory* (D =

1.66), were also examined by Lorenz. Lorenz found that, unlike his results for Le Corbusier's *Villa Savoye*, the results for both Rietveld and Behrens were consistent for the entire box-counting process, suggesting that 'even at first sight smooth modern architecture may offer complexity for smaller scales' (2012: 511).

There have been very few applications of the box-counting method to more recent architecture; the only published examples are by ourselves and form the basis for later chapters. Because we were interested in determining the lower practical limits of fractal dimensions for architecture, we examined the work of late twentieth-century Japanese Minimalist architect, Kazuyo Sejima (Vaughan and Ostwald 2008; Ostwald et al. 2009). Of a set of five of her houses built between 1996 and 2003, the fractal dimensions we developed using an early variation of this method range from $D = 1.192$ (*S-House*) to $D = 1.450$ (*Small House*). Minimalism, with its monochromatic finishes and unadorned surfaces would be expected to have a low D value and the significantly lower fractal dimension of Kazuyo Sejima's architecture supports this assumption. Another topic of interest to us at the time was the argument that high fractal dimensions were somehow more human (phenomenological, spiritual or accommodating) than low fractal dimensions. While any logical review of the argument would swiftly reject it (just look at the lower fractal dimensions for much vernacular housing), we felt that it was worthwhile to measure some famously abstract, post-representational designs that had been criticised as lacking human scale. Thus, we considered the architecture of John Hejduk and Peter Eisenman (Ostwald and Vaughan 2009a, 2013a). The results for Eisenman's famous *House* series elevations ranged from $D = 1.344$ (*House I*) to $D = 1.533$ (*House III*), with an average for the set $D = 1.419$. Five of John Hejduk's designs were also analysed, with the elevation results ranging from $D = 1.406$ (*House 4*) to $D = 1.519$ (*House 7*) with an average of $D = 1.472$. While this is considered in greater detail later in Chap. 9, it should be obvious that fractal dimensions are measures of characteristic complexity, regardless of any symbolic, semiotic or emotional cues present in a design.

3.3 Conclusion

As demonstrated in this chapter, the box-counting method is deceptively simple to apply, and this is why there are a growing number of examples of its application in urban and architectural analysis. Conversely though, the method has multiple complicating factors that have repeatedly undermined its usefulness and validity. These problems are readily apparent in the review of past results, where multiple applications of the method to the same façades (and even the same elevations representing these façades) have produced often-divergent results, and where a large number of completely counter-intuitive outcomes have been published. The three major concerns facing those seeking to use the method are about image representation standards, data pre-processing standards and methodological considerations.

The first of these concerns arises from the fact that the box-counting method measures information contained in an image. Obviously, if the image is a photograph then the information contained in it will be very different from the information in a line drawing. Shadows, textures and perspective are all part of the way in which we experience the world, but they also complicate the process of measurement to such a degree that they are typically removed from any consideration of questions of form. For example, unless a person was specifically interested in the visual impact of plants, trees or shadows, all of these features will completely dominate any analysis of visual complexity in a building façade. Furthermore, when studying a building, the question must be asked, what data is relevant? That is, which lines in a plan or elevation should be measured and why? These are all questions about representational standards and they are addressed in Chap. 4.

If a consistent rationale for the correct starting image information and representation can be developed, then the second challenge is to determine the correct way to present or prepare that data prior to mathematical or computational analysis. For example, how large should the starting image be, what line weights should architecture be depicted in, and how much space should be left around the image for a reasonable starting grid to be drawn. These seemingly trivial issues have repeatedly been identified as being responsible for incorrect estimations. Chapter 5 presents the results of a detailed investigation of the impact of these issues on architectural applications of the box-counting method, and then identifies the optimal image pre-processing standards to use for achieving consistent results.

Finally, once the starting image is correctly represented and prepared for analysis, then there are several features of the method itself that must be determined. The first of these is about the scale by which each successive grid reduces; that is, what is the correct scaling coefficient. The second is about the orientation (or starting point) for each successive reduction in grid scale. In the three worked examples in this chapter, a 2:1 ratio was used to halve the size of each grid and so subsequent grids fitted perfectly within the footprint or area of the previous one. But what if this isn't the case? What is the right or best ratio, not just the simplest? These factors are almost never mentioned in the past research in architecture and urbanism, but scientists and mathematicians have repeatedly affirmed that they will have an impact on the results. To quantify and determine this impact, and to identify the optimal approach to these factors, Chap. 5 also records the results of a calibration process for the method.

Chapter 4
Measuring Architecture

Before a researcher measures and analyses the form of a building, three seemingly simple questions need to be answered. *Why* is this building being analysed, *how* will its form be measured and *what* parts of the building will be measured? All three of these issues are interconnected and the answers must be well aligned to each other for the result to be meaningful. For example, for a practising architect or surveyor, the answer to the second question is seemingly straightforward; there are many standard ways of measuring the length, height and depth of a wall using rulers, tape measures or lasers (Watt and Swallow 1996; Swallow et al. 2004). However, for the scholar or researcher, the issue is more contentious, as these methods may not meet the needs of the first question (why) (De Jonge and Van Balen 2002; Stuart and Revett 2007). The often unstated assumption in architectural research is that the more accurate the measure, the better the result. However, as several researchers have demonstrated, this can provide a poor basis for testing a hypothesis (Frascari and Ghirardini 1998; Eiteljorg 2002). Thus, the 'how' question cannot be answered without first considering the 'why'. The third question is even more complex: what parts of the building should be measured? Because architecture operates across a range of scales—from the macro-scale of the city and the piazza, to the micro-scale of the doorjamb or the pattern on a wall tile—there is no simple answer to this question. Arthur Stamps, when considering this problem, notes that a building façade 'may be described in terms of its overall outline, or major mass partitions, or arrays of openings, or rhythms of textures' (1999, p. 85). He goes on to ask; '[w]hich of these many possible orderings should be used to describe' (1999, p. 85) a building?

These three questions, *why* to measure, *how* to measure and *what* to measure, are further complicated when practices in computational analysis are considered. Each of the main computational methods of formal and spatial analysis used in architecture relies on measuring representations of buildings or spaces. Thus, they derive data from orthographic projections (plans, elevations and sections), CAD models and photographic surveys. For two of the most established computational methods the rationale describing which part of a building plan to analyse and how it is

M.J. Ostwald and J. Vaughan, *The Fractal Dimension of Architecture*,
Mathematics and the Built Environment 1, DOI 10.1007/978-3-319-32426-5_4

undertaken is relatively straightforward. Space Syntax research typically analyses habitable space, being rigorous in its geometric mapping of lines of sight, visually defined space or permeability (access) in a plan (Hillier and Hanson 1984; Hillier 1996). Shape grammar researchers have a different but similarly meticulous way of extracting geometric and topological properties from a building plan before any analysis is undertaken (Stiny 1975; Knight 1992). While the logic underlying these approaches to measurement continues to be debated, there are accepted standards in each field. In contrast, several other approaches to computational research are more explicitly concerned with the distribution of detail in a building's form. These methods—including fractal analysis, semantic layering and distribution analysis by Zipf's law or Van der Laan septaves (Stamps 1999; Crompton and Brown 2008; Crompton 2012)—do not have such established standards for deciding which lines on a plan or elevation should be measured and why.

Fractal analysis is used to quantify the characteristic visual complexity of a building plan or elevation. Like architectural analysis using Zipf's law or Van der Laan septaves, it provides a measure of the distribution of information (lines or details) in an object across multiple scales. Despite early attempts to identify which parts of a building façade or plan should be the subject of fractal analysis (Bovill 1996), researchers have repeatedly noted that there is a lack of consistency in the field (Lorenz 2003; Ostwald et al. 2008). Moreover, it has been demonstrated that different measures can be derived from an analysis of the same elevation of a building, depending on which lines are chosen for consideration (Zarnowiecka 2002; Vaughan and Ostwald 2009b). Consider the example of Le Corbusier's *Villa Savoye*, a design which has been the subject of five separate applications of fractal analysis (Bovill 1996; Lorenz 2003; Wen and Kao 2005; Ostwald et al. 2008; Lorenz 2012). Across these five studies, the variation in the measurements recorded for the north elevation is in the order of 18 %. While there are subtle mathematical and methodological differences between the ways each of these studies have been undertaken, if the technique is valid then the results should be much closer.

The most obvious explanation for the 18 % anomaly is simply that computational methods like fractal analysis do not measure buildings; rather, they extract measures from representations of buildings. The primary representational media used for architectural analysis is the orthographic drawing and there are many variations in how a building can be represented in such a drawing. For example, Wen and Kao (2005) measured and analysed the lines present in a series of published design drawings of the *Villa Savoye*. In these drawings a range of graphic standards were used to represent the textural and material qualities of the architecture and to signal the function of certain spaces. For example, tiles were represented on the plan as a regular grid and the presence of timber boards by hatching on the drawing with parallel lines. Neither the size of the grid, nor the spacing of the lines, is a reflection of the real materials in the Villa or their dimensionality. Architects understand that such graphic conventions are symbolic of a material presence; they are not meant to be taken literally. In contrast, Bovill's (1996) analysis of the same building is derived from a tracing of a similar drawing wherein he chose to only delineate major changes in form. The difference between Bovill's

(1996) and Wen and Kao's (2005) representations of the Villa explains a large portion of the 18 % anomaly; they were measuring different lines on a drawing of the same building. But which lines, if any, are right?

Past research in fractal analysis has noted that without some consistent and reasoned rationale for selecting the particular lines in a building to analyse, the measurements extracted from that building are likely to be meaningless for any comparative purpose (Zarnowiecka 2002; Lorenz 2003; Ostwald et al. 2008; Lorenz 2009). The challenge then for fractal analysis is to determine which lines in a plan, section or elevation should be the measured, and why.

4.1 Introduction

This chapter is focussed on answering a two part question that is typically both unasked and unanswered in the majority of architectural applications of the box-counting method. That question is, which lines in an architectural representation are being measured and why? To answer this question, the chapter commences by describing the philosophical paradigm that typically governs architectural research, which is reliant on measuring. This paradigm, postpositivism, is then used to illuminate a common argument in architectural analysis about the misalignment between purpose and precision when measuring buildings. Postpositivism assists in delineating the essential values that must be present in a reliable research method. It has provided a basis for both established architectural research methods (Groat and Wang 2002) and for specialised research into the computational analysis of design (Gero 1998). Thereafter the chapter describes and illustrates five different ways of representing buildings in preparation for analysis. Each of these variations is illustrated with a different variation of a plan and elevation from Le Corbusier's *Villa Jaquemet*, and a calculation of the fractal dimension of that view. Through this process a conceptual framework containing five cumulative levels of representation is presented. The goal of this framework is to support decisions about which lines should be measured in an architectural image and for what purpose. Thus, the framework assists us to answer the three questions about measurement-based research raised at the start of this section—*why* to measure, *how* to measure and *what* to measure.

Before progressing, three points need to be made about the content of this chapter. First, in this chapter the word 'measuring' is taken to include any process which extracts numerical or geometric information from a building or representation of a building, whether drawings, models or photographs. In computational analysis it is common to talk about the processes of abstracting or translating information derived from the built environment into a graph or map; these are both types of measuring. Second, while parts of the philosophical discussion hereafter are relevant to all types of measuring (including, for example, the consideration of acoustic reverberation or temperature change) the majority of the chapter is more explicitly about the measurement of form. Third, the chapter provides some example measures which have been produced using the box-counting method, for each of the five representational permutations in the framework. While these values have been prepared in accordance with the standard system used throughout the book, their purpose here is to illustrate how the same method of measurement, when applied to the same elevation, but identifying different characteristics or elements of that façade, will produce different results.

4.2 Philosophical Foundations

In *Architectural Research Methods*, Linda Groat and David Wang suggest that a person who is seeking to study a particular object or phenomenon should commence by understanding the 'system of inquiry' they are operating within. At a superficial level, the concept of a system of inquiry encompasses the relationship between the research hypothesis being tested, the 'tactics of information gathering and analysis' being applied, and 'the practices of the researcher as s/he conducts the inquiry' (2002, p. 41). Thus, on one level, the system of inquiry is a framework of multiple parts, all of which are appropriate for a type of research and are consistently applied with an awareness of any limitations. But a system of inquiry is also broader than that; it refers to the philosophical or foundational values of the researcher. In this sense, the act of measuring a building, as a precursor to some form of analysis, is considered a postpositivist system of inquiry (Bechtel et al. 1987; Groat and Wang 2002; Sirowy 2012).

The English noun *positivism* is derived from the French words *positivisme* and *positif* which mean *that which has been learnt through experience* (Barnhart 2000). In the philosophy of science, the word positivism has come to be used to describe the belief that all legitimate systems of knowledge are derived from rigorous, logical and objective observations. Positivism, like empiricism, asserts that knowledge must be verifiable through the acquisition of appropriate evidence. While positivism was originally, during the Enlightenment at least, positioned in opposition to metaphysical modes of inquiry, by the late nineteenth-century it had not only become the dominant model of knowledge construction, but several variations, including logical positivism and neopositivism, refined its edicts in order to seek a higher level of rationalism or reasoning. However, in the twentieth-century a growing number of philosophers of science, led by Horkheimer (1947), Popper (1959) and Lakatos (1976), demonstrated that positivism, while a reasonable principle, was not so indisputable as its proponents maintained. Despite conflicting counter-arguments from Kuhn (1962) and Feyerabend (1975), by the latter half of the twentieth-century Popper's (1959) 'theory of falsification' had become the *cause célèbre* for social scientists and humanities researchers, who used it to justify the rise of action research, participatory research, design research and reflective practice, all modes of inquiry which sought to legitimise the observations of the individual above those of the collective. However, Popper never sought to promulgate such researcher-centric approaches. Instead his aim was threefold: to remind scientists that (i) they could not assume that they were always objective observers, (ii) outside the phenomenon they were investigating, or (iii) that their findings were necessarily universal.

In the aftermath of the breakdown in confidence in positivism, support for a more contingent or nuanced mode of scientific inquiry, one which acknowledged its limitations, became known as postpositivism. The postpositive stance accepts that the backgrounds, beliefs and conceptual frameworks of the researcher can and do have an influence on the results of the research. Nevertheless, it does not accept this fact as a reason to abandon the desire for objectivity, transparency and repeatability. Postpositivism could, therefore, be considered a contingent or realistic form of empiricism.

Postpositivist systems of inquiry seek to ensure that four standards are met: intrinsic validity, extrinsic validity, reliability and objectivity (Paul 2004). The first of these, intrinsic validity, is a reflection of the degree to which the primary theoretical frameworks and methods used will collectively produce a reasonable or truthful representation of the phenomena being studied. This issue is directly associated with ensuring an appropriate correlation between a method and its purpose. In regards to the measurement of buildings, it is about achieving an alignment between *what* is being measured and *why* is it being measured.

The second expectation of postpositivist systems of inquiry is that they possess extrinsic validity. This property, which is also known as generalizability or transposability (Kuhn 1962), determines whether 'the results of this study are applicable to the larger world' (Groat and Wang 2002, p. 36). In practice, two standard principles for achieving generalizability relate to sample size and benchmarking. In the first instance, the larger the sample size, the more likely a set of results is able to be extrapolated to suggest useful findings. The second is that research which is focussed on cases that are well known or well documented is more likely to be widely applicable.

The expectation of reliability means that the methods being used by a researcher should be both consistent and repeatable (Fellows and Liu 1997). As Groat and Wang observe, '[w]ithin the postpositivist paradigm the assumption is that the research methods would yield the same results if the study were conducted under the same conditions in another location or at another time' (2002, pp. 36–37). When measuring architecture, this standard applies to the selection of tools, to the way the tools are used and to what is being measured. All of these details should be recorded to ensure that any future measurements will be undertaken using the same parameters and thereby allow comparisons between studies to be constructed.

The final quality of a postpositivist system of inquiry is objectivity. This refers to the apparent neutrality of the method or the capacity of the researcher to reduce, control or limit any potential bias. The measuring of architecture can and should occur in an objective way if a researcher is clear about the limits involved in the particular tools being used and of any potential error rates and mitigation strategies.

4.3 Precision or Purpose

Past research into the process of measuring buildings has often instinctively adhered to the basic values of postpositivism. In particular, the issue of intrinsic validity has been repeatedly discussed in terms of the tension that exists between precision and purpose (Caciagli 2001; Eiteljorg 2002; Hebra 2010; Calter and Williams 2015). A classic example of this dilemma is the argument that a survey of the dimensions of the *Pantheon* in Rome using a laser scanner with 0.1 mm accuracy will produce a better analysis than a conventional manual survey with 100 mm accuracy. Consider this argument in the context of research which is, for example, seeking to examine the proposition that the plan and section of the *Pantheon* were designed to have the same radius dimension (Masi 1996). The purpose of such a study is to investigate evidence of design intent and, in particular, answer the question: did the architect consciously create a building where the plan and section are identically sized? Disregarding the possible existence of any written accounts of this intent, in a postpositivist sense some evidence for this proposition could legitimately be developed by measuring the historic building. However, a highly accurate measurement of the building is not necessarily better than a less accurate measure for this purpose. There are two reasons for this, one associated with the limits of construction techniques, the other with the practical problems of working with historic structures.

Roman stone masons worked with absolute dimensions—like *pollice*, *braccio*, *piede* and *canna* —which were derived from wooden copies of stone rods kept in each city in the empire (Kostof 1977). Moreover, the stone mason's tools (the square and the rule) were typically capable of around 50–100 mm accuracy for the repetitive production of elements in a quarry, although on-site, a different type of precision could be produced. For example, a mason could 'fit' two stones together with a gap of less than 2 mm if called upon to do so. However, this does not mean that every stone was produced with this level of precision, or even that the masons could measure this level of accuracy, only that they could produce a level of fit that was relatively precise. For this reason, Caciagli (2001) argues that it is fundamentally meaningless to measure a building to a higher level of precision (or tolerance) than was commonly available to the people who constructed the building. The second problem with measuring the *Pantheon* is simply that the building has changed over time. Not only was the *Pantheon* partially rebuilt on several occasions, but its foundations have also settled unevenly and its dome has slumped over time and been repositioned using secondary structures. Thus, in this case precision is largely irrelevant; a new, high quality measurement of the sectional geometry of the dome cannot be considered to provide any more evidence for this argument than an older survey using more limited technology. Indeed, much older measurements, regardless of how imprecise they may be, will always be better for testing this particular hypothesis.

Marco Frascari and Livio Volpi Ghirardini are highly critical of the process of producing supposedly precise measurements of buildings to locate particular geometric proportional systems (like golden sections) in historic and modern building plans. They too argue that it is meaningless to measure architecture without due consideration of the purpose of the process because, in 'metrical terms, every constructive part of building has its geometric order: masonry, in decimetres; wood carpentry, in centimetres; metal works, in millimetres. Every part is exactly approximate' (Frascari and Ghirardini 1998, pp. 68–69). This position, which maintains that precision should be relative to purpose, is reflected in several studies which identify the misuse of dimensional accuracy to suggest evidence for a proposition which is clearly not related to precision alone (Ostwald 2001b; Eiteljorg 2002). As Paul-Alan Johnson notes, '[p]recision *per se* is not enough no matter how satisfying it is for the analyst' (1994, p. 19). Ultimately, precision in architectural measurement must be relative to purpose; the postpositivist tenet of intrinsic validity. For example, James (1981, 1982) demonstrated, using increasingly fine observations of the chisel marks made by masons on the stones of *Chartres cathedral* (a system he called 'toichology'), that he could determine how many master stoneworkers were involved in the building's construction. Such an application requires very fine scale measurements of patterns. But when James measured the famous labyrinth pattern on the floor of the nave of *Chartres*, he was less concerned with the precise measure than with how many multiples of the Roman foot or hand it equated to. This is because the labyrinth was known to have been produced to such multiples. What this means is that, as Harrison Eiteljorg states, 'every project has its own needs for precision. Those needs should be carefully determined, explicitly stated, and properly met by the survey methods and procedures' (2002, p. 17). This message, derived from the need for intrinsic validity, is developed in the next section into a way of thinking about the relationship between measurement and representation.

4.4 Framework

Under the postpositivist paradigm a legitimate system of inquiry for investigating the formal properties of buildings must have a clearly aligned method of measurement and research purpose. Thus, the research question or hypothesis must be one for which measurements can provide useful evidence. While this may seem obvious, it is less common to observe that the particular architectural features being measured must also be appropriate for the research purpose. In computational analysis this is a question of representation or delineation. For example, it is possible to delineate the façade of a building, in a drawing or model, using many different combinations of lines or textures. However, in all of these cases the act of measuring is reliant on the conventions of representation. This is because, regardless of whether field dimensions are taken using a laser scanner or tape measure, the dimensions have to be transcribed into a representational system, be it a drawing, CAD or BIM model, for it to be analysed.

In order to accommodate this need, a framework is proposed wherein five cumulative levels of representation are defined and mapped against comparable research purposes (Table 4.1). This framework shows that, for instance, it is not reasonable to study the impact of material texture on a building when only the footprint of a plan is measured. Conversely, if the impact of planning decisions regarding site amalgamation is to be studied, then measuring the geometry of material textures in a façade will be counter-productive. This framework, aligning level of representation with research intent, is also inherently cumulative. That is, it relies on the fact that the building outline is required as a precursor to defining primary forms. Thereafter, secondary forms can only be added if the boundaries of the primary forms are already present, and so on. Thus, the framework is described in terms of what is added with each level of representation, as shown in Table 4.1.

Each of the five levels of representation are described in the following sections and illustrated using variations of an elevation and plan of the same building which are presented alongside the different fractal dimension measures derived from each. Both plans and elevations are used in the analysis, as an elevation can be considered to provide a measure of the geometric complexity of the building as measured from the exterior and the plan provides a measure of the complexity of the design as it is

Table 4.1 Levels of representation mapped against research purpose

Level	Representation	Research focus	Purpose
1	Outline	Building skyline or footprint	To consider major social, cultural or planning trends or issues which might be reflected in large scale patterns of growth and change in the built environment
2	+ Primary form	Building massing	To consider issues of building massing and permeability which might be a reflection of social structure, hierarchy, responsiveness (orientation) and wayfinding (occlusion)
3	+ Secondary form	Building design	To consider general design issues, where 'design' is taken to encompass decisions about form and materiality, but to not extend to concerns with applied ornament, fine decoration or surface texture
4	+ Tertiary form	Detail design	To consider both general and detail design issues, or where 'design' is taken to include not only decisions about form and materiality but also movable or tertiary forms and fixed furniture which directly support inhabitation
5	+ Texture	Surface finish and ornament	To consider issues associated with the distribution or zoning of texture within a design, or the degree to which texture is integral to design

inhabited (Ostwald 2011a). The example used to illustrate this framework is Le Corbusier's *Villa Jaquemet* in Pouillerel, Switzerland. Completed in 1907, this ornate, chalet-style house, which is stylistically reminiscent of the Arts and Crafts movement, gives little indication of the type of Modernist architecture Le Corbusier would later produce. This house was previously the subject of computational analysis where it was postulated that the particular lines chosen for analysis would have a dramatic impact on the results (Vaughan and Ostwald 2009b). The *Villa Jaquemet* is a four storey house with rough stone walls on the lower levels, and dressed stone around the windows and doors. The upper stories are clad in stuccoed timber, with exposed timber panelling on the highest sections of the walls. The tall, sloping roof has exposed carved timber supports and it is clad in dark tiles.

4.4.1 Level 1: Outline

The silhouette of a building—its elevation and associated ground plane—when drawn as one simple continuous line, is often referred to as a 'skyline' drawing. The analysis of skyline characteristics is common in urban design and town planning and there are many examples of this approach to considering the visual complexity of urban or architectural landscapes (Heath et al. 2000) including several applications of fractal analysis (Stamps 2002; Cooper 2003; Chalup et al. 2008). Less commonly, the plan form of a building can also be represented in this way, as a figure of the footprint of a building or as an outline of the roof plan (Brown and Witschey 2003; Frankhauser 2008), for the analysis of the structure of large urban neighbourhoods. In both cases, the consideration of silhouettes and footprints, the purpose is typically to examine the way in which particular types of construction create distinctive patterns, which in turn are thought to reflect the individual social and cultural characteristics of a region. This strategy is used where the large-scale patterns found in the built environment are thought to be a reflection of distinct differences between regions or groups.

If this approach is taken to the representation of the *Villa Jaquemet*, the silhouette retains much of the character of the aggregate geometry of the design, with the roof outline, showing its hips, gables and the two chimneys, being clearly discernible (Fig. 4.1). The curved brackets supporting the roof are shown with a permeable silhouette whereas other features, such as the windows, through which the sky cannot be seen, are not shown. The silhouette could also be represented without any permeable elements, and thereby depict only the outline of the building. This variation might be more suitable for analysing a neighbourhood-massing layout, however, as the present framework is focussed more on the analysis of architecture, some degree of detail has been included. Notably, this permeable silhouette variation has been used for analysing the fractal dimension of cityscape skylines (Chalup et al. 2008). While the *Villa Jaquemet* elevation prompts such minor methodological concerns, the plan is more straightforward, having an almost symmetrical footprint (Fig. 4.2).

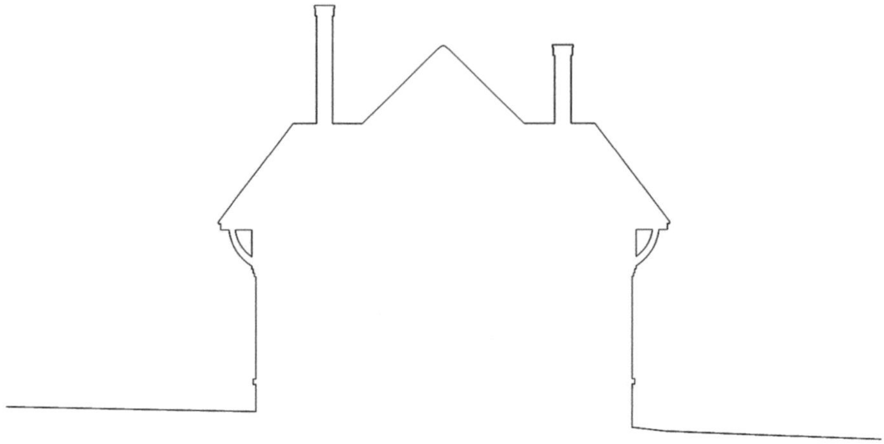

Fig. 4.1 *Villa Jaquemet*, elevation; showing silhouette. D = 1.084

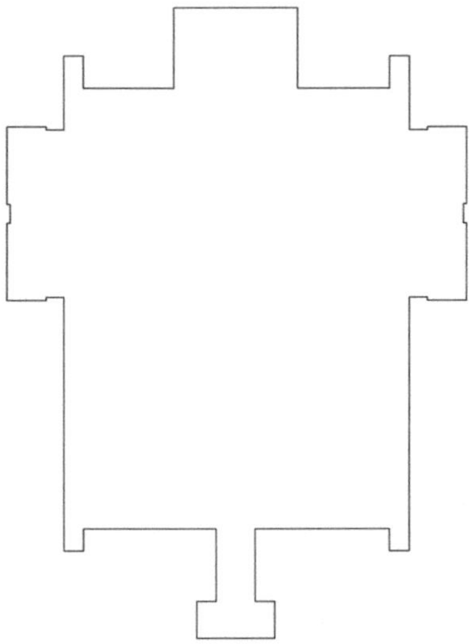

Fig. 4.2 *Villa Jaquemet*, plan; showing footprint. D = 1.076

4.4.2 Level 2: Outline + Primary Form

The second level of detail that could be considered for analysis is the formal massing of the building as a whole; what might be termed its primary form. This level is focussed on major formal gestures, not secondary forms, detail or ornament. The building is represented by the outline but now with the addition of massing elements, including openings. Smaller scale formal changes within these elements, such as individual stair treads or brick corbels, would not be included. All windows

Fig. 4.3 *Villa Jaquemet*, elevation; showing form and openings. $D = 1.348$

and doors are shown as portals but with no indication of fenestration or detail. Window openings of all sizes can be included at this stage as they represent a significant impact on the three-dimensional form of the building. In elevation specifically, gross changes in form, such as protruding walls, significantly advancing and receding elements (which measure greater than 250 mm) and the roof planes, are also delineated. Likewise in plan, the walls and major changes in floor level are shown. This level of representation was selected by Bovill (1996), for

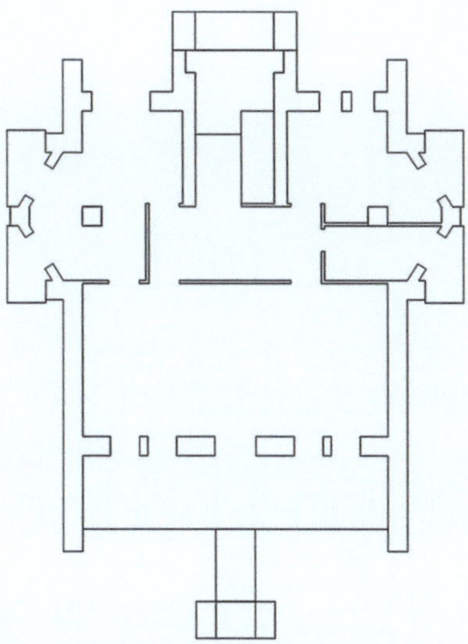

Fig. 4.4 *Villa Jaquemet*, plan; showing form and openings. $D = 1.256$

a fractal analysis of Frank Lloyd Wright's *Robie House* to gain a sense of the geometry of the major formal gestures in the plan.

The level of information in the *Villa Jaquemet* elevation increases at this stage as it has an articulated structural system which results in projecting walls and levels, all of which are shown in the representation of the elevation (Fig. 4.3) and the plan (Fig. 4.4). The windows, although actually quite detailed, are only shown as blank rectangles at this stage. In the elevation additional permeable elements, such as the foreground roof brackets, are now shown, whereas only the rear brackets appear against the skyline in the previous level of the framework.

4.4.3 Level 3: Outline + Primary Form + Secondary Form

In combination, the elements which make up the overall massing in a design along with major changes in materials could be considered secondary forms. By including secondary forms—in addition to the information previously provided by the outline and massing of the building—the primary geometric gestures that make up a design become measurable. In both plan and elevation, a single line separating surfaces should represent any changes in material. Basic mullions in doors and windows, stair treads and other elemental projections of a similar scale should be included in plan and elevation. Formal changes included in the drawing are more refined at this level and include any building elements which produce a change in surface level of greater than 25 mm. For example, the gutter and a fascia would be represented, but not the top lip of the gutter. These representational standards are very similar to those which have been used for the fractal analysis of Le Corbusier's *Villa Savoye* (Bovill 1996) and of an urban district in Istanbul (Cagdas et al. 2005).

Fig. 4.5 *Villa Jaquemet*, elevation; showing form, changes in material and mullions. $D = 1.425$

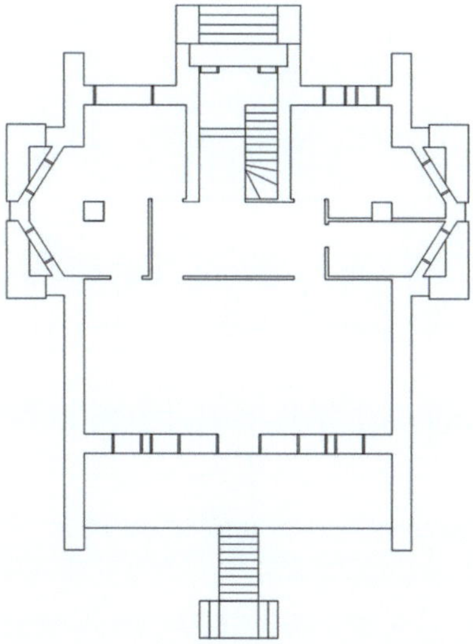

Fig. 4.6 *Villa Jaquemet*, plan; showing form, changes in material and mullions. $D = 1.310$

The *Villa Jaquemet* is constructed from a range of materials and the lines where one material ends and another begins are now represented. The difference between the rough stone base of the building, the timber panelling and the smooth walls are all clearly delineated. In elevation, the windows begin to show their detail in the top mullions (Fig. 4.5). In plan, the window frames and individual stair treads can now be seen (Fig. 4.6).

4.4.4 Level 4: Outline + Primary Form + Secondary Form + Tertiary Form

Once the form of a building has been defined (along with any secondary elements or changes in material needed to support that form) then various additional features must be added to more directly support the building's users. These tertiary forms, including doors, window panes and built-in furniture, are all critical to the inhabitation of a building, but are often simply assumed to be part of a design process. For example, windows are obviously represented in design analysis but what about the glass that is so integral to the window's function. Kitchens and bathrooms have built-in furniture and fittings which are often forgotten in architectural formal analysis. If a broad definition of design is being considered—that is, one that takes into account basic physical needs—then this level of detail is required. This level of detail has commonly been used in the analysis of regional and traditional housing (Bechoefer and Bovill 1994; Zarnowiecka 1998) and of architect designed housing (Ostwald and Vaughan 2010; Vaughan and Ostwald 2011). It could be argued that

Fig. 4.7 *Villa Jaquemet*, elevation; showing form, changes in material and mullions and outline of ornament. $D = 1.447$

this level represents design decisions that have clear consequences for inhabitation. As Chap. 6 reveals, this is also the level of representation that we have chosen for the majority of the research contained in the present book.

Drawings of the *Villa Jaquemet* at this level of representation include doors (but not door swings), glass in window panes, as well as beam-end details (Fig. 4.7). In the plan, kitchen and bathroom furniture is now clearly seen (Fig. 4.8).

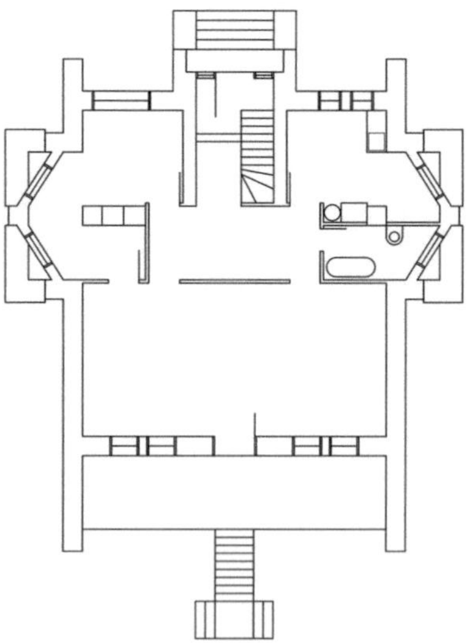

Fig. 4.8 *Villa Jaquemet*, plan; showing form, changes in material and mullions and outline of ornament. $D = 1.377$

4.4.5 *Level 5: Outline + Primary Form + Secondary Form + Tertiary Form + Texture*

The final level of representation is of surface texture or pattern. This level includes the repetitive surface geometry of a material (the grid marked by floor tiles, the parallel lines of floor boards or the distinctive wavy lines made by rows of roof tiles) or the patterns in ornamental tiles, wall-paper or applied decorations. In theory it could even include some level of representation of the grain in wood or marbling in polished stone. But it has to be acknowledged that it is rare for an architect to 'design' the pattern or geometry in a surface; more often a material is specified, and the grain chosen is indicative, rather than particular. Moreover, many of these textures are effectively invisible from a distance, or require very close observation to become apparent. This is why this last level of detail, while able to be represented, adds a new level of abstraction or artificiality to the process. It could even be argued that the major geometry of a design is complete before the materials, fabrics and colours are chosen. This does not mean that these surface or textural decisions are unrelated to the design process, but rather that they are no longer such clearly measurable geometric ones. Moreover, at the level of surface textures, the capacity to produce consistent results (in the way that postpositivist reasoning anticipates) is diminished by a growing number of peculiarities and singularities in the design and construction process.

Because fractal analysis operates across multiple scales of observation, this has led various researchers to include a high level of textural or ornamental information in some examples of Mayan architecture (Burkle-Elizondo 2001), Hindu Temples (Rian et al. 2007) and Islamic Mosques (Ediz and Ostwald 2012; Ostwald and Ediz 2015) In these studies, any ornamental textures or painted patterns are included as part of the geometry of the design, along with representations of materials of the walls and roof in elevation, and the floors in plan. Joye has even argued that this level of information in an elevation is critical to its fractal dimension, claiming that '[s]urface finishes and textures' are an 'important aspect of the visual richness of the architectural structure, which also influences perceived complexity' (2011, p. 822).

Le Corbusier used several different materials in the *Villa Jaquemet* and while he chose them specifically, he could not be said to have designed the precise geometry of the building texture. For example, the rough stone blocks which make up the primary walls have a distinct texture, but not a repetitive geometry. The rendered walls, timber linings and tiles do have a general geometric structure and a broadly consistent texture, but to measure them would be an exacting process (Fig. 4.9). The existing plans for the house do not indicate what floor surfaces were intended and so, for demonstration purposes, the house has been delineated in plan with tiled floors to the bathroom areas, stone to the entry foyer and timber boards for the rest (Fig. 4.10).

Fig. 4.9 *Villa Jaquemet*, elevation, showing form, changes in material, mullions and detailed ornament and materiality. $D = 1.606$

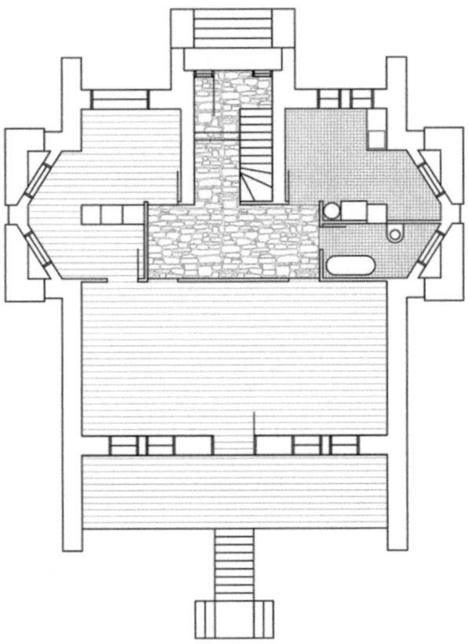

Fig. 4.10 *Villa Jaquemet*, plan, showing form, changes in material, mullions and detailed ornament and materiality. $D = 1.671$

4.5 Discussion

The particular question of how much texture to include when measuring a façade or plan is one of the more controversial ones in fractal analysis. Bovill (1996) argued that the geometric patterns produced by repetitive materials (like the horizontal lines of floorboards) should not be measured; a position which Joye (2011) has at least partially rejected. Bovill's position has been repeated by past researchers (Lorenz 2003; Ostwald et al. 2008) even though it has been acknowledged that inconsistent decisions regarding which lines to measure in a representation can have a major impact on the result (Vaughan and Ostwald 2009b).

Jadwiga Zarnowiecka in particular offers a balanced account of this issue. She originally disagreed with Bovill's proposition that, for example, the horizontal lines in a façade made by timber siding should be ignored. Zarnowiecka notes that if this is the case, then 'one must make a decision if a decorated top roof boarding is still a siding or a detail. Should this decision depend on the width of the planks being used in boarding?' (2002, p. 343). However, after measuring the difference between the elevation with planks, and without, she realised that the 'concentration of the lines on the façade' (2002, p. 344) changes the measured result even though they are not an important feature. When considering a simple regional house in Poland, Zarnowiecka notes that the addition of window mullions 'change[s] the results of the measurement, even though aesthetically they are quite meaningless' (2002, p. 344). Zarnowiecka's problem may be traced to the fact that measuring texture (to use the terminology of the current chapter) skews the results of the analysis, effectively making it unusable. These are arguments about intrinsic validity and the alignment between the representation used for measuring and the application of the measure.

Figure 4.11 charts the differing results for the images analysed in the present chapter and shows that although they are all of the same building, the fractal dimension increases with each additional layer of information included in the representation. Despite this pattern, the differences between the levels 1 and 2 of detail, and the levels 4 and 5, are responsible for the biggest changes in the chart. In contrast, there is a more stable zone in the results, for both plans and elevations, around the levels 3 and 4 of representation. The question of whether this is a characteristic of the particular house being examined, or of domestic architecture

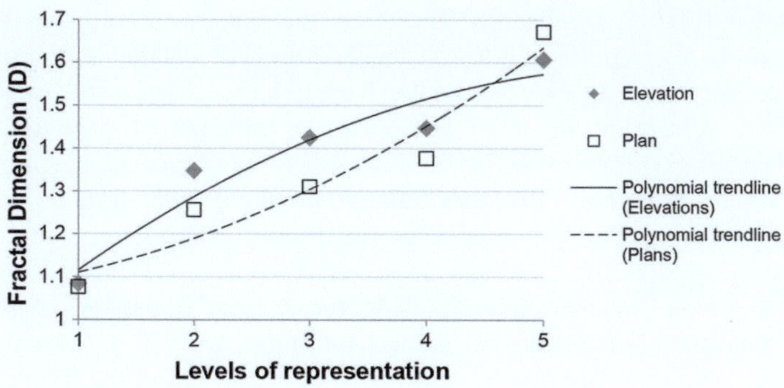

Fig. 4.11 Change in fractal dimension with increase in level of representation examined

more generally, is outside the scope of this chapter, but these patterns, along with the significance of the point at which the two trendlines cross, are worthy of future research.

4.6 Conclusion

Past researchers using fractal analysis have made evident the problems that arise when measuring the complexity of the same building represented with differing delineations. For example, Zarnowiecka (2002) notes discrepancies in the fractal dimension when the same elevations of vernacular houses in Poland were depicted with different materials, or drawn by different people. Lorenz (2003) studied one elevation of a vernacular Italian farmhouse which was represented in three different ways and found different fractal dimensions for each. Ediz and Ostwald (2012) analysed the *Süleymaniye Mosque* using the same elevation but with three different layers of detail—the form, the form plus ornament, and the form, ornament and material—obtaining different fractal dimensions for each layer included. By using the five-level framework presented in this chapter to align the type of representation being measured with the purpose of taking measurements, many of the problems noted in previous examples can be ameliorated.

At the start of this chapter, postpositivism was presented as a conceptual foundation for developing the framework; it also provides a useful basis for assessing the relative success of the proposal. Returning to the postpositivist principles, it can be seen that intrinsic validity provides the key rationale for the framework, where ideal levels of representation are correlated with the particular purpose of the analysis. For example, if the reason for a study (the *why*) is to investigate urban layouts, then only level 1 representations from the current framework may be required for analysis (the *what*). The extrinsic purpose of the framework is to facilitate comparable results between scholars by providing a set of rules, so that all data can be prepared for analysis in the same way, making outcomes more universally applicable. Therefore, the framework supports reliability and verifiability by reporting relevant methodological details and providing a reproducible list of delineation elements. Nevertheless, despite the success with each of these principles, the final postpositivist principle of objectivity is not entirely solved in this framework because, as described below, architectural design is a field where individual designers' intentions, along with different styles and scales, can produce a level of variance. Some designs that rely on straightforward surface treatments, like the buildings of Robert Venturi and Denise Scott Brown, will be able to be directly analysed using the present framework without any additional interpretation. However, not all buildings will fit so neatly into the system. Thus, anyone using this framework will need to report the limits involved in the particular techniques and representations they have chosen to ensure some degree of objectivity.

The problem with achieving objectivity in the current framework is that there will always be instances where more reasoned and considered decisions need to be made. For example, Frank Lloyd Wright specified that in the *Robie House,* all external brickwork would be horizontally raked, but with brick-coloured and filled vertical mortar joints, thereby giving the house an exaggerated horizontal appearance. Wright's decision challenges the distinction between tertiary detail and texture proposed in the present framework. In a similar way, Mario Botta detailed complex masonry bands of different colours in his façades, along with different types of brick-bonds, undermining the distinction offered in this chapter between tertiary form and texture. The framework provided in this chapter seeks to provide guidance about the appropriate use of representational standards for particular measuring purposes. However, as these examples indicate, there will always be exceptions that require additional consideration.

Chapter 5
Refining the Method

The last two decades has seen the publication of a growing body of research which uses the box-counting method to analyse architectural or urban forms. However, as stated in previous chapters, much of this research displays only a low level of awareness of the sensitivities or limits of the method. As a consequence, often widely varying results have been produced using the same mathematical approach and, in some cases, exactly the same images. The challenges associated with the accuracy and accountability of the box-counting method are threefold. First, determining in a consistent and reasoned manner, *the significant lines for analysis* in an architectural or urban drawing. Inconsistent representational standards will render the results of most studies meaningless. Second, an *optimal approach to preparing image data* for the box-counting approach is yet to be fully documented. There are many different ways of preparing an image for analysis and the consequences of these decisions have not previously been determined. Finally, the *impact of several key methodological variables* is largely untested. The first of these challenges was treated in Chap. 4. The present chapter is concerned with the second and third; it uses two different tests to determine the optimal image properties and methodological variables for fractal analysis. These tests were initially designed to develop confidence in the results of the method, but they also provide a deeper understanding of its qualities and limits (Ostwald and Vaughan 2013b).

The purpose of the first of the tests is to quantify the significance of various image properties and, in doing so, identify an optimal range for each. In the present context, image properties are characteristics of the architectural drawing being analysed. They include the resolution of the digital image, the line thicknesses used in the representation and the location of the image in the page or 'field' that is being considered. In order to determine which of these properties are significant, in this chapter we test seven 'elevations' and thirty-five variations of these properties, to

M.J. Ostwald and J. Vaughan, *The Fractal Dimension of Architecture*,
Mathematics and the Built Environment 1, DOI 10.1007/978-3-319-32426-5_5

produce 245 results for comparison. From these results we can not only determine the optimal image properties, but we can also begin to suggest the magnitude of errors that will arise from other settings. This test, its background and results, make up the majority of the first half of the present chapter.

The second half of the chapter examines two methodological factors: the ratio by which successive grids are reduced in size—the 'scaling coefficient'—and the position from which these grids are generated—the 'grid disposition'. To determine which variations of these methodological settings will produce the most accurate results we analyse nine classical fractal sets, each of which have a known or correct dimension. We then compare our estimated results using the box-counting method with the correct or calculated values, to identify the best variations of these parameters.

These two tests are conducted in parallel, rather than in sequence, because they require different types of test images and, for the most part, they examine unrelated issues. Thus, to combine the results of these tests, the final section of the chapter considers the two classic examples used for demonstrating the fractal analysis of architecture—an elevation from Le Corbusier's *Villa Savoye* and one from Frank Lloyd Wright's *Robie House*—applying the newly determined optimal settings for both image standards and methodological application. These two elevations have previously been the subject of multiple different measurements that we revisit in this section. The purpose of this final stage is not to criticise the past results—all of which have been undertaken specifically to illustrate or refine the method—but rather to suggest the extent to which basic variables may have an impact on the results of this method, and to plot the gradual improvements which have been made over time. The chapter concludes with a summary of the ideal settings for fractal analysis and a list of the information that is required for the verification of its results.

5.1 Introduction

In the years after mathematician Voss (1986, 1988) first demonstrated the use of the box-counting method, a growing number of scientific, engineering and medical researchers began to observe problems with both its accuracy and repeatability. In terms of its accuracy, Asvestas et al. (2000) found that for complex images, where $D > 1.8$, the box-counting method loses veracity and its results become both inconsistent and understated. In terms of the reliability of the method, it has been argued that the central problem with the box-counting method is that 'no step-by-step general procedure to use [it] has ever been written' (Buczkowski et al. 1998, p. 412). Multiple studies have confirmed that, for such a seemingly simple method, problems of accuracy and repeatability have plagued its application from the start (Xie and Xie 1997; Yu et al. 2005). Moreover, a lack of understanding of the role played by several methodological variables has exacerbated this situation (Camastra 2003; Jelinek et al. 2005).

Of the two major problems identified, repeatability is regarded as the most straightforward; it is solvable by clearly stating all of the parameters used in an application (Buczkowski et al. 1998). The more complex problem is accuracy. Computer scientists argue that four critical methodological variables—scale range, grid shifting, orientation of the grid and error characterisation—should be analysed and tested in every field where the method is applied to determine its limits (Da Silva et al. 2006). A similar point has been made in architectural applications of the method, which identify five key problematic variables (Lorenz 2003; Cooper and Oskrochi 2008; Ostwald et al. 2008). If all of these lists of factors are combined, they reveal eleven common variables that can be broadly divided into three categories—image pre-processing, data processing and post-processing (Table 5.1).

The four image pre-processing variables in Table 5.1 are all related to the way in which an image is prepared for analysis. The subject image must be produced and composed in such a way as to avoid adding 'noise' to the data. While the statistical validity of the fractal analysis method is largely reliant on the later data processing variables, image pre-processing factors also have the potential to cause substantial errors. These factors are particularly perplexing for people using the method for the first time, because it is possible to test four seemingly identical elevations, which are all derived from the same CAD file, but which, because of the way they have each been saved or positioned, will produce different results.

The four image pre-processing variables are divided into properties of the field (white space and image position) and those of the image (line weight and image resolution). To quantify the impact of these factors, in this chapter a series of test images are examined using a number of permutations of the relevant factor. By tabulating and plotting the *D* results for each of these permutations, their impact on each factor can be seen. Multiple test images are required to have confidence in any of the trends identified, and so for this study five house elevations were selected based on the typical range identified in past research. In addition, for comparative purposes two abstract shapes were added to the set of images. The rationale for the complete set of images is described later in the chapter. At the end of the process, the most stable data settings are identified, along with, in several cases, their limits and indications of error rates.

Of the five common data processing variables in Table 5.1—that is, those methodological settings which shape the way the procedure is undertaken—several have either been convincingly optimised in the past or rely on relatively straightforward decisions or parameters. For example, the ideal starting grid proportion (its $X \times Y$ number of cells) has been determined both intuitively and mathematically (Bovill 1996; Foroutan-Pour et al. 1999). Various 'rules of thumb' have also been proposed and tested for selecting the ideal size of the first and the last grid cells used in a set of calculations (Koch 1993; Cooper and Oskrochi 2008). However, despite this past research, the optimal settings for the scaling coefficient and grid disposition variables have only recently been demonstrated (Ostwald 2013). The process for determining these last two variables is the subject of the second test in this chapter.

Table 5.1 Variables in the box-counting method

Category	Sub-category	Variable	Description
Pre-processing	Field properties	White space	The proportions and dimensions of the field containing the image being analysed determine how much of the surrounding 'white space' is included in each calculation
	Field properties	Image position	The location of the image relative to the field has been theorised as having an impact on the calculated result
	Image properties	Line weight	More relevant for architectural analysis than for the consideration of data extracted from photographs, the thickness of the lines or points being analysed has been demonstrated as shaping the calculated result
	Image properties	Image resolution	The depth (dpi) of the source image is an indicator (along with the image size) of the potential quantity of information in the starting image. The less information present in an image (that is, the lower the dpi) the less accurate the calculation is likely to be
Processing	Grid properties	Scaling Coefficient (*SC*)	The ratio by which successive grids are reduced in size. The scaling coefficient determines how many grid comparisons are able to be used in the calculation of *D*, but it also determines how much extraneous white space is added with each set of comparisons
	Grid properties	Grid Disposition (*GD*)	The location from which successive grids are generated. Are successive grids positioned such that they share a common corner, edge, or centroid?
	Grid properties	Starting grid size	The dimensions of the largest cell in the starting grid. This effectively determines the upper limit (or largest scale) of the data being collected
	Grid properties	Starting grid proportion	The number of cells on each axis ($x \times y$) which make up the first grid. This variable shapes the usefulness of the data derived from the opening grid; if too few or too many cells are filled, the opening calculation is unlikely to be statistically close to the average
	Grid properties	Closing grid size	The dimensions of the smallest cell in the closing or last grid analysed. This effectively determines the lower limit

(continued)

Table 5.1 (continued)

Category	Sub-category	Variable	Description
			(or smallest scale) of the data being collected. This can be either pre-determined as part of the method, or it can be 'corrected' in the data post-processing stage
Post-processing	Statistical properties	Statistical Divergence (*SD*)	A means of moderating or managing the impact of two types of methodological biases (opening and closing divergence) to which the central *D* calculation in the box-counting method can be overly sensitive
	Statistical properties	Error characterisation	Instead of managing statistical divergence, an alternative approach is to record and explain the character of the data. Often presented in the form or a correlation coefficient (r) or coefficient of determination (r^2) of the complete data set, this is useful for supporting interpretation, but it does not respond to or correct the explicit flaws in the method

In order to determine which scaling coefficient and grid disposition variables will produce the most reliable result, in the second test we examine nine mathematical fractals with known dimensions. Using seven different scaling coefficients and the two most common grid disposition variables, this test produces 126 separate estimates of the *D* values of these geometric figures. The estimated results are then compared with the correct or ideal values that have been calculated mathematically. The grid ratios and dispositions that consistently generate the best results are therefore the ideal settings for these variables.

A combination of the results of the pre-processing test and the processing test are used to identify optimal settings for the method, either reducing its potential sensitivities or focusing its calculations into a more robust and reliable range.

5.2 Image Pre-processing Test

5.2.1 *Field and Image Properties*

Four types of image processing properties are significant for understanding the limits of the box-counting method. The first pair, white space and image position, are associated with the field on which the starting image is positioned and the relationship between the field and the image being analysed. The second pair of factors, line weight (the thickness of the lines which make up the image being analysed) and image resolution (the size and sharpness of the image), are properties of the image itself. Each of these four factors are discussed in detail in what follows before the test used to examine their impact on the method is described and the results presented.

5.2.2 Field Properties

The background on which the image being analysed is placed is called the field. This field comprises three components: white space, image space and empty space. The descriptor 'white space' refers to the region surrounding the image; 'image space' refers to the lines that make up the image itself; and 'empty space' is any region enclosed by the lines (Fig. 5.1). The image space and the empty space are effectively fixed quantities, but the initial amount of white space is determined when the image is positioned or cropped on the 'page' or 'canvas' prior to analysis. Why is this seemingly trivial feature so significant? Because, hypothetically, the more white space there is around an image the more the results of the calculation will be skewed by factors that are not intrinsic to the elevation or plan being

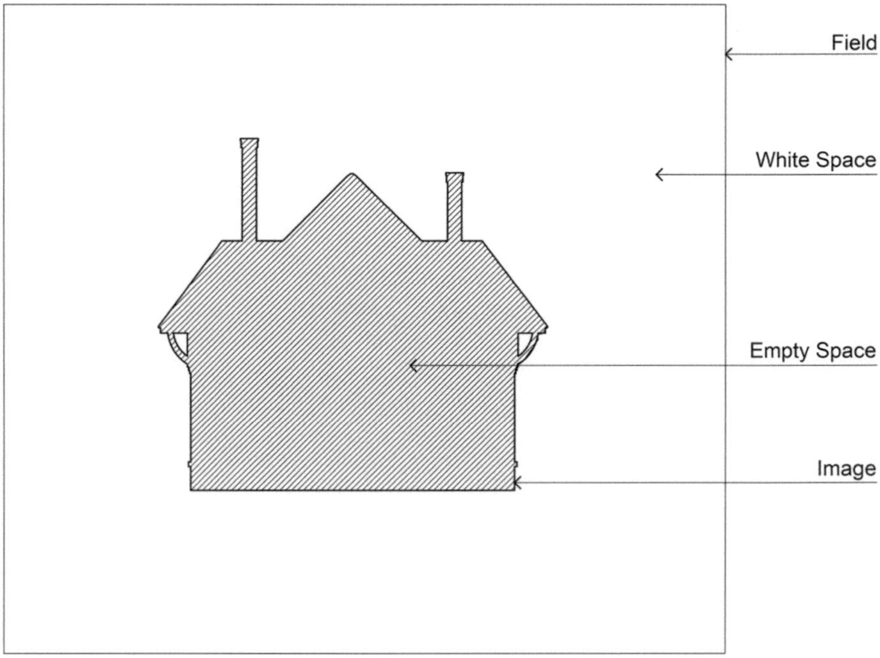

Fig. 5.1 Defining the parts of the image prior to analysis

analysed. Alternatively, if there is almost no white space (that is, the image is tightly cropped), then the first few grid comparisons may be statistically biased because every cell may have information in it. Figure 5.1 demonstrates the different field properties.

Just as the area of white space surrounding the image is thought to have an impact on the result, so too is the location of the white space relative to the image space (the image position). If, for example, the field is twice as large as the image on it, then the image could be positioned in a range of alternative locations on that field. If it is placed to the left side of the field, a large amount of white space will appear to the right. However, if the image space is primarily placed on the top right of the field, the white space on the lower left will be counted in a different iteration of the box-counting process; both architectural images are essentially the same, but may possibly result in different fractal dimensions.

Two other important properties of the field are its size and proportion. Size is measured in pixels (the length and breadth of the image) to accommodate different image densities. The field size is the first determinant of the practical limits of the analytical process. Ideally, the larger the field and image, the more grid comparisons may be constructed and the better the result. The proportion of the field is important because it determines the field's capacity to be neatly divided by grids. Because the box-counting method uses regular grids, it is obvious, but almost never stated, that the dimensions of the field should be multiples of the same figure. Thus, a field 1000 pixels high by 2000 pixels wide will accommodate several ideal starting grid configurations, including a 500 pixel grid (2×4 cells), a 250 pixel grid (4×8 cells) and a 200 pixel grid (5×10 cells). However, it has been demonstrated that the ideal starting proportion for the grid is a multiple of four on the shortest side (Foroutan-Pour et al. 1999). This starting configuration limits the volume of white space included in the first calculation and thus, reduces the need for post-processing corrections (see the discussion of statistical divergence later in this chapter). Thus, in the example given, the 250pixel grid (4×8 cells) is the optimal starting configuration for reducing errors, provided the image is large enough.

If the field does not have an ideal proportion, then it must be cropped or enlarged to achieve such a configuration. There are several variations of this process, depending on how much white space surrounds the image, but careful selection of field size and proportion avoids the need for this additional stage. Thus, we do not test this factor in the present chapter, because the optimal proportion has already been convincingly demonstrated.

5.2.3 Image Properties

Starting images for analysis may potentially be in colour (32 bit), greyscale (16 bit) or black and white (2 bit), but the analytical method only handles black and white lines or points. Data is either present in a grid cell (black) and can be unequivocally counted, or it isn't (white). Without exception, every application of the box-counting method in architectural or urban analysis which used photographs has relied on the application of multiple additional filters (often hidden in the software being used and possibly not obvious even to the researchers) to reduce greyscale and colour gradients to simple black and white lines and points. This factor also explains many of the false or grossly exaggerated results that have been reported in the past. To the human eye, a greyscale photographic image may seem to clearly represent a building form, but by the time the greyscale has been converted to a 2-bit image, it has lost all resemblance to the original.

One of the critical image variables for architectural analysis, line weight, relates to the thickness of the lines in the image being analysed. The box-counting method will incorrectly calculate the dimension of solid black sections of images and even thickened lines will be artificially counted twice, with each reduction in grid scale, leading to high error rates (Taylor and Taylor 1991; Chen et al. 2010). The standard solution to this problem is that all images must be pre-processed with edge detection software to convert them into lines 1 pixel wide (Chalup et al. 2009). For example, Chen et al. (2010) test six common edge detection methods for pre-processing images for the box-counting method; while they fail to identify a single ideal solution for all image densities, both the Sobel and Prewitt algorithms each produce consistent results for most images and perfect results for most line drawings. Alternatively, all images should be pre-processed in a CAD program to choose the finest practical line that the software can produce.

Too often in architectural computational analysis the size of the image being analysed is given as a metric measure; for example, '200 × 100 mm'. This description is often meaningless because it is the resolution of the image—its dpi (dots per inch)—and its size in pixels that is relevant, not its physical size. The same image, printed at the same physical size, will be very blurry at 75 dpi but very sharp at 500 dpi. Thus, the field size of a digital image must be understood as its length and breadth measured in pixels. The field size is important because it is the first determinant of the practical limits of the analytical process. The larger the field, the more grid comparisons may be constructed and the more accurate the result. However, increasing the image size multiplies the computer processing power needed and there are practical limits to all current software. The scale of the image on the drawing is irrelevant (1:100, 1:500 etc.) because the method calculates the visual complexity of the representation of the building, regardless of its size.

5.2.4 Test Description

Four image-processing factors (white space, image position, line thickness and image resolution) are examined in this section using seven test images. Between five and eleven permutations of each of the test images, tailored to the particular factor being considered, are each processed, producing a fractal dimension estimation ($D_{Est.}$) which is in turn, after an initial review of results, compared with a target result (D_{Targ}). The seven test images include five elevations of works by well-known architects and two artificial shapes. The elevations were selected because they represent a range of *D* values ranging from a very simple composition to a much more complex one. The *D* results for the elevations typically fall between 1.3 and 1.8, a span we tentatively call the 'architecture range' for the purposes of this chapter, because most buildings examined by past researchers have recorded results for visual complexity that fall between these values. However, to further test the limits of the method, two additional artificial 'elevations' were added to the set. The first of these is an empty square (like a blank elevation), which is expected to have the lowest result. Indeed, as the estimated results show, for some permutations the empty square was repeatedly measured as having a fractal dimension of around 0.998, which means that it is so minimal that it is no longer an image, but has become a 'dust' of points (Mandelbrot 1982) (Figs. 5.2 and 5.3). The second artificial image added to the set is a densely packed grid (suggesting a highly detailed elevation) which was intended to be within the higher part of the range, and in practice, always measured as the second highest result.

After the square, the image with the lowest *D* result is the south elevation of Kazuyo Sejima's *House in a Plum Grove* (2003), a typically minimalist elevation from this Japanese architect (Fig. 5.4). The next pair of elevations, which have similar levels of visual complexity, are the north elevation of Eileen Gray's *Tempe à Pailla* (1934) (Fig. 5.5) and the north elevation of Robert Venturi and Denise Scott Brown's *Vanna Venturi House* (1964) (Fig. 5.6). The west elevation of Le Corbusier's *Villa Savoye* (1928) (Fig. 5.7) is the next most complex and finally, the most complex elevation tested is the south elevation of Frank Lloyd Wright's *Robie House* (1910) (Fig. 5.8).

Fig. 5.2 "Square"

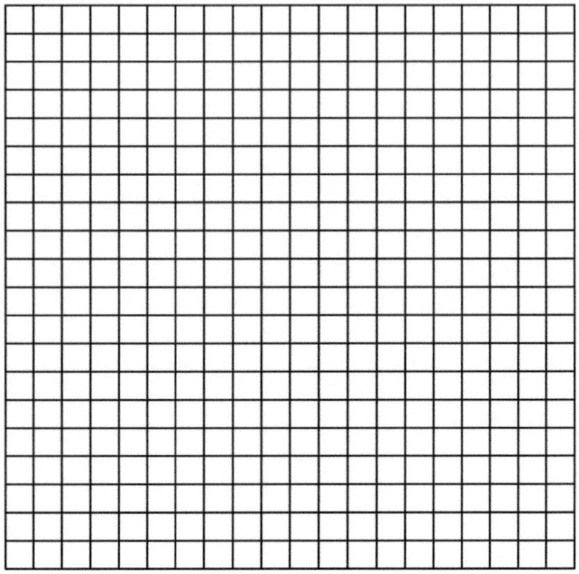

Fig. 5.3 "Grid"

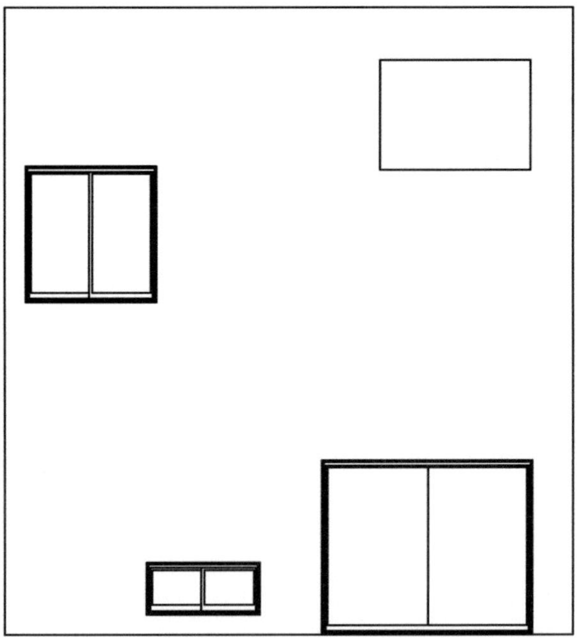

Fig. 5.4 *House in a Plum Grove,* south elevation, Kazuyo Sejima

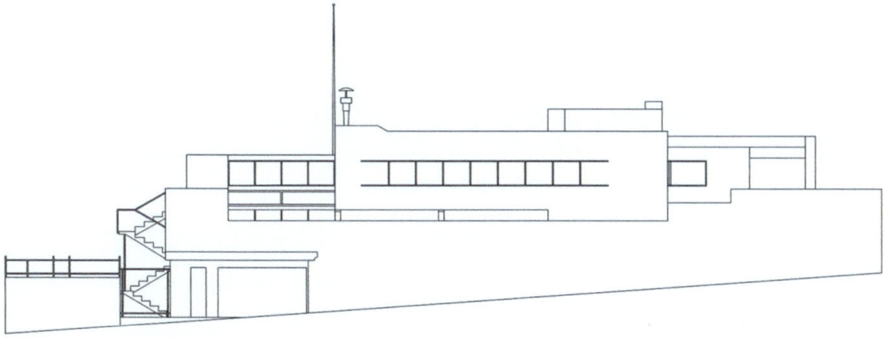

Fig. 5.5 North elevation, *Tempe à Pailla,* Eileen Gray

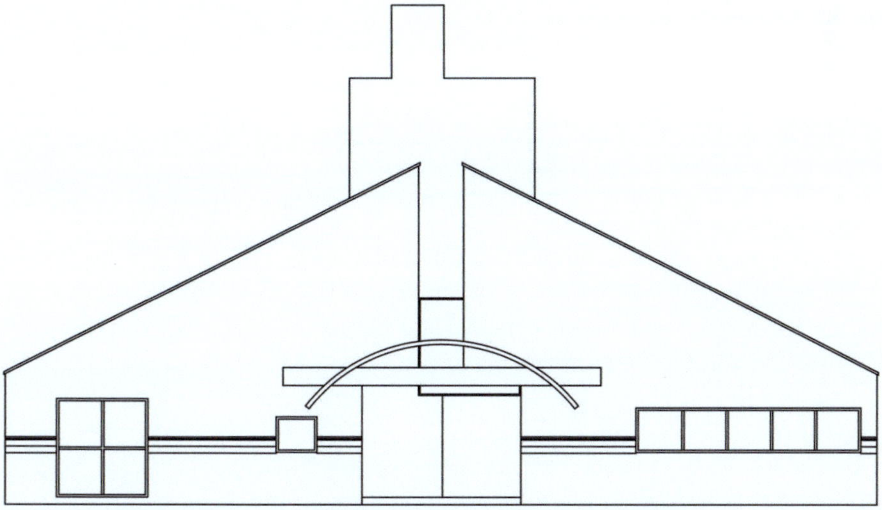

Fig. 5.6 *Vanna Venturi House,* "front" elevation, Venturi and Scott Brown

For each of the seven test images, a series of permutations were prepared in order to examine the four image-processing factors. For white space, nine permutations of each test image were prepared. Each involved gradually adding a controlled amount of white space around the same image (growing the field space, but keeping the image space the same). The first image tested had only a minimal amount of white space (that is, it was cropped very close to the elevation) and the incremental growth was determined for each test image by calculating the number of pixels equivalent to a given percentage of the shortest image dimension, then dividing this into two and adding that result to each side of the image, creating the final field (Fig. 5.9). This provides a consistent area, relative to the image space, for all of the test images. The percentage increments used are 0, 10, 20, 40, 50, 60, 70, 80, 90 and 100 % (Fig. 5.10).

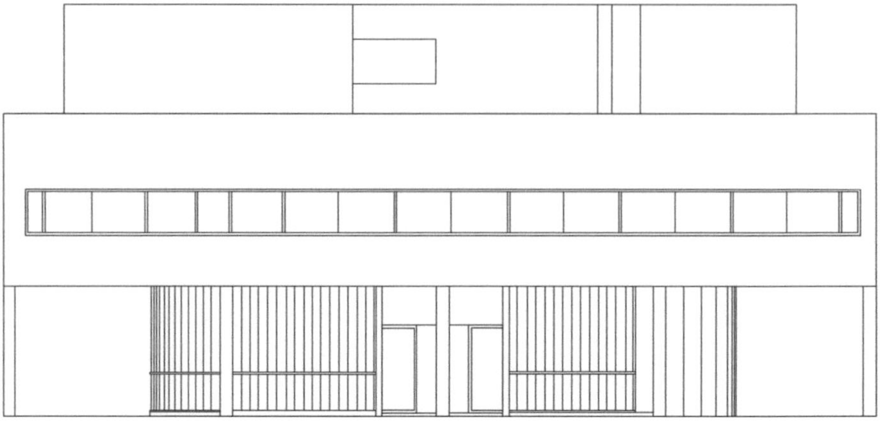

Fig. 5.7 *Villa Savoye,* west elevation, Le Corbusier

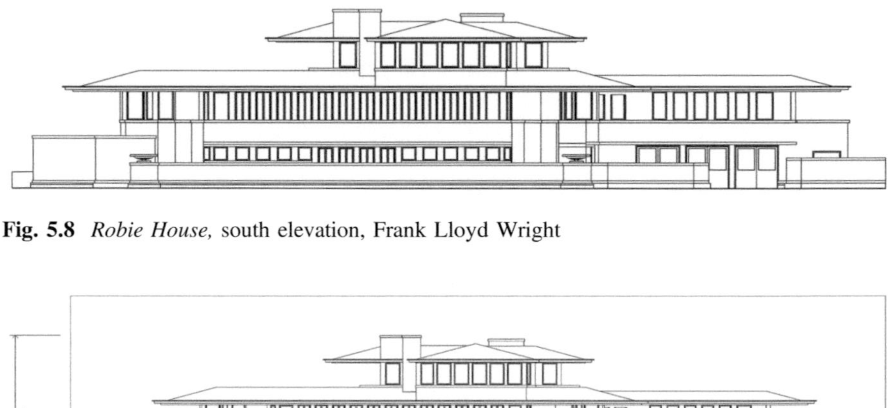

Fig. 5.8 *Robie House,* south elevation, Frank Lloyd Wright

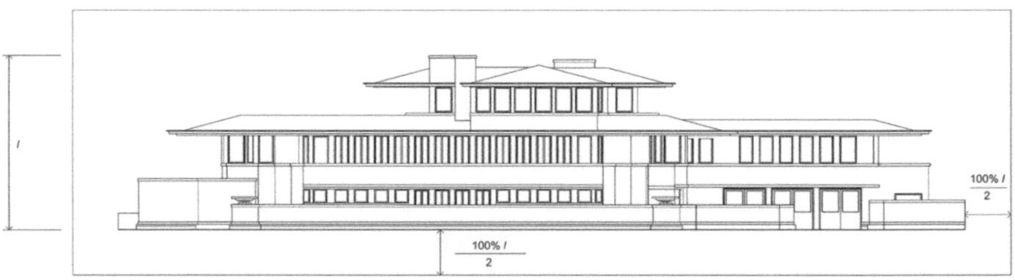

Fig. 5.9 Percentage white space increase

For the image position factor, the same image was placed on the same field but in one of nine different positions in that field, creating different relationships between the white space and the image space. The field size was determined for the initial image by adding 100 % white space (that is, taking the full length of the shortest side of the image, dividing this by two and adding this amount to each side of a centrally positioned image). Then, within this oversized field, the image was located in nine different positions, designated by a combination of left, centre or right and top, centre or base (Fig. 5.11).

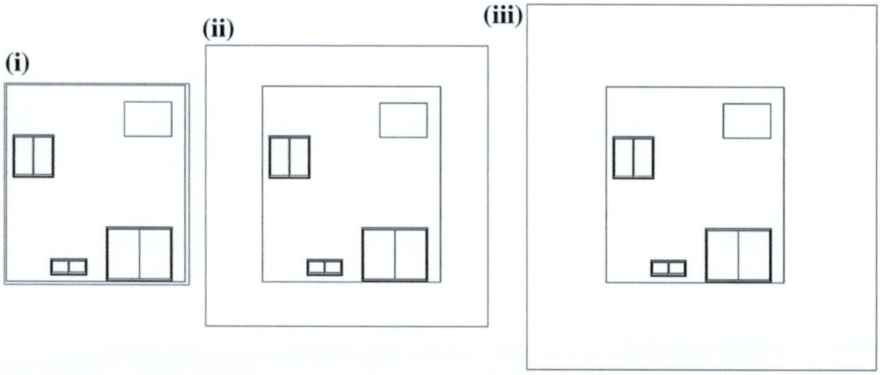

Fig. 5.10 Examples of white space incremental growth permutations for *The House in a Plum Grove.* (i) 0 (ii) 50 % (iii) 100 %

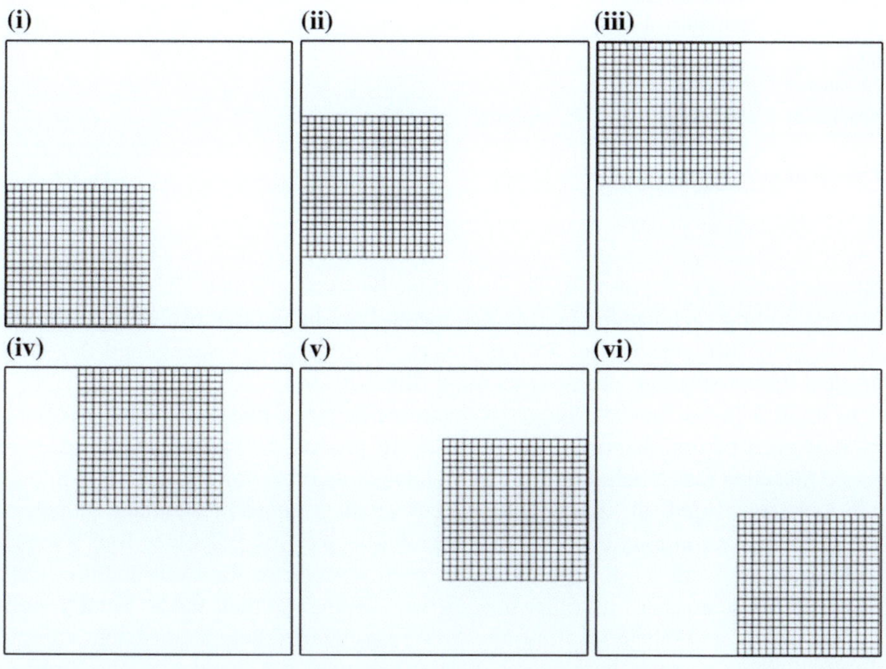

Fig. 5.11 Examples of image position permutations for the grid 'elevation'. (i) Left base (ii) Left centre (iii) Left top (iv) Centre top (v) Right centre (vi) Right base

The impact of line weight was examined by producing eleven permutations of each of the test images with different weights and calculating the results. The line weights tested are widths of 1, 10, 20, 30, 40, 50, 60, 70, 80, 90 and 100 point (Fig. 5.12). For the final factor, image resolution, the test images were saved at five

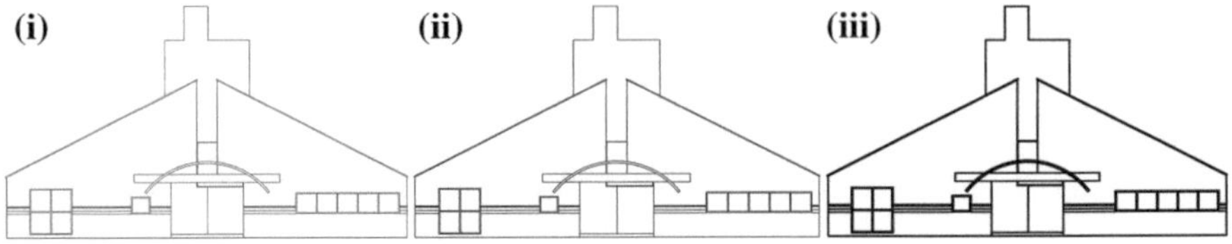

Fig. 5.12 Examples of line thickness permutations for the *Vanna Venturi House*. (i) 1 pt (ii) 50 pt (iii) 100 pt variations

Table 5.2 Summary of tests undertaken for image pre-processing standards

Factors examined	Description of permutations	Target for comparison	Test images (#)	Permutations (#)	Results (#)
White space	Increase white space: 0–100 %	50 % increase	7	10	70
Image position	Various image positions, 1 field	Centre field	7	9	63
Line thickness	Saved at line weights 1–100pt	1pt	7	11	77
Resolution	Saved at 75–175 dpi	175 dpi	7	5	35
Total number of results					245

different levels of compression. This was done by starting with a 175 dpi figure and then resampling each test image (bicubic method), reducing the resolution from 175 to 150, 125, 100 and, finally, 75 dpi. Resampling the image maintained the same physical dimensions but changed its pixel dimensions.

In total, using seven test images to examine between five and eleven permutations of each of four factors, 245 results were produced (Table 5.2). In order to ensure that each factor being tested was isolated from other variables, and its impact able to be measured, all other factors were set to a range of standard values or settings. For example, except for the test of line weights, all other line weights being analysed were set at 1 pt thickness. Next, except for the examination of the impact of image position, all other images were centred on their fields. Finally, both the line weight and image resolution tests were conducted with a consistent volume of white space around them which was determined by calculating the shortest dimension of each image in the set and adding 20 % of this length to each side of the image (defining the field). The data processing settings used for all calculations (for reasons which will become apparent later in the chapter) were a scaling coefficient of 1.41421, edge growth (top-left) grid disposition and no correction for statistical divergence.

5.2.5 *Data Analysis Method*

For each of the four factors being tested, the following steps are taken to interpret the results.

i. The D_{Est} results are tabulated and charted. Using this data, and informed by past, theorised ideal standards, a 'target' permutation is chosen for comparison. In two of the cases the theorised optimal setting matched the data, whereas in the other two the results were less differentiated and a range of possible targets, with similar outcomes were available. In this situation, the central setting in the range is chosen and the targets are used to assist the interpretation of the results, rather than as absolute indicators.
ii. The difference, expressed as a percentage, between the D_{Est} and D_{Targ} results is recorded. The average of these differences is then calculated, as is its correlation coefficient (r). The r value is an indicator of degree to which one body of data may be efficaciously compared to another. In this case, for all examples, the test charts the results of the permutations of D_{Est} against the target fractal dimension, D_{Targ}. For a perfect correlation, r should equal 1.0. The lower the result below 1.0, the less consistent the correlation. Despite this, because all of the tests in this paper are comparing variations of similar images, even the worst of the r results, 0.84, is relatively high.
iii. The tabulated data is then analysed using a scatter graph of D results and a distribution graph, charting results against the percentage gap ($D_{Est}-D_{Targ}$) to identify patterns and to quantify limits. This process clarifies the range of divergences from the target and assists to identify trends and quantify the average magnitude of errors. For these charts, linear and polynomial trend-lines are used to assist the analysis.
iv. The result with the highest percentage difference is identified; this is effectively the worst result or highest error. For evaluation purposes, anything with less than 20 % of this level of difference is considered to be within a stable or robust zone in the results.

5.2.6 *Results of the Pre-processing Test*

5.2.6.1 White Space

The theorised impact of white space has been one of the more contentious issues in fractal analysis, with many authors ignoring the issue and others suggesting various approaches to it (Bovill 1996; Cooper and Oskrochi 2008; Ostwald et al. 2008a). Amongst those who have considered the question there is a broad agreement that some white space around the image is necessary, but that too much will undermine the veracity of the method. An initial review of the results (Table 5.3) confirms this,

Table 5.3 Results of white space analysis

White space		Square	Sejima	Venturi	Gray	Le Corb.	Grid	Wright	Ave. % diff	r
0 %	D_{Est}	1.038	1.217	1.241	1.305	1.315	1.398	1.528	–	0.9818
	% diff	5.000	0.800	0.700	4.900	1.500	2.900	0.000	1.800	–
10 %	D_{Est}	1.017	1.242	1.237	1.286	1.304	1.401	1.506	–	0.9535
	% diff	2.900	3.300	0.300	3.000	2.600	3.200	2.200	2.430	–
20 %	D_{Est}	0.977	1.173	1.254	1.279	1.397	1.382	1.508	–	0.9620
	% diff	1.100	3.600	2.000	2.300	6.700	1.300	2.000	2.980	–
30 %	D_{Est}	0.991	1.196	1.259	1.283	1.328	1.388	1.557	–	0.9929
	% diff	0.300	1.300	2.500	2.700	0.200	1.900	2.900	1.920	–
40 %	D_{Est}	1.004	1.172	1.241	1.261	1.34	1.38	1.545	–	0.9884
	% diff	1.600	3.700	0.700	0.500	1.000	1.100	1.700	1.450	–
50 %	D_{Targ}	0.988	1.209	1.234	1.256	1.33	1.369	1.528	–	1.000
	% diff	*0.000*	*0.000*	*0.000*	*0.000*	*0.000*	*0.000*	*0.000*	*Target*	–
60 %	D_{Est}	0.998	1.196	1.255	1.278	1.333	1.403	1.53	–	0.9912
	% diff	1.000	1.300	2.100	2.200	0.300	3.400	0.200	1.580	–
70 %	D_{Est}	0.992	1.179	1.27	1.276	1.337	1.404	1.549	–	0.9854
	% diff	0.400	3.000	3.600	2.000	0.700	3.500	2.100	2.480	–
80 %	D_{Est}	0.974	1.237	1.268	1.237	1.354	1.387	1.52	–	0.9912
	% diff	1.400	2.800	3.400	1.900	2.400	1.800	0.800	2.180	–
90 %	D_{Est}	0.941	1.148	1.262	1.236	1.344	1.373	1.507	–	0.9712
	% diff	4.700	6.100	2.800	2.000	1.400	0.400	2.100	2.470	–
100 %	D_{Est}	1.036	1.248	1.261	1.242	1.331	1.463	1.488	–	0.8427
	% diff	4.800	3.900	2.700	1.400	0.100	9.400	0.400	2.980	–

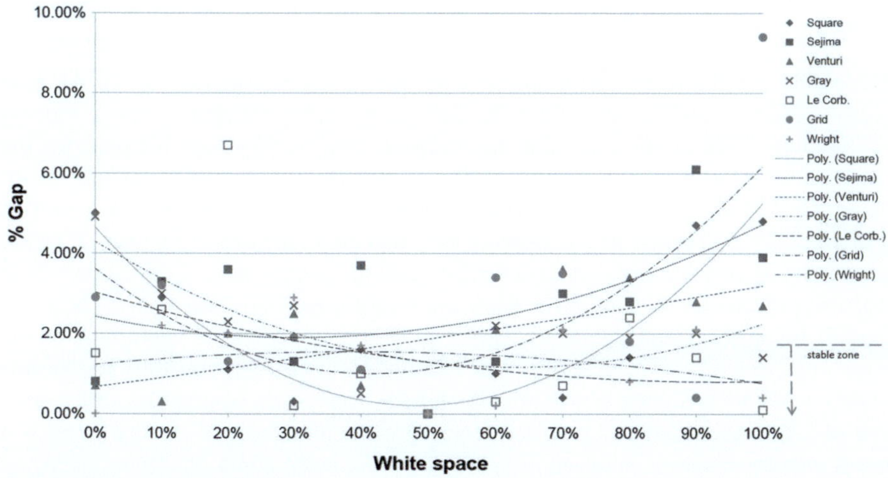

Fig. 5.13 Results for white space test

indicating that the most consistent sets were in the central part of the graph (between 30 and 60 % white space) and so the 50 % result was selected as the target. In this zone there was typically less than a 1.58 % average variation caused by the differing quantities of white space. Outside of this zone, while not consistent, the average growth in variation was in the order of 3.98 %, with isolated results up to 9.4 % (Fig. 5.13).

These results suggest that the best image pre-processing setting was either 40 or 50 % white space. The magnitude of errors caused by too little white space was relatively similar, commonly in the 2.2 % range, but for larger amounts of white space this grew to around 2.98 %, with higher trends indicated beyond that (>4 %). However, despite this, and taking into account the *r* results, it is also clear that within the 30 to 60 % range, white space has less impact on the results than previously suggested, with none of the average differences in that range being above 1.72 %.

Notably, in this set of results there is one test image which showed a much higher sensitivity to the changes in white space than any other. The abstract grid elevation had a low result of 1.1 % difference and a high of 9.4 %, more than double the typical range for the other test images. A more detailed test would be required to determine why such sensitivities occur in some images but not in others. As the focus of the present chapter is on optimising the method, rather than examining some of its occasional anomalies, this issue was not pursued.

5.2.6.2 Image Position

The results for the image position test were the least consistent of any of the four pre-processing variables examined in this chapter (as reflected in the *r* values) (Table 5.4). With no clear pattern, the centre-centre position was adopted for the target value (Fig. 5.14). However, with the highest percentage difference result being 11 %, the optimal zone was determined as any average difference result of less than 2.2 %, a range which none of the other permutations fell within. If, then, the results by position are considered relative to the centre, the only position which has a significant set of 'low' error rates is the centre-base position (2.5 %) with the top-left being the worst, (4.28 %). However, in combination the magnitude of the error rates, regardless of position, was relatively minor. One curiosity in this test relates to the *Villa Savoye* elevation, which had a very wide range of results, from a low of 1.1 % to a high of 11 % difference. For the remainder of the test images, a much smaller range of between 1.3 and 4 % was more common. Once again, an explanation for such an isolated anomaly is beyond the scope of the present test, but it is a reminder that some images are especially sensitive to the more extreme image factors.

5.2.6.3 Line Weight

The clearest trend in any of the results was for the line weight factor. It was readily apparent in the preliminary analysis stage that, as the line weight increases, so too does the calculated result (Table 5.5). The target line weight for this comparison was therefore the thinnest, 1 pt, as this cannot be counted multiple times in an analysis of the same box-counting grid; a 1 pt line is either emphatically inside or outside a 1 pt grid-shifted line, whereas a 20 pt thickness line can be partially inside (say, 8 pts) and partially outside (12 pts) a grid line, which means that it will be counted twice at that scale.

With the thinnest line as D_{Targ}, the D_{Est} value grows relatively steeply in the chart as the lines thicken, this is confirmed by the *r* values (Fig. 5.15). However, for both of the first two permutations, the line weights of 1 and 10 have identical results. Beyond the 10 pt line thickness, the more abstract test images—including the square, grid and Sejima elevation—display a rapid rate of increasing errors. Most of the other elevations—by Gray, Le Corbusier and Wright—tend to remain largely unchanged until the line thickness increases to 30 point whereupon the error rate increases slightly to around 3.4 % (20 % of 17) until the 50 point permutation is reached and the errors escalate.

When the complete set of line weight results is considered two things are apparent. First, for a sufficiently large starting image (say, 1 MB), as long as the lines being analysed are very thin (less than 10 pt in this case), the impact on the results is negligible. Second, once lines become marginally thicker, they have a heightened capacity to produce quite large errors. In particular, five of the test images showed errors of a magnitude of over 6 % for the thicker permutations,

Table 5.4 Results of image position analysis

Image position		Square	Sejima	Venturi	Gray	Le Corb.	Grid	Wright	Ave. % diff	r
Centre-Base	D_{Est}	1.036	1.179	1.279	1.288	1.342	1.462	1.524	–	0.9645
	% diff	0.000	6.900	1.800	1.500	1.100	0.100	3.600	2.500	–
Centre-Top	D_{Est}	1.048	1.201	1.281	1.251	1.402	1.488	1.516	–	0.9598
	% diff	1.200	4.700	2.000	2.200	7.100	2.500	2.800	3.550	–
Centre-Centre	D_{Targ}	1.036	1.248	1.261	1.273	1.331	1.463	1.488	–	1.000
	% diff	*0.000*	*0.000*	*0.000*	*0.000*	*0.000*	*0.000*	*0.000*	*Target*	–
Left-Base	D_{Est}	1.062	1.208	1.303	1.321	1.387	1.478	1.535	–	0.9625
	% diff	2.600	4.000	4.200	4.800	5.600	1.500	4.700	4.130	–
Left-Centre	D_{Est}	1.062	1.186	1.287	1.276	1.367	1.452	1.502	–	0.9694
	% diff	2.600	6.200	2.600	0.300	3.600	1.100	1.400	2.530	–
Left-Top	D_{Est}	1.077	1.229	1.287	1.295	1.441	1.504	1.527	–	0.9465
	% diff	4.100	1.900	2.600	2.200	11.000	4.100	3.900	4.280	–
Right-Base	D_{Est}	1.050	1.202	1.294	1.304	1.359	1.460	1.534	–	0.9142
	% diff	1.400	4.600	3.300	3.100	2.800	0.030	4.600	3.070	–
Right-Centre	D_{Est}	1.050	1.190	1.285	1.261	1.343	1.433	1.500	–	0.962
	% diff	1.400	5.800	2.400	1.200	1.200	3.000	1.200	2.470	–
Right-Top	D_{Est}	1.062	1.221	1.287	1.289	1.410	1.477	1.523	–	0.9673
	% diff	2.600	2.700	2.600	1.600	7.900	1.400	3.500	3.280	–

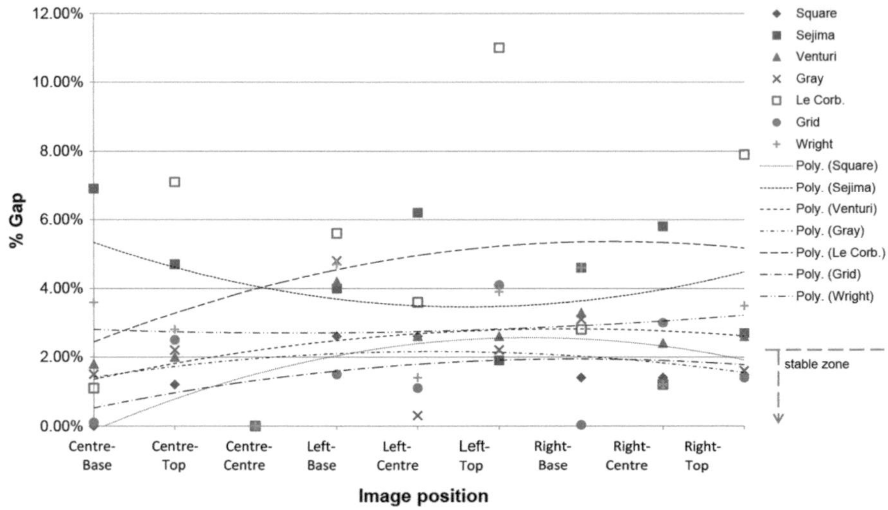

Fig. 5.14 Results for image position test

values which are consistently higher than for the other factors considered in this chapter.

Whereas in the previous two tests a particular image set showed a greater than expected sensitivity to the changes in settings, in this set all of the image results conformed to a relatively consistent pattern.

5.2.6.4 Image Resolution

A preliminary analysis of the results of the image resolution test shows that the higher the resolution the more convergent the results (Table 5.6). However, this is not a clearly linear trend, as is that of the line weight test, although it does broadly conform to the theorised ideal standard. For this reason, the 175 dpi version was selected as the target. Moreover, there is another compelling rationale for choosing this target: the larger the image the more grid comparisons may be undertaken by the software, and the more statistically viable the result. Thus, for this factor test, an additional piece of information, the number of grid comparisons possible, was also tabulated.

Despite the results generally worsening with lower image resolutions, some of the indicators for the 125 dpi permutation are superficially superior to those of the 150 dpi version. The former has both a lower percentage difference and a higher r value (both suggesting a better result), although it always has a lower grid-comparison value than both the 150 dpi and the target 175 dpi permutations. However, as none of the results are below the reduced error zone of 1 % (being 20 % of the highest difference 5), only the target value is considered to be the best

Table 5.5 Results of line thickness analysis

Line thickness		Square	Sejima	Gray	Venturi	Le Corb.	Grid	Wright	Ave. % diff	r
1 pt.	D_{Targ}	0.977	1.196	1.279	1.279	1.397	1.382	1.513	–	1
	% diff	*0.000*	*0.000*	*0.000*	*0.000*	*0.000*	*0.000*	*0.000*	*Target*	–
10 pt.	D_{Est}	0.977	1.196	1.279	1.279	1.397	1.382	1.513	–	1
	% diff	0.000	0.000	0.000	0.000	0.000	0.000	0.000	0.000	–
20 pt.	D_{Est}	0.977	1.216	1.279	1.279	1.397	1.382	1.513	–	0.9981
	% diff	0.000	2.000	0.000	0.000	0.000	0.000	0.000	0.290	–
30 pt.	D_{Est}	0.998	1.237	1.279	1.298	1.397	1.434	1.513	–	0.9853
	% diff	2.100	4.100	0.000	1.900	0.000	5.200	0.000	1.900	–
40 pt.	D_{Est}	1.015	1.25	1.279	1.311	1.413	1.458	1.513	–	0.9736
	% diff	3.800	5.400	0.000	3.200	1.600	7.600	0.00	3.090	–
50 pt.	D_{Est}	0.998	1.258	1.279	1.326	1.419	1.475	1.513	–	0.9633
	% diff	2.100	6.200	0.000	4.700	2.200	9.300	0.000	3.500	–
60 pt.	D_{Est}	1.068	1.285	1.280	1.328	1.430	1.496	1.543	–	0.9453
	% diff	9.100	8.900	0.100	4.900	3.300	11.400	3.000	5.810	–
70 pt.	D_{Est}	1.062	1.285	1.284	1.34	1.443	1.509	1.547	–	0.9449
	% diff	8.500	8.900	0.500	6.100	4.600	12.700	3.400	6.390	–
80 pt.	D_{Est}	1.062	1.303	1.286	1.351	1.442	1.522	1.551	–	0.9315
	% diff	8.500	10.700	0.700	7.200	4.500	14.000	3.800	7.060	–
90 pt.	D_{Est}	1.072	1.309	1.303	1.362	1.450	1.535	1.558	–	0.9336
	% diff	9.500	11.300	2.400	8.300	5.300	15.300	4.500	8.090	–
100 pt.	D_{Est}	1.072	1.309	1.303	1.362	1.450	1.535	1.558	–	0.9336
	% diff	9.500	11.300	2.400	8.300	5.300	15.300	4.500	8.090	–

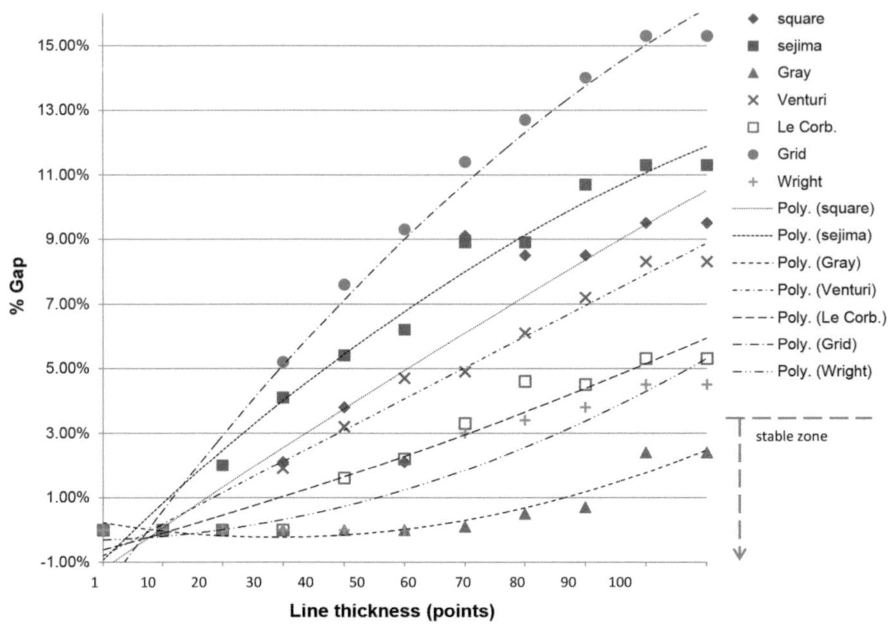

Fig. 5.15 Results for the line thickness test

option for resolution settings for image processing (Fig. 5.16). Another potentially misleading part of this set of results is that, under the influence of changing resolution, the errors range from 0.00 % to ±2.5 %, which might seem to indicate that resolution has little impact on the image. However, when additional tests were undertaken using low resolutions like 50 and 25 dpi as well as higher resolutions up to 275 dpi (Table 5.7), the software was not able to return results for all of the images. This is because at the lower resolutions the images became so blurry that the software was unable to detect their full extent and in some cases (25 dpi) it was unable to detect the presence of an image at all. Such instances are coded with an 'x' in Table 5.7. At the higher resolutions, particularly 250 and 275 dpi, the amount of information processing required was too high for the software and the calculations could not be completed. These are shown as 'xx' in Table 5.7. Thus, image resolution has clear practical upper and lower limits, beyond which a result often simply cannot be produced.

5.2.7 *Discussion*

The first of our two tests reveals that two of the pre-processing options for images have clear trends: line weight and image resolution. In both cases, even seemingly minor changes in image standards can result in errors of up to 18 %, completely

Table 5.6 Results of image resolution analysis

Image resolution		Square	Sejima	Gray	Venturi	Le Corb.	Grid	Wright	Ave. % diff	r
75 dpi	D_{Est}	0.953	1.169	1.226	1.230	1.343	1.353	1.465	–	0.8414
	% diff	4.500	5.000	3.700	3.600	1.100	0.600	3.800	2.200	–
	#Grids	14	14	14	17	16	14	12	–	–
100 dpi	D_{Est}	0.998	1.178	1.213	1.273	1.340	1.376	1.51	–	0.9715
	% diff	0.000	4.100	5.000	0.700	1.400	2.900	0.700	1.800	–
	#Grids	15	15	15	17	17	15	12	–	–
125 dpi	D_{Est}	0.982	1.207	1.246	1.270	1.368	1.353	1.515	–	0.9974
	% diff	1.600	1.200	1.700	0.400	1.400	0.600	1.200	0.900	–
	#Grids	16	15	15	18	17	16	13	–	–
150 dpi	D_{Est}	0.998	1.207	1.253	1.235	1.346	1.372	1.496	–	0.9886
	% diff	0.000	1.200	1.000	3.100	0.800	2.500	0.700	1.600	–
	#Grids	16	16	16	19	18	16	14	–	–
175 dpi	D_{Targ}	0.998	1.219	1.263	1.266	1.354	1.347	1.503	–	1.000
	% diff	*0.000*	*0.000*	*0.000*	*0.000*	*0.000*	*0.000*	*0.000*	*Target*	–
	#Grids	17	16	16	19	18	17	14	–	–

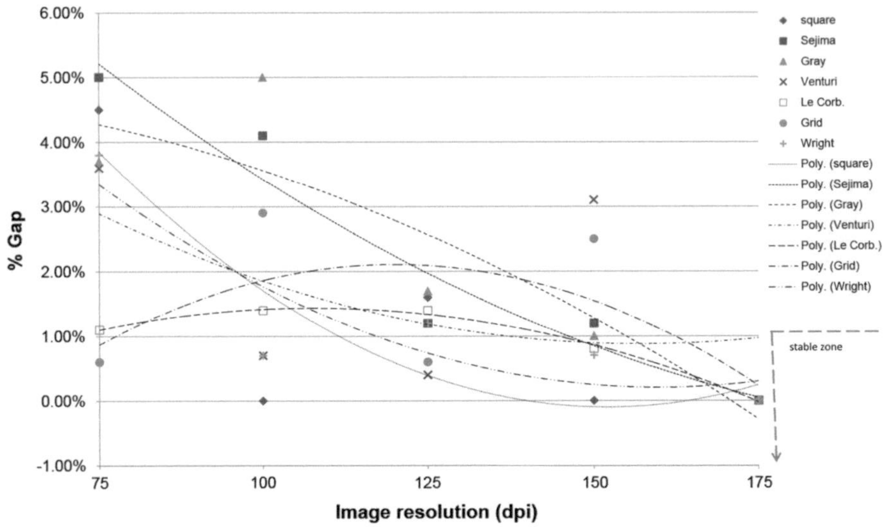

Fig. 5.16 Results for image resolution test

undermining the veracity of the calculations made using these settings. However, in both cases the use of particular standards or settings identified in this chapter will limit possible errors to less than 1 %.

The results for variations in white space and image position are less compelling. For the former case, a relatively robust zone is identified (between 40 and 50 % white space) where divergent results are minimised, while outside of that zone they increase. Nevertheless, a much larger set of tests would be required to discern a clear pattern. The results for the image position factor are even less consistent, with no position, in comparison to centre placement, offering a persistent reduction in possible errors. While this supports the default practice found in much past research of using a central position, it does not necessarily provide evidence against other practices. Ultimately, though, the magnitude of errors relating to both white space and image position is relatively minor (averaging between 1.42 and 2.98 % for area and 1.46 and 4.28 % for position) although more extreme permutations produce much larger errors.

5.3 Image Processing Test

There are five types of data processing factors that need to be understood in order to usefully apply the box-counting method. All of these factors are associated with the grids chosen for the analysis, either with the relationship between successive grids, or their proportionality or limits. There are also two post-processing approaches—

Table 5.7 Extended set of results for image resolution

Image resolution		Square	Sejima	Gray	VSB	Le Corb.	Grid	Wright	Ave. % diff
25 dpi	D_{Est}	x	x	x	x	x	0.292	1.014	–
	% diff	–	–	–	–	–	105.500	48.900	51.470
	#Grids	–	–	–	–	–	11	8	–
50 dpi	D_{Est}	−0.050	0.870	0.726	0.866	0.857	0.841	1.229	–
	% diff	100.300	−86.700	−71.500	−86.600	−85.700	−83.600	−122.500	−62.380
	g–c	13	12	12	15	12	13	10	–
200 dpi	D_{Est}	0.998	1.212	1.264	xx	1.343	1.357	1.501	–
	% diff	0.00	0.700	0.100	–	1.100	1.000	0.200	0.520
	#Grids	17	17	17	–	17	17	14	–
225 dpi	D_{Est}	1.008	1.229	1.267	xx	1.351	1.339	1.493	–
	% diff	1.000	1.000	0.400	–	0.300	0.800	1.000	0.750
	#Grids	18	17	17	–	19	18	15	–
250 dpi	D_{Targ}	1.037	1.222	1.274	xx	xx	1.352	1.499	–
	% diff	3.900	0.300	1.100	–	–	0.500	0.400	1.240
	#Grids	18	17	17	–	–	18	15	–
275 dpi	D_{Est}	xx	1.221	xx	xx	xx	xx	1.506	–
	% diff	–	0.200	–	–	–	–	0.300	0.250
	#Grids	–	18	–	–	–	–	15	–

to either improve the statistical reliability of the results or conversely to report their limits—that are best described in the same context. Each of these factors is described hereafter in this section, prior to focussing on just two of these which are the subject of a more detailed test to determine the optimal scaling coefficient and grid disposition variables for the method.

5.3.1 *Image Processing Factors*

5.3.1.1 Grid Disposition

The grid disposition variable describes the point of origin from which successive grids are generated. This in turn determines where white space is added to the calculation and, in combination with the scaling coefficient, how much white space is added. The two most common grid disposition variations are edge-growth and centre-growth. Edge-growth typically generates the first grid from a corner point of the field, say the top left-hand corner, and white space is then successively 'grown' or 'cropped' to the right and base of the field to form a suitable starting proportion. Depending on the degree to which each successive grid is reduced, further white space may be added to, or removed from, the right edge and base of the field for each comparison. Centre-growth uses the centroid of the image as the point of origin for each successive grid and draws or crops white space equally from around all four sides of the field. A variation of centre-growth uses the centroid as a point of origin, but rotates the grid with each scale reduction, a process that past research suggests has negligible impact on accuracy (Da Silva et al. 2006).

5.3.1.2 Scaling Coefficient

Architectural and urban applications of the box-counting method conventionally present it as starting with the largest grid and gradually reducing its size for each subsequent comparison. The ratio between one grid and the next is the 'scaling coefficient' (*SC*). For example, Bovill (1996) describes halving each successive grid, a scaling coefficient of 2:1, so that the larger grid is double the size of the smaller. The scaling coefficient has a direct impact on two factors: the number of possible mathematical comparisons that can be made of detail in an image, and the amount of white space around the image that is included in each comparative calculation. The lower the scaling coefficient, the larger the number of comparisons that can be constructed and, by implication, the more accurate the result (Roy et al. 2007; PourNejatian and Nayebi 2010). However, the lower the scaling coefficient, the more variable the white space included with each comparison, undermining the accuracy of the result.

Consider the examples of Bovill (1996) and Sala (2002), each of whom uses a scaling coefficient of 2:1, effectively the only practical value that does not add any white space to the calculation. This is because successive iterations of the grid have the same external dimension and therefore include exactly the same quantity of white space. However, the 2:1 ratio only allows them to produce around three scale grids for comparison before the cells become too small. The difficulty with this is that it potentially takes at least five comparative scales for the error rate to be reduced to ±25 % (Chen et al. 1993). To achieve a result of ±5 % accuracy for the same image, Meisel and Johnson (1997) suggest that between fifteen and twenty comparative scales may be required, and to achieve a ±0.5 % error rate anything between fifty and 125 comparative scales is potentially necessary. Thus, the choice of scaling coefficient is a balance between maximising the number of grid comparisons available and minimising the variable growth of white space included in each calculation (Roy et al. 2007). One solution to the scaling coefficient variable is to use a ratio of $\sqrt{2}$:1 (approximately 1.4142:1, the ratio of a diagonal to the side of a square) which increases the number of grid comparisons while moderating the variable amount of white space to a tight zone. Scaling coefficients of less that 1.4 will produce more comparative results, but will cyclically vary the amount of white space included in each calculation.

5.3.2 *Managing Limits*

When the log-log chart is plotted, the slope of the line—its fractal dimension—is determined by the data points generated by the mathematical comparison between detail in cells at different scales. The slope of the line is the average of the set of points, but like any average, not all points in the set will be close to that value (Fig. 5.17). Statistical divergence (*SD*) refers to the degree to which certain data points do not fit neatly in a set but still participate in the calculation of its average. There are three types of statistical divergence in the box-counting method, which is why past researchers have tended not to immediately resort to calculating r (correlation coefficients) or r^2 (coefficient of determination) values to examine the validity of a trend line. There is no consistency in how these three are named, but here we will call them 'opening', 'central' and 'closing' divergence (Fig. 5.18).

Opening divergence occurs in the first few grid comparisons for one of three common reasons. First, because the proportionality of the opening grid is poor; second, because excessive white space surrounds the image; third, because the image fills the entire first grid. All of these problems are associated with poor starting field and image settings. Central divergence occurs in the 'stable', middle part of the graph and it represents an inconsistent shift in detail in the image itself (meaning that the image is a multi-fractal). Such a shift is not an anomaly; it is an important property of the image. Closing divergence occurs when the analytical grids have become so small that they are mostly counting empty space within the image (Chen et al. 1993). Opening and closing divergences are flaws which can be minimised or controlled in various ways. Central divergence is a quality of the image itself, representing the scale at which the characteristic irregularity begins to

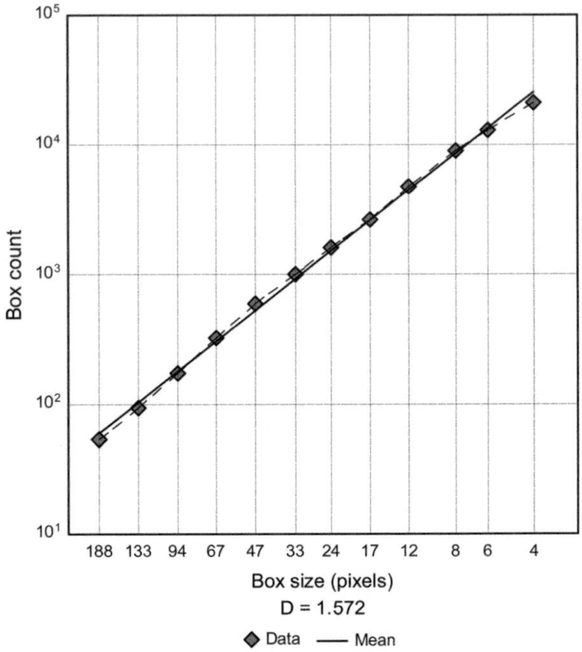

Fig. 5.17 Example of a log-log chart with a high degree of correlation between data and average

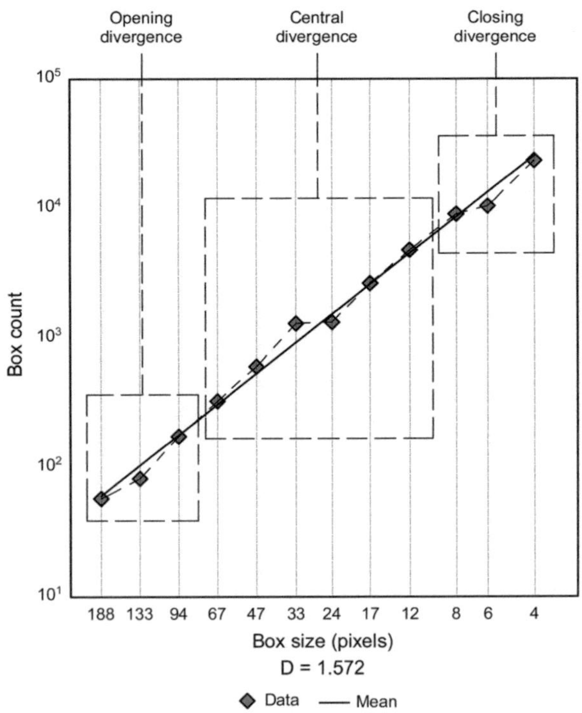

Fig. 5.18 Example of a log-log chart identifying the three zones of potential statistical divergence

break down. Some software allows for the tactical removal of particular points in the 'central' range, but such a process alters the measured character of the object, so it should be avoided unless the user has a clear reason for making such a decision.

While central divergence is critical for the calculation, opening and closing divergences can be controlled. For example, past research suggests that an ideal proportion for the opening field and associated first or largest grid cell is 0.25 l, where l is the length of the shortest side (Foroutan-Pour et al. 1999). Conversely, the smallest grid that should be considered has a cell size of 0.03 l (Koch 1993; Cooper and Oskrochi 2008). By using these two standards, the impact of opening and closing divergences is mitigated, but there is no way of managing or understanding the impact of grids which are either larger or smaller than these rules accommodate.

Another way to approach this problem is to post-process the results to control the extent to which divergence is allowed. For example, to limit the impact of opening divergence, the overall result for all grid comparisons is first calculated and then the first data point is removed and the result recalculated. If the difference between the original and the revised result is greater than a particular threshold level (*SD*), then the first point is removed. Then the process is repeated for the second point and potentially for the third, if a large enough data set is available. The same process also occurs with the last point or two in the line, to limit the impact of closing diversity.

The ideal *SD* value is relative to two factors: the accuracy of the other variables in the method and the purpose of the analysis. In the first instance, there is no need to choose an *SD* of 0.5 % (a value which will remove data points which deviate from the average by more than 0.5 % relative to the log-log result), if the accuracy of data produced by the scaling coefficient is at best 10 %. In the second instance, for example in architecture, human visual perception will readily differentiate between dimensions with around 4 % difference (Stamps 2002; Ostwald and Vaughan 2010; Vaughan and Ostwald 2010b), so for some limited purposes, an *SD* of less than 4 % may be unnecessary. However, Westheimer's (1991) research into the capacity of the human eye to differentiate between different types of fractal lines (mathematical 'random walks') finds that a less than 1 % variation is readily detectable by the human eye and mind for similar objects. Thus, if there are two similar forms (say two different elevations of the Villa Savoye), the human eye and mind is likely to be able to detect which one is more visually complex, even if the difference is only in the order of 1 %. However, if the images are stylistically dissimilar (say an elevation of the *Villa Savoye* and one of the *Robie House*), then human perceptions will readily identify the more complex image, though it has a much lower ability to determine how much more complex it is. For this reason, the purpose of the research will determine the correct level of *SD* to use, between 4 and 0.5 %.

5.3.3 Test Description

In the second test in this chapter, images of nine well-known fractal sets are analysed using the box-counting method to compare the results, estimated using various scaling coefficient and grid disposition variables, with the correct result for each fractal set. The Hausdorff Dimension for each of these nine geometric forms has been widely documented (Mandelbrot 1982; Voss 1988; Górski et al. 2012). The nine test cases, listed in order of increasing visual complexity, are as follows: the Koch Snowflake, the Terdragon Curve, the Apollonian Gasket, the Minkowski Sausage, the Sierpinski Triangle, the Sierpinski Hexagon, the Fibonacci Word, the Pinwheel Fractal and the Sierpinski Carpet (Figs. 5.19, 5.20, 5.21, 5.22, 5.23, 5.24, 5.25, 5.26 and 5.27). All of these cases are deterministic fractal sets; so-called because they have mathematically calculable fractal dimensions (D_{Calc}). The process of comparing a series of permutations of a system against an object with a known set of properties is conventionally called 'calibration'. Thus, this test calibrates the box-counting method to determine which scaling coefficient and grid disposition settings will produce an estimated result (D_{Est}) that consistently comes closest to the actual dimension of these fractals.

Because the scaling coefficient and grid disposition variables are the only ones being tested, all of the other settings are standardised as follows.

Fig. 5.19 Koch Snowflake, $D_{(\mathrm{calc})} = 1.2619$

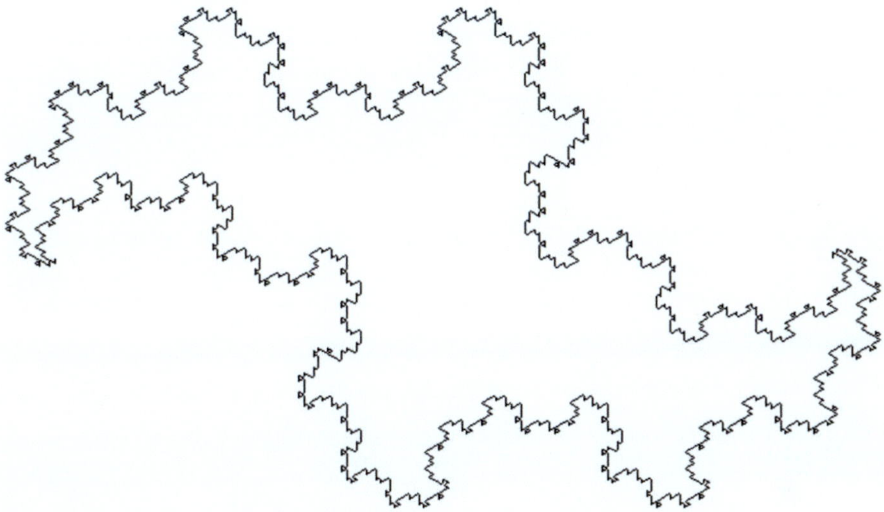

Fig. 5.20 Terdragon Curve, $D_{(calc)} = 1.2619$

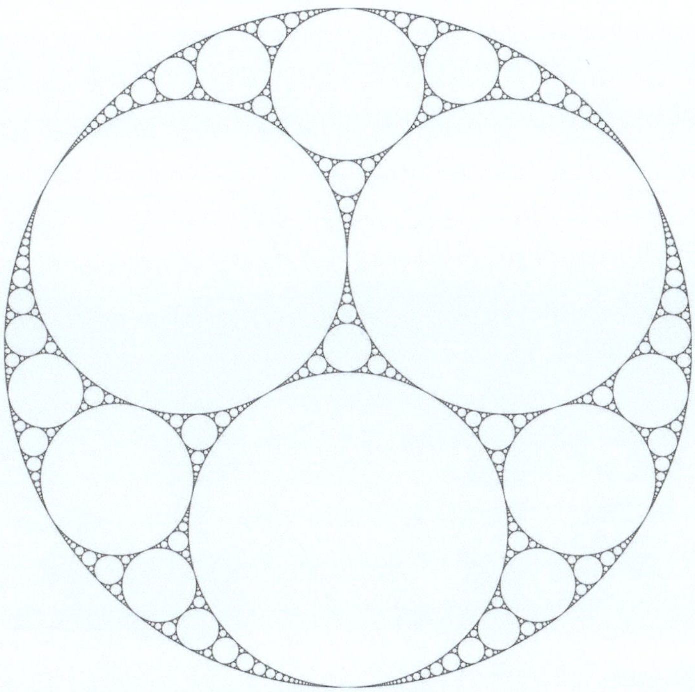

Fig. 5.21 Apollonian Gasket, $D_{(calc)} = 1.3057$

Fig. 5.22 Minkowski Sausage, $D_{(calc)} = 1.5000$

Fig. 5.23 Sierpinski Triangle, $D_{(calc)} = 1.5849$

Fig. 5.24 Sierpinski Hexagon, $D_{(\text{calc})} = 1.630$

Fig. 5.25 Fibonacci Word, $D_{(\text{calc})} = 1.6379$

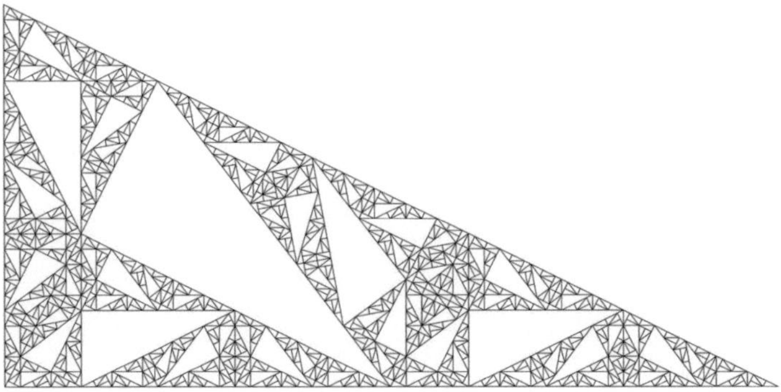

Fig. 5.26 Pinwheel Fractal, $D_{(calc)} = 1.7227$

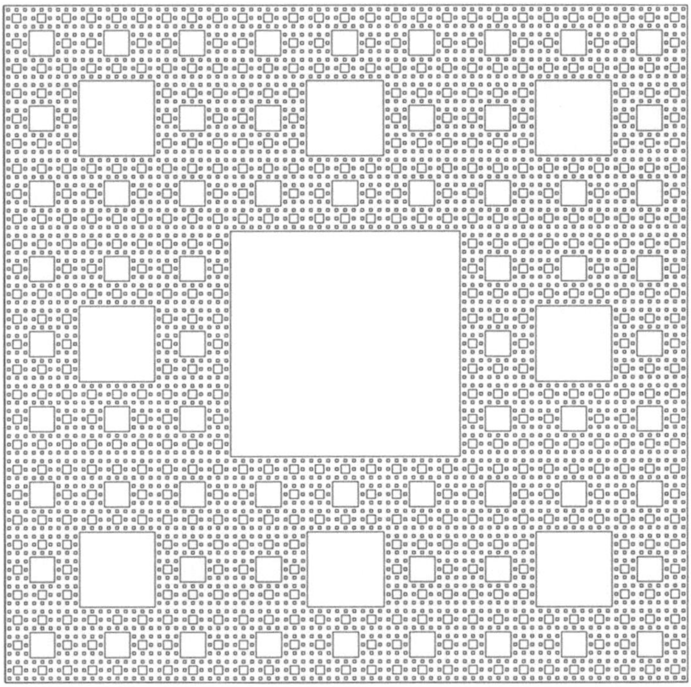

Fig. 5.27 Sierpinski Carpet, $D_{(calc)} = 1.8928$

i. All of the images are placed on a similarly sized field with the base of each image conforming to a similar range of widths (1700–1900 pixels) to accommodate their different shapes.
ii. All images are positioned in the centre of the field, with 40 % white space surrounding them and with an image resolution is 175 dpi.

Table 5.8 The number of grids for comparison generated by each scaling coefficient (*SC*) tested

SC	2	1.8	1.65	1.4142	1.3	1.2	1.1
# Grids	7	9	10	14	19	27	51

iii. Each image is pre-processed using Sobel edge detection (50 % black/white p-gradient—100 % contrast) and all lines are reduced to a width of one pixel.
iv. The starting grid configuration in the y-axis is four cells high.
v. Statistical divergence and error characterisation strategies and settings are not used because deterministic fractal geometric sets all have, by definition, data that conform to the average.

Because all of the starting field and image settings are identical, the number of grids generated for comparison is consistent, ranging from a low of 7 (for the 2:1 ratio) to a high of 51 (for the 1.1: 1 ratio) (Table 5.8). The purpose of standardising the test images in this way is to ensure that the only factors shaping the results are the combination of grid disposition and scaling coefficient.

Each of the nine images is analysed using seven commonly-used scaling coefficients that are common in the field and the two standard variations of the grid disposition parameter. The scaling coefficient variants are 2:1, 1.8:1, 1.65:1, 1.4142:1 (approximately $\sqrt{2}$:1), 1.3:1, 1.2:1 and 1.1:1. The two grid disposition variants tested are 'edge-growth' or 'centre-growth'. Thus, in total, 126 D_{Est} results are generated and then compared with the 9 target D_{Calc} figures.

5.3.4 Data Analysis Method

The first set of results (Table 5.9) is for edge-growth, using the seven scaling coefficient settings. The second set (Table 5.10) is for the same range of scaling coefficients but for centre-growth. In order to determine which of the combinations of these variations is optimal, two processes, each with two variations, are followed.

For the first process, a comparison is constructed between the D_{Calc} and D_{Est} results, using r values. Because the calibration process effectively compares two measures of the same data, high r values are anticipated. This process is undertaken for two variations of the data, first the complete set (*All*) of nine determinable fractal images and second a more limited set (*Arch.*), which is the range of D wherein most architectural results have previously been recorded (that is, where $1.3 < D < 1.8$). This latter set is more significant for architectural and urban analysis than for several other fields, like astronomy or geology, which have sought to calibrate the method across different portions of the range. The highest result for the correlation coefficient in the Tables 5.9 and 5.10 is then charted for comparison producing an r^2 value with an indicator of the gradient of the linear trend-line generated by this data set and a location where the trend crosses the y-axis (Fig. 5.28).

Table 5.9 D_{Est} results for nine fractal test sets, using edge-growth grid disposition and seven standard *SC* settings

SC	Fractal set									*r*	
	Koch snowflake	Terdragon curve	Apollonian gasket	Minkowski sausage	Sierpinski triangle	Sierpinski hexagon	Fibonacci word	Pinwheel fractal	Sierpinski carpet	*All*	*Arch.*
2	1.271	1.291	1.378	1.466	1.515	1.637	1.572	1.717	1.734	0.970	0.937
1.8	1.296	1.318	1.361	1.47	1.503	1.623	1.585	1.706	1.766	0.983	0.955
1.65	1.282	1.30	1.357	1.475	1.514	1.624	1.576	1.708	1.773	0.986	0.961
1.4142	1.274	1.309	1.353	1.486	1.519	1.628	1.595	1.730	1.813	0.988	0.965
1.3	1.279	1.310	1.351	1.478	1.507	1.614	1.58	1.714	1.768	0.985	0.962
1.2	1.278	1.321	1.355	1.485	1.487	1.614	1.594	1.725	1.758	0.975	0.944
1.1	1.283	1.324	1.374	1.479	1.494	1.612	1.585	1.726	1.762	0.975	0.934
D_{Calc} =	1.2619	1.2619	1.3057	1.500	1.5849	1.630	1.6379	1.7227	1.8928	–	–

Table 5.10 D_{Est} results for nine fractal test sets, using centre-growth grid disposition and seven standard *SC* settings

SC	Fractal set									*r*	
	Koch snowflake	Terdragon curve	Apollonian gasket	Minkowski sausage	Sierpinski triangle	Sierpinski hexagon	Fibonacci word	Pinwheel fractal	Sierpinski carpet	*All*	*Arch.*
2	1.263	1.314	1.377	1.466	1.515	1.611	1.572	1.716	1.765	0.979	0.945
1.8	1.266	1.309	1.365	1.466	1.506	1.614	1.562	1.677	1.742	0.983	0.958
1.65	1.272	1.307	1.363	1.460	1.508	1.617	1.583	1.718	1.764	0.981	0.951
1.4142	1.278	1.305	1.357	1.474	1.512	1.622	1.591	1.721	1.802	0.987	0.961
1.3	1.279	1.315	1.355	1.470	1.507	1.602	1.576	1.705	1.752	0.984	0.963
1.2	1.281	1.323	1.353	1.483	1.493	1.609	1.586	1.728	1.759	0.977	0.947
1.1	1.283	1.326	1.347	1.479	1.494	1.612	1.585	1.726	1.762	0.978	0.952
D_{Calc} =	1.2619	1.2619	1.3057	1.500	1.5849	1.630	1.6379	1.7227	1.8928	–	–

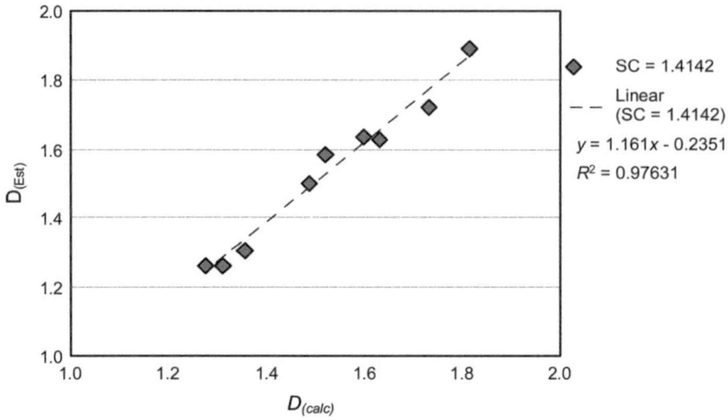

Fig. 5.28 The optimal result set; edge-growth grid disposition and scaling coefficient of 1.4142

The second approach to identifying the optimal set of results is to construct a table of percentage variations between the D_{Calc} and D_{Est} results for each of the grid disposition variables and the seven standard scaling settings (Table 5.11 and 5.12). Thereafter, an average result for the percentage difference for each scaling coefficient, and for both the complete set (*All*) and the common range for architecture (*Arch.*) is calculated.

While the raw data used for both comparisons (by *r* and by percentage difference) is the same, the approaches emphasise different things: the former is a reflection of the most statistically valid results, and the latter is the best average results. Furthermore, both of these data sets are assessed against the complete set of results (*All*) to gain an overall appreciation of the results, but also against the range wherein architecture is most likely to occur (*Arch.*).

5.3.5 *Results and Discussion*

Across the four tables of results, the two testing processes (*r* and % differences) for two sets of data (*All* and *Arch.*) produced eight indicators of the optimal combination of variables. Six of these eight indicators identified the scaling coefficient of 1.4142:1 ($\sqrt{2}$:1) as the best setting for that variable, and all of the indicators confirm that the edge-growth setting is superior.

How then, do these results compare with the trends suggested in past research from other disciplines? It has been argued that, while not a direct linear relationship, the higher the D_{Calc} value, the less accurate the D_{Est} result is likely to be (Asvestas et al. 2000). Similarly it has been concluded that the box-counting method underestimates the fractal dimension for higher levels of image complexity (Li et al. 2009; Górski et al. 2012). The present results confirm that the box-counting method marginally overestimates the very lowest order of results (where $D_{\text{Calc}} < 1.3$) by

Table 5.11 Comparison between D_{Calc} and D_{Est} results by percentage difference (% diff.) for edge-growth disposition and seven standard *SC* settings

SC	Fractal set									Mean % diff.	
	Koch snowflake	Terdragon curve	Apollonian gasket	Minkowski sausage	Sierpinski triangle	Sierpinski hexagon	Fibonacci word	Pinwheel fractal	Sierpinski carpet	*All*	*Arch.*
2	0.91	2.91	7.23	−3.40	−6.99	0.70	−6.59	−0.57	−15.88	−2.41	−1.60
1.8	3.41	5.61	5.53	−3.00	−8.19	−0.70	−5.29	−1.67	−12.68	−1.89	−2.22
1.65	2.01	3.81	5.13	−2.50	−7.09	−0.60	−6.19	−1.47	−11.98	−2.10	−2.12
1.4142	1.21	4.71	4.73	−1.40	−6.59	−0.20	−4.29	0.73	−7.98	−1.01	−1.17
1.3	1.71	4.81	4.53	−2.20	−7.79	−1.60	−5.79	−0.87	−12.48	−2.19	−2.29
1.2	1.61	5.91	4.93	−1.50	−9.79	−1.60	−4.39	0.23	−13.48	−2.01	−2.02
1.1	2.11	6.21	6.83	−2.10	−9.09	−1.80	−5.29	0.33	−13.08	−1.76	−1.85
Average % diff.	1.85	4.85	5.56	−2.30	−7.93	−0.83	−5.40	−0.47	−12.51	–	–

Table 5.12 Comparison between D_{Calc} and D_{Est} results by percentage difference (% diff.) for centre-growth disposition and seven standard *SC* settings

SC	Fractal Set									Mean % diff.	
	Koch snowflake	Terdragon curve	Apollonian gasket	Minkowski sausage	Sierpinski triangle	Sierpinski hexagon	Fibonacci word	Pinwheel fractal	Sierpinski carpet	*All*	*Arch.*
2	0.11	5.21	7.13	−3.40	−6.99	−1.90	−6.59	−0.67	−12.78	−2.21	−2.07
1.8	0.41	4.71	5.93	−3.40	−7.89	−1.60	−7.59	−4.57	−15.08	−3.23	−3.19
1.65	1.01	4.51	5.73	−4.00	−7.69	−1.30	−5.49	−0.47	−12.88	−2.29	−2.20
1.4142	1.61	4.31	5.13	−2.60	−7.29	−0.80	−4.69	−0.17	−9.08	−1.51	−1.74
1.3	1.71	5.31	4.93	−3.00	−7.79	−2.80	−6.19	−1.77	−14.08	−2.63	−2.77
1.2	1.91	6.11	4.73	−1.70	−9.19	−2.10	0.10	0.53	−13.38	−1.44	−1.27
1.1	2.11	6.41	4.13	−2.10	−9.09	−1.80	−5.29	0.33	−13.08	−2.04	−2.30
Average % diff.	1.27	5.22	5.39	−2.89	−7.99	−1.76	−5.11	−0.97	−12.91	–	–

between 2 and 5 % while, at the other end of the scale (where $D_{\mathrm{Calc}} > 1.7$) it begins to underestimate by up to 12.91 % for the most complex images. Overall, the average difference for the complete set of nine images and for all scaling coefficients is between 1.91 and 2.19 % less than the correct values. All of this confirms the findings of past attempts to calibrate the method for different ideal ranges.

A second observation is that the difference between edge-growth and centre-growth variables is minimal (<1 %) and that the $\sqrt{2}$:1 scaling coefficient is only marginally better than some of the others, including the 1.3:1 variation (<2 %). This might imply that the optimal settings identified are simply a result of 'noise' in a test that relies on a small starting sample. However, the calibration process is, by definition, concerned with fine gradations and past examples of fractal analysis have demonstrated that seemingly small differences in the starting data can be very significant. While a much larger set of images would produce more statistically reliable results, there is a sufficiently high correlation coefficient for the optimal combination of variables to have confidence in the present outcome.

A bigger question is, why doesn't the method, even using deterministic fractals as subjects, provide better estimates? One of the reasons is that deterministic fractals are infinitely deep or scalable in a simulated environment (the computer), whereas the box-counting method works on *images* of deterministic fractals. Thus, because the image is always less detailed than the fractal it represents, the estimated results tend to be less than the calculated ones.

Finally, some readers may find it worrisome that the best settings still cannot deliver results which are closer to the mathematical ideal. Actually, what is more important is that the method delivers consistent results, provided all factors have been controlled. This quality of consistency is more significant than the capacity to replicate absolute, but abstract, values, because D results are most useful for comparative purposes.

5.4 Revisiting the *Robie House* and the *Villa Savoye*

The *Robie House* and the *Villa Savoye* are two canonical works of twentieth-century architecture, the former representing the transition of the Arts and Crafts movement into early Modernism and the latter being regarded as the epitome of the Modern Movement (they are analysed in depth later in Chaps. 7 and 8). Bovill (1996) chose an elevation from each of these iconic buildings to demonstrate how the box-counting approach could be used to quantify the difference between the visual properties of each. Sala (2000), Lorenz (2003), Ostwald et al. (2008a), and Ostwald (2013) have each repeated variations of this analysis, the latter two in order to improve the reliability of the method and the former focussed entirely on the *Robie House*.

Starting with the *Robie House* (see Fig. 5.8), Bovill (1996) undertook three comparative calculations (over four scales with $SC = 2.0$) for the south elevation, recording D results of 1.645, 1.485 and 1.441; average $D = 1.520$. Lorenz (2003)

Table 5.13 Comparing the D_{Est} results for elevations of the *Robie House* and the *Villa Savoye*

Elevation		Calculated result								
		Bovill (1996)		Lorenz (2003)		Ostwald et al. (2008)		Ostwald (2013)		New
		D_{Est}	% diff.	D_{Est}	% diff.	D_{Est}	% diff.	D_{Est}	% diff.	D_{Est}
Robie House		1.520	−5.08	1.570	−0.8	1.62	4.92	1.572	0.12	1.5708
Villa Savoye		1.3775	−9.57	1.306	−16.72	1.53	5.68	1.492	1.88	1.4732
% diff. (Robie to Savoye)		14.25%	–	26.4%	–	9.0%	–	8.0%	–	9.76%
Settings	*SC*	2:1	–	2:1	–	1.41:1	–	1.4142:1	–	1.4142:1
	G#	3	–	4	–	11	–	15	–	13
	SD	–	–	–	–	–	–	3%	–	2%

used the same image over the same range of scales and the same *SC* value but produced an overall calculation of $D = 1.57$. Ostwald et al. (2008a) undertook eleven comparative calculations over twelve scales of grid with $SC = 1.41$ and using image pre-processing to convert all lines to a width of one pixel; the result was $D = 1.62$. Ostwald's (2013) result of $D = 1.572$ was produced using an *SC* of 1.4142:1, and the same elevation image, with a base width of 2900 pixels wide, centred in a 3000×1500 pixel field, with edge-growth grid disposition and a 50 % Sobel-gradient. *SD* was also used with a 3 % threshold. When we use all of the optimal settings identified in this chapter for both data processing and pre-processing, a new result is produced, $D = 1.5708$.

For the *Villa Savoye* elevation (see Fig. 5.7), Bovill (1996) undertook two comparative calculations (over three scales of grid with $SC = 2.0$) for the west elevation leading to the results $D = 1.42$ and $D = 1.33$; average $D = 1.3775$. Lorenz undertook three comparative calculations (over four scales with $SC = 2.0$) replicating Bovill's first two results and adding one additional result $D = 1.17$. Ostwald et al. (2008a) undertook eleven comparative calculations (over twelve scales with $SC = 1.41$) to achieve a result of $D = 1.53$ and a revised version (Ostwald 2013) calculated the result for the west elevation of *Villa Savoye* as $D = 1.492$, using $SC = 1.4142{:}1$, with edge-growth grid disposition, a 50 % Sobel-gradient and $SD = 3$ %. For the present book, using the new optimal image pre-processing factor settings, a revised calculation was produced, $D = 1.4732$.

The set of results for these two elevations is compared, for each case, by calculating the percentage difference between the past result and the most recent result (Table 5.13). In addition to this, the percentage difference determined for each set of past results between the *Robie House* and *Villa Savoye* elevations is also recorded. In effect, this process could be viewed as developing a record of the growing accuracy and accountability of the method, with the results slowly improving from a maximum 16.72 % difference to around 0.12 %. While there is no perfect result in this process, the estimations can continue to be improved in this way.

5.5 Conclusion

Previous attempts to refine the box-counting method have noted that it is both more subtle and more complex than most users realise. However, by using the optimal settings identified in this chapter, for both the pre-processing and processing stages, it is possible to produce a set of results that are both consistent and accountable. When combined with a detailed understanding of the potential error rates associated with various facets of the method, and with a clearly stated and consistent set of image standards (governing which lines in an elevation or plan are chosen for analysis and why), the box-counting method will provide highly credible results.

Through the results of the two tests presented in this chapter, and the discussion of the complete set of variables at play in the method, it is possible to propose a

Table 5.14 Optimal variables and settings (assuming an image size of between 2 MB and 3 MB at 175 dpi)

Category	Variable	Optimal setting	Notes
Pre-processing	White space	40–50 % increase	40–50 % white space around a starting image produces the most consistent result although the potential for errors is also reduced across the 30–60 % white space spectrum
	Image position	Centre-Centre	The more centred the image, the more consistent the set of results. Both Centre-centre and Centre-base are appropriate positions
	Line weight	1 pt	The thinner the line, the better the result. In practice, all images should be converted into lines of 1 pixel width using either Sobel or Prewitt edge detection algorithms (with a 50/50 black/white threshold, leading to 100 % contrast) as a precursor to analysis
	Image resolution	175 dpi	In principle, the higher the resolution and the larger the field, the better the result. However, within the limits of current computing power, 175 dpi consistently produced high quality results, while lower resolutions gradually lost accuracy
Processing	Scaling coefficient (*SC*)	1.4142:1	$\sqrt{2}$:1 (or 1.4142:1) produced the best balance between varying levels of white space being included in the calculations while generating enough grids for comparison to achieve a statistically viable data set
	Grid disposition	Top left	Edge-growth (top left-hand corner as point of origin) is the optimal setting although the centre-growth variable generates results with a similar level of accuracy
	Starting grid size	0.25 l	The short side (l) of the field should be divisible by four (0.25 l) to generate the starting grid proportion and cell size. If using statistical divergence correction techniques this setting may be less useful because the algorithm will determine the usefulness of starting and closing grid permutations

(continued)

Table 5.14 (continued)

Category	Variable	Optimal setting	Notes
	Starting grid proportion	$4 \times X$	The shortest side of the field should be divisible by 4 for the starting grid. It is also suggested that the image, on the field, should be sized in such a way that it has detail in more than half but ideally not all of the cells in the starting grid
	Closing grid size	0.03 l	If using error characterisation reporting, then the lowest grid cell size should be 0.03 l where l is the length of the shortest side of the field. If using statistical divergence correction techniques the closing grid size is determined by the spread of results, not an artificial limit
Post-processing	Statistical Divergence (*SD*)	1 %	Generally use *either* Statistical Divergence or Error Characterisation, not both For architectural analysis, a 1 % *SD* value is potentially sufficient for most analyses, although for some specific purposes higher or lower values might be more appropriate
	Error characterisation	r^2	Generally use *either* Statistical Divergence or Error Characterisation, not both Reporting the coefficient of determination can be useful for understanding and communicating the limits of the data

'best practice' model or 'standard method' for measuring the fractal dimensions of architectural and urban forms (Table 5.14). While developments will continue to refine these standards, at the very least researchers using this approach should begin to methodically record the variables they have used to calculate their results.

Part II
Analysing Architecture

Chapter 6
Analysing the Twentieth-Century House

Whereas Part I of this book provides a detailed review of the intricacies of the box-counting method, Part II presents an application of this method to a specific project: a review of the formal complexity of eighty-five canonical designs. This chapter describes the method used for this larger research project, its rationale, limitations and scope. It includes a detailed discussion of the stages in the research process and the use of calculated and derived measures to characterise and compare the properties of individual designs and sets of designs. It concludes with a short, hypothetical example of the way the results of this research will be visualised and interpreted throughout the later chapters.

6.1 Introduction

The two-dimensional variation of the box-counting method can be used to measure the fractal dimension of a wide range of objects, as represented in images, including such diverse forms as arterial networks, forest perimeters and star clusters in galaxies. For every application of the method there are two different aspects of the experimental design that must be considered and described. The first, the optimisation of factors which are innate to the mathematical basis of the approach, is contained in Part I of the present book, whereas the second, a considered approach to using that method to interrogate, interpret and understand the object of study, is contained in this chapter.

We commence by describing the methodological approach we have employed for the primary research project that is the subject of the remainder of this book. The first step in this process involves delineating the rationale for the selection of the data source and type and a justification of the interpretation of the source material. A comprehensive list of all of the designs that are analysed in Part II is then

M.J. Ostwald and J. Vaughan, *The Fractal Dimension of Architecture*,
Mathematics and the Built Environment 1, DOI 10.1007/978-3-319-32426-5_6

provided. Thereafter, the representational standards employed to prepare the data for analysis are recorded (in accordance with the framework in Chap. 4) and the software and the data processing variables we used are also reported, followed by a summary table of these factors.

In the second part of this chapter, the stages of the research method are described. These involve identifying and coding the plans and elevations, followed by the analysis process for each individual building, then for the set of five buildings and an optimal sub-set of buildings. The last stages include deriving comparative indicators of the consistency of the data (or lack thereof). To provide the reader with an understanding of how the results appear in later chapters, a hypothetical example set of data is tabulated, charted and discussed. Thereafter, other potential applications of the method (which are explored later in the book) are outlined prior to the chapter's conclusion.

6.2 Research Description

6.2.1 Data Selection

Canonical houses of the twentieth-century are the focus of the research undertaken in Part II of this book. Houses are an ideal subject for an application of fractal analysis because, as a type, they possess similar scale, program and materiality. While a study of commercial, urban or religious designs is also possible using this method, none of these building types have the same potential for producing consistent and, within reason, statistically valid results for comparisons with other designs. Famous houses, rather than traditional or project homes, have been chosen because, by definition, they have been extensively researched in the past and thereby offer an opportunity for comparing the quantitative results derived from the present study with past qualitative interpretations. Indeed, in the later chapters in this book we use our measured results to test specific arguments about these houses that have been suggested by historians, theorists and critics.

The difficulty with studying well-known works in isolation is that even with the benefit of the methodological refinements described in the previous chapters, fractal analysis can be sensitive to quite minor variations in the forms being measured. Thus, in Chap. 5 when we analysed elevations using different settings for line weight and white space, there were isolated outliers in the results that were difficult to explain. Nevertheless, across the complete set of data there was a very high degree of consistency. To overcome the problem of the outlier, sets of five houses by the same architect or practice, and including an acknowledged masterwork of design, were selected for analysis throughout the remainder of this book. For example, the set of five Prairie Style designs by Frank Lloyd Wright includes the *Henderson House*, *Tomek House*, *Evans House* and *Zeigler House* as well as

Wright's canonical *Robie House*. In many cases the masterwork is the last house in the sequence: the set of five of Le Corbusier's Modernist designs concludes with the *Villa Savoye,* and the Mies van der Rohe and Richard Meier sets culminate, respectively, with the *Edith Farnsworth House* and the *Douglas House*. In a few cases the masterwork was completed earlier in a sequence of works. Examples of this include Robert Venturi's *Vanna Venturi House* and Eileen Gray's *E.1027*, both of which are the second houses in their sets.

To maintain a level of consistency across each set of houses several guiding parameters were chosen. The first of these was the general goal that no more than ten years should separate the earliest design in a set from the last. Of the seventeen sets chosen, there are only three exceptions to this rule: Mies van der Rohe, Venturi and Scott Brown and the 'late' Murcutt set. The second parameter was that preference was given for single houses (rather than pavilions or estates) and in all cases this was met. The third was that completed works rather than unbuilt projects were chosen, in an attempt to ensure a similar level of design development. Three of the seventeen sets include at least one unbuilt project although in most cases the unbuilt projects were developed and documented in such detail that they are, in a mathematical sense, indistinguishable from the data derived from constructed designs. The fourth parameter was that houses with a relatively tight geographic distribution were preferred within their sets, to limit the impact of climate on the form of the house. Only the last house in the Mies van der Rohe set (*Edith Farnsworth*) breaks this pattern, being in North America whereas the rest are in Europe. Where it was not possible to identify an ideal set of houses that conformed to all of these parameters, compromises were made, which explains the exceptions listed above. Where there was a need to vary the broad parameters of the research, the decisions are discussed in each of the chapters.

The number five, as the standard size of a set of designs for this research, was determined by considering the minimum data pool to allow for the use of simple statistical methods. If each house has four elevations and at least one floor plan and a roof plan, then there are at least thirty separate results in a set of five houses from which various simple indicators (average, median, and standard deviation) of patterns in the data can be derived. Certainly, this sample size is too small to extrapolate the results to say anything globally meaningful about housing, but it is sufficient to offer observations about movements in twentieth-century design, or specific interpretations of famous buildings or an individual architect's works. This is especially the case when the results of each set are combined together to consider larger factors. Thus, across the complete group of houses examined in this book, we chose the sets in such a way that they could be combined into stylistic groupings which could then be compared. For example, Chap. 7 analyses three sets of five Modernist houses by Le Corbusier, Eileen Gray and Mies van der Rohe. Chap. 10 considers two sets of five Post-Modern houses, the first by Robert Venturi and Denise Scott Brown and the second by Frank Gehry. Another consideration in constructing the larger sets of works was to compare different stages in the same architect's career. For example, Chap. 7 compares a set of five of Le Corbusier's Arts and Crafts style houses with five of his Modernist works. In a similar way, Chap. 8 examines three stylistic shifts in the career of Frank Lloyd Wright: the Prairie Style houses, Textile-block houses and Usonian houses.

Some final considerations regarding the complete set of works being analysed in this book relate to their geographic and historic distribution and to the question of gender. If we had simply selected our great domestic designs from those identified in architectural histories of the twentieth-century, we would have found ourselves with works produced by male, European and North American architects, working predominately prior to the 1960s. Instead, we sought to provide a wider geographic distribution of works along with more recent designs and to expand the set to include, where the parameters allowed, female architects. These decisions shaped several of the architects and practices chosen, including, in Chap. 11, Japanese, late twentieth-century designers, Kazuyo Sejima, Ryue Nishizawa, Yoshiharu Tsukamoto and Momoyo Kajima and Australian Regionalist architects, Glenn Murcutt and Peter Stutchbury. In this way the research makes at least some concessions to geographic, cultural and gender diversity that are not always apparent in other selections of canonical works.

6.2.2 Data Scope

For the analysis, eighty-five houses were identified and divided into seventeen sets, with five designs in each. The years in which these houses were constructed or designed (if unbuilt), their locations (if they were constructed) and the number of plans or elevations produced for analysis in the present book are recorded in Tables 6.1, 6.2, 6.3, 6.4 and 6.5. Nine of the designs are unbuilt projects, sixteen were constructed in Europe, thirty-five in the United States of America, ten in Japan and fifteen in Australia. Four of the designers are female, and twelve male.

6.2.3 Data Source and Type

The most productive images to use for the fractal analysis of an architectural design are plans and elevations. While perspective images of buildings might seem to more closely replicate the experience of a person viewing a building, no two people experience precisely the same perspective. The height of the viewer, the geometry of their optic system (the distance between their eyes) and its acuity (how clear, deep or precise their vision is) all change the volume and character of geometric information experienced by a person. In contrast, although elevations and plans are less realistic, they offer a universal system of representation that can be

Table 6.1 Pre-modern and modern sets

Architect	Set	House	Year	Location	Elevations	Plans
Le Corbusier (Charles-Édouard Jeanneret-Gris) (1887–1965)	Pre-modern	*Fallet*	1905	Switzerland	4	4
		Jaquemet	1907	Switzerland	4	5
		Stotzer	1907	Switzerland	4	5
		Jeanneret-Perret	1912	Switzerland	4	4
		Favre-Jacot	1912	Switzerland	4	4
	Modern	*Ozenfant*	1922	France	3	5
		Cook	1926	France	2	4
		Weissenhof-Siedlung	1927	Germany	4	5
		Stein-de Monzie	1928	France	4	5
		Savoye	1931	France	4	3
Eileen Gray (1879–1976)	Modern	*Small house for an engineer*	1926	Unbuilt Project	4	3
		E.1027	1929	France	4	3
		Four storey Villa	1930	Unbuilt Project	4	5
		Tempe a Pailla	1934	France	4	3
		House for two sculptors	1934	Unbuilt Project	4	3
Mies van der Rohe (1886–1969)	Modern	*Wolf*	1927	Poland	4	4
		Lange	1930	Germany	4	4
		Esters	1930	Germany	4	4
		Lemke	1933	Germany	4	2
		Edith Farnsworth	1951	USA	4	2
Total		20			77	77

Table 6.2 Frank Lloyd Wright sets

Architect	Set	House	Year	Location	Elevations	Plans
Frank Lloyd Wright (1867–1959)	Prairie style	*Henderson*	1901	USA	4	4
		Tomek	1907	USA	4	4
		Evans	1908	USA	4	3
		Zeigler	1910	USA	4	3
		Robie	1910	USA	4	4
	Textile-block	*La Miniatura*	1923	USA	4	4
		Storer	1923	USA	4	4
		Freeman	1923	USA	4	3
		Ennis	1923	USA	4	2
		Lloyd-Jones	1929	USA	4	4
	Usonian	*William Palmer*	1950	USA	4	2
		Reisley	1951	USA	4	3
		Chahroudi	1951	USA	3	2
		Dobkins	1953	USA	4	2
		Fawcett	1955	USA	3	2
Total		15			58	46

Table 6.3 Late modernist, Avant-Garde and abstraction sets

Architect	Set	House	Year	Location	Elevations	Plans
John Hejduk (1929–2000)	Avant-Garde and Abstraction	*House 1*	1954	Unbuilt project	4	2
		House 4	1954	Unbuilt project	4	3
		House 5	1962	Unbuilt project	4	2
		House 6	1962	Unbuilt project	4	4
		House 7	1963	Unbuilt project	4	8
Peter Eisenman (1932–)	Avant-Garde and abstraction	*House I*	1968	USA	4	3
		House II	1970	USA	4	3
		House III	1971	USA	4	3
		House IV	1971	USA	4	3
		House VI	1976	USA	4	3
Richard Meier (1934–)	Avant-Garde and abstraction	*Smith*	1967	USA	4	4
		Hoffman	1967	USA	4	3
		Saltzman	1969	USA	4	4
		Douglas	1973	USA	4	5
		Shamberg	1974	USA	4	3
Total		15			60	53

Table 6.4 Post-modern sets

Architect	Set	House	Year	Location	Elevations	Plans
Robert Venturi (1925–) and Denise Scott Brown (1931–)	Post-modernism	*Beach*	1959	Unbuilt project	4	2
		Vanna Venturi	1964	USA	4	3
		Vail	1975	USA	4	5
		Delaware	1983	USA	4	3
		Long Island	1990	USA	4	4
Frank Gehry (1929–)	Post-modernism	*Familian*	1978	USA	4	3
		Gunther	1978	USA	4	4
		Spiller	1980	USA	4	5
		Wagner	1984	USA	4	3
		Norton	1984	USA	4	4
Total		10			40	36

independently validated and used to construct reasonable comparisons between buildings. The vast majority of all computational techniques for investigating architecture rely on plans and elevations, and this includes almost all past applications of fractal analysis listed in Chap. 3. Therefore, for the purposes of this research, the basic formal properties of these buildings are assumed to be sufficiently encapsulated in their 335 elevations and 290 plans.

Table 6.5 Japanese minimalist and Australian Regionalist sets

Architect	Set	House	Year	Location	Elevations	Plans
Kazuyo Sejima (1956–) Ryue Nishizawa (1966–)	Minimalism	*Y*	1994	Japan	4	4
		M	1994	Japan	4	3
		S	1996	Japan	4	3
		Small	2000	Japan	4	5
		Plum Grove	2003	Japan	4	4
Atelier Bow-Wow (est. 1992), Yoshiharu Tsukamoto (1965–) and Momoyo Kajima (1969–)	Minimalism	*Ani*	1997	Japan	4	4
		Mini	1998	Japan	4	4
		Shallow	2002	Japan	4	5
		Gae	2002	Japan	4	4
		Juicy	2005	Japan	4	5
Glenn Murcutt (1936–)	'Early' career Regionalism	*Marie Short*	1975	Australia	2	
		Nicholas	1980	Australia	4	3
		Carruthers	1980	Australia	4	3
		Fredericks	1982	Australia	4	3
		Ball-Eastaway	1983	Australia	4	2
	'Late' career Regionalism	*Magney*	1984	Australia	2	
		Simpson-Lee	1994	Australia	4	2
		Fletcher-Page	1998	Australia	4	2
		Southern Highland	2001	Australia	4	2
		Walsh	2005	Australia	4	2
Peter Stutchbury (1954–)	Regionalism	*Verandah*	2004	Australia	4	4
		Billabong	2005	Australia	4	2
		Beach	2006	Australia	4	3
		Paddock	2007	Australia	4	2
		Invisible	2007	Australia	4	3
Total		25			100	78

For each building being analysed we commenced by sourcing an authoritative set of working drawings or measured drawings. All of the elevations and plans were scanned and then traced using *ArchiCAD* (Graphisoft) software. Provided the plans and elevations matched, and overall dimensions were consistent, these documents were accepted as the primary data. If there were inconsistencies, the dimensions on the drawings or surveys were used to reconcile the views and then design drawings and photographs were employed as a final means of interpreting and correcting any discrepancies. For unbuilt works, final published or archived design drawings were accepted as the primary source of information. If only incomplete sets of drawings were available, we reconstructed missing views using information from other drawings and views (including sections or axonometrics) along with any photographs or models of the houses. In some situations where the design has since been altered—as is the case with Le Corbusier's *Maison-Atelier Ozenfant*, where the original sawtooth roof has been removed and replaced with a flat roof—we analysed the original design, not any later alterations.

For most designs we were able to produce four elevations for analysis; however, in the case of the small number of designs with triangular plans, it was more practical to develop three elevations. Furthermore, one design (*Villa Cook*) has shared 'party-walls' with adjacent buildings, meaning that only two elevations are available for analysis. Whether there are four, three or two elevations, they are regarded as collectively constituting what might be called the 'perimeter' or 'boundary' elevations of the building, that is, the formal expression of the building which is visible from points which are external to the structure and are largely perpendicular to the dominant geometry of the plan. Internal courtyard elevations that are completely surrounded by built form are not part of this analysis. In addition, some building plans are sufficiently complex that small surfaces may not be visible on any elevation or may be distorted. Such hidden elevations are also excluded from the analysis.

For the house plans, we traced and analysed the ground floor and any upper, habitable levels (which may include functional spaces in an attic level) and the roof plan. Basements were only drawn and analysed if they were clearly designed to contain habitable floor space. This means that plant or boiler rooms and wood storage spaces, which were on a separate floor plan and served no other habitable function, were excluded from the analysis.

6.2.4 Data Interpretation

The fractal analysis of the elevation of a building measures its characteristic visual complexity, that is, the level of detail or formal information that is typically visible across all scales of observation of the façade. This measure could also be considered a reflection of the functional or habitable qualities of its interior because the location of windows and doors, along with the modulation of walls, roofs and balconies, are all potentially expressions of function.

Plan analysis requires a different interpretation of the meaning of the fractal dimension of architecture. The fractal analysis of a building plan measures the formal and spatial complexity of a design, not as it can be seen in its totality, but as it can be experienced through movement or inhabitation (Ostwald 2011a). While an elevation is potentially close to the experience of viewing a façade (albeit through a telephoto lens and using perspective correction), the plan view assumes that part of the building has been completely removed to reveal a more abstract spatial relationship within, one which is never really seen in this way but is experienced. Past research has demonstrated that this experience of space and form, as a reflection of the social structure implicit in a building, is a significant property of a building (Hillier and Hanson 1984; Hillier 1996).

The roof plan poses a different dilemma for interpretation, as it is shaped by both the façade expression of a building and its interior planning. In theory, it is neither clearly separate from the set of elevations and plans nor does it fit with either set perfectly. Nevertheless, it is notable that with a few exceptions, relatively few roof plans resemble their elevations so much as they resemble their internal plans. This is because roof plans are typically either a product of expediency (weather-proofing the form of the plan) or a by-product of other decisions about massing and expression. Therefore, despite the roofscape being described in various architectural primers as the 'fifth façade' of a building, in this book we treat it as a special type of plan.

6.2.5 Data Representation

For all of the plans and elevations, for every building analysed, we digitally traced the building outline and primary, secondary and tertiary forms, but excluded any material representation. This is in accordance with the level 4 framework outlined previously in Chap. 4. In practice this meant that all tracing was undertaken using single lines, with no textures or infills. All changes in form and between materials were depicted using a single line separating the surfaces. As the delineated images represent a 'real' view of the building, dotted lines indicating hidden surfaces and forms were not shown. We typically included in the traced representation any building elements that would produce a change in surface level of more than 1 cm. Thus, we would draw a gutter and a fascia (if shown), but not the top lip of the gutter. In plans, we traced any change in floor material with a single line to divide them, but, as in the elevations, did not indicate the material. Any built-in furniture, such as bathroom items and built-in benches were delineated with a simple outline of the furniture item. For the representation of doors and windows in elevations, the main frame, plus any secondary sash details or mullions were included, but not secondary leadlight or ornate moulds and joinery. Glass was depicted as opaque unless otherwise stated. In plan, doors were all drawn open at 90°, while all windows were depicted closed. No 'swings' were shown on doors or windows.

There are many architectural graphic conventions for representing voids, stair runs and roofs in plan but none of these could be used for the present analysis. To assist us to decide how to represent the form of a plan we imagined what the building would really look like if we had sliced off its upper part, just below each ceiling line, and were looking down into it. In a multi-level house this was especially significant because all stair treads up to the top of the level or floor being depicted in a plan were drawn. Again, no hidden detail lines (representing forms that are not visible) were shown. However, any details that would be seen through void spaces and on the roof plan any features (such as lower roof levels) that would be seen below the roof, were all shown. Using a similar logic, it should be obvious that we did not depict dotted 'roof-lines above' on a plan, or landscape contour lines, paths or paving. The edge of a building was the limit of the drawing, with engaged steps and balconies included, but not site works. If the building was connected to another in part or whole, we drew this part of the elevation as a blank wall. If in elevation, the garden walls were a clear extension of the building form, and were integral to the visual appearance of a house, they were retained. Any internal construction details which would not be visible when built were not depicted. No vegetation, shadows or other entourage elements or textures were included in the analysis.

While acknowledging that an elevation and a plan are both artificially abstracted views, the aim of every decision taken in the re-drawing and re-representation of the buildings being analysed was to limit the impact of artificial graphic conventions and to standardise the representation such that reasonable comparisons can be made between the buildings.

6.2.6 Data Processing

All of the results in Part II of this book were produced using software to undertake both the box-counting procedure and the calculation of fractal dimensions. In the early stages of our research we used *Benoit*, a commercially available program, to calculate fractal dimensions but soon found that it wasn't flexible enough to test all of the methodological variations that we wanted to examine as part of our goal to fully understand and then optimise the method. Thereafter, our team prepared various modules using *MATLAB* for fractal analysis before settling on the *Python* programing language to author our own software, which was specifically adapted for architectural drawings. That software, called *ArchImage*, was then tested and refined through several iterations. *ArchImage* (Version 1.16) was used for all of the calculations undertaken for this book.

6.2.7 Data Settings

The data representation, pre-processing and processing standards used are summarised in Table 6.6. The specific pre-processing standards, including image position, line weight, white space and image depth were all standardised using prior to importing the files into *ArchImage* for analysis.

6.3 Research Method

There are six stages to the standard research method we used, each of which are described in detail over the following sections. These stages are:

Table 6.6 Methodological settings used in Part II of this book (unless otherwise specified)

Category	Variable	Our setting	Notes
Representation	Graphic presentation	Level 4 of the framework	The level 4 (Outline + Primary form + Secondary form + Tertiary form) standard in the representational framework is adopted for all of the results (unless otherwise stated)
Pre-processing	White space	50 % increase	This is an optimal setting as identified previously
	Image position	Centre-centre	This is an optimal setting as identified previously
	Line weight	1 pt	This is an optimal setting as identified previously
	Image resolution	125 dpi	In principle, the higher the resolution and the larger the field, the better the result. However, within the practical limits of hardware and software, 125 dpi was the highest resolution available for all images given their starting sizes
Processing	Scaling coefficient (*SC*)	1.4142: 1	This is an optimal setting as identified previously
	Grid disposition (*GD*)	Edge-growth:	This is an optimal setting as identified previously
	Starting grid size	0.25 l	This is an optimal setting as identified previously
	Starting grid proportion	$4 \times X$	This is an optimal setting as identified previously
	Reporting accuracy	4 decimal points	Most past research reported results to 2 decimal points only, whereas our own research typically reported to 3 decimal places. However, from that past experience we have decided to report data for our findings typically to 4 decimal places
	Edge sensitivity	0.05	Associated with Sobel edge detection algorithm
Post-processing	Statistical divergence (*SD*)	0.25 % (0.03 l)	*SD* was the preferred method but for some alternative applications a closing grid dimension of 0.03 l was more appropriate.
	Error characterisation	r^2	While the importance of error characterisation is mitigated by the use of statistical divergence techniques, we report some amalgamated data using this approach

i. Identifying and coding data (plans and elevations).
ii. Analysis of each building in isolation to determine D values and derive average values for elevations and plans and a 'composite' average for the entire building.
iii. Analysis of a set of buildings to determine mean results for elevations and plans as well as an 'aggregate' result for the entire set. Mean, median and standard deviation measures are developed across the set for interpreting the data.
iv. Analysis of an optimal sub-set of buildings.
v. Comparative analysis of all results.
vi. Presentation of results.

In the context of this research, the 'set' is a group of five houses designed by the same architect (or partnership) and defined by a particular style. Within each set we mathematically partition the three most visually similar houses and define it as the 'optimal sub-set'.

6.3.1 *Identifying and Coding Data*

The naming and coding procedure for all of the raw data (elevations and plans) analysed in this book is as follows.

i. The elevations of each house are either numbered ($E_{1\text{-}4}$) or designated by orientation ($E_{N,E,S,W}$), in accordance with the conventions used in past research.
ii. The ground floor plan is numbered zero (P_0) and any floors above ground level are numbered consecutively from 1 ($P_{1,}$ $P_{2,\ldots}$). If one or more basement levels, are present in a design, they are designated with negative integers ($P_{-1,}$ $P_{-2,\ldots}$).
iii. The roof is separately labelled (P_R) although it is grouped with the plan set for determining average values.

6.3.2 *Analysis of Each Building*

Results are directly calculated for each view of a particular house, along with some derived measures from this raw data (Table 6.7). The following steps describe the way a series of fractal dimension results for elevations and plans are determined and then combined together to create a 'composite' value for a building. While the focus of the present book is on houses, these same steps are also appropriate for the analysis of any building type.

i. Each elevation is measured using *ArchImage* software to determine its fractal dimension ($D_{E1\text{-}4}$ or $D_{EN\text{-}W}$).
ii. The average D_E value for the house is determined (μ_E). This value is a measure of the typical level of visual complexity observable in the exterior of the house.

Table 6.7 Summary of definitions relating to the analysis of an individual house

Abbreviation	Meaning	Explanation
D	Fractal dimension	Fractal dimension (D) is a measure of the formal complexity of a design and the consistency with which it is distributed across all scales of a design. For an image, D is a value between 1.0 and 2.0. The fractal dimension of an architectural elevation is D_E. The fractal dimension of an architectural plan is D_P
D_E	D for a specific *elevation*	
D_P	D for a specific *plan*	
μ_E	Average D for the visible *elevations* of a building	The average of a set of values (the sum of the values divided by their number). It is expressed here as a 'population' average (μ) because the findings of this research are generally not extrapolated to comment on anything other than the actual houses being analysed The average D for all of the elevations of a single building is μ_E. This value reflects the typical level of characteristic formal complexity visible in the exterior of a design The average D for all of the plans of a single building is μ_P. This value reflects the typical level of characteristic complexity present in the spatial and formal properties of the interior and the expression of its roof The 'composite' result, μ_{E+P}, is the average for the plans and elevations of a single building
μ_P	Average D result for the habitable *plans* of a building	
μ_{E+P}	Average D result for all of the plans and elevations of a building	

Most of the buildings analysed in the present book include four elevations in the calculation of μ_E but there are exceptions which are noted in the text.

iii. Each plan is measured using *ArchImage* software to determine its fractal dimension ($D_{P\#}$ or D_{PR}).

iv. The average of the $D_{P\#}$ and D_{PR} results for the house is determined (μ_P). This value is a measure of the typical level of formal complexity present in the spatial arrangement of the plan and its corresponding exterior expression in the roof.

v. The $D_{E1\text{-}4}$ (or $D_{EN\text{-}W}$) and $D_{P\#\text{-}PR}$ results for the house are combined into an average for the entire house (μ_{E+P}). This is a composite measure of the typical level of characteristic complexity present in the building.

6.3.3 Analysis of a Set of Buildings

The form of a single building might be shaped by the specific needs of clients, the complexity of a site, or even material or budgetary constraints. Because the present research is focussed on the analysis of broader design trends in an architect's body of work or across a stylistic movement, it is more meaningful to consider the characteristic formal complexity present in multiple buildings that have been designed by a single architect or practice and within a similar time scale and spread of locations. Thus, the most important results in this study are derived from the analysis of sets of buildings (Table 6.8). The results of sets are signified in this research by the presence of curly brackets {…}. The following steps describe the process for combining a series of fractal dimension measures derived from a set of five individual buildings to develop an 'aggregate' result for the set:

i. The five μ_E results in the set are averaged together to create an aggregate result for the set ($\mu_{\{E\}}$) which is a measure of the typical level of characteristic visual complexity present across all façades in the set.
ii. The median fractal dimension for all elevations in the set is calculated for comparative purposes ($M_{\{E\}}$). The median is the value of the midpoint in the set of data, or if there is no midpoint, it is the average of the centre two.
iii. The standard deviation of the elevations in the set is determined ($std_{\{E\}}$).

Table 6.8 Summary of definitions relating to the analysis of a set of houses

Abbreviation	Meaning	Explanation
$\mu_{\{E\}}$	Average *D* for a set of *elevations*	The average *D* result for all of the elevations in a set of buildings is $\mu_{\{E\}}$. This value is a measure of the typical level of characteristic formal complexity visible across the exterior of a set of designs The average *D* result for all of the plans of a single building is $\mu_{\{P\}}$. This value is a measure of the typical level of characteristic complexity present in the spatial and formal properties of the interiors of a set of buildings The average *D* result for all of the plans and elevations of a set of buildings is $\mu_{\{E+P\}}$
$\mu_{\{P\}}$	Average *D* for a set of *plans*	
$\mu_{\{E+P\}}$	Average *D* for a set of *elevations* and *plans*	
$M_{\{E\}}$	Median *D* for a set of *elevations*	In some sets of data, outliers skew the results in a particular way. A comparison between the mean *D* result for the set ($\mu_{\{E\}}$ or $\mu_{\{P\}}$) and the median *D* result for the same set ($M_{\{E\}}$ or $M_{\{P\}}$) allows for the detection of statistical outliers and identification of the direction of skew in the data. This combination is useful for being able to identify if a particular building, or group of elevations or plans, have unduly influenced the final result
$M_{\{P\}}$	Median *D* for a set of *plans*	
$std_{\{E\}}$	Standard deviation for a set of *elevations*	The standard deviation of a set is a measure of its distribution or dispersion relative to the mean. The higher the standard distribution, the more divergent the results The standard deviation of a set of buildings is calculated for both elevations ($std_{\{E\}}$) and plans ($std_{\{P\}}$)
$std_{\{P\}}$	Standard deviation for a set of *plans*	

iv. The five μ_P results are combined to create an aggregate result $\mu_{\{P\}}$ which is a measure of the typical level of formal complexity present in and experienced throughout the interior of the set of the architect's works.
v. The median fractal dimension for all plans in the set is calculated for comparative purposes ($M_{\{P\}}$).
vi. The standard deviation of the plans in the set is determined ($std_{\{P\}}$).
vii. The five μ_E and μ_P results are combined to create an aggregate value ($\mu_{\{E+P\}}$). This value measures the typical level of characteristic complexity present across the entire set of plans and elevations. When examining a large number of sets of works by different architects, this information can provide a general comparative value.

6.3.4 Analysis of a Sub-set of Buildings

Where the set of five or more houses could not perfectly fulfil the original selection criteria—for example, some of the works were unbuilt, or were designed over a longer time period—the identification of a sub-set that is more indicative of the work can be beneficial. The results of optimal sub-sets are signified in the reporting by the presence of square brackets, [...], and the houses within this group are identified in the data tables with an asterisk (*) (Table 6.9). Sub-sets are handled as follows:

Table 6.9 Summary of definitions relating to the analysis of an optimal sub-set of houses

Abbreviation	Meaning	Explanation
$\mu_{[E]}$	Average D for the optimal sub-set of *elevations*	In a set of five works, the three with the closest μ_{E+P} results are identified as the optimal sub-set. They comprise the core, or most stable and consistent part of the data The average of the elevations of the three most consistent houses is $\mu_{[E]}$ The average of the plans of the three most consistent houses is $\mu_{[P]}$ The average of the elevations and plans of the three most consistent houses is $\mu_{[E+P]}$
$\mu_{[P]}$	Average D for the optimal sub-set of *plans*	
$\mu_{[E+P]}$	Average D for the optimal sub-set of *elevations* and *plans*	
$M_{[E]}$	Median D for the optimal sub-set of *elevations*	The average ($\mu_{[E]}$ or $\mu_{[P]}$) and median ($M_{[E]}$ or $M_{[P]}$) D values for the optimal subset may also be compared to confirm the level and direction of skew present in the data. Ideally, this should be far less than for the equivalent full set of the data
$M_{[P]}$	Median D for the optimal sub-set of *plans*	
$std_{[E]}$	Standard deviation for the *elevations* in an optimal sub-set	The standard deviation of a set is a measure of its distribution or dispersion, relative to the mean. The higher the standard distribution, the more divergent the results The standard deviation of the optimal sub-set of buildings is calculated for both elevations ($std_{[E]}$) and plans ($std_{[P]}$)
$std_{[P]}$	Standard deviation for the *plans* in an optimal sub-set	

i. Within each set of houses, the three with the closest μ_{E+P} results are identified as an optimal sub-set.
ii. The average of the D_E results of the optimal sub-set is calculated ($\mu_{[E]}$). This is a measure of the characteristic visual complexity present across the most consistent façades of the set.
iii. The median fractal dimension for all elevations in the optimal sub-set is calculated for comparative purposes ($M_{[E]}$).
iv. The standard deviation for the elevations in an optimal sub-set is calculated ($std_{[E]}$).
v. The average of the D_P results of the optimal sub-set is calculated ($\mu_{[P]}$). This is a reflection of the typical formal complexity present in and experienced throughout the most consistent plans in the set of an architect's works.
vi. The median fractal dimension for all plans in the optimal sub-set is calculated for comparative purposes ($M_{[P]}$).
vii. The standard deviation of the plans in the optimal sub-set is determined ($std_{[P]}$).

6.3.5 Comparative Analysis

The difference between two fractal dimensions is defined as the *Range* (R) and it can be expressed in two different ways. First, the difference can be measured in terms of absolute fractal dimensions (R_D) where the subtraction of one D result from another determines the positive or negative difference in terms of D. Alternatively, because the fractal dimension of an image is necessarily between 1.00 and 2.00, the difference between two dimensions can also be expressed as a percentage ($R_{\%}$). Importantly, while we use both expressions, $R_{\%}$ is simply R_D with the decimal point moved two places to the right. In general we tend to use $R_{\%}$ for considering large sets of results whereas R_D is more commonly used for individual comparisons (Table 6.10). Range is handled as follows.

i. The range between the highest and the lowest D_E result in an individual house is calculated ($R_{E\ (D \text{ or } \%)}$).
ii. The range between the highest and the lowest D_E results in a set of houses is calculated ($R_{\{E\}\ (D \text{ or } \%)}$).
iii. The range between the highest and the lowest D_E results in an optimal sub-set of houses is calculated ($R_{[E]\ (D \text{ or } \%)}$).
iv. The range between the highest and the lowest D_P result in an individual house is calculated ($R_{P\ (D \text{ or } \%)}$).
v. The range between the highest and the lowest D_P results in a set of houses is calculated ($R_{\{P\}\ (D \text{ or } \%)}$).
vi. The range between the highest and the lowest D_P results in an optimal sub-set of houses is calculated ($R_{[P]\ (D \text{ or } \%)}$).

Table 6.10 Summary of definitions of data used for comparative purposes

Abbreviation	Meaning	Explanation
R_D	Range between the highest and lowest results expressed as a value of D	The difference between two sets of D results is the range (R). Because the maximum D value for an image is 2.0 and the minimum practical value is 1.0, R can be expressed as either an absolute value of D, or as a percentage (% = 100 × R); this is signalled in the subscript annotation of R
$R_{\%}$	Range between the highest and lowest results expressed as a percentage	
$R_{E\ (D\ \text{or}\ \%)}$	Range between the highest and lowest D_E results in a single building	A low R_E value implies that each of the elevations in a single building have a high degree of formal similarity. Conversely, a high R_E value suggests that at least one of the elevations, in terms of formal similarity, diverges from the others
$R_{P\ (D\ \text{or}\ \%)}$	Range between the highest and lowest D_P results in a single building	
$R_{\{E\}\ (D\ \text{or}\ \%)}$	Range between the highest and lowest D_E results in a set of buildings	The range between sets of results is significant for interpreting the relationship between elevations or plans across the complete set of buildings. Similarly, the mean for the combined plans and elevations of a building may be compared across the set
$R_{\{P\}\ (D\ \text{or}\ \%)}$	Range between the highest and lowest D_P results in a set of buildings	
$R_{\{\mu E+P\}\ (D\ \text{or}\ \%)}$	Range between the highest and lowest μ_{E+P} results in the set of buildings	
$R_{[E]\ (D\ \text{or}\ \%)}$	Range between the highest and lowest D_E results in the optimal sub-set of buildings	The comparative range results for optimal sub-sets are calculated automatically and reported although they only become especially significant if $R_{\{\mu E+P\%)}$ >5 %
$R_{[P]\ (D\ \text{or}\ \%)}$	Range between the highest and lowest D_P results in the optimal sub-set of buildings	
$R_{[\mu E+P]\ (D\ \text{or}\ \%)}$	Range between the highest and lowest μ_{E+P} results in the optimal sub-set of buildings	

vii. The range between the highest μ_{E+P} and the lowest μ_{E+P} result in a set is calculated ($R_{\{\mu E+P\}\ +\ (D\ \text{or}\ \%)}$). This measure reflects the degree of diversity within the $\mu_{\{E+P\}}$ result. If there is a relatively high degree of diversity (>5 %) then the optimal subset of results may be a better determinant of the properties of an architect's work ($R_{[\mu E+P]\ +\ (D\ \text{or}\ \%)}$).

6.3.6 *Interpretation of Results*

If you imagine that you have a set of images laid out before you, how similar might they appear in terms of their relative visual complexity and how would you describe this verbally to someone else? For the purpose of more intuitively relating the comparative results to various theorized relationships between buildings or architects works, we found it useful to map the mathematical results to some indicative

Table 6.11 Qualitative descriptors used for ranges

Range (%)	Qualitative descriptors
<1	'Identical' or 'indistinguishable'
2–5	'Very similar' or 'alike'
6–10	'Similar' or 'close'
11–20	'Comparable' or 'some correspondence'
>21	'Unrelated' or 'dissimilar'

qualitative descriptors (Table 6.11). This practice is purely qualitative, but we will use these descriptors consistently in the discussion sections over the following chapters.

6.3.7 Presentation of Results

The results for each set of five designs are first presented in a pair of tables. For example, the results for a hypothetical set of houses are in Tables 6.12 and 6.13.

Table 6.12 displays the D values for every elevation and plan of each house in the set, along with average results for each house's elevations and plans. Thereafter the table records average, median and standard deviation results for both the overall set and the optimal sub-set. At the base of the table composite results for each house are recorded (being the average of both elevations and plans) as well as aggregate results for both the overall set and the optimal sub-set.

Table 6.13 of data associated with each set of results contains comparative measures, expressed as either a range of D or as a percentage difference. For both plans and elevations, the range within individual houses and across both the overall set and the optimal sub-set is recorded. At the base of the table the range between the highest and lowest composite results (combined plans and elevations for each house) are reported for both the overall set and the optimal sub-set.

The results contained in Table 6.12 are also charted in a combined line and bar graph for every set. The vertical y-axis of this chart is the fractal dimension, while the horizontal x-axis is the set of houses. Each house has a vertical bar graph above it indicating the range of D values for both plans and elevations ($D_{E1\text{-}4}$ or $D_{EN\text{-}W}$ and $D_{P\#\text{-}PR}$). An overlaid line graph connects the mean results for both elevations (μ_E) and plans (μ_P), and a horizontal line records the mean value for the sets of both elevations ($\mu_{\{E\}}$) and plans ($\mu_{\{P\}}$) and the associated medians ($M_{\{E\}}$ and $M_{\{P\}}$). The D value for the roof in each case is indicated on the vertical bar with a triangle (Fig. 6.1).

This summary chart (Fig. 6.1) is an ideal starting point for interpreting the data, although specific measures contained in the tables will assist in developing a more detailed or nuanced reading of what it means. For instance, the Example Set could begin to be interpreted in the following way. The elevations of the *Smith*, *Stone* and *Koch* houses have a similar range ($R_{E\%}$) of between 4.35 and 4.45 %, meaning that

Table 6.12 Results for the example set

Example houses		Smith	Stone*	Koch	Heather*	Slate*	Opt. [*]	Set {…}
Elevations	D_{E1}	1.5650	1.5090	1.5995	1.4855	1.5053		
	D_{E2}	1.5505	1.5180	1.5975	1.5015	1.5734		
	D_{E3}	1.5495	1.5400	1.5565	1.5170	1.5095		
	D_{E4}	1.5215	1.5535	1.5555	1.5085	1.5095		
	μ_E	1.5466	1.5301	1.5773	1.5031	1.5244		
	$\mu_{[E]}/\mu_{\{E\}}$						1.5192	1.5363
	$M_{[E]}/M_{\{E\}}$						1.5095	1.5308
	$std_{[E]}/std_{\{E\}}$						0.0245	0.0325
Plans	D_{P0}	1.4765	1.4760	1.4750	1.4200	1.4150		
	D_{P1}	–	1.4650	–	–	1.4980		
	D_{PR}	1.4245	1.4575	1.4910	1.4500	1.4785		
	μ_P	1.4505	1.4662	1.4830	1.4350	1.4638		
	$\mu_{[P]}/\mu_{\{P\}}$						1.4575	1.4606
	$M_{[P]}/M_{\{P\}}$						1.4613	1.4700
	$std_{[P]}/std_{\{P\}}$						0.0287	0.0278
Composite	μ_{E+P}	1.5146	1.5027	1.5458	1.4804	1.4985		
Aggregate	$\mu_{[E+P]}/\mu_{\{E+P\}}$						1.4945	1.5084

*Signifies the optimal sub-set

Table 6.13 Comparative values for the example set

Example houses		Smith	Stone*	Koch	Heather*	Slate*	Opt. [*]	Set {…}
Elevations	R_{ED}	0.0435	0.0445	0.0440	0.0315	0.0680		
	$R_{E\%}$	4.3500	4.4500	4.4000	3.1500	6.8000		
	$R_{[ED]}/R_{\{ED\}}$						0.0435	0.0315
	$R_{[E\%]}/R_{\{E\%\}}$						4.3000	3.1000
Plans	R_{PD}	0.0520	0.0185	0.0160	0.0300	0.0830		
	$R_{P\%}$	5.2000	1.8000	1.6000	3.0000	8.3000		
	$R_{[PD]}/R_{\{PD\}}$						0.0185	0.0830
	$R_{[P\%]}/R_{\{P\%\}}$						1.8000	8.3000
Composite	$R_{[\mu E+PD]}/R_{\{\mu E+PD\}}$						0.0315	0.0680
	$R_{[\mu E+P\%]}/R_{\{\mu E+P\%\}}$						3.1000	6.8000

*Signifies the optimal sub-set

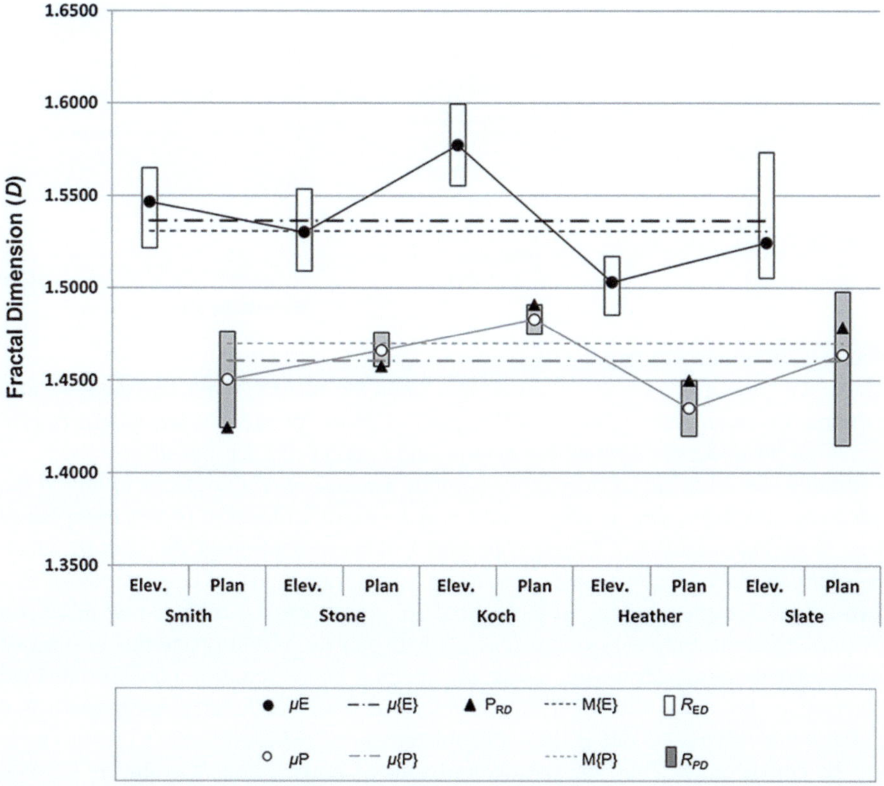

Fig. 6.1 Results for the example set

there is a consistent level of difference between the most and least complex elevations in these houses. This might seem to suggest the existence of a pattern but the elevations of the *Koch House* are more visually complex (μ_E = 1.5773) than those of the remainder of the set by 4.1 % ($\mu_{\{E\}}$ = 1.5363). The plans of the *Stone* and *Koch* houses are the most consistent ($R_{P\%}$ = 1.6 % and 1.8 %), whereas the plans of the *Slate House* are the least ($R_{P\%}$ = 8.3 %). In two cases (*Smith* and *Stone*) the roof plans are the least complex in the house, whereas in the other three, the roof is amongst the more complex plans in the house. Overall, only a single plan (for the roof of the *Koch House*) is more complex than any individual elevation (Elevation 1 of the *Heather House*). If we then examine the aggregate fractal dimension for works by this 'example' architect we can see that the aggregate of the optimal sub-set ($\mu_{[E+P]}$ = 1.4945) is slightly lower (1.39 %) than the aggregate for the whole set ($\mu_{\{E+P\}}$ = 1.5084) with the *Koch House* skewing the results higher than the mean for this architect. Is the *Koch House* larger or more complex? Did the architect experiment with a different façade expression, fenestration or roofline in the *Koch House*? Is the *Koch House* on an exposed site necessitating additional sun and wind screening devices? Without resorting to a more detailed review of the specifics of the *Koch*

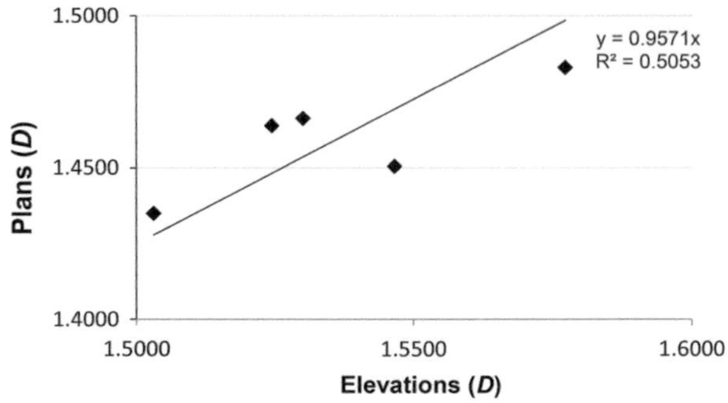

Fig. 6.2 Formal coherence results for the example set

House we cannot answer these questions. Thus the results, while informative in a numeric or quantitative way, still require a level of qualitative interpretation involving an understanding of the history and context of each building.

Finally, for each set of houses we provide a measure of the extent to which the plans and elevations are visually related or have a similar level of formal complexity (Fig. 6.2). This measure, 'formal coherence', is an indication of the way in which the plans in a set of houses correlate to the elevations, and the consistency of this correlation. A high R^2 value in the formal coherence chart indicates that there is a greater degree of similarity or consistency between the visual properties of a set of plans and elevations. However, unlike this method's more common use in statistical analysis, in this case the level of correlation cannot be directly interpreted as a qualitative assessment; rather, it is a characteristic of the architect's signature style and its consistency. Thus for the example set of houses, the R^2 value is 0.5053, which implies at best a medium level of formal coherence, however the result does emphasise the pattern in this work, that the elevations are more complex than the plans in almost all cases.

6.4 Additional Applications of the Method

Whereas the method described thus far is applied consistently to every set of houses analysed throughout this book, several secondary variations are also used in different chapters to test particular claims about buildings, architects or styles. The hypotheses underlying these secondary applications are detailed in Chap. 1, and described in the chapters in which they occur, but here we will summarise the different methodological factors.

For the first of these variations, in Chap. 7, the fractal dimension data derived from the Modernist house sets is augmented with additional information about the address and cardinal orientation of each house. This process involves clustering the fractal dimension data in accordance with information about orientation (north, south, east and west) and address (public or private access). Chap. 8 uses a large series of perspective views, generated at intervals along a path, to measure the changing levels of visual complexity experienced by a person moving through space. This is a rare demonstration of a case wherein orthographic images (plans and elevations) are not ideal for testing specific claims about space and form. Chap. 9 measures the fractal dimensions of a series of non-cardinal elevations, generated by rotating the view point of a building around its axis, and thereby comparing how a building's expression changes from different positions. The properties of three houses by three architects (Eisenman, Hejduk and Meier) are compared with this way. In Chap. 10, fractal dimension measures derived from a Modernist and a Post-Modernist set of houses (respectively from Le Corbusier and Venturi and Scott Brown) are compared. Specifically, each measure of D is augmented with data relating to the permeability of the façade it is derived from, allowing all ten houses to be classified into a sector-map to differentiate the combined formal and functional expressions of the designs. Finally, Chap. 11 compares measures derived from representations of façades with two different characteristics: transparent and opaque windows and openings. Whereas in all of the rest of the cases presented in this book windows are assumed to be opaque and louvres and doors closed, in the methodological variant in Chap. 11, the standard approach is compared with results from a transparent representation, for both elevations and perspective images.

6.5 Conclusion

The large number of measures presented in this chapter might seem complex at first, but there are really only two basic things being measured: the fractal dimensions of elevations and plans. These can then be combined together across an individual house, across a sub-set of three houses or across a complete set of five houses. To compare the various measures derived in this way, the difference or range between the results is then determined. If the range is relatively small, then the houses, plans or elevations are visually similar. If the range is large, then they are dissimilar, and the equivalent results for the sub-set are considered to see if the range is reduced. Ultimately, in most cases, a small range implies a degree of consistency in the way a designer works, even though various site- and program-specific differences might occasionally confound the data.

Chapter 7
The Rise of Modernity

In this chapter, twenty houses, divided into four sets, are measured and analysed using fractal dimensions. The first two sets are designs by Le Corbusier, respectively his Pre-Modern houses in Switzerland and his early Modern works in France and Germany. The third set is of five designs by Eileen Gray for sites in southern France, and the last set contains five houses by Mies van der Rohe in Germany, Poland and the USA.

At the very start of his career, Le Corbusier designed a series of Arts and Crafts style houses that seemed to mimic the strategies and techniques found in Swiss vernacular architecture. Five of these Pre-Modern works make up the first set of designs analysed in this chapter. The five are the Villas *Fallet*, *Jaquemet*, *Stotzer*, *Jeanneret-Perret* and *Favre-Jacot*. While for many years architectural historians were aware of these works but largely ignored them, they are today regarded as important precursors to Modernism and as marking the period when Le Corbusier turned away from ornamentation and towards a fascination with pure form. The second set analysed in this chapter features five of Le Corbusier's famous white-rendered, proto-Modernist designs. It comprises the *Maison-Atelier Ozenfant*, *Weissenhof-Siedlung Villa 13* and the Villas *Cook*, *Stein-de Monzie* and *Savoye*. For the purposes of this research, all ten designs by Le Corbusier were digitally reconstructed using published measured and design drawings (Ando 2001; Park 2012). Notably, two of these houses do not possess four visible elevations. The *Maison-Atelier Ozenfant*, sited on an almost triangular block, only has three primary elevations and the *Villa Cook*, which shares walls with two existing buildings, has only two visible elevations.

The houses of Eileen Gray, a contemporary of Le Corbusier, have been described as 'seminal examples of the spirit of the Modern Movement' (Garner 1993: np). The five designs by Gray that make up the third set in this chapter are *Small House for an Engineer, E.1027, Four Storey Villa, Tempe à Pailla* and *House for Two Sculptors*. Two of these houses were built in southern France, while the other three are unrealised projects. The inclusion of the unbuilt projects presents a challenge for the research, because they have a lower level of resolution than the

M.J. Ostwald and J. Vaughan, *The Fractal Dimension of Architecture*,
Mathematics and the Built Environment 1, DOI 10.1007/978-3-319-32426-5_7

constructed works. However, such is the importance of Gray's architecture that her unbuilt designs have also been subjected to repeated analysis using a range of methods. For this reason, we have elected to include these three unbuilt projects, most likely designed, like the built projects, for sites in southern France. The digital models used for the analysis of Gray's architecture have been adapted both from her originals drawings (Hecker and Müller 1993; Constant and Wang 1996) and from later reconstructions of her work (Constant 2000). The design for the *Four Storey Villa*, the design was never finalised, and so Gray's last set of drawings, modified to take into account her annotated corrections, were used for the analysis. Gray's archived documents for the *Small House for an Engineer* contain only three finished elevations; the fourth was constructed for the present research using dimensions extrapolated from the other views along with images of the final model produced of that house.

Five houses by Mies van der Rohe make up the last set measured in this chapter. These designs span the period from Mies's first realisation of his functionalist ideals to their refinement in one of his most famous works. The set includes Mies's *Wolf*, *Lange*, *Esters*, *Lemke* and *Farnsworth* houses. The digital models used for the fractal analysis were derived from published archival and measured drawings (Vandenberg 2003; Puente and Puyuelo 2009).

In the final section of the chapter, a set of data derived from the Modernist houses of Le Corbusier and Gray is used to search for patterns in the relationship between the visual complexity of an elevation and its orientation and address. If form does follow function in these designs, then a pattern should be visible for one or other of these site-related factors when the fractal dimension data is clustered in this way.

7.1 Functionalist Modernism

In the early years of the twentieth-century, the combination of widespread growth in industrialisation and the availability of new materials meant that, possibly for the first time in history, buildings were no longer being made in the same way as the majority of other constructed objects. While the Arts and Crafts tradition still dominated architectural production, other disciplines were taking advantage of highly efficient industrial techniques and processes. After the outbreak of the first world war in Europe, a shortage of human resources and an urgent demand for everything from clothing and guns to packaged food and machinery broadened the gap between conventional architectural practice and other major economies of production. In the aftermath of the war, with the need for large-scale housing at its peak, and under pressure from a burgeoning trend of car ownership, architecture was facing a crisis. At that time, '[t]he architect, along with every other specialist, had to heed these changed circumstances' (Kostof 1985: 696). However, despite this urgency, there was no single concerted response from architects. Instead, multiple designers, often working in isolation from one another, began to formulate theories that would help architecture to renounce its past, historically-dominated traditions and embrace not only new materials and processes, but a new aesthetic expression for the era. Rather than relying on applied decoration and symbolism, this new style sought to derive its character from contemporary concerns, processes and materials. In practice, the buildings proposed by these architects tended to emphasise stark geometric forms, often finished in a flat, white render and with details that evoked industrial production techniques. These designs rejected historic cellular and hierarchical interior spatial strategies in favour of open plans, roof terraces and integrated garages.

German architecture, led by the Bauhaus school and under the direction of Walter Gropius, was an early promoter of such 'functionalist' ideals. Their approach was informed by the *Russische Ausstellung*, the Constructivist exhibition of 1922 in Berlin, which had challenged the expressionist tendencies and professional complacency of German architects (Risebero 1982). By 1927, Mies van der Rohe (assisted by Lilly Reich) had organised the Second International Exposition of the Deutscher Werkbund in Stuttgart. This was not only an exhibition of drawings and models but also involved the construction of a new suburb, the *Weissenhof Siedlung*. Houses for this suburb were commissioned from leading modernist architects, including J.J.P. Oud, Le Corbusier, Walter Gropius and Peter Behrens.

In parallel with the efforts of the Bauhaus, the published theories of Viollet-le-Duc, Louis Sullivan and Adolf Loos inspired the spread of the Modern movement. In France, the built works of Le Corbusier provided a physical expression of his published theory of a *machine à habiter*—a machine for living. Modernism also spread to the USA in the early 1930s, influenced by an exhibition organized by Henry-Russell Hitchcock, Philip Johnson and Alfred Barr. Entitled the 'International Style', the exhibition 'was huge in the architecture world … even though comparatively few people came' (Hitchcock and Johnson 1995: 15). It documented different strands of Modernism across several nations and the name, International Style, soon came to represent the entire movement, even though historians have, in later years, tended to describe it more thematically as Functionalist Modernism.

Despite flourishing in many countries, and dominating architectural education in a similarly diverse range of locations, by the 1960s the Modern movement was heavily criticised for its failure to come to terms with a growing number of social and cultural problems. Furthermore, the extent to which the icons of Modernity were actually shaped by industrial and functional influences had been repeatedly questioned and exposed as inherently fallacious. Under attack for its apparent ignorance of regional values and environmental conditions, and lambasted for its inability to respond to sensitive cultural and historical settings, Modernism soon fell out of favour. Nevertheless, it has been argued that the influence of Modernism 'never really died; it did not even fade away, but transformed itself under several guises … After the 1960s there was a greater plurality of architectural expression, engendered by a world ever-increasingly aware of itself' (Khan 2009: 212). Thus, as later chapters in this book reveal, the values and concerns of Modernism found their way, albeit in a more nuanced form, into the Regionalist and Minimalist architectural movements, and in a more mannered way, in the work of the architectural *Avant-Garde* of the 1960s and 1970s.

7.2 Le Corbusier

Charles-Édouard Jeanneret-Gris was born in Switzerland in 1887 and later changed his name to Le Corbusier, before becoming a French citizen in 1930. In 1914, at a critical juncture in his career, he developed a concept for a modern structural system that would allow for both a flexible response to function and a clear programmatic expression. This system, the *Maison Dom-Ino*, laid the conceptual foundation for Le Corbusier's 'white period' of the 1920s, where he famously developed his 'five points for a new architecture' and its extrapolation into a proposal for 'machines for living'. The majority of Le Corbusier's works from 1914 onwards are clear examples of either Functionalist or Rationalist Modernism. However, prior to that time he designed five houses in Switzerland. These surprisingly ornate, chalet-style buildings are reminiscent of the Arts and Crafts movement and give little indication that they are the work of the future Modernist visionary. But despite their outward differences from his later work, and the neglect by earlier scholars of Le Corbusier, more recent historians and critics have repeatedly suggested that they are an important precursor to Le Corbusier's designs of the 1920s (Von Moos 1979). Both these early chalet-style buildings and his later Modernist designs are analysed in the following sections.

7.2.1 Pre-modern Houses (1905–1912)

Le Corbusier began his formal training in design at the Art School of Le Chaux-de-Fonds in Switzerland. Under the tutelage of Charles L'Eplattenier, students of the school worked to develop a style that would evoke and complement the physical features of their local Jura region. After studying the geometry of nature, the students were asked to design in such a way as to 'make the structural laws of nature visible and to express them in clear and universal geometric patterns' (Von Moos 1979: 4). In 1905, as part of his education, Le Corbusier began to apply these lessons in the design of a house in Pouillerel, a small enclave on the outskirts of Le Chaux-de-Fonds. After completing that work, the *Villa Fallet*, in 1907 he travelled across Europe furthering his studies in art and architecture. It was while he was based in Vienna that he designed two more houses for sites in Pouillerel, the villas *Jaquemet* and *Stotzer*. Then, after working for Peter Behrens in Germany and travelling through Greece and Turkey in 1911, Le Corbusier returned to a teaching position at his former school. The following year he completed both a house for his parents, the *Villa Jeanneret-Perret* and a large family home situated in the town of Le Locle, the *Villa Favre-Jacot*. Table 7.1 provides example elevations and plans for this set.

The first of the five Arts and Crafts houses, the 1905 *Villa Fallet*, is often described as an architectural representation of the geology and ecology of the local area. The stone base of the building is typically interpreted as depicting the lower mountains in the Jura landscape, while the 'windows and door-frames [are] of

Table 7.1 Le Corbusier, Pre-Modern set, example elevations and floor plans (level 4 representation—not shown at uniform scale)

Villa Fallet

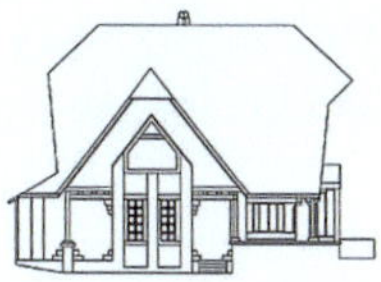

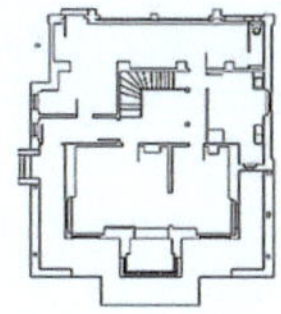

Villa Stotzer

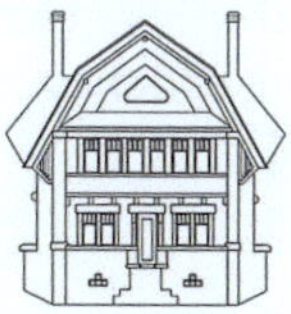

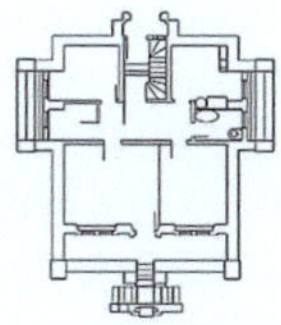

Villa Jaquemet

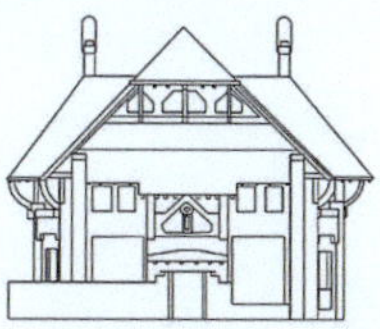

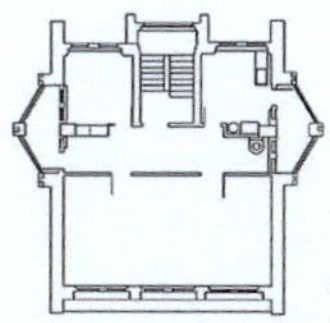

Villa Jeanneret-Perret

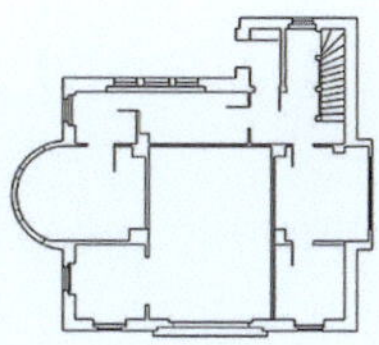

Villa Favre-Jacot

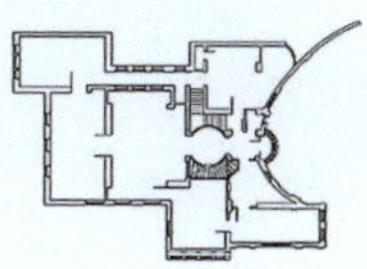

dressed stone … being chiselled to produce a textured surface' (Baker 1996: 52). The upper storeys of the *Villa Fallet* are sheathed in decoratively carved, stucco-faced timber. The sloping roof has exposed carved timber supports and it is clad in dark tiles, suggesting, once again, a section through the landscape. As Charles Jencks notes, '[i]ts base of heavy rusticated limestone blocks represents bedrock. … Then above these two layers of geology springs life, above all the pine tree' (2000, 31–32).

The 1907 *Villa Stotzer*, which was designed for Albert Stotzer-Fallet, allegedly 'displays the early signs of a reassessment of the historicist idiom' (Baltanás 2005: 19). The third house, the *Villa Jaquemet* was designed for Jules Jaquemet-Fallet and it is similar in planning and appearance to the *Villa Stotzer* (Fig. 7.1). These two villas share a common program, which required that they appear as single houses from the street, but actually contain two separate apartments. Furthermore, the initial designs for both houses were more ornate than the completed buildings, with financial restrictions forcing Le Corbusier to remove some of the planned decorative elements.

Le Corbusier's 1912 design for his parents' home, the *Villa Jeanneret-Perret*, is a departure from the first three works in several subtle ways. This two-storey family home has large unadorned exposed windows in its plain white-rendered walls and a dark tiled roof that lacks the dramatic pitched form of the earlier houses.

Fig. 7.1 *Villa Jaquemet*, perspective view

Completed in the same year, the *Villa Favre-Jacot* continues Le Corbusier's early exploration of neo-classical façade expression in an otherwise vernacular idiom. It too has mostly plain, white walls and a non-chalet-shaped roof clad in dark tiles. The external ornamentation that was so significant for the first three houses is, in the last two villas, much reduced and classicized using geometry more than texture to achieve its visible expression. Jencks argues that these two houses signal a move away from regional forms and materials and towards a more formally rich 'composition with cubic and cylindrical volumes' (2000: 195). Thus, while the last two works are less decorative or crafted, they have an increased reliance on formal modelling.

7.2.2 Pre-modern Houses, Results and Analysis

In the set of Le Corbusier's Pre-Modern works, the lowest average elevation result is for the *Villa Jaquemet* (μ_E = 1.3788) and the highest is found in the *Villa Favre-Jacot* (μ_E = 1.5143), leading to a range for the set of $R_{\{E\%\}}$ = 23.27. The median elevation result is 1.4352 and the standard deviation is 0.0750. Results for the plans show the lowest average is found in the *Villa Stotzer* (μ_P = 1.2911) while the *Villa Favre-Jacot* again has the highest (μ_P = 1.3601), as it did for the elevation average. The range for the set of plan results is $R_{\{P\%\}}$ = 28.23, the median is 1.3307 and the standard deviation is 0.0771, meaning that the plan data is only slightly more distributed from the average than it is for the elevation data. Notably, the roof is the least complex of any plan view in every case (Tables 7.2 and 7.3, Fig. 7.2).

These results indicate that there is more diversity (or less consistency) in the plans than in the elevations, although the average plan results are all lower than the average elevation results for each house. Furthermore, a 23 % difference across both sets of plans and elevations is not especially close. However, when all of the plans and elevations are examined together, the aggregate result is $\mu_{\{E+P\}}$ = 1.3714. Thus, while both the separate ranges for elevations and plans show a more substantial gap (23.27 and 28.23 % respectively), the composite range is much reduced, $R_{\{\mu E+P\%\}}$ = 10.65, an outcome which means that these five houses, as complete objects (merging both plans and elevations), have a more consistent level of visual complexity than either the elevations or plans in isolation.

The optimal sub-set comprises the earliest three houses, the villas *Fallet*, *Jaquemet* and *Stotzer*. The aggregate result for the sub-set is $\mu_{[E+P]}$ = 1.3441, a figure which is only marginally less than the overall set. However, the composite range is more substantially lowered, from 10.65 % to less than 4 % ($R_{[\mu E+P\%]}$ = 3.22), supporting the notion that these three houses make up a very distinct group. Indeed, the villas *Jaquemet* and *Stotzer* share a similar design brief and similar modulation and massing, leading to a tight range between the two of $R_{E\%}$ = 3.60. With a μ_E value of 1.3856 the elevations of the *Villa Stotzer* have a slightly higher level of visual complexity than the *Villa Jaquemet* (μ_E = 1.3788) but otherwise there are few differences. Furthermore, for both houses, the south

Table 7.2 Le Corbusier, Pre-Modern set, results

Houses		Villa Fallet*	Villa Jaquemet*	Villa Stotzer*	Villa Jeanneret-Perret	Villa Favre-Jacot	Opt. [*]	Set {…}
Elevations	D_{E1}	1.4383	1.4132	1.3379	1.4016	1.4674		
	D_{E2}	1.4794	1.4707	1.4796	1.5154	1.5328		
	D_{E3}	1.3494	1.3159	1.3602	1.4320	1.5479		
	D_{E4}	1.3945	1.3152	1.3648	1.5001	1.5285		
	μ_E	1.4154	1.3788	1.3856	1.4623	1.5192		
	$\mu_{[E]}/\mu_{\{E\}}$						1.3933	1.4322
	$M_{[E]}/M_{\{E\}}$						1.3797	1.4352
	$std_{[E]}/std_{\{E\}}$						0.0620	0.0750
Plans	D_{P-1}	1.3310	1.3422	1.2524	1.3101	1.3420		
	D_{P0}	1.3712	1.3542	1.3950	1.3158	1.3805		
	D_{P1}	1.3510	1.3403	1.3304	1.3968	1.4304		
	D_{P2}	–	1.3180	1.3297	–	–		
	D_{PR}	1.2080	1.1555	1.1481	1.1885	1.2874		
	μ_P	1.3153	1.3020	1.2911	1.3028	1.3601		
	$\mu_{[P]}/\mu_{\{P\}}$						1.3019	1.3127
	$M_{[P]}/M_{\{P\}}$						1.3307	1.3307
	$std_{[P]}/std_{\{P\}}$						0.0788	0.0771
Composite	μ_{E+P}	1.3654	1.3361	1.3331	1.3825	1.4396		
Aggregate	$\mu_{[E+P]}/\mu_{\{E+P\}}$						1.3441	1.3714

Table 7.3 Le Corbusier, Pre-Modern set, comparative values

Houses		Villa Fallet*	Villa Jaquemet*	Villa Stotzer*	Villa Jeanneret-Perret	Villa Favre-Jacot	Opt. [*]	Set {…}
Elevations	R_{ED}	0.1300	0.1555	0.1417	0.1138	0.0805		
	$R_{E\%}$	13.0000	15.5500	14.1700	11.3800	8.0500		
	$R_{[ED]}/R_{\{ED\}}$						0.1644	0.2327
	$R_{[E\%]}/R_{\{E\%\}}$						16.4400	23.2700
Plans	R_{PD}	0.1632	0.1987	0.2469	0.2083	0.1430		
	$R_{P\%}$	16.3200	19.8700	24.6900	20.8300	14.3000		
	$R_{[PD]}/R_{\{PD\}}$						0.2469	0.2823
	$R_{[P\%]}/R_{\{P\%\}}$						24.6900	28.2300
Composite	$R_{[\mu E+PD]}/R_{\{\mu E+PD\}}$						0.0322	0.1065
	$R_{[\mu E+P\%]}/R_{\{\mu E+P\%\}}$						3.2200	10.6500

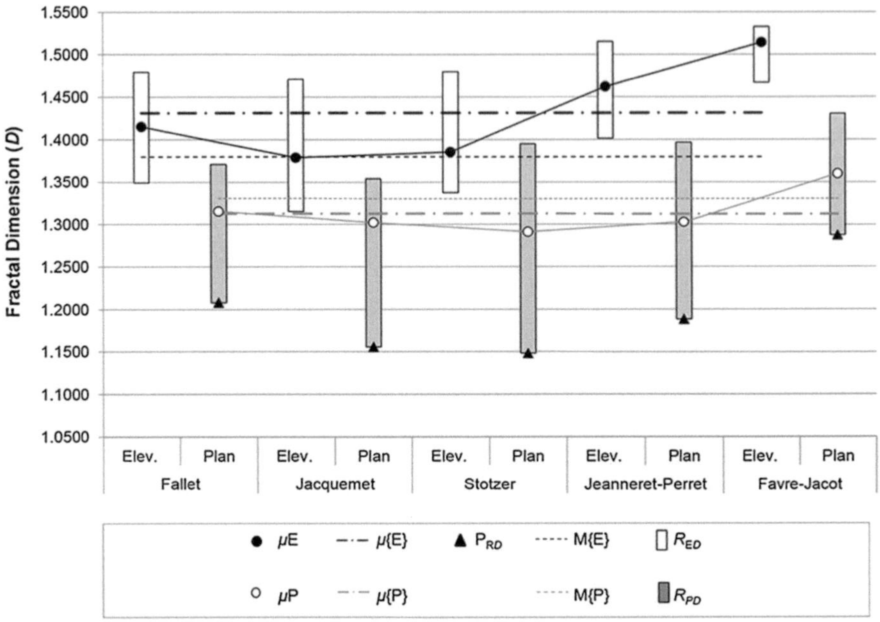

Fig. 7.2 Le Corbusier, Pre-Modern set, graphed results

elevation is the most visually complex (for the *Villa Stotzer*, D_{E2} = 1.4796 and for the *Villa Jaquemet*, D_{E2} = 1.4707). For both houses the east and west elevations are almost mirror images, which is reflected in the almost identical *D* results: the range between the east and west elevations for the *Villa Jaquemet* is $R_{\{E\%\}}$ = 0.07 for the *Villa Jaquemet* and $R_{\{E\%\}}$ = 0.46 for the *Villa Stotzer*.

As previously noted, historians have tended to divide this group of five houses into two sets, based on their visual character and planning approach. This position is confirmed by the fractal analysis results, where the first three villas (*Fallet*, *Jaquemet*, *Stotzer*) have both lower and more consistent composite results ($1.3331 < \mu_{E+P} < 1.3654$), while the later two, *Jeanneret-Perret* and *Favre-Jacot*, are higher and have a larger range (μ_{E+P} = 1.3825 and 1.4396 respectively). According to Geoffrey Baker, the *Fallet*, *Stotzer* and *Jaquemet* houses share an essential quality where 'their massing is very powerful' and their 'surface and structure, and the vigour of the forms … all foreshadow Le Corbusier's later work' (1996: 56). Of the *Jeanneret-Perret* and *Favre-Jacot* houses, Baker remarks that these signal a change in Le Corbusier's design style. It is the last of these, the *Villa Favre-Jacot*, that has the highest overall fractal dimension result of these five Pre-Modern houses (μ_{E+P} = 1.4396) and the most complicated planning (reflected in the highest average plan dimension (μ_P = 1.3601). Considering its advancing and receding elevation planes, curved walls, complex fenestration and expressed columns, the high *D* values for the elevations are readily understood ($1.4674 < D_E < 1.5479$). This result also signals an important lesson about interpreting the fractal dimension of a building: a design which features a large amount of surface decoration on relatively simple walls *may*, depending on the level of representation chosen for the analysis, have a much lower fractal dimension than a more formally articulated, but completely unornamented, work.

7.2.3 Modern Houses (1923–1931)

In the decade after completing the *Villa Favre-Jacot*, Le Corbusier was given the opportunity to experiment with the use of structural concrete, an experience that was to shape his 1914 proposal for the *Maison Dom-Ino*. By 1916 he had become close friends with the cubist painter Amédée Ozenfant, an influence that Kenneth Frampton credits as encouraging Le Corbusier to embrace both the 'machine aesthetic of Purism' and the abandonment of 'existing types' (1992: 152). All of these influences found their way into Le Corbusier's proposition of five strategies for an architecture that would reflect the technological and social spirit of its era. Developed in the early 1920s, published in the journal *L'Esprit Nouveau* and later collated in *Vers Une Architecture* (1923), these five strategies are: elevation of the building on *pilotis* (slender columns), *plan libre* (open planning), *fenêtre en longueur* (horizontal strip windows), *façade libre* (a façade expression which is free from traditional structural constraints) and the inclusion of a *toît jardin* (roof garden). These features are found (at least in part) in all of the houses in the set studied here, and are visible in their most refined form in the *Villa Savoye* (Curtis 1986).

Le Corbusier was given his first opportunity to apply his new theory of design in the *Maison-Atelier Ozenfant*. Located in Paris and constructed in 1923, this building was intended to function as both home and studio for Amédée Ozenfant, and is regarded as reflecting Le Corbusier's purist attitude to creativity (Willmert 2006). Set on a steep corner location, in an urban area which features multiple artists' studio houses, its unusual footprint was designed to accommodate the tight, five-sided site. The three-storey, white-rendered masonry structure was designed with large glazed areas and a prominent, industrial saw-tooth roof.

The *Villa Cook* in Boulogne-sur-Seine was constructed in 1926 and described by Le Corbusier as 'the true cubic house' because, as Gans observes, 'plan, section and elevation all derive from the same square and in reference to one another' (2000: 66). The house is a four-storey structure of white rendered masonry, which shares party walls with neighbouring buildings to the east and west, meaning it has only two façades. The house is one of the first designs in which Le Corbusier used all of his five strategies, starting with the slender *pilotis* on the ground floor, the unrestricted room layouts and elevation treatments (*plan libre* and *façade libre*) with horizontal windows running the full width of the north façade (*fenêtre en longueur*) and finally, the upper level, outdoor living space (*toît jardin*).

The *Villa Stein-de Monzie* is sited on a narrow block in the suburbs of Vaucresson. The unusual domestic brief was for a house and studio for Gabrielle de Monzie and her daughter, to be shared with Michael and Sarah Stein. The house has

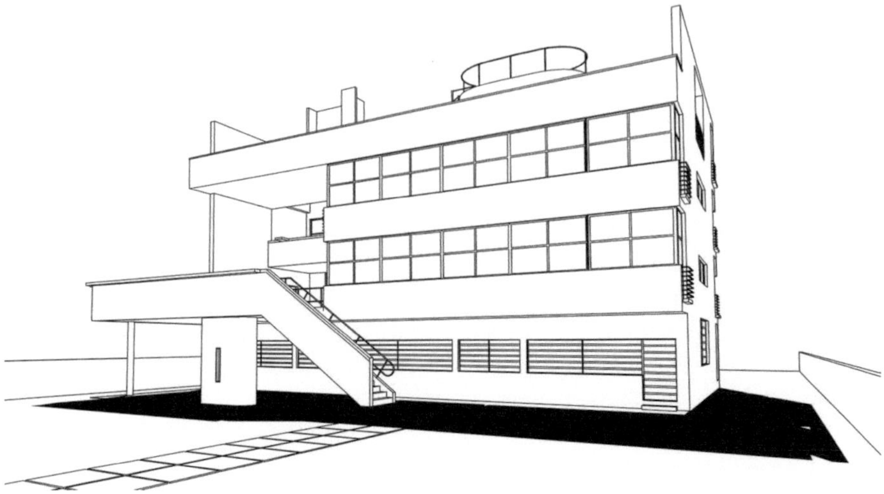

Fig. 7.3 *Villa Stein-de Monzie*, perspective view

been described as the 'most monumental and luxurious of Le Corbusier's houses of the 1920s' (Curtis 1986: 79). Constructed of white plaster-rendered masonry and finished to look machine-made, with this four-storey house Le Corbusier modified his application of the five strategies. The house has a free façade and horizontal strip windows, but the *pilotis* are set within the interior, giving the building a more traditional ground floor exterior expression. The *toît jardin* also differs in this house, arranged over multiple, open levels, hence giving the house its other name, '*Les Terraces*' (Fig. 7.3).

In the late 1920s Le Corbusier and Pierre Jeanneret were invited to produce two residential designs for the Deutscher Werkbund exhibition at Stuttgart. *Villa 13* was designed as a prototype for suburban row housing and was constructed on a corner block in the *Weissenhof-Siedlung* estate. Described as 'a dynamic and complex articulation of the basic cube form' (Park 2012: 124) the long, narrow, four-storey house demonstrates all of Le Corbusier's five strategies. In particular, the *fenêtre en longueur* can be found on the north and south elevations.

The *Villa Savoye*, also known as '*Les heures Claires*', was constructed in 1928 and is sited in Poissy, France. Unlike the other houses in this set, this villa is located in an open field in a rural landscape, which meant that Le Corbusier could design a house 'in the round', with all four elevations of the building being carefully articulated. Designed as a weekend house for Pierre and Emilie Savoye, this is a predominantly two-storey building of white-rendered masonry, with an extensive roof-garden making up a third level. The *Villa Savoye* is widely regarded as the most complete and refined application of Le Corbusier's 'Five Points of a New Architecture' and is one of the world's masterworks of design (Meier 1972) (Fig. 7.4). Table 7.4 provides example elevations and plans for this set of Le Corbusier's houses.

Fig. 7.4 *Villa Savoye*, perspective view

7.2.4 Modern Houses, Results and Analysis

The fractal dimension measurements taken from all five of Le Corbusier's Modernist houses show that the lowest average elevation is found in the *Weissenhof-Siedlung Villa 13* ($\mu_E = 1.3685$) and the highest is in the *Maison-Atelier Ozenfant* ($\mu_E = 1.4611$), defining a range of $R_{\{E\%\}} = 30.31$ for all elevations of the set. The median for all elevations is 1.4450 and the standard deviation is 0.0686, the data being slightly more clustered than his Pre-Modern works. The highest individual elevation result is in the *Villa Stein-de Monzie* ($D_{E2} = 1.5400$). Likewise, the highest plan average is also found in the *Villa Stein-de Monzie* ($\mu_P = 1.3808$), however the highest individual plan is the first floor of the *Villa Savoye* ($D_{P1} = 1.4216$) and the lowest plan average is from the *Villa Cook* ($\mu_P = 1.2889$). The range of all the plans is $R_{\{P\%\}} = 25.24$. The median for all plans is 1.3535 and the standard deviation is 0.0608, once again more consistent than his Pre-Modern houses (Tables 7.5 and 7.6, Fig. 7.5).

Superficially at least, these results suggest that there is a considerable difference in the levels of formal complexity present in both the elevations and plans of these Modern houses. However, the data is complicated by several factors, including two almost blank walls in the *Weissenhof-Siedlung Villa 13* (meant to be shared with future adjacent developments) and some similarly blank roofs in three of the houses. The difference between the median and the average (coupled with the range outcomes) in both the plan and elevation sets confirm the presence of a small number of very low results in several houses which had shaped the overall results.

When all of the plan and elevation results are considered together, the aggregate is $\mu_{\{E+P\}} = 1.3825$ and the composite range is $R_{\{\mu E+P\%\}} = 8.00$, a result which places the five houses within a similar scale of visual complexity. Intuitively, this

Table 7.4 Le Corbusier, Modern set, example elevations and floor plans (level 4 representation—not shown at a uniform scale)

Maison-Atelier Ozenfant	
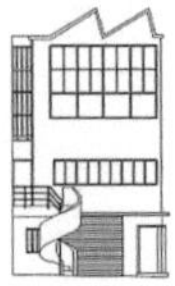	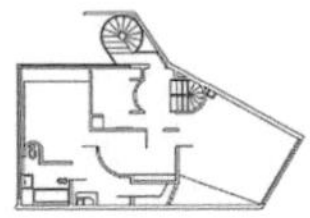
Villa Cook	
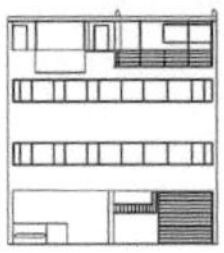	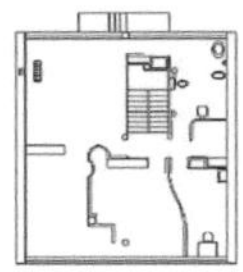
Villa Stein and de Monzie	
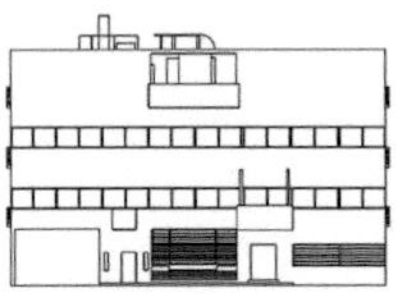	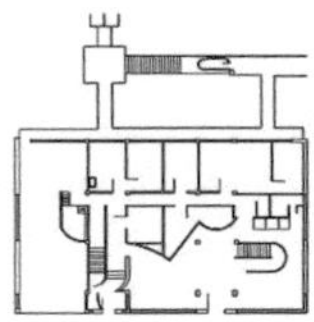
Weissenh of-SiedlungVilla 13	
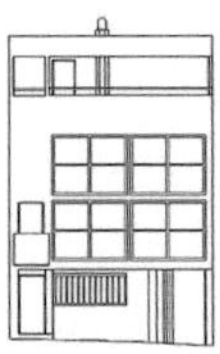	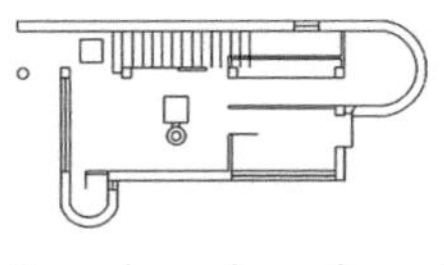
Villa Savoye	
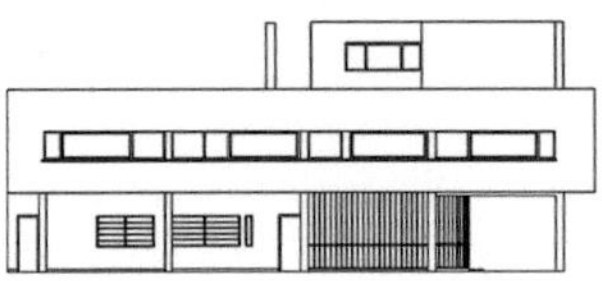	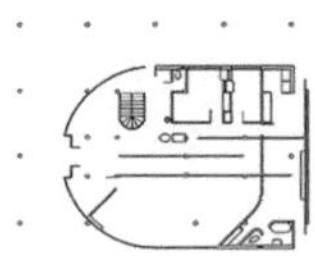

Table 7.5 Le Corbusier, Modern set, results

Houses		Ozenfant*	Cook*	Weissenhof-Siedlung*	Stein-de Monzie	Savoye	Opt. [*]	Set {…}
Elevations	D_{E1}	1.4927	1.4681	1.4076	1.4637	1.4916		
	D_{E2}	1.4002	1.3960	1.4441	1.5400	1.3961		
	D_{E3}	1.4905	–	1.3852	1.4450	1.4732		
	D_{E4}	–	–	1.2369	1.3782	1.4694		
	μ_E	1.4611	1.4321	1.3685	1.4567	1.4576		
	$\mu_{[E]}/\mu_{\{E\}}$						1.4135	1.4352
	$M_{[E]}/M_{\{E\}}$						1.4076	1.4450
	$std_{[E]}/std_{\{E\}}$						0.0780	0.0686
Plans	D_{P0}	1.3610	1.2795	1.2851	1.4019	1.2944		
	D_{P1}	1.4001	1.3559	1.3490	1.4065	1.4216		
	D_{P2}	1.3332	1.3511	1.3735	1.3928	–		
	D_{P3}	1.3389	–	1.3366	1.4041	–		
	D_{PR}	1.4030	1.1692	1.2619	1.2985	1.3699		
	μ_P	1.3672	1.2889	1.3212	1.3808	1.3620		
	$\mu_{[P]}/\mu_{\{P\}}$						1.3284	1.3449
	$M_{[P]}/M_{\{P\}}$						1.3440	1.3535
	$std_{[P]}/std_{\{P\}}$						0.0619	0.0608
Composite	μ_{E+P}	1.4025	1.3366	1.3422	1.4145	1.4166		
Aggregate	$\mu_{[E+P]}/\mu_{\{E+P\}}$						1.3617	1.3825

Table 7.6 Le Corbusier, Modern set, comparative values

Houses		Ozenfant*	Cook*	Weissenhof-Siedlung*	Stein-de Monzie	Savoye	Opt. [*]	Set {…}
Elevations	R_{ED}	0.0925	0.0721	0.2072	0.1618	0.0955		
	$R_{E\%}$	9.2500	7.2100	20.7200	16.1800	9.5500		
	$R_{[ED]}/R_{\{ED\}}$						0.2558	0.3031
	$R_{[E\%]}/R_{\{E\%\}}$						25.5800	30.3100
Plans	R_{PD}	0.0698	0.1867	0.1116	0.1080	0.1272		
	$R_{P\%}$	6.9800	18.6700	11.1600	10.8000	12.7200		
	$R_{[PD]}/R_{\{PD\}}$						0.2338	0.2524
	$R_{[P\%]}/R_{\{P\%\}}$						23.3800	25.2400
Composite	$R_{[\mu E+PD]}/R_{\{\mu E+PD\}}$						0.0658	0.0800
	$R_{[\mu E+P\%]}/R_{\{\mu E+P\%\}}$						6.5800	8.0000

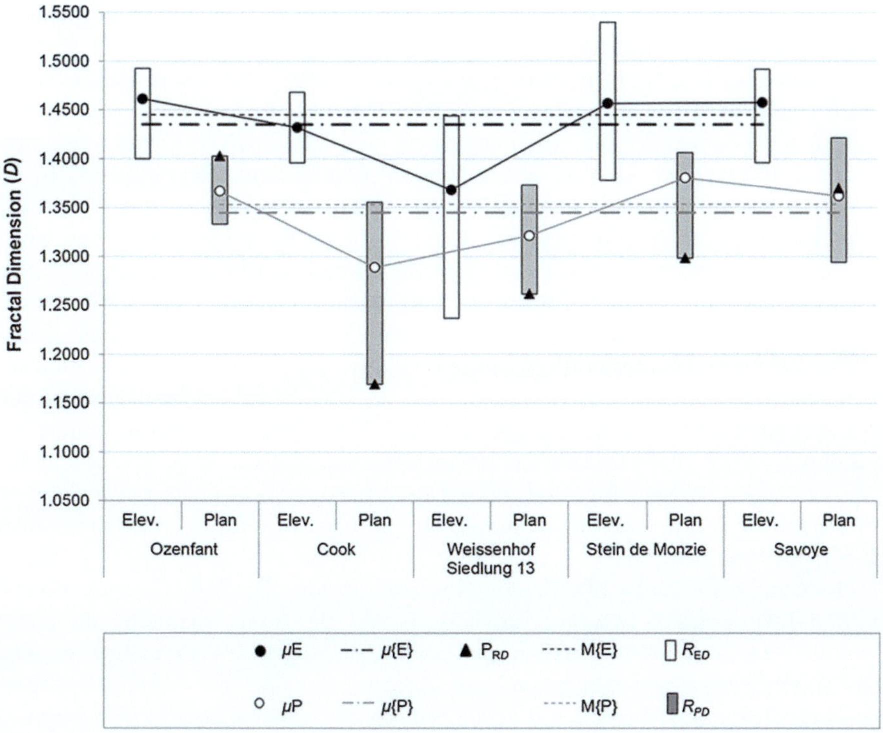

Fig. 7.5 Le Corbusier, Modern set, graphed results

supports the conventional understanding that these houses are part of a sustained attempt by Le Corbusier to apply his design theory to multiple works in a consistent way. However, even the most cursory examination of these designs shows that they vary in their formal expression in response to constraints of site and program.

The *Maison-Atelier Ozenfant*, the *Villa Cook* and *Weissenhof-Siedlung Villa 13* were identified as the optimal sub-set. Considering only these three houses, the aggregate result for all plans and elevations is reduced to $\mu_{[E+P]} = 1.3617$ and the composite range is reduced to $R_{[\mu E+P\%]} = 6.58$. This is not a substantial reduction over the full set, reinforcing the notion that they are visually similar but not identical.

The overall results for this set of houses by Le Corbusier are partially compromised by the fact that neither the *Villa Cook* nor the *Maison-Atelier Ozenfant* possesses four elevations. Furthermore, two of the elevations of the *Weissenhof-Siedlung Villa 13* are exposed to view, but were originally intended to be blank party walls, leading to some very low results for that house. Despite this, the data does reflect several key properties of Le Corbusier's architecture. For example, the fractal dimension of the roof plans, typically a low result for most architects, are generally higher in his set ($1.1692 < D_{PR} < 1.4030$) than they are for his earlier Arts and Crafts style works. This is most likely due to the roof gardens featured in the latter houses. Furthermore, while the roof dimension for any set of plans is usually the lowest overall value, the roof plan of the *Villa Savoye* (D_{PR} = 1.3699) is higher than the average of its floor plans (μ_P = 1.3620). This result reinforces the visual significance of the roof garden in the *Villa Savoye*.

The most unusual roof plan result is for the *Maison-Atelier Ozenfant* (D_{PR} = 1.4030). The roof space of this studio is treated somewhat differently by Le Corbusier, where he created an artist's space which, 'when viewed from the inside … appears to be an illuminated cube dissolved by northern light. However, when seen from the outside, the profile of skylights accentuated by a projecting cornice defines the top of the building' (Park 2012: 8). It is this particular topography, made up of two sawtooth skylight windows, each made of twenty individually glazed panels, that gives the interior its remarkable quality of light and also adds significant visual complexity to the form of the roof. Conversely, the results for the *Villa Cook* do not support Le Corbusier's suggestion that it is a 'true cubic house' where the complexity of the plans and elevations are reflective of each other. The results for the plans (μ_P = 1.2889) and elevations (μ_E = 1.4412) show only a low level of correspondence (R = 15.23 %) although the presence of only two elevations may be complicating this issue.

Producing consistently high fractal dimension results, The *Villa Stein-de Monzie* (μ_{E+P} = 1.4145) has a complex set of plans and elevations, including the south elevation (D_{E2} = 1.5400) which is the highest of all the houses in the set. The unusual terracing of the roof garden over three levels influences the appearance of the south elevation, a feature that is not seen in any other houses in this group.

The minimal detailing of the east and west elevations of the *Weissenhof-Siedlung Villa 13* is also reinforced by the results (μ_{E+P} = 1.3422). The west wall in particular could be considered a party wall, with the addition of four simple windows, and it has the lowest fractal dimension of all elevations of this set (D_{E4} = 1.2369). The east wall is less visually complex, but actually has a higher fractal dimension than expected, due to the presence of *pilotis* and the *toît jardin*.

Designed as a freestanding, sculptural object, the similarity of all fractal dimensions for the elevations of the *Villa Savoye* ($R_{E\%}$ = 9.55) is not surprising. The lower result for the 'south' elevation (D_{E2} = 1.3961) can be traced to the reduced number of small windows that appear on the ground floor in comparison with all other elevations of this building. If the other three elevations of the *Villa Savoye* were considered as a type of optimal set, then the range would be reduced to 2.22 %. Ultimately, the *Villa Savoye* produces the highest composite result (μ_{E+P} = 1.4166) of any house in this set. While Bovill (1996) may be right to suggest that the level of visual detail present in the elevation, under a close scale of observation, is very low, the opposite is true when the elevations are considered in their totality.

The complete results for Le Corbusier's Modern set also present some quantifiable indicators that his five strategies of design, and in particular the *toît jardin* and *pilotis*, directly contribute to the visual character of his architecture. An additional feature found in all of the five houses that increases their visual complexity, is Le Corbusier's use of individually-framed glazing panels to form each fenestration unit. These frames consistently generate a much finer-grained level of detail than most critics acknowledge in the design.

7.2.5 Comparing the Pre-modern and Modern Houses

When comparing the results derived from Le Corbusier's Pre-Modern and Modern houses some interesting aspects become apparent, helping to explain several conflicting arguments that historians and critics have offered about the work. In particular, apart from the *Villa Favre-Jacot*, which has the highest *D* result ($\mu_{E+P} = 1.4396$) of the ten houses analysed, the results for the early works are typically no more complex than those of the 1920s (Table 7.7). All the houses fall within a similar range, $1.3331 < \mu_{E+P} < 1.4396$, of just over 10 % visual difference. This is of interest because of the number of scholars who have suggested that the Pre-Modern work may prefigure Le Corbusier's later stylistic and compositional strategies, and of those who see the *Villa Favre-Jacot* (with the highest result) as being a bridging work to the later Modern houses (von Moos 1979; Jencks 2000). Both claims are supported by the data.

Comparing the aggregate values for both sets, the result for the Modern houses ($\mu_{\{E+P\}} = 1.3825$) is slightly higher than for the Pre-Moderns ($\mu_{\{E+P\}} = 1.3714$), with a difference of 1.11 %. Looking just at the elevations, the overall value for the Moderns ($\mu_{\{E\}} = 1.4352$) is also slightly higher than the Pre-Moderns ($\mu_{\{E\}} = 1.4322$), with a range between the two averages of 0.03 %. Some of the individual elevations also produce strikingly similar results, including the counter-intuitive pairing of the *Villa Jaquemet* ($D_{E1} = 1.4132$) and the *Villa Cook* ($D_{E2} = 1.4142$). When the plans are compared, the aggregate of the Modern set ($\mu_{\{P\}} = 1.3449$) is actually 3.2 % higher than for the Pre-Modern plans ($\mu_{\{P\}} = 1.3127$). Thus, despite the development of the *plan libre* approach, the level of complexity is very similar. However, there has been a shift from cellular planning (reliant on small rooms) to open planning, but now with voids, ramps and screens of a similar level of geometric complexity within the open plans. On an individual level, some of the plans from the different sets also display similar levels of measured correspondence. For example, the Villas *Jaquemet* and *Cook* are not only similar in elevation, but also in plan, with floor and roof plans offering similar results, particularly the first floor ($D_{P1} = 1.3559$) of the *Villa Cook* and the ground floor ($D_{P0} = 1.3542$) of the *Villa Jaquemet*.

Table 7.7 μ_{E+P} and $\mu_{\{E+P\}}$ values for all ten Le Corbusier houses analysed

Villas (1905–1912)	μ_{E+P}	Villas (1922–1928)	μ_{E+P}
Fallet	1.3654	Ozenfant	1.4025
Jaquemet	1.3361	Cook	1.3366
Stotzer	1.3331	Stein-De Monzie	1.4145
Jeanneret-Perret	1.3825	Weissenhof–Siedlung13	1.3422
Favre-Jacot	1.4396	Savoye	1.4166
$\mu_{\{E+P\}}$	1.3714	$\mu_{\{E+P\}}$	1.3825

Using this new data it is possible to explain the apparently counter-intuitive view that the later Modern works are either more visually complex than, or similar to, the chalet-style works. The critical distinction that is needed to understand this position is one between ornament and form. The first three Pre-Modern houses have relatively simple forms, but are heavily modelled with decorative elements. The next two are somewhat bare of decorative detailing, relying instead on a more elaborate, neo-classical planar modelling to express interior spatial relations, a strategy which is repeated, albeit with a different aesthetic expression, in the Modern houses.

In effect, the decoration and detail that characterises the early works gives way to an increasing reliance on formal modelling in the later ones. This shift of visual impact from decoration to formal modelling provides a viable explanation for why some scholars have observed a degree of consistency in the works, while others see differences. Mathematically, the ten houses have a high degree of consistency in the distribution of their complexity across multiple scales, but the compositional elements that generate this complexity shift between the two sets of houses.

7.3 Eileen Gray

Born in County Wexford, Ireland, in 1879, Eileen Gray studied at the Slade School of Art in London and at the Colarossi and Julien Académies in Paris, completing her formal education in 1905. Gray spent her early career as an artist and furniture designer. In 1915 she opened a lacquerwork gallery with the Japanese artist Suguwara in London; she relocated this gallery to Paris in 1917. By 1922 Gray was dealing mostly in her own designs for furniture and rugs and had developed a reputation as a designer of interiors (Rykwert 1971). It was at this time that Gray's work first began to be favourably received by architects of the Modernist and de Stijl movements and she was encouraged by this to design a number of small buildings. While most of Gray's architectural works remain unrealised, many are sufficiently well documented to demonstrate that, according to Elizabeth Murphy, they display a 'full and original understanding of the language' of the Modern movement (1980: 306). Gray's architectural designs were widely exhibited and published during her lifetime and there has been a renewed interest in her work since the late 1970s.

Having come to architecture from a background in art and object design, the interiors and furnishings in Gray's houses formed an integral part of her work. Gray's architectural forms have been said to suggest an 'overlapping of the architectonic outer skin' of the house, 'with a shell consisting of the individual furniture and fixtures' (Hecker and Müller 1993: 161). Gray's architecture expresses the Modernist ideal of a machine-like building and her planning shows a clear understanding of the differing functions of private and public spaces within a house. However, in a sense, Gray's real design skill lay in her ability to reinforce the importance of humanity in a movement otherwise fixated on function. This focus endowed her buildings with a sense of place and a poetry of purpose that is conspicuously absent from other Modernist works of the era. Murphy argues that Gray 'revolted against the over mechanisation of things to the exclusion of emotion' (1980: 306). Her understanding of the formal and spatial requirements for human inhabitation is reflected in her interior designs and in the flexibility she built into her architecture. Both of Gray's completed works are constructed from concrete, stone, steel and glass. They feature intersecting white concrete planes, glazed bands in fine steel frames and tubular steel balustrades. Her houses are set on local stone bases, a detail which seems to connect each structure inextricably to the site. While her unbuilt projects do not detail the proposed construction method, it is apparent from her drawings and models that she intended most of the building materials to be similar to those in her constructed works (Table 7.8).

7.3.1 Modern Houses (1926–1934)

Gray's first detailed architectural design was most likely the *Small House for an Engineer*, an unbuilt project for a site in the south of France. This elevated, split-level house and office was probably not for an actual client; rather, it is likely an early exploration of Gray's architectural ideas. The design was evidently re-worked by Gray, with two versions of it in her archives. In one version, the ground floor is smaller and more enclosed, while in the other, it covers a much larger area and opens out into the gardens. The latter version, which is more developed and refined, has been selected for the present analysis. Stefan Hecker and Christian Müller observe of this house that, while the ground floor is more traditional in its planning, the upper floor 'reveals itself astonishingly as an interpretation of Le Corbusier's five points for modern architecture' (1993: 42).

Gray's first built work, *E.1027* or *Maison en Bord de Mer*, is located in Roquebrune on the Côte d'Azur, France and was completed in 1929. Sources describe the house as being designed either by Gray for herself, or for Jean Badovici, an architect who encouraged Gray to design houses as well as interiors and furniture. The name *E.1027* is an oblique reference to Gray's and Badovici's joint inhabitation and relationship. It is literally derived from their partially coded and rearranged initials; respectively, E, J(10), B(2) and G(7). *E.1027* is a two-storey house set on a steep hillside beside the Mediterranean Sea. It is Gray's best-known work and, according to Constant and Wang, at that time 'no other architect had produced anything comparable' (1996: np). Featuring elongated, white concrete walls and overhangs, with steel balustrades and strip windows, the house has a maritime character that is appropriate to its seaside location (Fig. 7.6). Importantly, Gray designed everything: the exterior, the interior and the furniture. Each room has easy access to the outside, and because of the many operable windows, doors and shutters, it has been argued that *E.1027* 'may count among the first convincing attempts to adapt Modernist forms to a hot climate' (Hecker and Müller 1993: 59). *E.1027* has been described as demonstrating a desire to resist the common Modernist approach of individually expressing or celebrating every element in a

Table 7.8 Gray set, example elevations and floor plans (level 4 representation—not shown at a uniform scale)

Small House for an Engineer	
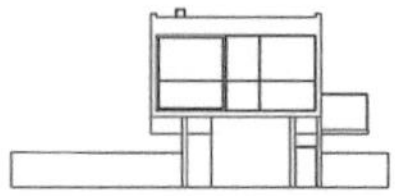	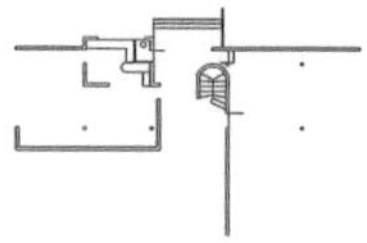
E.1027	
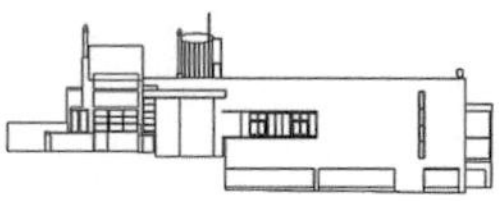	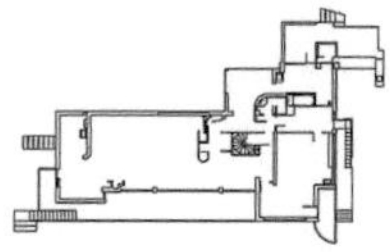
four storey villa	
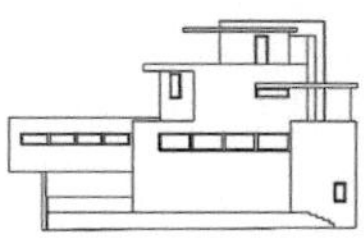	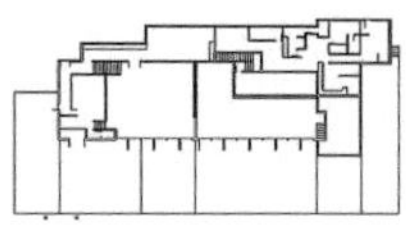
Tempe à Pailla	
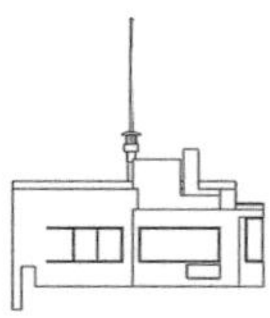	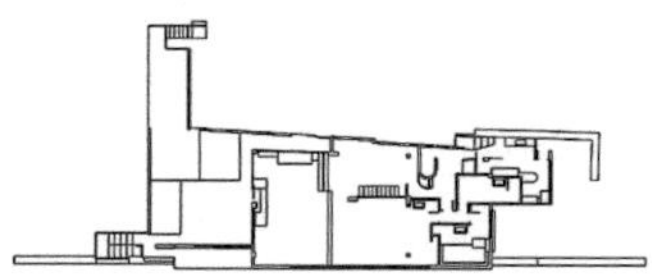
House for Two Sculptors	
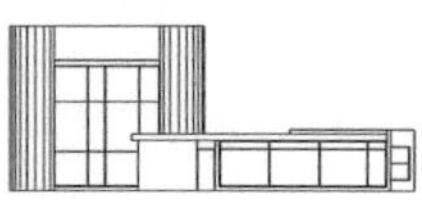	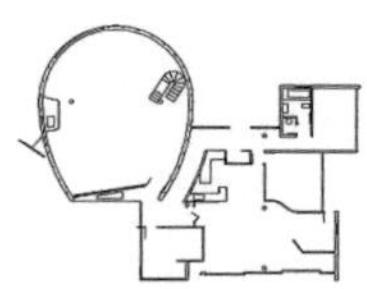

Fig. 7.6 *E.1027*, perspective view

design. Instead, distinctions between architectural space, form and furnishings are blurred, with all three seeming to flow together. Furthermore, 'Gray's fascination with opacity and indecipherability led her to focus on the surface of elements, their colours, textures, and reflective qualities, rather than their profiles, modelling, or placement in a legible space' (Constant and Wang 1996: 107). In this way individual elements are not expressed formally, but instead become part of a larger textual composition.

The Four Storey Villa, an unbuilt project from the mid 1930s, is Gray's largest building design. This bulky home is set in an unspecified rural area and, with its large wall planes, strip windows and maritime feel, has strong similarities to *E.1027*. The plan includes a gallery and bar in a large internal void and, on the roof terrace, a gym, sunbathing terrace, shower and dressing room.

Gray's *Tempe à Pailla* is sited in Castellar, France, on a long, narrow, hillside block with distant views over the sea and mountains. Gray designed it for her own residence in 1934, with rooms for her maid and chauffeur. Constructed five years after *E.1027*, Hecker and Müller state that '[t]he design reveals itself as a continuation of well-tried concepts. The result is more mature, although less spectacular than the first building' (1993: 120). The *Tempe à Pailla* is a two-storey residence with living areas and bedrooms on the main floor (constructed from concrete elements) and cellars, garage and chauffeur's quarters below in an exposed stone base. Garner records that like *E.1027*, the *Tempe à Pailla* possesses a 'clean, uncluttered linearity, with flat overhanging roofs, simple column props and tubular balustrades, terraces and walkways, long louvred windows and large picture windows' (1993: 33).

The *House for Two Sculptors* is an unbuilt project from 1934 that may have been designed for the brothers Jan and Joël Martel. The design features a dynamic, curved, two-storey studio intersecting with a single-level, rectilinear residence. The program of the two volumes seems to explore the relationship between public and private spaces. There are two versions of this project and the present analysis is focussed on the second version, in which the studio is more curvilinear. The form of this design has been described as an 'oval shape' that 'is cut up into crescent-shaped parts, which are then displaced vertically, one against the other' (Hecker and Müller 1993: 162). Unlike Gray's other unbuilt projects, where a complex relationship to the ground has been signalled in the drawings and models, the *House for Two Sculptors* is designed on a nondescript, flat site.

7.3.2 Gray, Results and Analysis

The lowest elevation average is found in the *Small House for an Engineer* (μ_E = 1.2697) and the highest in *E.1027* (μ_E = 1.4223), leading to a range of $R_{\{E\%\}}$ = 22.53. For the entire set of elevations the average is $\mu_{\{E\}}$ = 1.3589, the median is 1.3711 and the standard deviation is 0.0655 (that is, slightly more clustered that the data for Le Corbusier's elevations). Results for the plans show that the lowest average is also for the *House for Two Sculptors* (μ_P = 1.2699) and the highest average is for *E.1027* (μ_P = 1.3427) with the range for the plans being $R_{\{P\%\}}$ = 22.10. For the entire set of plans the average is $\mu_{\{E\}}$ = 1.3079, the median is 1.3337 and the standard deviation is 0.0681. When all of the plans and elevation results are considered together, the aggregate is $\mu_{\{E+P\}}$ = 1.3353. While the separate ranges for both elevations and plans show a fairly wide gap, the composite range is much closer with $R_{\{\mu E+P\%\}}$ = 6.51 (Tables 7.9 and 7.10, Fig. 7.7).

The optimal sub-set comprises the later three houses, the *Four Storey Villa*, *House for Two Sculptors* and *Tempe à Pailla*. The aggregate result for the optimal sub-set is $\mu_{[E+P]}$ = 1.2801, a figure which is slightly higher than the overall value. However the composite range result is dramatically lowered to $R_{[\mu E+P\%]}$ = 2.28. Thus, when each house is taken as a whole object, the later works are very similar in their level of visual complexity.

The results for the *Small House for an Engineer* are unusual, as the fractal dimensions for the elevations (1.2445 < D_E < 1.3260) are equal to, or even marginally lower than the dimensions of the plans (1.2642 < D_P < 1.3584). This means that all the elevations and plans for this building share a similar level of visual complexity, whereas most houses from this era show a higher complexity for elevations. Perhaps, as this was one of Gray's first building designs after a career creating objects and furniture, she may have devoted equal attention to the interior planning and the exterior form.

E.1027, as the most visually complex house in the set, has higher results, in both individual plans and elevations, than most other results for the five buildings. Furthermore, the results for the sets of plans and elevations for *E.1027* are in a very

Table 7.9 Gray set, results

Houses		Small house	E-1027	Four storey villa*	House for two sculptors*	Tempe à Pailla*	Opt. [*]	Set {…}
Elevations	D_{E1}	1.3260	1.3827	1.3069	1.3990	1.3734		
	D_{E2}	1.2582	1.4698	1.4292	1.4426	1.3045		
	D_{E3}	1.2502	1.4261	1.2935	1.3781	1.3687		
	D_{E4}	1.2445	1.4104	1.3536	1.3912	1.3687		
	μ_E	1.2697	1.4223	1.3458	1.4027	1.3538		
	$\mu_{[E]}/\mu_{\{E\}}$						1.3675	1.3589
	$M_{[E]}/M_{\{E\}}$						1.3711	1.3711
	$std_{[E]}/std_{\{E\}}$						0.0472	0.0655
Plans	D_{P-1}	–	–	–	–	1.3291		
	D_{P0}	1.2703	1.3337	1.3566	1.3605	1.3665		
	D_{P1}	1.3584	1.3713	1.3699	1.2592	–		
	D_{P2}	–	–	1.3644	–	–		
	D_{P3}	–	–	1.3339	–	–		
	D_{PR}	1.2642	1.3230	1.2333	1.1899	1.1503		
	μ_P	1.2976	1.3427	1.3316	1.2699	1.2820		
	$\mu_{[P]}/\mu_{\{P\}}$						1.1928	1.3079
	$M_{[P]}/M_{\{P\}}$						1.3315	1.3337
	$std_{[P]}/std_{\{P\}}$						0.3832	0.0681
Composite	μ_{E+P}	1.2817	1.3881	1.3379	1.3458	1.3230		
Aggregate	$\mu_{[E+P]}/\mu_{\{E+P\}}$						1.2801	1.3353

Table 7.10 Gray set, comparative values

Houses		Small house	E-1027	Four storey villa*	House for two sculptors*	Tempe à Pailla*	Opt. [*]	Set {…}
Elevations	$R_{\mathrm{E}D}$	0.0815	0.0871	0.1357	0.0645	0.0689		
	$R_{\mathrm{E}\%}$	8.1500	8.7100	13.5700	6.4500	6.8900		
	$R_{[\mathrm{E}D]}/R_{\{\mathrm{E}D\}}$						0.1491	0.2253
	$R_{[\mathrm{E}\%]}/R_{\{\mathrm{E}\%\}}$						14.9100	22.5300
Plans	$R_{\mathrm{P}D}$	0.0942	0.0483	0.1366	0.1706	0.2162		
	$R_{\mathrm{P}\%}$	9.4200	4.8300	13.6600	17.0600	21.6200		
	$R_{[\mathrm{P}D]}/R_{\{\mathrm{P}D\}}$						0.2196	0.2210
	$R_{[\mathrm{P}\%]}/R_{\{\mathrm{P}\%\}}$						21.9600	22.1000
Composite	$R_{[\mu\mathrm{E+P}D]}/R_{\{\mu\mathrm{E+P}D\}}$						0.0228	0.0651
	$R_{[\mu\mathrm{E+P}\%]}/R_{\{\mu\mathrm{E+P}\%\}}$						2.2800	6.5100

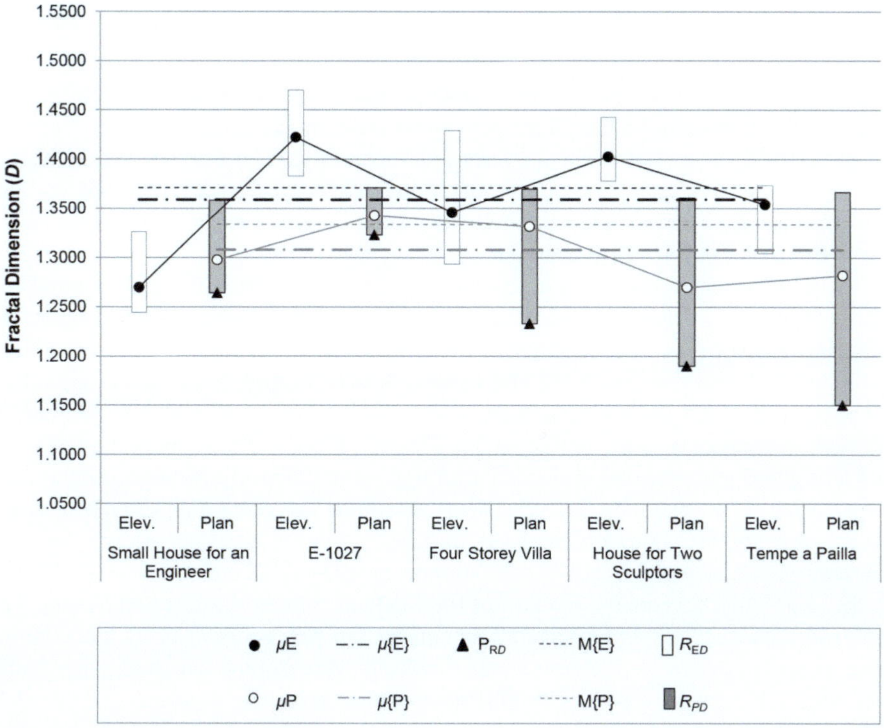

Fig. 7.7 Gray set, graphed results

tight range, which may reflect the amount of time Gray invested into designing every facet of this house. The high level of complexity in the roof plan (D_{PR} = 1.3427) is an outcome of both her use of terracing and her fascination with articulating three-dimensional form.

The complete set of results for Gray's architecture does not display a strong consistency, something which is possibly due to the inclusion of unbuilt works, or to the fact that she was only gradually developing her architectural skills at the time. Like Le Corbusier's Pre-Modern houses, Gray's suggest not so much an architect with a fixed style, but one whose style was evolving. There is, however, a clear trend in the graph which shows that the uppermost result for the plans of each house are all in a close range (1.3584 < D_P < 1.3713). Perhaps this confirms the view that Gray's sense of interior planning and spatial organisation was already more advanced and stable when she began these works, while the exterior forms continued to be refined with each project.

7.4 Mies van der Rohe

Ludwig Mies was born in Germany in 1886 and learnt the basics of architecture while working with his father, a master stonemason, in Aachen. Later extending his surname to Mies van der Rohe (a reference to his mother's name), he attended trade school in Aachen and become a meticulous draftsman, a skill that led to his employment by several talented architects and designers including Bruno Paul, Peter Behrens and Hendrick Berlage.

In 1913 Mies established an independent practice in Berlin, producing a series of traditionally-styled houses; in the years that followed he became a central figure in the *Novembergruppe* and was involved in the design journal *G*, both of which promoted Modern theories of art and culture. It was through these two groups that he began to be well known for a number of theoretical projects, which he promoted through exhibitions and publications (Blaser 1965). These unbuilt designs, including two glass-walled high-rise buildings, were stridently Modern in their use of materials, open planning and crystalline detailing. The aesthetic promise of these works was to be eventually realised in Mies's ground-breaking design for the German pavilion at the Barcelona Exhibition of 1929 (Blake 1966).

In 1930 Mies became the director of the Bauhaus, but the Fascist and Nationalist political agenda of the era rejected Modernism—in part because of its association with Constructivism and communism—making Mies's position at the Bauhaus untenable (Hochman 1990). Despite Mies's appeals to the Gestapo to save the school, it was forced to close and Mies immigrated to the USA, where he became head of the Armour Institute of Technology's architecture school in Chicago. It was while he held this position that he produced many of the buildings that he is now famous for. One of the first of these was a weekend house for Edith Farnsworth. Colomina (2009) suggests that Mies's understanding of the *Farnsworth House* as an idealised pavilion was to shape much of his commercial architecture as well. Mies went on to produce several famous university buildings including *Crown Hall* at the Illinois Institute of Technology, along with the *Seagram Building* in New York (Speyer 1968). The structural expression of all of these works echo the simple pavilion house in its desire to express a higher unity of space, where internal and external areas are perceived as part of a greater whole (Tegethoff 1985). It has been considered especially significant that Mies 'saw the architect, foremost of all, as an apolitical artist concerned with beauty and Platonic universals' (Kostof 1985: 701).

This brief summary of Mies, his life and legacy, is typical of that found in many texts; however, very few mention that at the start of his career he produced an important series of houses. These designs, typically orthogonal and flat-roofed forms with masonry walls, have often been left out of the Miesian oeuvre because they do not possess the structural purity and transparency of the *Farnsworth House* or the *Barcelona Pavilion*. However, these houses are critical early attempts to produce Modern architecture in a country that had a growing industrial economy but was still mired in a social system derived from the previous century. Thus, although these houses do not have the structural clarity or neo-Platonic purity of his later works, they do show Mies's development as a designer, and his early attempts to use geometry to create a contemporary expression of space and form.

7.4.1 Modern Houses (1930–1951)

The *Wolf House* is the first of a series of Modernist brick houses that demonstrate Mies's early commitment to creating an open-planned design with a glazed building envelope, but using more traditional materials and respecting the constraints of the European family structure of the era. This house was designed in conjunction with Lilly Reich for Erich Wolf, an executive in the textile industry. Occupying the top of a hill on a narrow sloping site in Gubin, Poland, this three-storey, flat-roofed structure appears as a series of terraces and rectangular forms, their physical arrangement expressed 'in a stepped play of brick volumes, jutting planes and protruding chimneys' (Puente and Puyuelo 2009: 70). Unlike the machine-like, white-rendered finish found in the Modernist houses of Gray and Le Corbusier, the Wolf house is constructed of a fastidiously laid, Flemish-bond, unpainted brickwork finished with a flush vertical coping that gives the wall a strong planar appearance.

The *Lange* and *Esters* houses are two separate residences on adjacent properties in the German city of Krefeld, where both were completed in 1930 (Figs. 7.8 and 7.9). The design and construction processes for these two houses occurred in parallel and using the same palette of materials. They are similar in appearance to the *Wolf House* in their use of neat dark brickwork, however these two houses have copper copings atop their planar walls. The significant advance found in these houses is the use of a steel structural system. They are said to be 'among the first modern buildings to free brick from its load-bearing function' (Zimmerman 2006: 33). The steel structure allowed Mies to use larger window and door openings, and

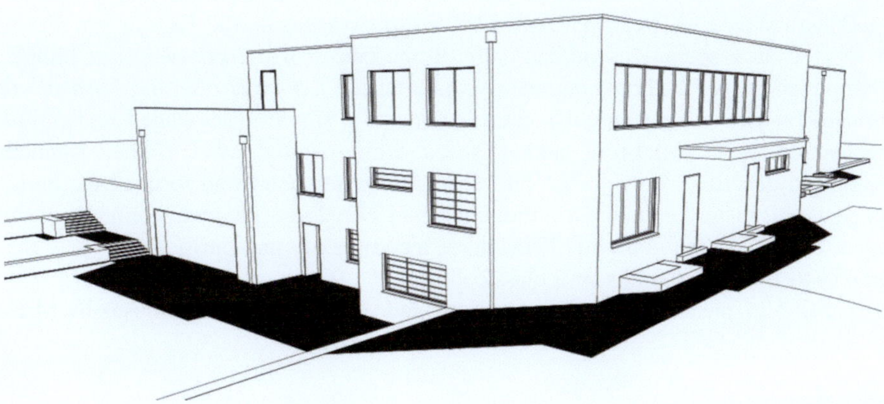

Fig. 7.8 *Lange House*, perspective view

Fig. 7.9 *Esters House*, perspective view

the steel framing for these units lends the two houses a distinctly sleek and functional appearance. However, despite their innovative exteriors, the interior planning of the houses is much more conventional, relying on a type of hierarchically divided, cellular planning which is more reminiscent of stately homes from the previous century (Ostwald and Dawes 2013a). It has been argued that these two houses were 'first repressed by the architect himself and subsequently suppressed by his apologists' (Kleinman and Van Duzer 2005: 12), and thus they have, despite their innovations and aspirations, received relatively little scholarly attention.

The *Lemke House*, from 1933, was designed as both a home and gallery and is sited on the shores of the lake Obersee in Berlin. More modest in its scope and scale than the previous three works, it has been described as a 'footnote' (Schulze and Windhorst 2012: 160) in Mies's early career, as it is effectively the last design in a series prior to his more famous work in the USA. The *Lemke House* is a single-storey, flat-roofed residence with an L-shaped plan. However, unlike the previous works, here the brickwork is of a paler hue, the masonry has an English-bond finish and the walls are capped in stone. The windows are also larger than those in the previous houses, with some walls almost entirely glazed, further reinforcing the notion that it is an important precursor to the *Farnsworth House*.

Designed as a weekend retreat for Dr. Edith Farnsworth, the *Farnsworth House* is located on a secluded woodland site on the banks of the Fox River in Illinois. Completed in 1951, almost eighteen years after the *Lemke House*, the *Farnsworth House* consists of a rectangular patio leading to a single glass-walled rectangular enclosure, with a flat-roof and exposed white-painted steel frame. Kenneth Frampton describes the rigorous and unforgiving geometry and form of the house as elevating it 'to the status of a monument' (1992: 235). The twin elements of the open terrace and the glass box house have a complexity and purity about them that is akin to the formal traits of a classical temple.

Table 7.11 presents example elevations and plans of the set of houses by Mies.

Table 7.11 Mies set, example elevations and floor plans (level 4 representation—not shown at a uniform scale)

Wolf House

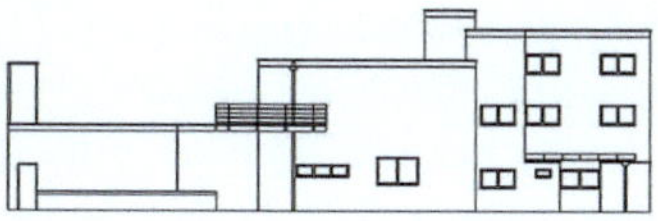
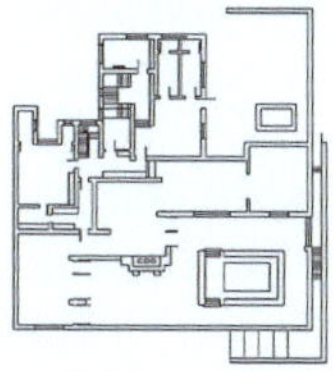

Lange House

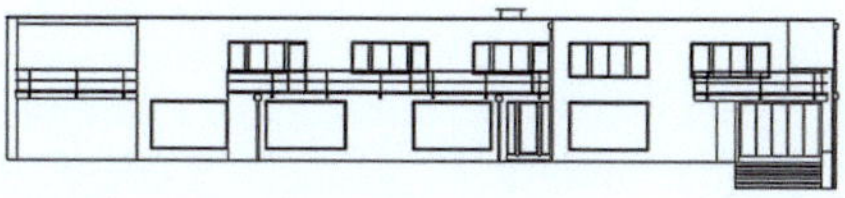
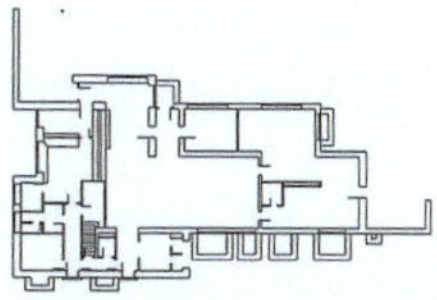

Esters House

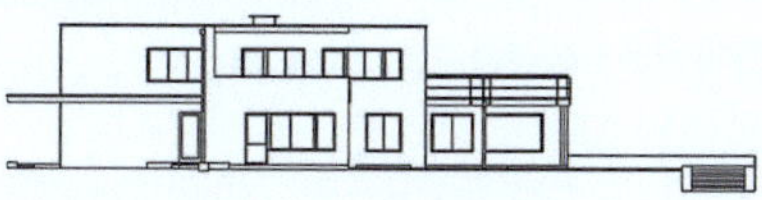
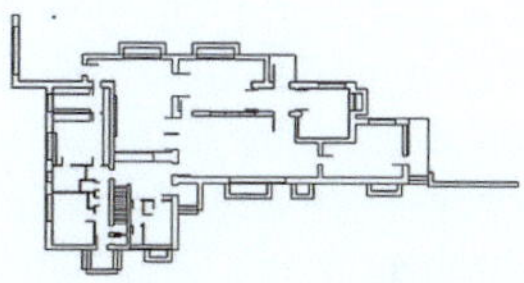

Lemke House

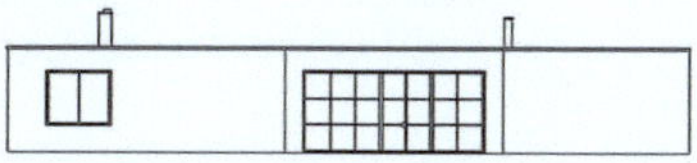
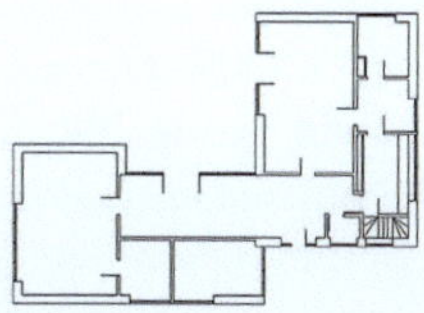

Farnsworth House

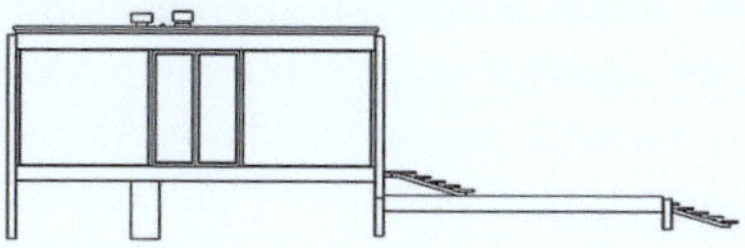
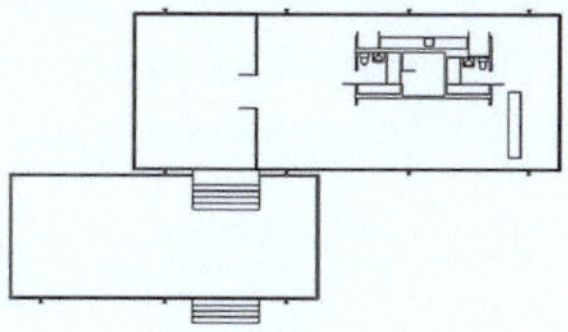

7.4.2 Mies van der Rohe, Results and Analysis

Results from the fractal dimension calculations of all five houses in the Mies set show the lowest elevation average is for the *Lemke House* (μ_E = 1.3471) and the highest for the *Lange House* (μ_E = 1.4533), while the *Esters House*, sometimes considered to be the Lange House's 'twin sister', has the highest individual elevation (D_{E2} = 1.5099). The range for elevations, averaged by house, across the set is $R_{\{E\%\}}$ = 20.70, the median for these elevations is 1.4129 and the standard deviation is 0.0569 (the most tightly clustered about the average of the three Modernists in this chapter). Not only does the *Lange House* have the highest elevation average, it also has the highest plan average (μ_P = 1.3542), and the *Lemke House* also has lowest plan average (μ_P = 1.2463). The range of all the plans is $R_{\{P\%\}}$ = 25.01, the median for all plans is 1.3290 and the standard deviation is 0.0875 (the least tightly clustered about the average of the three architects plans in this chapter). When all of the plan and elevation results are considered together, the aggregate is $\mu_{\{E+P\}}$ = 1.3659 and the composite range is $R_{\{\mu E+P\%\}}$ = 9.03, a result which places the five houses within a similar scale of visual complexity (Tables 7.12 and 7.13, Fig. 7.10).

The *Wolf*, *Lange* and *Esters* houses are the optimal sub-set. Considering only these three works, the aggregate result for all plans and elevations is reduced to $\mu_{[E+P]}$ = 1.3840 and the composite range is reduced by more than half of the complete range to $R_{[\mu E+P\%]}$ = 4.52 reinforcing the notion that they are very similar in character. Indeed, these three earlier houses form a clear group, all being large-scale, three storeys high and the most detailed of the five analysed. This visual complexity is at its highest in the *Lange* and *Esters* houses, the results also confirming that these are indeed 'twin' houses, as they are sometimes described. They both produce several top results: with the highest average elevations ($\mu_{E(Lange)}$ = 1.4533, $\mu_{E(Esters)}$ = 1.4456), plans ($\mu_{P(Lange)}$ = 1.3542, $\mu_{P(Esters)}$ = 1.3337) and highest south elevations ($D_{E2(Lange)}$ = 1.5007, $D_{E2(Esters)}$ = 1.5099).

As a second group, the later *Lemke* and *Farnsworth Houses* are both simple, small-scale, one-storey houses, although their materiality and planning are different. Their fractal dimension results are generally lower than the earlier houses; the plans in particular are all less visually complex than the median of the entire set. That the *Lemke* has the lowest average elevation and plan dimensions is unsurprising, although the fractal dimensions for the *Farnsworth House* are slightly higher than anticipated. This design of 'abstract simplicity' (Zimmerman 2006: 63) actually has a very clearly articulated structure that the method measures, along with a stair detail in which every element, however minimal, is expressed in the design.

Table 7.12 Mies set, results

Houses		Wolf *	Lange*	Esters *	Lemke	Farnsworth	Opt. [*]	Set {…}
Elevations	D_{E1}	1.4450	1.4663	1.4056	1.3361	1.4371		
	D_{E2}	1.3787	1.5007	1.5099	1.3665	1.4429		
	D_{E3}	1.4452	1.3799	1.4202	1.3267	1.3904		
	D_{E4}	1.3029	1.4663	1.4465	1.3590	1.3868		
	μ_E	1.3930	1.4533	1.4456	1.3471	1.4143		
	$\mu_{[E]}/\mu_{\{E\}}$						1.4306	1.4106
	$M_{[E]}/M_{\{E\}}$						1.4451	1.4129
	$std_{[E]}/std_{\{E\}}$						0.0578	0.0569
Plans	D_{P-1}	–	1.3854	1.3398	–	–		
	D_{P0}	1.4232	1.3899	1.3869	1.3122	1.3182		
	D_{P1}	1.4018	1.3859	1.4060	–	–		
	D_{P2}	1.2984	–	–	–	–		
	D_{PR}	1.1731	1.2554	1.2019	1.1803	1.2096		
	μ_P	1.3241	1.3542	1.3337	1.2463	1.2639		
	$\mu_{[P]}/\mu_{\{P\}}$						1.3373	1.3168
	$M_{[P]}/M_{\{P\}}$						1.3857	1.3290
	$std_{[P]}/std_{\{P\}}$						0.0851	0.0875
Composite	μ_{E+P}	1.3585	1.4037	1.3896	1.3003	1.3443		
Aggregate	$\mu_{[E+P]}/\mu_{\{E+P\}}$						1.3840	1.3659

Table 7.13 Mies set, comparative results

Houses		Wolf *	Lange*	Esters*	Lemke	Farnsworth	Opt. [*]	Set {…}
Elevations	R_{ED}	0.1423	0.1208	0.1043	0.0398	0.0561		
	$R_{E\%}$	14.2300	12.0800	10.4300	3.9800	5.6100		
	$R_{[ED]}/R_{\{ED\}}$						0.2070	0.2070
	$R_{[E\%]}/R_{\{E\%\}}$						20.7000	20.7000
Plans	R_{PD}	0.2501	0.1345	0.2041	0.1319	0.1086		
	$R_{P\%}$	25.0100	13.4500	20.4100	13.1900	10.8600		
	$R_{[PD]}/R_{\{PD\}}$						0.2501	0.2501
	$R_{[P\%]}/R_{\{P\%\}}$						25.0100	25.0100
Composite	$R_{[\mu E+PD]}/R_{\{\mu E+PD\}}$						0.0452	0.0903
	$R_{[\mu E+P\%]}/R_{\{\mu E+P\%\}}$						4.5200	9.0300

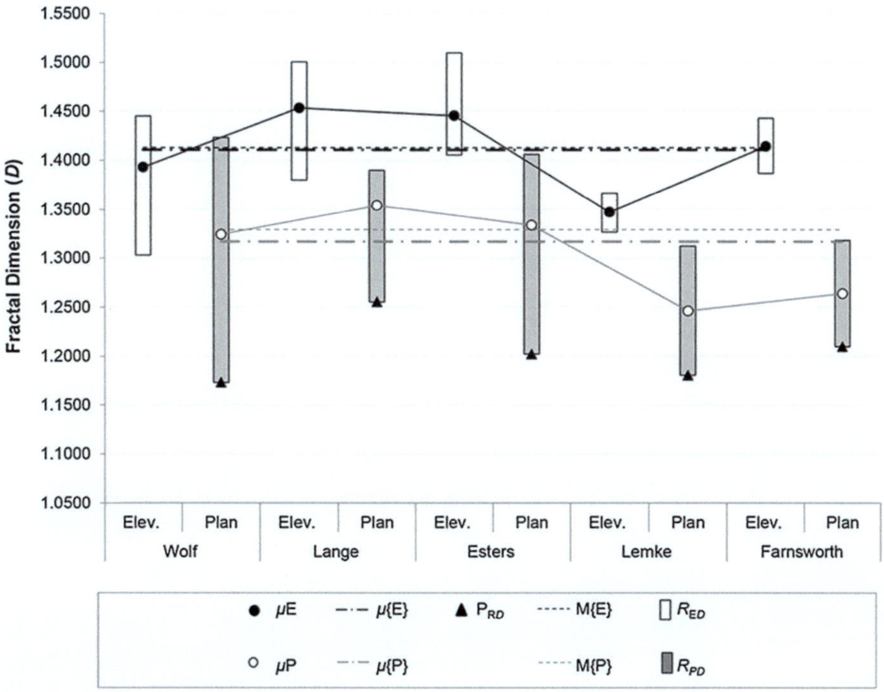

Fig. 7.10 Mies set, graphed results

7.5 Comparison of the Three Modernists

Comparing the aggregate results of the three Modernist architects, it can be seen that of the three, Le Corbusier's buildings are generally the most complex ($\mu_{\{E+P\}}$ = 1.3825), Gray's are the least ($\mu_{\{E+P\}}$ = 1.3353) and Mies van der Rohe's are midway between the two ($\mu_{\{E+P\}}$ = 1.3659). The formal coherence graphs for the three sets of houses show that the correlation between the formal properties of the plans and elevations is strongest in the work of Mies (R^2 = 0.6179) and weakest in Le Corbusier (R^2 = 0.1028) (Figs. 7.11, 7.12, 7.13 and 7.14). This means that Le Corbusier's elevations are heavily modelled, expressing functional properties of the interior including responding to climate and address, while his interior plans are relatively open and less intricately defined. In contrast, Mies's plans and elevations are more similar in their modelling. The orthogonal modulated exteriors, with no ornament or visible roofline, are visually reminiscent of the plans of these same houses, which are also orthogonal compositions, lacking finer detail. Conversely, Gray's architecture displays a strong inverse correlation, meaning that her plans are consistently more detailed and formally rich than her elevations.

While the size of the agglomerated data set is not sufficient to produce a reliable analysis of trends, the linear indicators offer a valuable way of visualising how these

Fig. 7.11 Le Corbusier, pre-Modern set, formal coherence graph

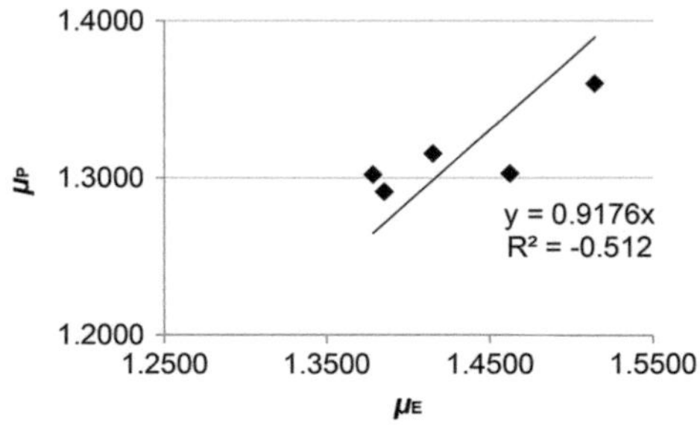

Fig. 7.12 Le Corbusier, Modern set, formal coherence graph

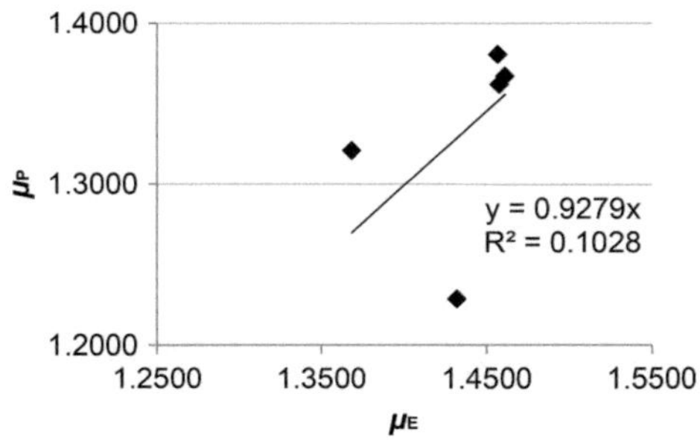

Fig. 7.13 Gray set, formal coherence graph

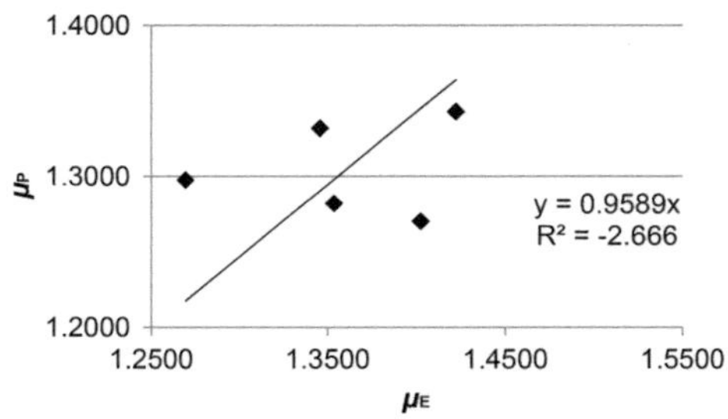

Fig. 7.14 Mies set, formal coherence graph

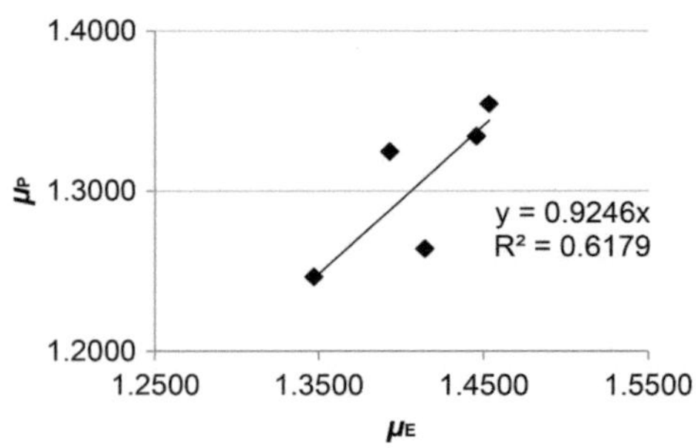

architects' works evolved over the course of their five projects. For example, just considering elevations, visual complexity is relatively constant across the works of Le Corbusier, it falls slightly across the set by Mies, and it rises more noticeably over time in the work of Gray (Fig. 7.15). Le Corbusier's five strategies for a Modern architecture were already, despite later refinements, well developed before these five houses were produced, perhaps accounting for their stability, while Mies was pursuing a type of distillation of form across his set, which is reflected in his results. For the plans of the sets of houses, both Gray and Mies show a definite simplification in planning over time, while Le Corbusier's is the only one that grows, culminating in the richly modelled roof terraces and vertical circulation planning of the *Villa Savoye* (Fig. 7.16). Finally, when elevations and plans are combined, Mies continues to show a clear trend towards increasing simplification, minimalism or purity, whereas the designs of Le Corbusier and Eileen Gray each increase throughout the period (Fig. 7.17).

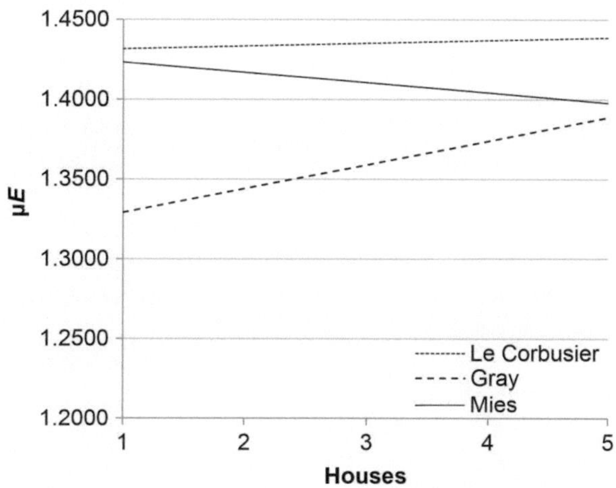

Fig. 7.15 Le Corbusier, Gray and Mies sets, linear trendline data for elevations (μ_E)

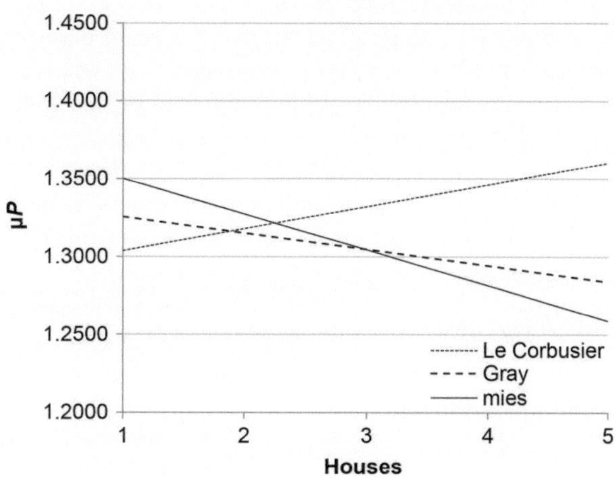

Fig. 7.16 Le Corbusier, Gray and Mies sets, linear trendline data for plans (μ_P)

Fig. 7.17 Le Corbusier, Gray and Mies sets, linear trendline data for composite results (μ_{E+P})

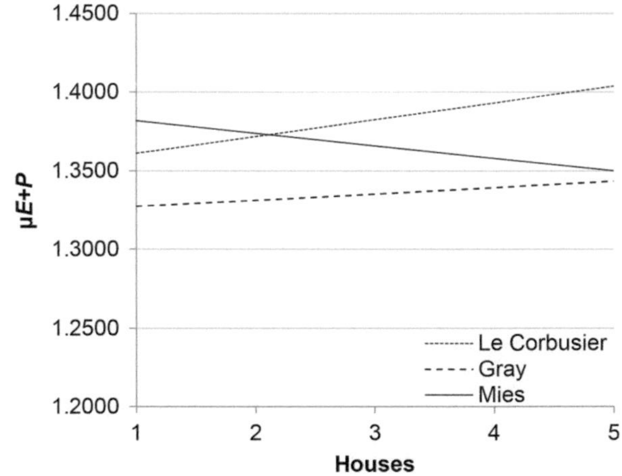

7.6 Testing 'Form Follows Function'

A common assumption in architectural design is that the functional expression of a building façade is typically shaped by a combination of environmental and programmatic conditions (Feininger 1956; Grillo 1975; Jones 1992; Box 2007; Frederick 2007). These characteristics are not necessarily all present in all architecture, but they are apparent in many buildings (Kruft 1994; Ching 2007). It has been proposed that this is particularly true in the case of domestic architecture where a façade is formed 'in response to the resources, climate and topography of a particular region' (Moore et al. 1974: 71). In this view, the form of each house expresses something about its capacity to accommodate human activities and attitudes within a distinct context and region. Thus, the façade of a house can be understood as a reflection of two primary factors: orientation (its siting) and address (its approach). The first of these two factors, orientation, is related to the impact of the environment on a design (Leatherbarrow 2000). For instance, a house may be shaped to respond to the movement of the sun, either to restrict its consequences or to capture its energy. The second factor is concerned with the way in which a building addresses a visitor, or the manner in which a design differentiates its public and private façades (Venturi 1966; King 2005). This is because the design of a façade generally shifts to acknowledge points of entry and signal rights of access.

Despite the repeated assertions that these two assumptions about façade expression are normative in architecture, there is no quantitative data to support the efficacy of this position or even to test whether or not it is true in the case of particular architects who are renowned for their functionalist values. In this section we propose a unique application of fractal analysis to test whether or not these two functional factors are broadly expressed in architectural form. This application of the method augments some of the data generated earlier in this chapter by adding additional information pertaining to the orientation and address of the façades. The focus of this alternative variation of the method is the elevations (not the plans or roof plans) of ten Modern houses by Le Corbusier and Eileen Gray. In total, thirty-seven elevations from these houses are each augmented in two different ways and then analysed to see if there is any pattern in the way the elevations have been designed with respect to site and address. If there is a pattern, then this set of data supports the general hypothesis that *the form of a functional façade is shaped by a combination of its orientation and address.*

7.6.1 Orientation and Approach

To augment the fractal dimension data to take into account orientation, each elevation is coded in accordance with its position relative to the cardinal points of the compass (North, South, East and West). Not only is the differentiation of elevations using this nomenclature common practice in architecture, but a determination of orientation, by way of magnetic bearings, is a universal system that can potentially be used for comparisons between most buildings. There are, however, several practical considerations in determining the orientation of an elevation. First, only four categories of orientation are used in the present work. While it might be possible to subdivide orientation by angle (within a 360° array) very few architectural drawings record this information and it is not available for most of the projects being studied. This also means that when an elevation is not clearly oriented towards a cardinal point (for example, it is facing 20° west of north), it is placed into the closest possible category (in this example, north). This procedure works well for all of the houses studied here except the *Villa Savoye*, which is set at almost exactly 45° to north, meaning that the façade conventionally labelled 'north elevation' could also arguably be labelled 'east elevation'. But because this elevation is always described in the literature as the 'north' we have repeated this classification for consistency.

This data-augmentation approach is also most appropriate for dwellings that are both orthogonal and freestanding, because it assumes that a house may be described using a set of four elevations. If a house needs fewer than four elevations to describe it (say it has a triangular plan) this classification method will be less useful.

While it might be imagined that houses designed for uniformly flat, rural sites would be strongly shaped by their orientation, for the majority of houses the strategic siting options are much more limited and the impact of orientation tends to be ameliorated by the importance of addressing a street and providing acoustic and visual privacy for its inhabitants. This is because the majority of sites and designs have a single obvious public face or 'address' and a single private face. This means that the primary factor shaping the design of a typical urban or suburban façade is more likely to be related to the presentation of the house to the street (and the associated impact of positioning internal spaces appropriately with respect to that street) than to the passage of the sun. This implies that, perhaps more so than orientation, patterns should be discernible in the way in which dwellings orient their public and private façades, especially in the case of designs for dense urban environments. Therefore, for the second augmentation method, the fractal dimension data for each elevation is coded to reflect its provision of access to the building. This type of physical accessibility is typically understood as being different for non-inhabitants—visitors or the public—and for inhabitants (Hillier and Hanson 1984), with access for non-inhabitants in the 'public' or 'front' façade of the building, while access for inhabitants is provided through its 'private' or 'back' façade. However, while the method for coding fractal dimensions using orientation provided a universal system—magnetic bearings—the second approach is concerned with local and more intuitive or relative determinations. For this reason, here the front is defined as the public or street address of a building, which most often also contains its formal entry. Once the front is defined, most of the remainder of the elevations are described in relation to the front. Thus, the elevation that is facing in the opposite direction becomes the back elevation. The remaining two elevations, in a predominantly orthogonal or rectilinear plan, are the sides. These also tend to be distinguished from each other by their relationship to the front. In particular, they are typically called the left or right side of the house, a relative determination made with respect to a viewing point perpendicular to the front elevation.

There are several issues to consider in identifying the front elevation. First, as just noted, the front is typically the location of the formal entry. Not all people will necessarily use this entry in a large house and in a more modern house the garage might partially replace this entry for everyday use. However, the majority of houses still have a formal entry for visitors and it is frequently signalled in some way by the positioning of a porch or by the siting of windows or paths. Moreover, the formal entry is normally, but again not always, sited in relation to the primary elevation that addresses the street. There are some exceptions, including corner sites, but for the majority of houses the designation is relatively clear.

7.6.2 Method and Hypothesised Results

Two sets of five houses analysed earlier in this chapter—the Modern set of Le Corbusier and the set of Eileen Gray—are analysed in this section using the following process.

i. The *orientation* of each elevation is recorded, either from the original drawings or other means (including the use of Google Earth and photographic observations).
ii. The *approach* to each of the ten houses is identified from a review of plans, elevations, photographs and descriptions. Once this front elevation is identified, the remainder of the elevations are classified relative to it (back, left and right).
iii. The D_E result for each individual elevation is coded into one of four orientation categories (N, S, E, W) and one of four approach categories (F, B, L, R).
iv. The coded data for each set of five houses is tabulated and charted to seek patterns in the relationship between elevation complexity, orientation and approach.

Before looking at the results of this method, it is useful to consider what pattern in the data might be anticipated depending on whether the visual expression of an elevation is shaped by orientation or by address. For example, if an architect applied a consistent set of design strategies to similar scale projects in similar geographic regions over a relatively short timeframe, it might be anticipated that a pattern could be uncovered in the work. Furthermore, if all five houses by the same architect are on rural or 'green-field' sites (without nearby neighbours) and within a similar geographic region (say southern France), then it might be anticipated that there would be some consistency between the complexity of a house's façade and its orientation. In such an example, the southern elevation would typically feature more windows and balconies to capture warmth and light, while the western elevation would be relatively unadorned to shelter it from the afternoon sun and winter winds. The northern and eastern elevations would be between these two extremes. When these hypothesised conditions are charted, the results would show a marginal rise in visual complexity from north to east and then a sharp rise to the southern elevation before a uniform fall to the lowest set of results, the west elevation (Fig. 7.18).

For houses on more urban sites it is unlikely that the orientation will produce such a pattern. Instead, all other things being equal, a pattern should be evident in the approach chart. For example, for buildings that face a busy urban street and have side elevations facing neighbouring houses (typically in close proximity) and a single rear elevation (to a private courtyard or garden) the following might be an expected pattern. The front elevation has a middle level of relative visual complexity, reflecting the desire for natural light from the street, and the positioning of some formal areas (foyer, dining or home office) toward the busier side of the property. The left and right side elevations would have little formal modelling, as they would have neither outlook nor need for shelter. The back or rear elevation would have the highest level of visual complexity as it would contain the private

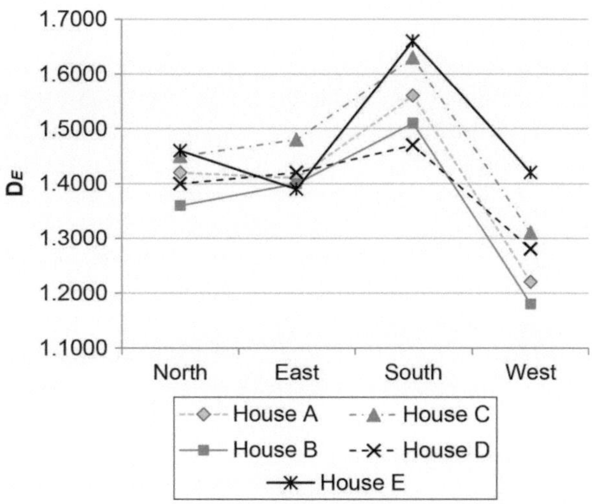

Fig. 7.18 Results for a hypothetical set of idealised houses on rural sites in southern France

Fig. 7.19 Example, hypothetical set of idealised houses on urban or suburban sites

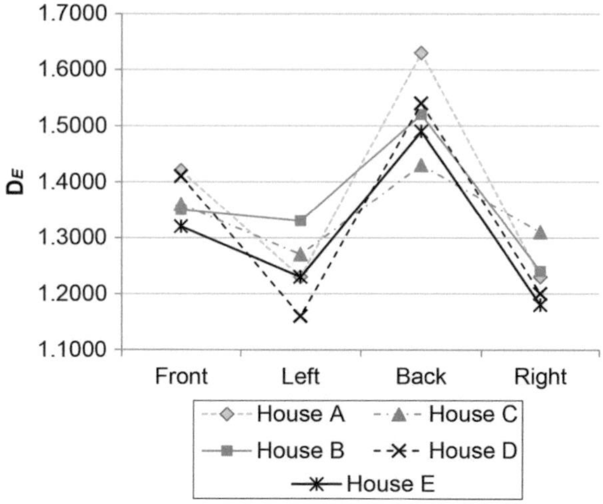

spaces (bedrooms, living rooms) that require natural light and ventilation, along with any balcony spaces and more extensive connections to the landscape or yard (Fig. 7.19).

7.6.3 Results and Discussion

The tabulated and charted orientation data for Le Corbusier's Modern houses feature several interesting results or trends (Table 7.14, Fig. 7.20). First, for his freestanding urban houses (the *Villa Stein-De Monzie* and the *Weissenhof-Seidlung Villa 13*) the south elevation is the most complex and the west is the least complex. This result mirrors the hypothesised outcome outlined in the previous section. In contrast, the *Villa Cook* and the *Maison-Atelier Ozenfant* have party walls and unusual siting which may explain their lack of consistency, while the *Villa Savoye* is a completely freestanding house with relatively little differentiation across its façade. Three houses—*Cook*, *Ozenfant* and *Savoye*—display a secondary trend, with all southern elevations having a similar level of visual complexity. For the second set of results, divided by address or approach, the data for Le Corbusier's houses does not display a clear pattern, although, apart from the *Villa Stein-de Monzie*, the back elevations are less visually complex than the front elevations, which is not as hypothesised in the previous section (Fig. 7.21).

The tabulated and charted orientation data for Gray's houses suggests a partial pattern for three of the houses (*E.1027*, *House for Two Sculptors*, *Four Storey Villa*) wherein the south is more visually complex than the north (Table 7.15, Fig. 7.22). Furthermore, with one exception, the south elevation is more complex than both the east and west. The exception is the *Tempe à Pailla*, where the west has an unusually high level of visual complexity. Thus, Gray's designs follow a pattern similar to the

Table 7.14 Augmented results for Le Corbusier

Houses	D_E	Orientation	Approach
Maison-Atelier Ozenfant	1.4927	North	Right
	1.4905	East	Front
	1.4002	South	Left
Villa Cook	1.4681	North	Front
	1.3960	South	Back
Villa Stein-de Monzie	1.4637	North	Left
	1.445	East	Front
	1.5400	South	Back
	1.3782	West	Right
Weissenhof-Siedlung Villa 13	1.4076	North	Back
	1.3852	East	Right
	1.4441	South	Front
	1.2369	West	Left
Villa Savoye	1.4916	North	Front
	1.4694	East	Left
	1.3961	South	Back
	1.4732	West	Right

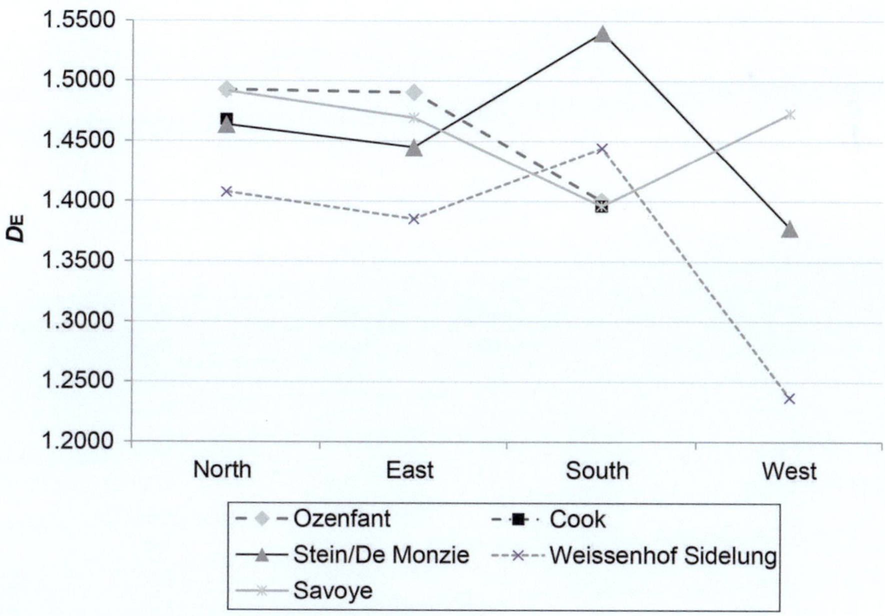

Fig. 7.20 Orientation results for Le Corbusier

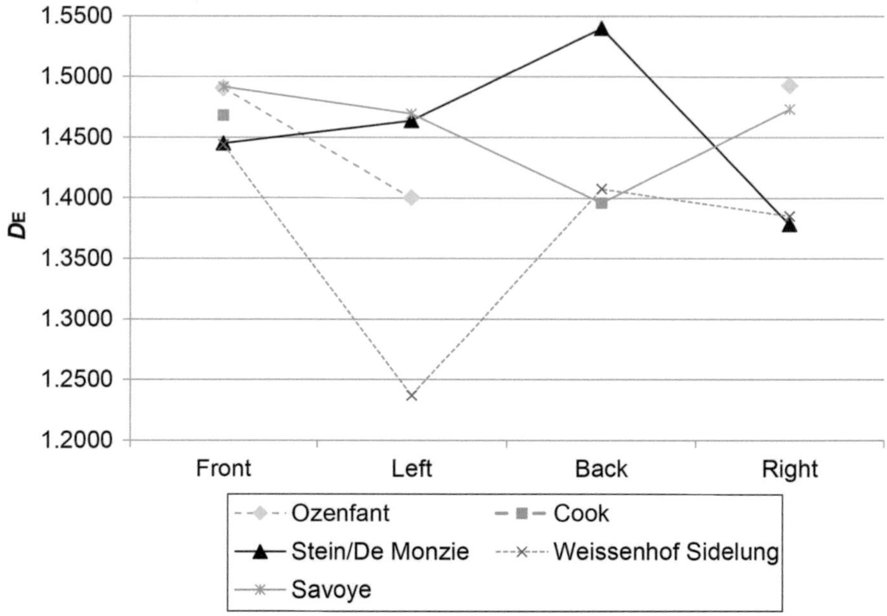

Fig. 7.21 Approach results for Le Corbusier

Table 7.15 Augmented results for Gray

Houses	D_E	Orientation	Approach
Small house for an engineer	1.3260	North	Front
	1.2502	East	Back
	1.2582	South	Left
	1.2445	West	Right
E.1027	1.3827	North	Front
	1.4261	East	Left
	1.4698	South	Back
	1.4104	West	Right
House for two sculptors	1.3990	North	Back
	1.3781	East	Right
	1.4426	South	Front
	1.3912	West	Left
Tempe à Pailla	1.3734	North	Front
	1.3045	East	Right
	1.3485	South	Back
	1.3687	West	Left
Four storey villa	1.3069	North	Left
	1.2935	South	Back
	1.4292	East	Right
	1.3536	West	Front

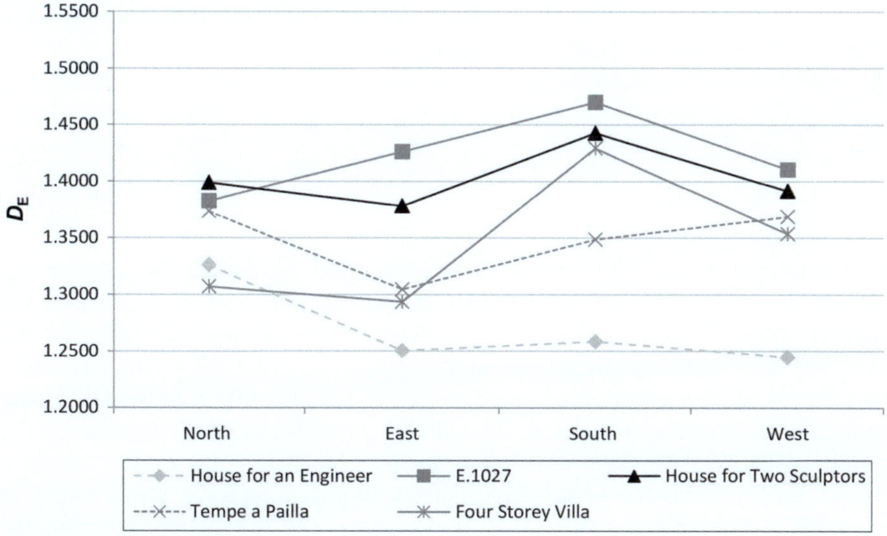

Fig. 7.22 Orientation results for Gray

hypothesised results outlined in the previous section in terms of orientation. The approach chart for Gray (Fig. 7.23) is primarily of interest because it so closely resembles her orientation chart. With three of the five houses designed for non-urban sites, and two for idealised non-urban sites, Gray evidently adopted a strategy wherein the northern elevation is typically the front and the southern, the

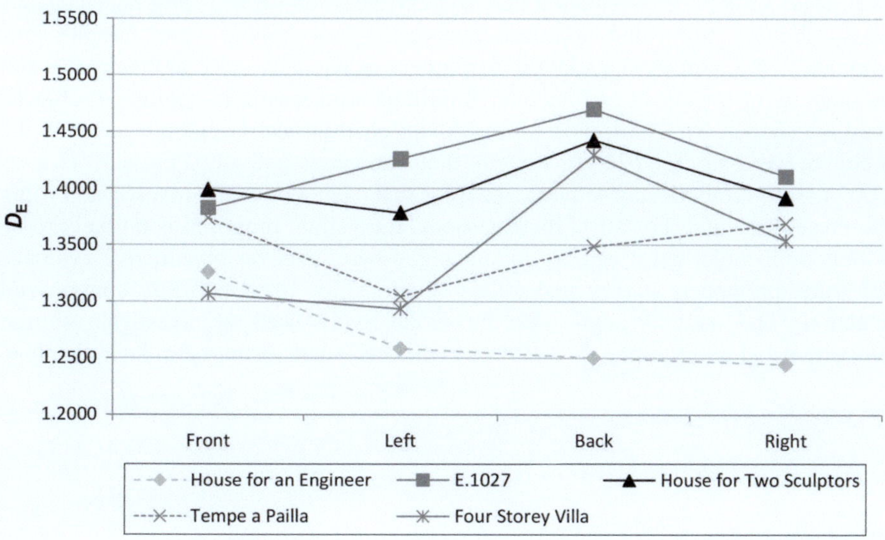

Fig. 7.23 Approach results for Gray

rear. There is one reversal of this strategy, in her *House for Two Sculptors*, and while the *Four Storey Villa* doesn't comply with this trend in its coding, the data is still a relatively close match to it.

7.7 Conclusion

In this chapter fractal dimensions have been used to explain the relationship between Le Corbusier's Pre-Modern and Modern designs, and to demonstrate the visible impact of his five principles on the plans and elevations of the latter set. For Gray, the significance of the plan is emphatically clear, with the results showing a high degree of complexity, detail and layering in her plan forms, in contrast to her more minimal exterior expressions. For Mies, a curious outcome is revealed in the review of his lesser-known works. Specifically, the data suggests that his unadorned, starkly geometric façades have a distribution of complexity that is similar to his cellular, hierarchical plans. Such observations and interpretations about the designs of these three famous European Modernists are not solely a result of the mathematical method used, but of the careful interpretation of the data using both quantitative and qualitative means.

Finally, the alternative application of the fractal analysis method was primarily used for demonstration purposes, not to prove or disprove specific aspects of the claim that form follows function. Unfortunately for the Modernists, many scholars have already demonstrated that in most ways, their forms did not truthfully or transparently express their underlying functions. However, even in the simple study presented in this chapter it is interesting to note how consistently Gray designed her façades to respond to orientation and address, and how unwavering her decisions about aligning siting and approach features in her designs were. Despite this, ultimately the proposed variation can never adequately accommodate the full complexity of a work of architecture. Buildings are necessarily contingent objects; they are shaped by a multitude of forces and each project is different, even if the architect has a strong vision or strategy that transcends individual projects. Houses also possess symbolic and semiotic qualities that cannot be easily investigated using the present method. For all of these reasons, the variant proposed is unlikely to be widely applicable, but it can, as demonstrated here, uncover quantitative evidence for some previously poorly understood relationships between form, context and function. The method might also be productively used to investigate similar properties and relationships in the other architects' works featured in later chapters.

Chapter 8
Organic Architecture

Frank Lloyd Wright, one of the world's most famous architects, designed more than 300 houses during his almost seventy-year career. While historians and critics have repeatedly classified his architecture as'Organic Modernism', during his lifetime he was actually responsible for developing several distinct stylistic sub-sets of this movement. This chapter examines five freestanding houses from each of three distinct stylistic phases in Wright's career: the Prairie Style, Textile-block and Usonian works. The fractal dimensions of the plans and elevations of these fifteen designs are first calculated and compared within their respective stylistic sets, and then across the complete group of works. This process provides a series of mathematical measures of the changing levels of formal complexity found in Wright's architecture throughout his career. The last part of the chapter takes a different approach to Wright's architecture, demonstrating a novel application of fractal analysis to measuring the changing visual complexity of the experience of walking through the *Robie House*. This application examines sequential perspective views to provide a measure of spatio-visual experience and to test a well-known argument about Wright's domestic architecture.

The plans, elevations and three-dimensional models of Wright's architecture used in this chapter were all digitally reconstructed from his original working drawings reproduced by Storrer (2006) and Futagawa and Pfeiffer (1984, 1985a, b, c, 1987a, b). Where Wright altered a particular house during construction, or only an incomplete set of working drawings was available, the measured drawings of the Historic American Buildings Survey, supplemented with photographs of the houses, were used to digitally reconstruct the designs.

M.J. Ostwald and J. Vaughan, *The Fractal Dimension of Architecture*,
Mathematics and the Built Environment 1, DOI 10.1007/978-3-319-32426-5_8

8.1 Organic Modernity

Even before the emergence of Functionalism in Europe in the 1920s, a different approach to the use of contemporary materials and construction methods was evolving in the United States. In the last two decades of the nineteenth-century Louis Sullivan broke from the tradition of designing buildings for load-bearing masonry construction, and in doing so abandoned the carved and decorated neo-classical façades that typically resulted from this approach. Instead, Sullivan designed a series of tall buildings that used steel-framed structures clad in thin masonry façades with large window openings. While these buildings were still decorated, often with Sullivan's characteristic stylized botanical traceries, they nevertheless signalled a clear departure from the formal and aesthetic practices of previous eras. Sullivan is today credited with developing the maxim 'form follows function', and it was in his Chicago office that a young Frank Lloyd Wright would be given his first major commissions. Wright would go on to be the most important proponent of 'Organic Architecture', a style which Alan Hess and Alan Weintraub describe as a type of 'Modern architecture that engaged both contemporary machinery and the ageless natural landscape' (2006: 6).

As is often the case with portmanteau titles that are used to artificially group a set of works, descriptions of the formal properties of Organic Modernity vary considerably (Joedicke 1997; Kuhlman 2008). John Farmer and Kenneth Richardson, noting this confusion, suggest that the movement should simply be associated with architecture that has a 'freer geometrical approach' (1996: 124). Indeed, buildings with a curvilinear appearance are still often described as 'organic' despite having none of the other properties of this movement or, as demonstrated in a previous chapter, any connection to nature. Furthermore, the most famous early works of Organic Modernism are Wright's Prairie style houses, designs characterized by their horizontal, rectilinear forms (Nute 2008). Indeed, the common thread binding the architecture of Organic Modernity together is not form, but an underlying set of philosophical values. Thus, Hess and Weintraub explain that what organic buildings 'have in common is the concept of seeing a building's design, structure, use, and life as an organic thing—that is, as a thing that grows from the germ of an idea into a fully articulated, variegated and unified architectural artifact' (2006: 6).

In the 1900s the Prairie School of architecture developed the poetic and conceptual concerns that would eventually define the Organic tradition in America and by the 1920s, and with the emergence of Modernism in Europe, the values of Organic architecture and Modernity began to be conflated. In 1931 Wright's first book was published in a series entitled 'Modern Architecture'. Other authors in this series included Philip Johnson, Bruno Taut, Henry-Russell Hitchcock, Catherine Bauer, Lewis Mumford and Alfred Barr. Wright used his book in this series to position himself as the founder of Modernity and, regardless of whether his 'self-promotion as fountainhead' (Levine 2008: xix) was reasonable or not, there are clear connections between his early theories of Organic design and the later pronouncements of the European Modernists. For example, 'the Aristotelian ideal of an organic whole' was a recurring theme in Modernity, and even in 'the theories of architects who never used natural forms in their designs' (Kuhlmann 2008: 40). However, the key difference between organic and functionalist strains of Modernity can be found in the fact that the Organic movement 'had sources not only in science but in poetic thinking too' (Farmer and Richardson 1996: 124). This was especially the case for Wright, whose goals and aspirations were similar to those of his European counterparts, but believed that there was a different path to achieving these outcomes. For example, whereas Le Corbusier presented his architecture using the rhetoric of science and manufacturing, Wright frequently spoke of more spiritual values, including the poetry of the land and the importance of familial social structures.

While Organic architecture is often considered synonymous with the works of Wright, there were several other famous proponents of the style in North America including Rudolph Schindler, Richard Neutra, Bruce Goff and John Lautner. In Europe, Alvar Aalto, Reima Pietilä, Hugo Haaring, Hans Scharoun, Frei Otto, Michel de Kerk and Piet Kramer all used principles that were similar to those of the American organic school. In Brazil, Oscar Niemeyer employed related theories and formal expressions, and Marion Mahoney Griffin and Walter Burley Griffin carried Wright's teachings to Australia. Organic architecture continues as a movement today, with architects Bart Prince, Kendrick Bangs Kellogg, Helena Arahuette and Gregory Burgess producing designs with a clear lineage to this style.

8.2 Frank Lloyd Wright (1867–1959)

With a career spanning from 1886 to 1959 and almost 500 completed buildings to his name, Frank Lloyd Wright could be considered the epitome of the successful American architect. Brash, opinionated and controversial, Wright was nevertheless infinitely talented, industrious and inventive. His life was fraught with drama, both professional and personal, and he was forced to reinvent himself on several occasions as his fortunes rose and fell.

Wright began his architectural education working for Joseph Silsbee and then found a like-minded mentor when Louis Sullivan employed him from 1888 to 1893. Sullivan, a key figure in the development of large-scale buildings in Chicago in the late 1800s, espoused 'an emphatic rejection of any autonomous form in building which failed to take account of function and construction' (Von Seidlein 1997: 326). Sullivan's position, that the form of a building should be derived from its purpose, remained a persistent theme throughout much of Wright's career.

After Wright and Sullivan parted ways, Wright struggled to develop his own identity as a designer, producing works which were both 'eclectic and experimental' (Storrer 2006: 18). But this situation soon changed as, from the beginning of the twentieth-century, Wright's personal views became more clearly resolved and so too his designs grew more consistent. The first sustained application of Wright's new design ethos was in his Prairie Style works, which were realised between 1900 and 1910. Primarily a domestic architectural type, the Prairie style was 'the *built* manifestation of [Wright's] reformist social program for the betterment of a growing middle class' (Alofsin 1994: 35). In these works, the majority of which were actually constructed for relatively wealthy clients, Wright demonstrated a considered language of modern design that shaped each building across all scales, from its formal modelling to its detail and ornament.

Despite the success of the Prairie Style, Wright's interest in this approach began to wane after 1910 and in 1911 he had two portfolios of his Prairie work published, an event which, in hindsight, seems to signal the end of that stage of his career. His next decade was spent on formal experimentation, including developing new approaches to construction. With an existing interest in Japanese art and architecture, Wright spent the years between 1915 and 1922 first designing and later overseeing the construction of the *Imperial Hotel* in Tokyo. After returning to the USA, Wright moved to Los Angeles and set about testing a new concrete construction system. This so-called 'textile-block' method featured interlocking, modular patterned blocks, in stark contrast to the 'arts and crafts' inspired materials and methods of his previous domestic architecture, but reminiscent of the 'Mayan-revival' work he had produced in Japan. Not including the *Hollyhock House*, a transitional work that is sometimes grouped with these designs, Wright completed only five houses using the textile-block construction system. The first of these was the highly patterned *La Miniatura*, designed for Alice Millard in 1923, and the last was the less ornate, but more extensively modelled *Lloyd Jones House*, which was built in Tulsa in 1929.

In the 1930s, as North America descended into a deep financial depression, Wright maintained an income by setting up the Taliesin fellowship. It was during this time that he re-visited several ideas he had developed previously in his Prairie and Textile-block houses, but reformulating them to be more suitable for the fiscally constrained era. He called this new approach 'Usonian' architecture, and the majority of the sixty Usonian houses that were eventually built were completed between 1935 and 1955. However, with only one of these houses constructed, Wright was offered a commission that would lead to what is perhaps his greatest work. Designed in 1935, the *Kaufmann House*, known as *Fallingwater*, is one of the world's best-known buildings. This remarkable dwelling is sited in a wooded valley and perched above a waterfall, its modern terraces stacked and cantilevered like geological extrusions from the hill behind.

Buoyed by the success of *Fallingwater* and granted increased opportunities by the improving economy, Wright soon returned to his Usonian ideas. The Usonian houses were intended to be quintessentially American, suburban, homes. Spiro Kostof describes them as being driven by Wright's 'romantic, transcendental vision' (1985: 740), but often featuring abstract geometric forms—'hexagons and piercing points, jagged fragmentation [and] scaly surfaces'—that were successfully controlled by 'Wright's geometric command and his unfaltering sense of scale' (1985: 740). While continuing to produce Usonian houses for the remainder of his career, Wright's last great civic work, commenced in 1943, was the white spiralling form of the *Solomon R Guggenheim Museum* in New York.

During his lengthy career Wright pioneered many architectural strategies for domestic design. The houses selected for analysis in this chapter are drawn from three distinct periods in his body of work and are often described as representing the early, middle and late stages of his career. The first five of Wright's early house designs analysed in this chapter are from his Prairie Style, the next set are from Wright's mid-career, Textile-block period and the last group are a sub-set of the Usonian period, called the 'triangle-plan' houses.

8.3 Five Prairie Style Houses (1901–1910)

Wright described his Prairie Style architecture as being inspired by the long, flat reaches of the American plains, leading to the design of houses with similarly strong horizontal lines, wide overhanging eaves and low-pitched roofs. Hess and Weintraub declare it an undeniably Modern approach that was nevertheless 'rooted in the American Midwest and its progressive political and intellectual landscape' (2006: 12). In the Prairie houses Wright developed a formal vocabulary or grammar that sought to express its underlying geometric structure at every turn.

The five Prairie Style houses selected for analysis in this chapter were constructed between 1901 and 1910. Four of the five are in the state of Illinois and the fifth is in Kentucky. The first design is the *Henderson House*, which was constructed in 1901 in Elmhurst, Illinois. It is a wooden, two-storey structure with plaster-rendered elevations (Fig. 8.1). The *Tomek House*, from 1907 in Riverside, Illinois, is also a two-storey structure, although it includes a basement, and it is sited on a large urban lot. It is finished with pale, rendered brickwork, dark timber trim and a red tile roof. The *Robert W. Evans House* in Chicago, Illinois, features a formal planning diagram wherein the 'basic square' found in earlier Prairie Style houses is 'extended into a cruciform plan' (Thomson 1999: 100). The house is set on a sloping site and possesses a plan similar to one Wright proposed in 1907 for a 'fireproof house for $5000'. The *Zeigler House* in Frankfort, Kentucky has a similar plan to the *Evans House*. Designed as a home for a Presbyterian minister, this two-storey house is sited on a small city lot and was constructed while Wright was

Fig. 8.1 *Henderson House*, perspective view

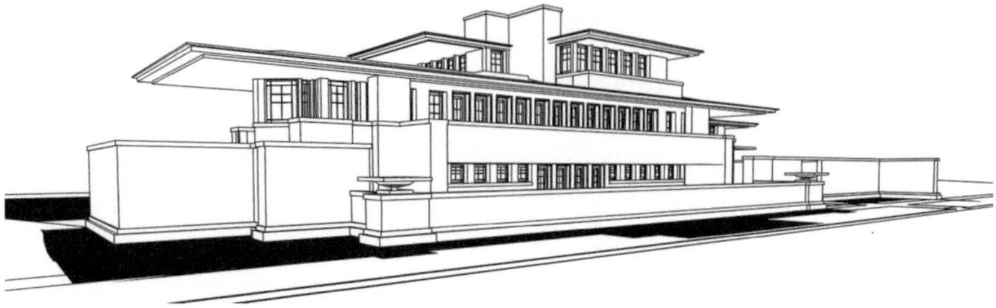

Fig. 8.2 *Robie House*, perspective view

in Europe. After a decade of development and refinement, the consummate example of the Prairie Style, the *Robie House* was constructed in Chicago, Illinois in 1910. Designed as a family home, the three-storey structure fills most of its tight corner site (Fig. 8.2). Unlike many of Wright's other houses of the era, the *Robie House* features a façade of exposed Roman bricks with horizontal raked joints giving it an expression which leads Anthony Alofsin to describe it as 'a startling image of sliding parallel horizontal masses hugging the ground' (1994: 36). Importantly, these five houses span the period between the first publication of Wright's Prairie Style, in the *Ladies Home Journal* in 1901, and what is widely regarded as the pinnacle of this approach, the *Robie House*. Table 8.1 provides example elevations and plans for this set of Wright's houses.

8.3.1 Prairie Style Houses, Results and Analysis

In the set of Wright's Prairie works, the *Zeigler House* has the lowest average elevation result ($\mu_E = 1.4448$) while the highest is in the *Evans House* ($\mu_E = 1.5473$). The median elevation result is 1.4991, the average is 1.4979 and the standard deviation is 0.0432. Thus, these results represent a consistent, non-skewed outcome. The data for the plans shows the lowest average is the *Henderson House* ($\mu_P = 1.3270$) while the *Tomek House* has the highest ($\mu_P = 1.3787$). The median for the set of plan results is 1.3783, the average is 1.3579 and the standard deviation is 0.0734. Thus, there is a slight negative skew to the plan data, caused by the roof results in every case. Despite the averages being very consistent, the ranges for each house are more diverse, and in the case of the elevations, often reflect particular site conditions (Tables 8.2 and 8.3, Fig. 8.3).

The entire set of five houses has a close, or comparable range of complexity across all twenty elevations ($R_{\{E\%\}} = 13.67$), although the degree of complexity present across all thirteen plans is less close ($R_{\{P\%\}} = 26.82$). This situation is further amplified when the individual house results are considered. For example, the *Zeigler House* has a remarkably tight range of fractal dimensions in elevation ($R_{E\%} = 1.57$) which suggests the four elevations of this house are virtually identical

Table 8.1 Wright, Prairie Style set, example elevations and floor plans (level 4 representation—not shown at a uniform scale)

Henderson House

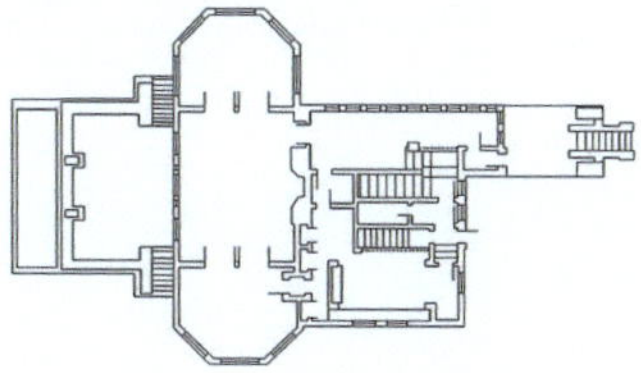

Tomek House

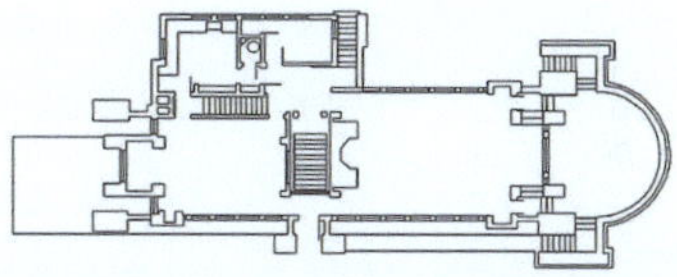

Evans House

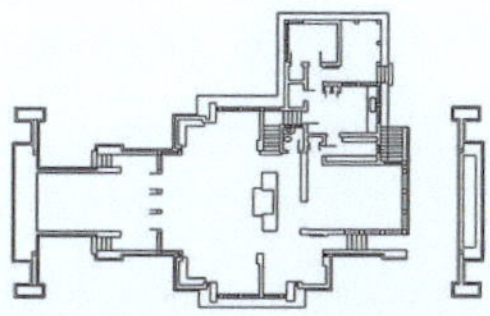

Zeigler House

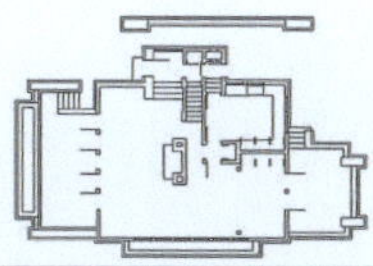

Robie House

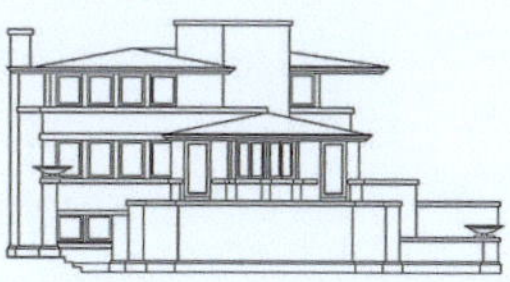
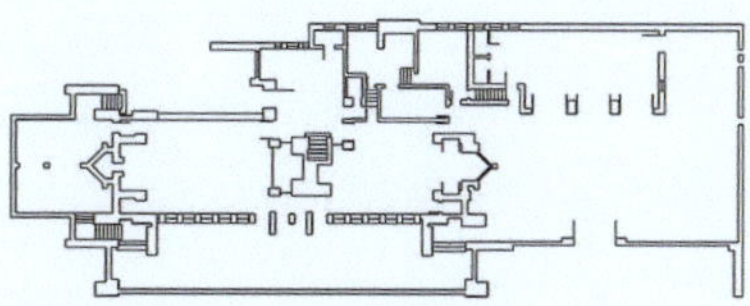

Table 8.2 Wright, Prairie Style set, results

Houses		Henderson*	Tomek*	Evans	Zeigler	Robie*	Opt. [*]	Set {…}
Elevations	D_{E1}	1.5255	1.5103	1.5592	1.4442	1.5174		
	D_{E2}	1.5177	1.4885	1.5709	1.4542	1.5708		
	D_{E3}	1.4910	1.4342	1.5254	1.4385	1.4785		
	D_{E4}	1.5072	1.4799	1.5337	1.4424	1.4677		
	μ_E	1.5104	1.4782	1.5473	1.4448	1.5086		
	$\mu_{[E]}/\mu_{\{E\}}$						1.4991	1.4979
	$M_{[E]}/M_{\{E\}}$						1.4991	1.4991
	$std_{[E]}/std_{\{E\}}$						0.0342	0.0432
Plans	D_{P-1}	1.3001	1.4448					
	D_{P0}	1.4499	1.3902	1.4307	1.4170	1.3385		
	D_{P1}	1.3763	1.3721	1.3817	1.3802	1.4220		
	D_{P2}	–	–	–	–	1.3984		
	D_{PR}	1.1817	1.3077	1.3147	1.2295	1.3066		
	μ_P	1.3270	1.3787	1.3757	1.3422	1.3664		
	$\mu_{[P]}/\mu_{\{P\}}$						1.3574	1.3579
	$M_{[P]}/M_{\{P\}}$						1.3742	1.3783
	$std_{[P]}/std_{\{P\}}$						0.0758	0.0734
Composite	μ_{E+P}	1.4187	1.4285	1.4738	1.4009	1.4375		
Aggregate	$\mu_{[E+P]}/\mu_{\{E+P\}}$						1.4282	1.4318

Table 8.3 Wright, Prairie Style set, comparative values

Houses		Henderson*	Tomek*	Evans	Zeigler	Robie*	Opt. [*]	Set {…}
Elevations	R_{ED}	0.0345	0.0761	0.0455	0.0157	0.1031		
	$R_{E\%}$	3.4500	7.6100	4.5500	1.5700	10.3100		
	$R_{[ED]}/R_{\{ED\}}$						0.1366	0.1367
	$R_{[E\%]}/R_{\{E\%\}}$						13.6600	13.6700
Plans	R_{PD}	0.2682	0.1371	0.1160	0.1875	0.1154		
	$R_{P\%}$	26.8200	13.7100	11.6000	18.7500	11.5400		
	$R_{[PD]}/R_{\{PD\}}$						0.2682	0.2682
	$R_{[P\%]}/R_{\{P\%\}}$						26.8200	26.8200
Composite	$R_{[\mu E+PD]}/R_{\{\mu E+PD\}}$						0.0188	0.0729
	$R_{[\mu E+P\%]}/R_{\{\mu E+P\%\}}$						1.8800	7.2900

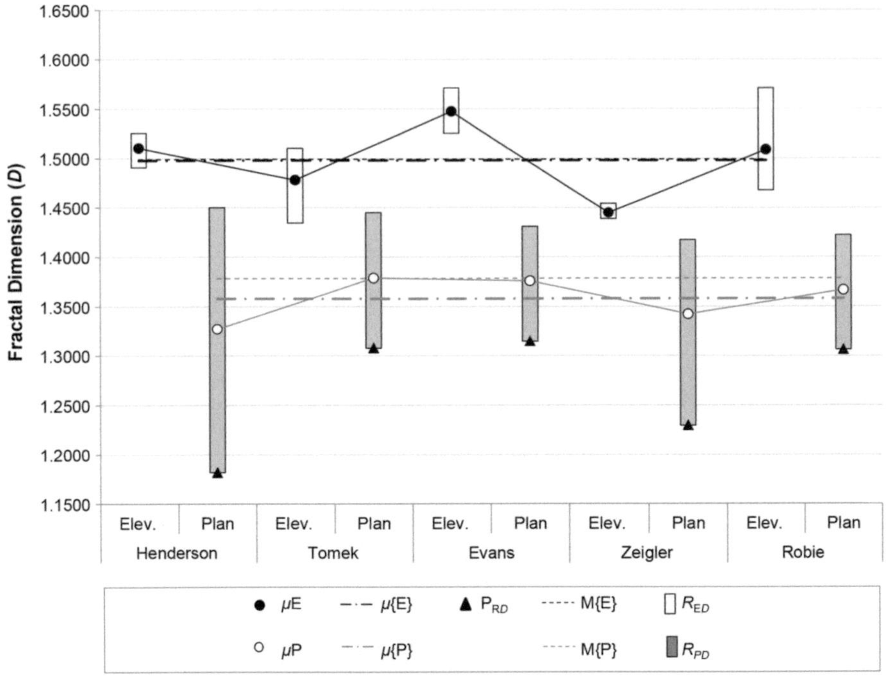

Fig. 8.3 Wright, Prairie set, graphed results

in their level of visual complexity. However, the range for the plans of the *Zeigler House* ($R_{P\%}$ = 18.75), while offering more visual correspondence than the range for the entire group of five Prairie Style houses, is less consistent, once again as a result of the roof plan.

The graphed data shows that the *Henderson House* has the most diverse results in terms of plan forms, but is amongst the most consistent for façade treatment. The set of fractal dimensions for the elevations ($1.4910 < D_E < 1.5255$) corresponds with the even distribution of detail on the exterior of the house, where each elevation has around fifteen windows and a similar level of wall detailing. In contrast, each level of the plans serves a different function, with the most complex being the ground floor (D_{P0} = 1.4499), which includes flexible living spaces and outdoor terracing. The roof plan is the least complex (D_{PR} = 1.1817), reflecting the fundamental simplicity of the layout of the *Henderson House*.

The *Tomek House* is the only one in the set of Prairie Style designs to have an overlapping level of visual complexity present in the plans and elevations. This occurs because one of the elevations has very little detail in it, and one of the plans has a particularly high level of detail. Specifically, the east elevation (D_{E3} = 1.4342) has very limited detail, being dominated by a typical Prairie Style, externally-expressed, wide chimney, leaving little space for fenestration or any of the types of details found in the other elevations. In contrast, the entry-level plan has a higher fractal dimension (D_{P-1} = 1.4448) as it includes the additional details of the stonework mouldings that Wright used in many of his Prairie houses to anchor them, visually and symbolically, to the ground.

The *Evans House* and the *Robie House* share a similar pattern of results, both with complex elevations and a very similar set of outcomes for their plans. The *Evans House* elevations are more consistent, in terms of complexity ($R_{E\%}$ = 4.5), than the *Robie House* ($R_{E\%}$ = 10.3), and the *Evans House* results for elevations all fall within the range of the *Robie* (1.4677 < D_E < 1.5708). In a like manner, the maximum plan dimension of the *Evans House* is similar to that of the *Robie* (D_{P0} = 1.4307 and D_{P1} = 1.4220 respectively) and the minimum plan dimension is also similar for the two houses (D_{PR} = 1.3147 and D_{PR} = 1.3066 respectively).

Overall, the results show that the elevations of Wright's Prairie houses are generally more complex than the plans. Furthermore, the median and the average of all elevations are almost identical. Thus, despite some wider ranges existing in individual houses, which are often the result of specific siting and planning factors, several results derived from Wright's Prairie Style works do demonstrate a remarkably consistent level of visual expression.

8.4 Five Textile-Block Houses (1923–1929)

Appearing as imposing, ageless structures, several of which have been compared to pagan temples, the five Textile-block houses were typically constructed from a double skin of pre-cast, patterned and plain exposed concrete blocks, held together by Wright's patented system of steel rods and grout (Table 8.4). Occasional bands of ornamented blocks punctuate the otherwise plain square masonry grid in the façades of these houses and for each client a different pattern was created for these ornamental highlight features. Despite their apparent difference in appearance to the Prairie Style houses, Hess and Weintraub state that 'every aspect of the LA homes followed organic principles' (2006: 38).

The first of the five houses, the *Millard House* or '*La Miniatura*', as it came to be known, is in Pasadena and it was constructed in 1923 (Fig. 8.4). This house is the only one of the Textile-block works not to feature a secondary structure of steel rods. Reflecting on *La Miniatura*, Wright wrote that in this project he 'would take that despised outcast of the building industry—the concrete block—out from underfoot or from the gutter—find a hitherto unsuspected soul in it—make it live as a thing of beauty textured like the trees' (Wright 1960: 216–217). The second of the Textile-block designs, the *Storer House* from 1923, is a three-storey residence in Hollywood with views across Los Angeles. The *Samuel Freeman House* in Los Angeles is regarded as the third Textile-block house and the first to use mitred glass in the corner windows; all of the previous works in this style have solid corners. The *Samuel Freeman House* is a two-storey, compact, flat-roofed house made from both patterned and plain textile-blocks and with eucalyptus timber detailing. The next completed design, the 1924 *Ennis House*, is probably the most famous of

Table 8.4 Wright, Textile-block set, example elevations and floor plans (level 4 representation—not shown at a uniform scale)

Millard House	
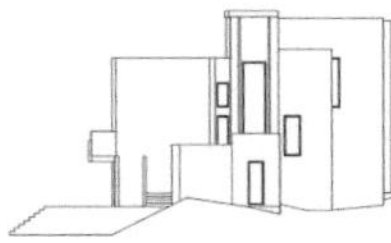	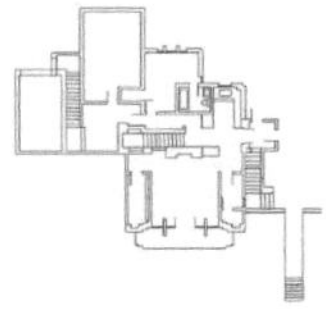
Storer House	
	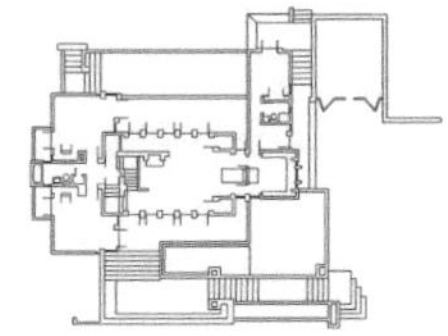
Freeman House	
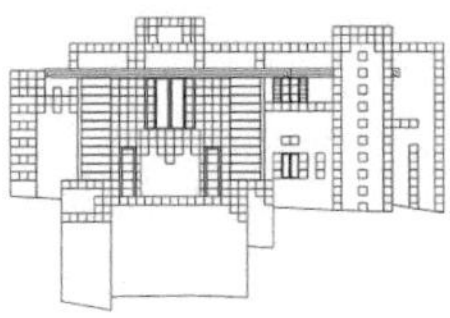	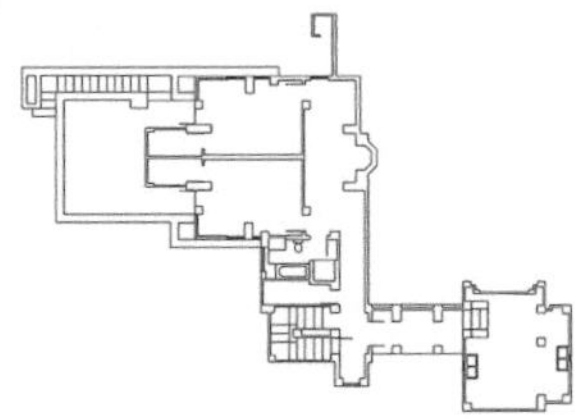
Ennis House	
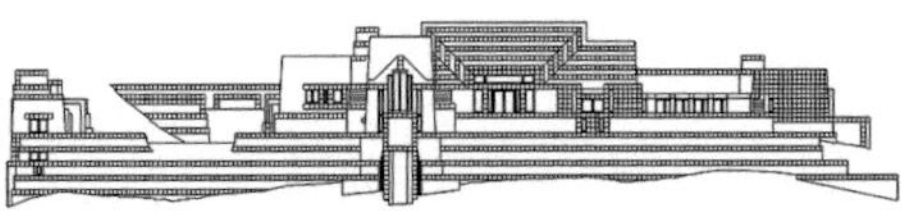	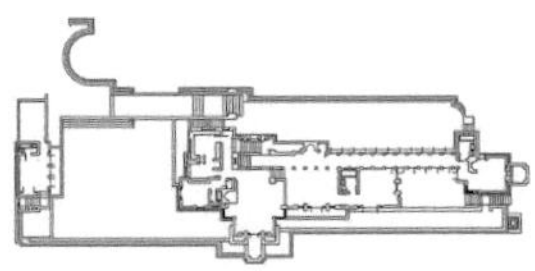
Lloyd-Jones House	
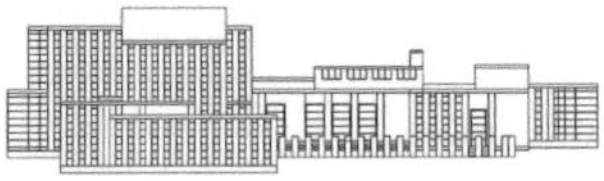	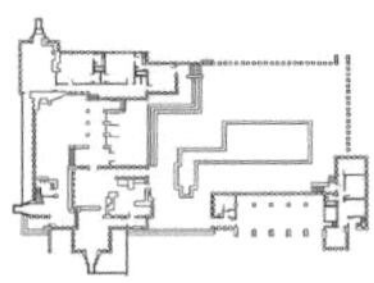

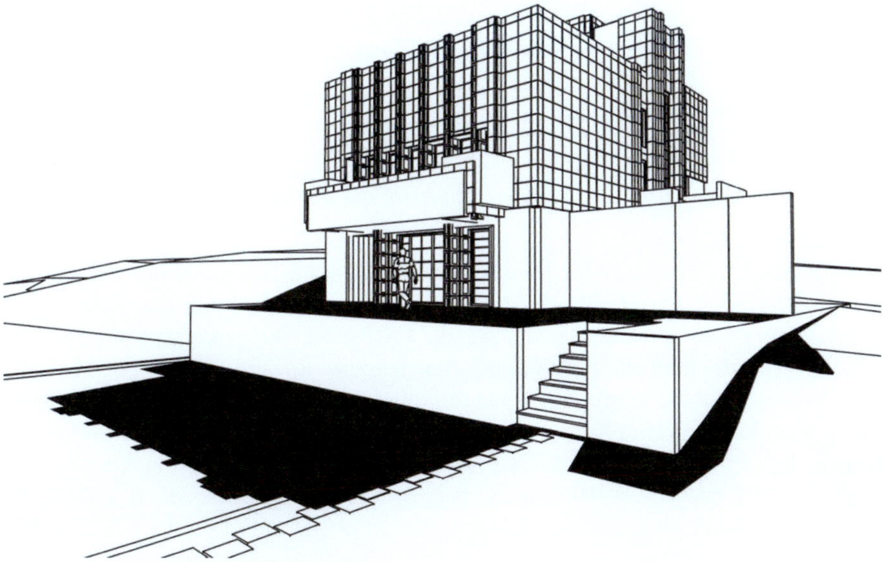

Fig. 8.4 *Millard House*, perspective view

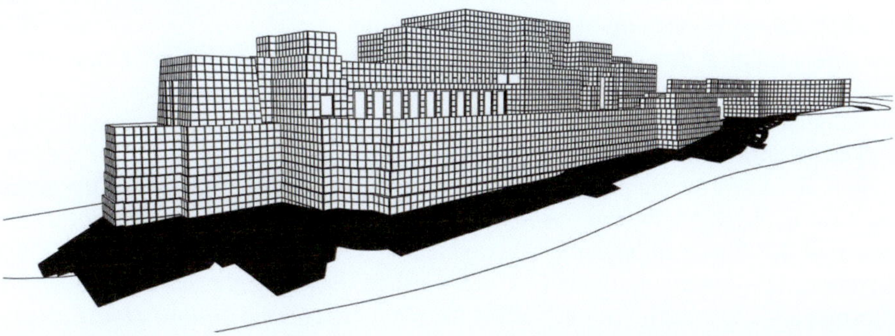

Fig. 8.5 *Ennis House*, perspective view

Wright's works of this era (Fig. 8.5). It is regarded as both ambitious and enigmatic, being described as 'looking more like a Mayan temple than any other Wright building except [the] *Hollyhock House*' (Storrer 1974: 222). The *Ennis House* is conspicuously sited overlooking Los Angeles and is made of neutral-coloured blocks with teak detailing. Some of the windows feature art glass designed by Wright as a geometric abstraction of the form of wisteria plants. The final design in this set, the *Lloyd-Jones House*, is in Tulsa, Oklahoma. It is the only non-Californian Textile-block house. Designed for Wright's cousin, it is a large home with extensive entertaining areas and a four-car garage. The *Lloyd-Jones House* is notably less ornamental than the others in the sequence, with Wright rejecting richly decorated blocks 'in favor of an alternating pattern of piers and slots' (Frampton 2005: 170). Table 8.4 Provides example elevations and plans of this set.

8.4.1 Results and Analysis of Wright's Textile-Block Houses

Among Wright's Textile-block houses, the earliest in the set, the *Millard House* has the lowest average elevation result (μ_E = 1.3942), while the highest is found in the *Lloyd-Jones House*, the latest of the set (μ_E = 1.5906). The average elevation result is 1.4996 and the median, 1.5006, with the standard deviation being 0.0925. Thus, there is very little skew in the results, although the deviation in the data is higher than it was for the Prairie Style works. Results for the plans identify that the lowest average is also found in the *Millard House* (μ_P = 1.3379) while the *Ennis House* has the highest (μ_P = 1.4810). The average for the set of plan results is 1.4055, the median is 1.4127 and the standard deviation is 0.0557 (Tables 8.5 and 8.6, Fig. 8.6).

The entire set of results of the seventeen Textile-block plans are not closely related enough that they can be considered 'comparable' ($R_{\{P\%\}}$ = 21.46). The range of complexity across all twenty elevations ($R_{\{E\%\}}$ = 32.62) is even wider. The optimal set is made up of the *Storer*, *Ennis* and *Lloyd-Jones* houses, and while this new grouping does not affect the range results for the plans, the range of the sixteen elevations of these three houses ($R_{[E\%]}$ = 19.95) reveals a 12.67 % reduction, signalling a much higher level of visual correspondence within the sub-set.

The *Millard House* results are lower than expected, with all those for the elevations and most for the plans falling below their respective averages for the entire set. One explanation for this result is that the textured, ornamental blocks in the façade of this house all have the same pattern and in a level 4 representation of the elevation, this texture is treated as one surface, lowering the anticipated result. However, this approach does not affect the planning and the *Millard House* results do generally show a simpler plan form. The *Storer House* fits neatly into the overall results, with the median and average for all of the Textile-block set falling within its range of results for both plans and elevations. As expected, most roof plans provide the lowest result for each house in the Textile-block set; however, for the *Freeman House* it is the first floor which has the least visual complexity (D_{P1} = 1.3799). The explanation for this anomaly is found in the fact that this house was designed with terraced levels, and the view of the roof therefore includes a roof garden and the rooflines below, making the visual complexity of this plan higher than might be expected (D_{PR} = 1.3901).

The *Ennis House* could be considered the most complex of the Textile-block set, having the highest results for all plans (D_{P0} = 1.4955 and D_{PR} = 1.4664), and with its north (D_{E1} = 1.6130) and south elevations (D_{E2} = 1.6390) being the most visually complex of the set. However, the average elevation result for the *Lloyd Jones House* is the highest overall. The elevations for this house have a high fractal

Table 8.5 Wright, Textile-block set, results

Houses		Millard	Storer*	Freeman	Ennis*	Lloyd-Jones*	Opt. [*]	Set {…}
Elevations	D_{E1}	1.4420	1.5389	1.3603	1.6130	1.5947		
	D_{E2}	1.4786	1.5543	1.5125	1.6390	1.5589		
	D_{E3}	1.3434	1.5111	1.4666	1.4900	1.6105		
	D_{E4}	1.3128	1.4395	1.4868	1.4417	1.5983		
	μ_E	1.3942	1.5110	1.4566	1.5459	1.5906		
	$\mu_{[E]}/\mu_{\{E\}}$						1.5492	1.4996
	$M_{[E]}/M_{\{E\}}$						1.5566	1.5006
	$std_{[E]}/std_{\{E\}}$						0.0668	0.0925
Plans	D_{P0}	1.4078	1.4497	1.3964	1.4955	1.4465		
	D_{P1}	1.3801	1.4330	1.3799	–	1.4228		
	D_{P2}	1.2826	1.4311	–	–	1.4158		
	D_{PR}	1.2809	1.4024	1.3901	1.4664	1.4127		
	μ_P	1.3379	1.4291	1.3888	1.4810	1.4245		
	$\mu_{[P]}/\mu_{\{P\}}$						1.4055	1.4055
	$M_{[P]}/M_{\{P\}}$						1.4127	1.4127
	$std_{[P]}/std_{\{P\}}$						0.0557	0.0557
Composite	μ_{E+P}	1.3660	1.4700	1.4275	1.5243	1.5075		
Aggregate	$\mu_{[E+P]}/\mu_{\{E+P\}}$						1.4650	1.4591

Table 8.6 Wright, Textile-block set, comparative values

Houses		Millard	Storer*	Freeman	Ennis*	Lloyd-Jones*	Opt. [*]	Set {…}
Elevations	R_{ED}	0.1658	0.1148	0.1522	0.1973	0.0516		
	$R_{E\%}$	16.5800	11.4800	15.2200	19.7300	5.1600		
	$R_{[ED]}/R_{\{ED\}}$						0.1995	0.3262
	$R_{[E\%]}/R_{\{E\%\}}$						19.9500	32.6200
Plans	R_{PD}	0.1269	0.0473	0.0165	0.0291	0.0338		
	$R_{P\%}$	12.6900	4.7300	1.6500	2.9100	3.3800		
	$R_{[PD]}/R_{\{PD\}}$						0.2146	0.2146
	$R_{[P\%]}/R_{\{P\%\}}$						21.4600	21.4600
Composite	$R_{[\mu E+PD]}/R_{\{\mu E+PD\}}$						0.1582	0.1582
	$R_{[\mu E+P\%]}/R_{\{\mu E+P\%\}}$						15.8200	15.8200

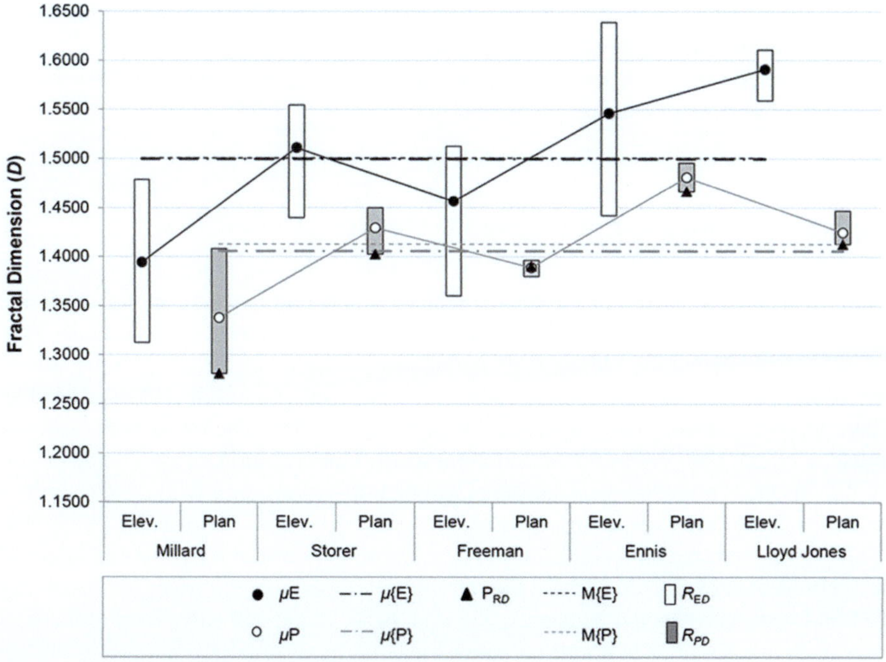

Fig. 8.6 Wright, Textile-block set, graphed results

dimension due to the window framing used by Wright in this building, where each panel of glass is framed to match the blockwork. This house also has an unusual result for its plans and elevations, which are distinctly different, where the elevations ($D_{\mu\{E\}} = 1.5906$) are far more complex than the plans ($D_{\mu\{P\}} = 1.4245$). For all of the other houses in this set there is some overlap between the complexity of the plans and elevations.

Overall, the results for the Textile-block set confirm that they are a series of visually complex dwellings where there is a broad relationship between the expression of both plans and elevations. Furthermore, over the six-year period between the first to the last, the complexity of the house designs increased. This result partially confirms the typical descriptions of these houses provided by historians, who argue that Wright's architecture became more visually complex, heavy and ornate throughout this time, largely as a property of the decoration embedded in the blocks. However, some historians disagree with this, suggesting that in the *Lloyd-Jones House* Wright moved away from the 'primitivism' or 'Mayan-revivalism' found in the first four to produce a much simpler formal expression. For example, Alofsin argues that as Wright 'responded to the incipient International Style he simplified his surface patterns, a shift that marked the end of his primitivist phase' (1994: 42). Yet, the total level of formal complexity in the work did not fall; instead, the level of ornamental detail fell in the final house, whereas the formal modelling reached its most articulated expression. This interpretation of the data supports the views of those critics and historians who see the *Lloyd-Jones House* as triggering a shift from vertical to horizontal modelling, rather than being less ornamental in its expression (Sweeney 1994).

8.5 Five Triangle-Plan Usonian Houses (1950–1956)

More than twenty years passed before Wright developed his third major sequence of domestic works, the Usonian houses. For Wright, the Usonian house was intended to embrace the elements of nature and make them 'integral to the life of the inhabitants', It was also to be truthful in its material expression; 'glass is used as glass, stone as stone, wood as wood' (Wright 1954: 353). The archetypal Usonian house is effectively 'a simplified and somewhat diluted Prairie house characterized by the absence of leaded glass and the presence of … very thin wall screens with a striated effect from wide boards spaced by recessed battens' (Hoffmann 1995: 80). While there were multiple variations on the Usonian house, the five works featured in the present analysis are all based on an underlying equilateral triangular planning grid and were constructed between 1950 and 1956. Plans and elevations of the five houses in this set are presented in Table 8.7.

One of the earliest of the triangle-plan Usonian houses, the 1950 *Palmer House*, is located in Ann Arbour, Michigan. It is a two-storey brick structure set into a sloping site, with wide, timber-lined eaves, giving the viewer an impression of a low, single-level house (Fig. 8.7). The brick walls include bands of patterned, perforated blocks in the same colour as the brickwork. The repeatedly-scaled triangle motif in the Palmer house plan has made it the subject of multiple fractal studies (Eaton 1998; Joye 2006; Harris 2007).

The *Reisley Residence*, from 1951, a single-level home with a small basement, is constructed from local stone with timber panelling and it is set on a hillside in Pleasantville, New York. In contrast, the *Chahroudi House*, from the same year, was built on an island in Lake Mahopac, New York, and constructed using Wright's desert masonry rubblestone technique, with timber cladding and detailing. Wright originally designed the cottage as the guest quarters for the Chahroudi family home; although only the cottage was built and subsequently served as the primary residence. The *Dobkins House* was built for Dr. John and Syd Dobkins in Canton, Ohio. This small house is constructed from brick, with deeply raked mortar joints. Unlike the *Robie House*, for the *Dobkins House* the mortar colour was chosen to contrast with the bricks in the vertical as well as the horizontal joints. Finally, the 1955 *Fawcett House* had an unusual brief for Wright: to design a home for a farming, rather than a suburban, family. The house is set on the large, flat expanse of the Fawcett's farm in Los Banos, California. The single-storey house is constructed primarily of grey concrete block with a red gravel roof.

Table 8.7 Wright, Usonian set, example elevations and floor plans (level 4 representation—not shown at a uniform scale)

Palmer House

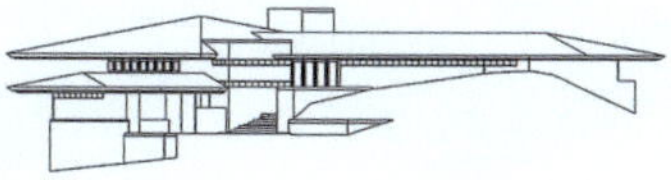
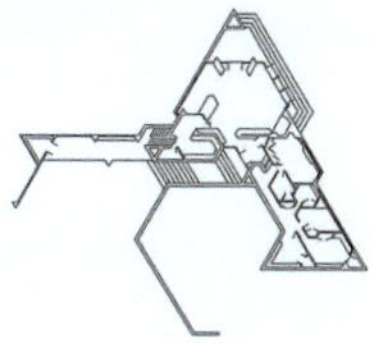

Reisley House

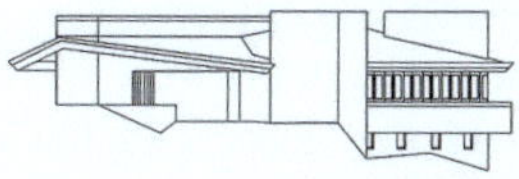
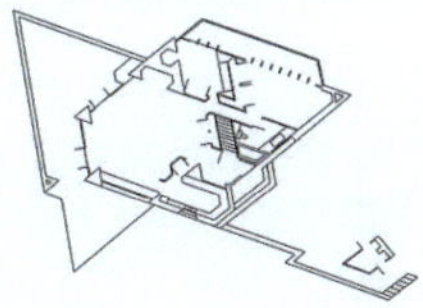

Chahroudi House

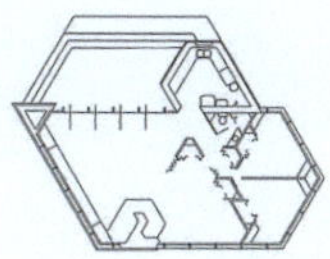

Dobkins House

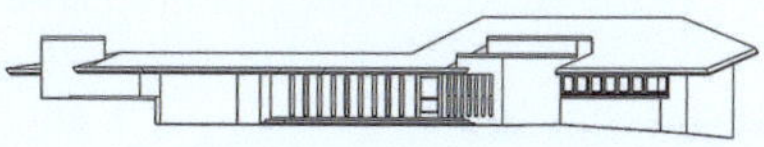
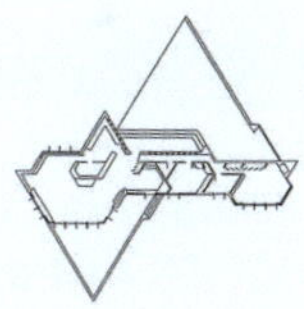

Fawcett House

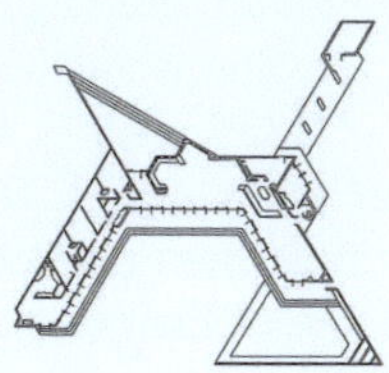

Fig. 8.7 *Palmer House*, perspective view

8.5.1 Usonian Houses, Results and Analysis

The data for Wright's Usonian houses indicates that the lowest average elevation is found in the *Reisley House* (μ_E = 1.3982) and the highest in the *Fawcett House* (μ_E = 1.4719). The highest individual elevation result is also from the *Fawcett House* (D_{E2} = 1.5575) and the complete set of results from this house suggest it is complex in both plan and elevation. The median for all elevations in the Usonian set is 1.4297, the average is 1.4350 and the standard deviation is 0.0560. The highest plan average is from the *Fawcett House* (μ_P = 1.3997), however the highest individual plan is the ground floor of the *Palmer House* (D_{P0} = 1.4412) and the lowest plan average is found in the *Dobkins House* (μ_P = 1.3105). The range of all the plans ($R_{\{P\%\}}$ = 20.12) is in a close percentile band to that of the elevations ($R_{\{E\%\}}$ = 22.00). The median for all plans is 1.3687 and the standard deviation is 0.0634. In four of the five cases the roof is the least complex plan (Tables 8.8 and 8.9, Fig. 8.8).

The aggregate average for all plans and elevations is $\mu_{\{E+P\}}$ = 1.4032 and the composite range is $R_{\{\mu E+P\%\}}$ = 7.39. Considering only the optimal sub-set—the *Reisley*, *Chahroudi* and *Dobkins* houses—the aggregate result is reduced to $\mu_{[E+P]}$ = 1.3828 and the composite range is reduced to $R_{[\mu E+P\%]}$ = 2.66. This is a substantial change over the full set, supporting the notion that while the optimal subset is, by virtue of its definition, the tightest grouping of results, these often display very high levels of correspondence.

The *Palmer House* ground floor plan (D_{P0} = 1.4412) is higher than the mean of all the elevations in the set ($\mu_{\{E\}}$ = 1.4350), and in the *Fawcett House* all plan results (1.3839 < D_P < 1.4155) are higher than the mean of the other plans of the Usonian set ($\mu_{\{P\}}$ = 1.3480). These two houses, along with the *Dobkins House*, all have ground floor plans which share corresponding levels of visual complexity with at least one of their elevations. The other two houses in the Usonian set, the *Reisley* and *Chahroudi*, present a ground floor fractal dimension which is very close to, but not as complex, as the least complex of their elevations.

Due to the triangular nature of the planning system that Wright employed with these houses, the *Chahroudi House* and the *Fawcett House* have only three elevations in representational form. This lesser number of sources does not, however, appear to affect the results, as the two houses are still typical when compared with

Table 8.8 Wright, Usonian set, results

Houses		Palmer	Reisley*	Chahroudi*	Dobkins*	Fawcett	Opt. [*]	Set {…}
Elevations	D_{E1}	1.4802	1.3865	1.4328	1.4596	1.3991		
	D_{E2}	1.4461	1.3710	1.4529	1.3375	1.5575		
	D_{E3}	1.4642	1.4086	–	1.5359	–		
	D_{E4}	1.4018	1.4265	1.4045	1.3745	1.4591		
	μ_E	1.4481	1.3982	1.4301	1.4269	1.4719		
	$\mu_{[E]}/\mu_{\{E\}}$						1.4173	1.4350
	$M_{[E]}/M_{\{E\}}$						1.4086	1.4297
	$std_{[E]}/std_{\{E\}}$						0.0538	0.0560
Plans	D_{P-1}	–	1.2968	–	–	–		
	D_{P0}	1.4412	1.3687	1.3973	1.3810	1.4155		
	D_{PR}	1.2875	1.3256	1.2908	1.2400	1.3839		
	μ_P	1.3644	1.3304	1.3441	1.3105	1.3997		
	$\mu_{[P]}/\mu_{\{P\}}$						1.3286	1.3480
	$M_{[P]}/M_{\{P\}}$						1.3256	1.3687
	$std_{[P]}/std_{\{P\}}$						0.0568	0.0634
Composite	μ_{E+P}	1.4202	1.3691	1.3957	1.3881	1.4430		
Aggregate	$\mu_{[E+P]}/\mu_{\{E+P\}}$						1.3828	1.4032

Table 8.9 Wright, Usonian set, comparative values

Houses		Palmer	Reisley*	Chahroudi*	Dobkins*	Fawcett	Opt. [*]	Set {…}
Elevations	R_{ED}	0.0784	0.0555	0.0484	0.1984	0.1584		
	$R_{E\%}$	7.8400	5.5500	4.8400	19.8400	15.8400		
	$R_{[ED]}/R_{\{ED\}}$						0.1984	0.2200
	$R_{[E\%]}/R_{\{E\%\}}$						19.8400	22.0000
Plans	R_{PD}	0.1537	0.0719	0.1065	0.1410	0.0316		
	$R_{P\%}$	15.3700	7.1900	10.6500	14.1000	3.1600		
	$R_{[PD]}/R_{\{PD\}}$						0.1573	0.2012
	$R_{[P\%]}/R_{\{P\%\}}$						15.7300	20.1200
Composite	$R_{[\mu E+PD]}/R_{\{\mu E+PD\}}$						0.0266	0.0739
	$R_{[\mu E+P\%]}/R_{\{\mu E+P\%\}}$						2.6600	7.3900

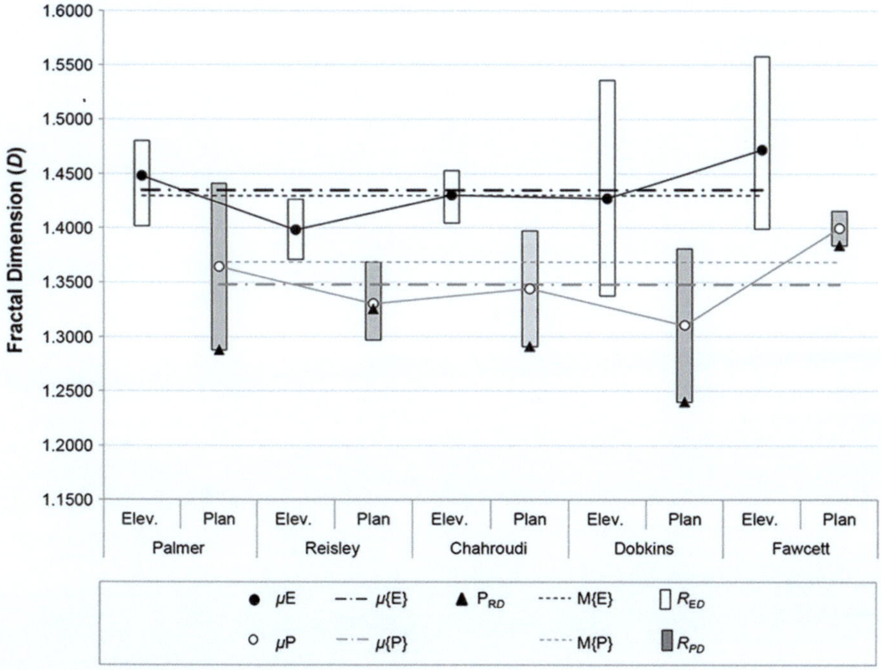

Fig. 8.8 Wright, Usonian set, graphed results

the others. Indeed, the *Chahroudi House* provides balanced results with the extent of the *D* values falling neatly above and below the mean in the case of both plans and elevations.

8.6 Comparing the Three Sets

Comparing the aggregate results developed from the analysis of three periods of Wright's architecture, it can be seen that, of the three, the Textile-block designs are generally the most visually complex ($\mu_{\{E+P\}} = 1.4591$), the Usonians are the least ($\mu_{\{E+P\}} = 1.4032$), and the Prairie Style houses are midway between the two ($\mu_{\{E+P\}} = 1.4318$). The formal coherence graphs for the three sets of houses demonstrate that the correlation between the formal properties of the plans and elevations is very weak in the Prairie houses ($R^2 = 0.0849$) and considerably stronger in both the Textile-block ($R^2 = 0.6729$) and Usonian sets ($R^2 = 0.6902$) (Figs. 8.9, 8.10 and 8.11). This means that there is a greater level of disparity or difference between the open plan interiors and the detailed and decorated exteriors of the Prairie Style houses, than is found in either of the other styles. In contrast, the sculptural geometric modelling of the exterior of the Textile-block houses is matched more

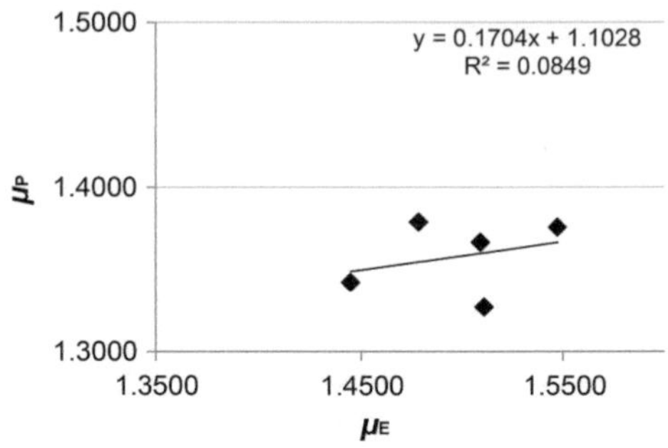

Fig. 8.9 Wright, Prarie Style set, formal coherence graph

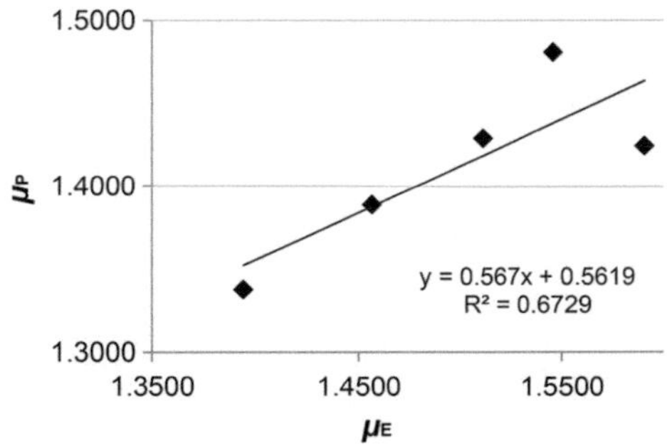

Fig. 8.10 Wright, Textile-block set, formal coherence graph

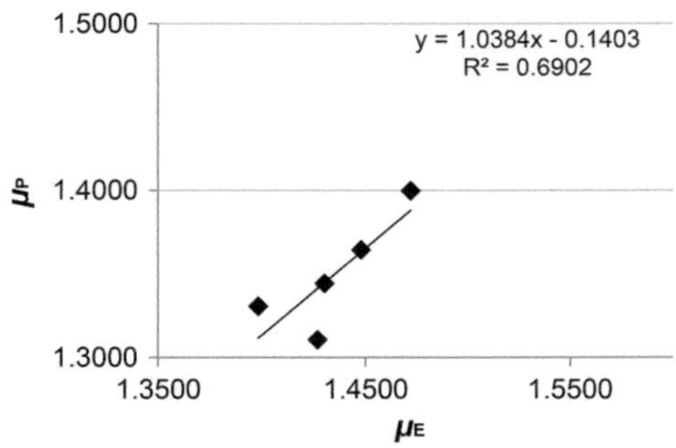

Fig. 8.11 Wright, Usonian set, formal coherence graph

closely with their textured and labyrinthine interior planning. While still utilising a relatively open interior in the main spaces of the Textile-block houses, Wright's designs for them includes complicated recesses, staggered layouts and outdoor terracing. The Usonian houses display the highest degree of formal coherence. Planning in these houses is open and straightforward despite the lack of right-angles in many of the triangular grid plans, and the exteriors, designed for simple construction, have a level of visual complexity that is similar to that of the plans of these same houses.

Linear trends extrapolated from the data offer another way of understanding how Wright's architecture evolved. For example, just considering elevations, visual complexity is relatively flat or constant across both the Prairie Style and Usonian

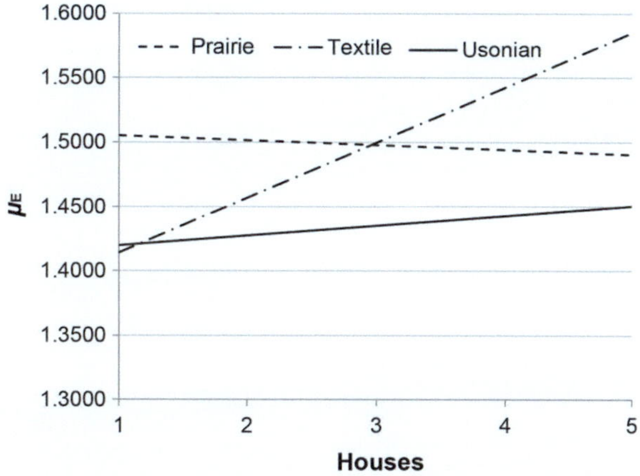

Fig. 8.12 Wright's Prairie Style, Textile-block and Usonian sets, linear trendline data for elevations (μ_E)

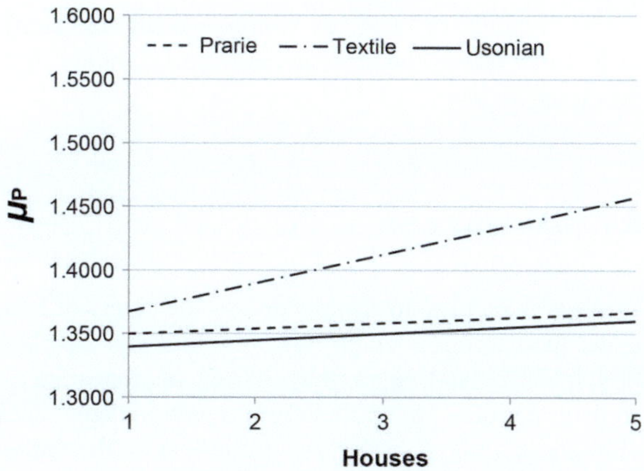

Fig. 8.13 Wright's Prairie Style, Textile-block and Usonian sets, linear trendline data for plans (μ_P)

houses, while it rises more noticeably over time in his Textile-block works (Fig. 8.12). For the plans of the sets of houses, a similar pattern occurs, with the Prairie Style and Usonian houses remaining similar in their complexity (the former falling slightly and the latter rising slightly) with only the Textile-block set showing a marked increase over time (Fig. 8.13). Finally, when elevations and plans are combined, these trends are confirmed, with the Prairie Style house trendline almost flat from the *Henderson House* in 1901 to the *Robie House* in 1910 (Fig. 8.14). The Usonian houses increase in complexity slightly over time from the *Palmer House* in 1950 to the *Fawcett House* in 1955. The dramatic increase in the results for the Textile-block houses might be somewhat surprising, considering that the blocks of the *Millard House* (1923) are highly textured, and these ornate features seem to decrease in complexity over the period, concluding with the un-patterned *Lloyd-Jones House* in 1929. However it must be remembered that the analytical method used throughout this book does not include the ornamental patterns that are present in some textile-blocks in this set of houses. Such details are also not visible until

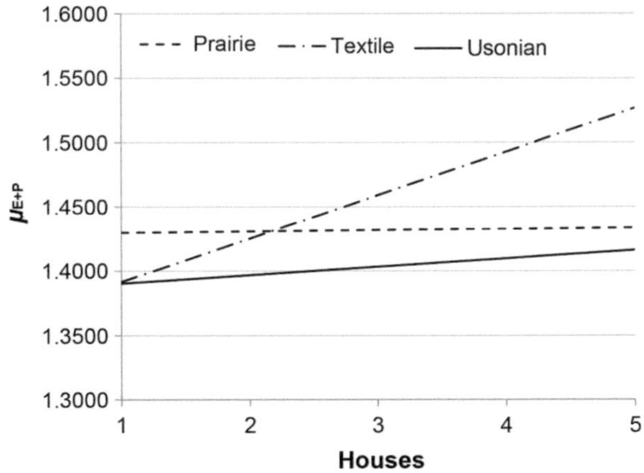

Fig. 8.14 Wright's Prairie Style, Textile-block and Usonian sets, linear trendline data for composite (μ_{E+P}) results

one is in relatively close proximity—so that the analysis of these houses is of their formal modelling and changes in material, rather than the intricacies of individual, decorated finishes.

8.7 Wright's Style, Perceived and Measured

When architectural historians review the career of Frank Lloyd Wright, the majority agree that, throughout his life, his buildings displayed a consistent set of design principles, even though they varied in appearance across a number of distinct stylistic periods. Historians tend to acknowledge such obvious visual and stylistic differences while focusing on similarities in the underlying tactics and theories that shaped Wright's work. For example, Robert Sweeney concedes that Wright's ability to 'renew himself repeatedly throughout his career' (1994: 1) is a characteristic of his approach, but argues that it does not change his underlying values. David De Long supports this view when he proposes that, over time Wright 'was able to retain allegiance to earlier principles while arriving at markedly different conclusions' (1994: xii). Kenneth Frampton similarly maintains that there is a constant thread throughout Wright's work which is related to a modular system of planning and construction which 'varied according to local circumstance' (2005: 178). Robert McCarter, who claims that Wright's Usonian houses are derivations of his Prairie house ideals, argues that Wright's architecture represents a cyclical pattern of 'continuous reinvention or rediscovery of the same fundamental principles' (1999: 249). Finally, Donald Hoffmann supports this position confirming that 'the language of Wright's buildings continued to change, but the logic did not; once he grasped the principles, his work no longer evolved' (1995: 52). As evidence for this assertion Hoffman quotes Wright himself stating that 'I am pleased by the thread of structural consistency I see inspiring the complete texture of the work revealed in my designs and plans, … from the beginning, 1893, to this time, 1957' (qtd. in Hoffman 1995: 52).

While the above quotes provide interpretations of the reasons why Wright's architecture remained so consistent, when looking at the data for Wright's Prairie Style, Textile-block and Usonian periods, the scholarly view is generally supported by data derived from visual complexity. By comparing the average complexity for the elevations of Wright's houses over these three periods, the largest variation is 6.4 %, between the Usonian and Prairie Style elevations. Starting with the first of Wright's stylistic periods analysed, the Prairie style, the houses have an average fractal dimension for all elevations of $\mu_{\{E\}} = 1.4979$. This level of complexity remains virtually unchanged in Wright's Textile-block period ($\mu_{\{E\}} = 1.4996$) before decreasing slightly in the Usonian period ($\mu_{\{E\}} = 1.4350$). In plan form, Wright appears to have gone full circle in complexity starting from the Prairie Style ($\mu_{\{P\}} = 1.3579$), then increasing in the Textile-block houses ($\mu_{\{P\}} = 1.4055$), before returning to the lower levels in the Usonian houses ($\mu_{\{P\}} = 1.3480$).

In his earliest writings, Wright set down some 'propositions' regarding a method for creating an American architecture. On the subject of creating a unique character for a building, he stated that 'I have endeavoured to establish a harmonious relationship between ground plan and elevation of these buildings considering the one as a solution and the other as an expression of the conditions of a problem of which the whole is a project' (Wright 1908: 158). Robert MacCormac emphasises this view, arguing that Wright 'saw the architectural relationship between plan and section' (2005: 131) as a fundamental principle. John Sergeant supports MacCormac, suggesting that Wright's three-dimensional, non-symmetrical grid planning allows for a connection between the plan and the elevation of his buildings on both a practical and an experiential level. Sergeant further states that this was achieved by Wright during his Prairie style period where 'a vocabulary of forms was used to translate or express the grid at all points—the solid rather than pierced balconies, planters, bases of flower urns, clustered piers, even built-in seats were evocations of the underlying structure of a house' (Sergeant 2005: 192), In the Usonian houses, Sergeant maintains that 'Wright's skill lay in the perfect coordination of horizontal and vertical systems to manipulate the[ir] character' (2005: 197).

Early observations arising from the present data do not necessarily support the idea that Wright's architecture always displays a similar set of formal measures for plans and elevations. The results show that the elevations in general have higher fractal dimensions than the plans for all fifteen houses. Conceptually, such a result might be a particular characteristic of Wright's architecture, but it is more likely a reflection of the minimum scale and dimensionality of rooms required to accommodate human inhabitation, that is, the primary forces shaping the plan. In contrast, elevations are shaped by materials, outlook, environmental effects and privacy needs, all of which may occur across a wider range of scales. When the results are examined more closely, it can be seen that for all medians and averages, the elevations are indeed generally more complex than plans, particularly in the Prairie Style houses where this difference is distinct, with a change of 12.08 % between the $M_{\{E\}}$ and $M_{\{P\}}$ and only one elevation dimension lower than any plan. Although not so divergent, the average and median are also higher for the elevations than the plans for the Textile-block houses (8.79 % between the $M_{\{E\}}$ and $M_{\{P\}}$) and for the Usonians (6.10 % between the $M_{\{E\}}$ and $M_{\{P\}}$). However, for these last two sets some elevation and plan results overlap, so when scrutinised individually, it would appear that some of the houses do share a level of complexity in plan and elevation. This data only partially supports the views of Sergeant (2005) and MacCormac (2005).

Comparing sets of fractal dimensions is one method of considering the similarities between the elevations and plans of a building. Another is to use the range results to determine the level of correspondence between the visual complexity of the elevations and of the plans. The composite, or overall ranges for each set show the Prairie Style ($R_{\{\mu E+P\%\}} = 7.29$) and Usonian ($R_{\{\mu E+P\%\}} = 7.39$) houses to have a similar level of difference across all plans and elevations. However the range is higher ($R_{\{\mu E+P\%\}} = 15.82$) for the Textile-block set: more than double that of the other two.

8.8 Measuring Spatio-Visual Experience

Carl Bovill proposes that architecture is necessarily produced through the manipulation of rhythmic forms. He expands this idea to argue that fractal geometry allows the development of a 'quantifiable measure of the mixture of order and surprise' (1996: 3) in architecture and, moreover, that this measure reveals the essence of a building's formal composition. For Bovill, '[a]rchitectural composition is concerned with the progression of interesting forms from the distant view of the façade to the intimate details' (1996: 3). Both the historical and the methodological facets of Bovill's argument are worth considering in more detail.

In terms of his historical observation, it is possible to take a contrary position to Bovill's view about architectural composition and argue that, with only a few exceptions, the desire to capture the viewer's interest, by creating a progressive sequences of details, has not been a major goal in any established architectural theory since Ancient Rome (Kruft 1994). One of the exceptions relates to late twentieth-century phenomenologically-inspired theories of design that were critical of both Modern and Post-Modern architecture for failing to respond to the full range of human sensory needs (Norberg-Schulz 1980). However, these phenomenologists were also highly dismissive of designs that relied solely on formal manipulation to maintain visual interest, what they characterised as the ocular-centric fetish of architecture (Pallasmaa 2005). More conventionally though, throughout the history of architecture there have been many major movements that have completely rejected the concept that changing levels of detail in a building should be maintained or indeed emphasized. For example, Ancient Greek architects used elaborate geometric strategies (like *entasis* in columns) to artificially correct the visual changes that occur when a building is viewed from different distances and viewpoints. An equally strong view on this issue is seen in Renaissance architecture which was designed to be appreciated from a singular, neo-Platonic, perspective viewpoint. Some of the world's most admired and respected buildings do not possess a cascade of detail, and many that do would be considered too decorative for contemporary tastes.

The more important issue Bovill alludes to relates to the method of measuring changing fractal dimensions in response to the shifting position of the viewer. He effectively asks: why don't we measure the fractal dimensions of perspective views of buildings? The practical answer to this question—which is the same one given earlier in the present book and also offered by Bovill (1996)—is that it is impossible to meaningfully compare the measures derived from different buildings unless there is a common orthographic basis for them. But what if we are not interested in a comparison between buildings, but of the way a specific building is visually experienced from different positions in space? This is where Bovill's suggestion leads to an interesting alternative application of the fractal analysis method.

The human eye reads the world through a type of perspective lens, which is why it is impossible to actually experience an elevation in the same way that it is drawn. The problems of parallax ensure that in the 'real world' no two lines are ever, perceptually at least, parallel, we see them as converging. Thus, while plans and elevations are universal modes of representation and this is why they are useful, they do not replicate the way we view the world. But what if we wanted to measure the visual experience of a person as they viewed a building, or even as they moved through it? Admittedly, such a measure would provide at best an approximation of the visual experience of an individual with a certain height, visual acuity and facial dimensions (since the distance between the eyes has an impact on the way we view space), but for some purposes, such an estimate might be very useful. In this final section of the chapter, we offer several alternative ways of thinking about the fractal analysis of architecture and then demonstrate one of them using the *Robie House* as an example.

8.8.1 Alternative Perspective-Based Approaches

In this section, five variations of perspective-based applications of fractal analysis are proposed. These variations are based on a series originally described by Ostwald and Tucker (2007). Each relies on a different combination of viewpoints, perspective planes and picture planes. These variations also introduce the role of the cone of vision, something conspicuously lacking from past attempts to use fractal analysis for considering photographs of buildings. Furthermore, in what follows, for the sake of simplicity, the methods are described for structures which are primarily orthogonal in plan, although the principles are the same regardless of the geometry of the plan.

The first variation is called 'fixed position, one-point perspective'. It requires that a fixed viewpoint—one with the centreline of the eye at right angles to the dominant surface of a façade and with no correction for parallax—is used to

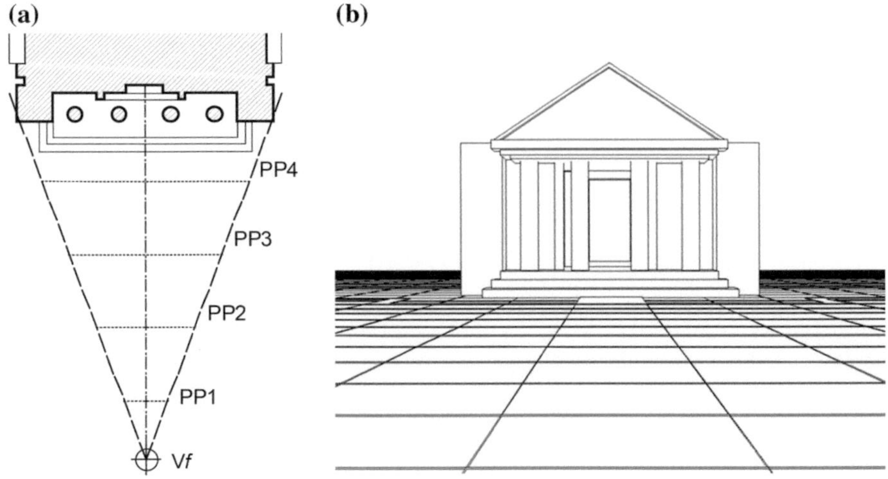

Fig. 8.15 Fixed position, one-point perspective alternative

construct the image that will be analysed (Fig. 8.15). The resultant image is effectively a one-point perspective view of a building façade. This variation has the advantage of using a consistent rule for setting up the image composition; at right angles to the façade and a certain distance from it, based on the dimensions of the building being considered and determined by a defined cone of vision, view limits and eye height. Thus, with these four additional measures recorded, this variation could be used to compare different buildings, if there was a sound reason to do so using this approach.

The second variation is 'fixed position, multi-point perspective'. It requires that a fixed viewpoint—with the eye *not* at right angles to the dominant surface of a façade and with no correction for parallax—be used to construct the image that will be analysed (Fig. 8.16). Thus, the image being generated for analysis is, for an orthogonal design, predominately a two-point perspective view (or three if it is tall enough). This variation has the challenge that there is no clear rule for setting the viewpoint angle relative to the façade, even though the image is more 'natural' than the first variation proposed because the fixed, one-point view is relatively artificial in its framing and is rarely how anyone experiences a building for any length of time. Thus, with some additional defined elements including angles and dimensions, this method could be repeated for different buildings, although a clear logic describing the rationale for the elements would need to be developed.

The third alternative is 'variable or sequential position, one-point perspective', which commences by drawing a line at right angles to the dominant surface of a façade and then dividing this line into a number of equal-length segments. The end of the line, furthest from the façade, is then used to position the eye and generate the first image for analysis. Then the second last segment of the line is used to locate a new viewing position, generating an image that is slightly closer to the façade. The

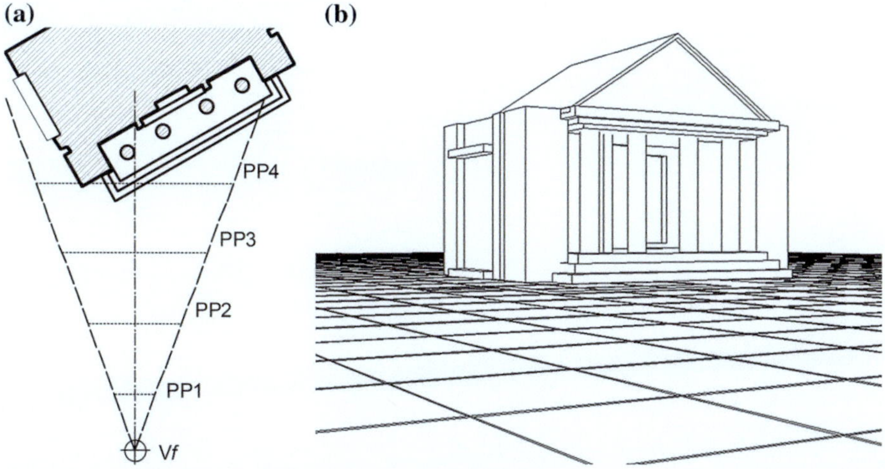

Fig. 8.16 Fixed position, two-point perspective alternative

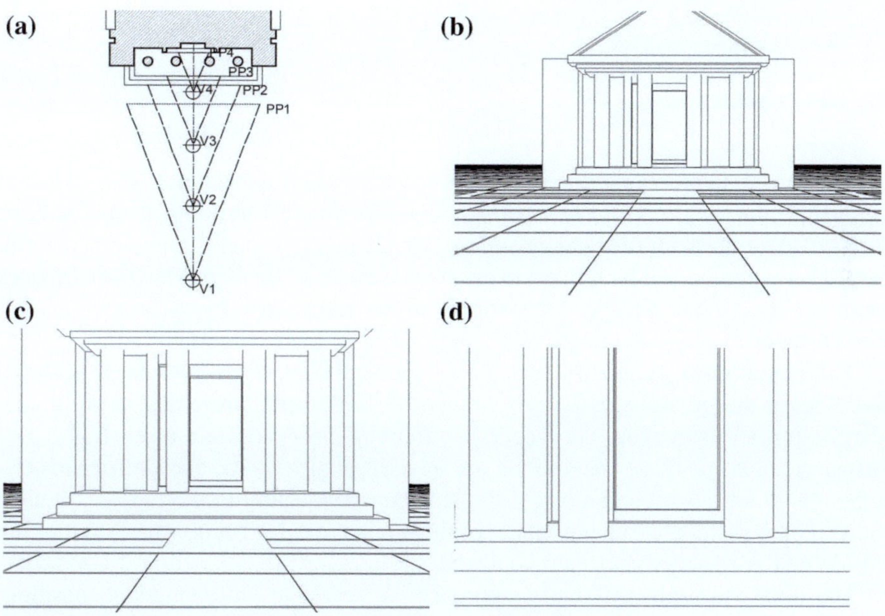

Fig. 8.17 Variable position, one-point perspective alternative

third view is created from the next closer segment and so on (Fig. 8.17). At all times, the eye is at right angles to the dominant surface of the façade, there is no correction for parallax, and the cone of vision of the eye (or its high-acuity zone) determines the extent of the façade that is analysed at each step. This also means

(a) (b)

(c) (d)

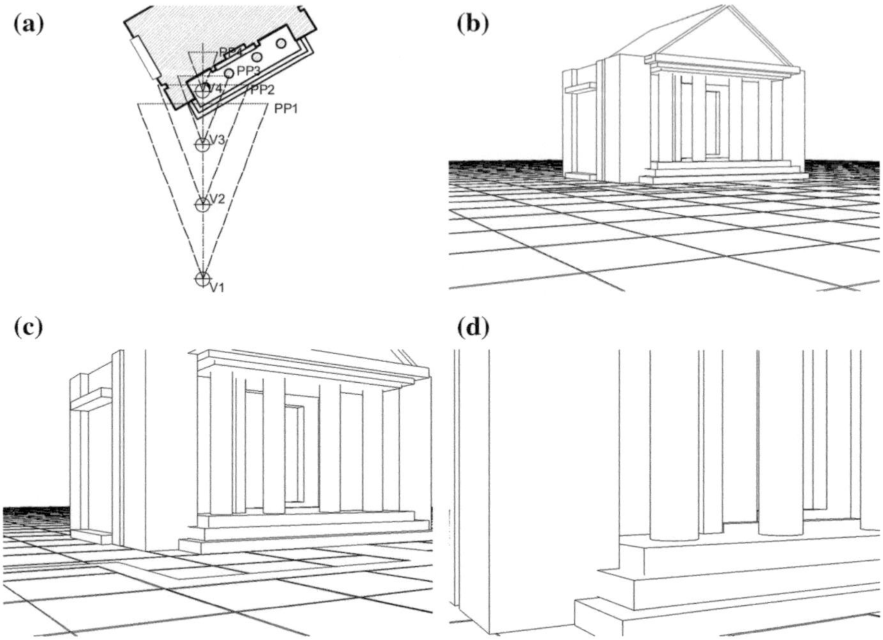

Fig. 8.18 Variable position, two-point perspective alternative

that, with each iteration a reduced portion of the façade is considered. This variation is close to the way a human would visually experience a building if they walked directly towards a façade and examined its changing visual complexity at each step. This variation can be refined to establish a range of standard viewing distances along a line to the façade, allowing it to be repeatable for a wide range of circumstances.

The penultimate option is 'variable or sequential position, two-point perspective'. As in the previous variation, a range of viewpoints positioned along a line, starting further away from the façade and moving closer to it are used (Fig. 8.18). However, none of these viewpoints are at right angles to the façade's geometry, although all are positioned along a single vector connecting to the façade. At each viewpoint the standard cone of vision of the human eye determines the extent of the façade that is depicted and then analysed.

The final and most flexible approach is 'variable or sequential position, multi-point perspective'. In this version, a distinct path is identified—to or through a building—that is relevant for the assessment of that design. At evenly spaced intervals along this path viewpoints are established for the generation of perspective images (Fig. 8.19). At each viewpoint the cone of vision of the human eye determines the extent of the building that is recorded. This is the closest of any of the variations to measuring the visual experience of a person approaching or using a building. It suggests that people rarely view buildings along a single vector and

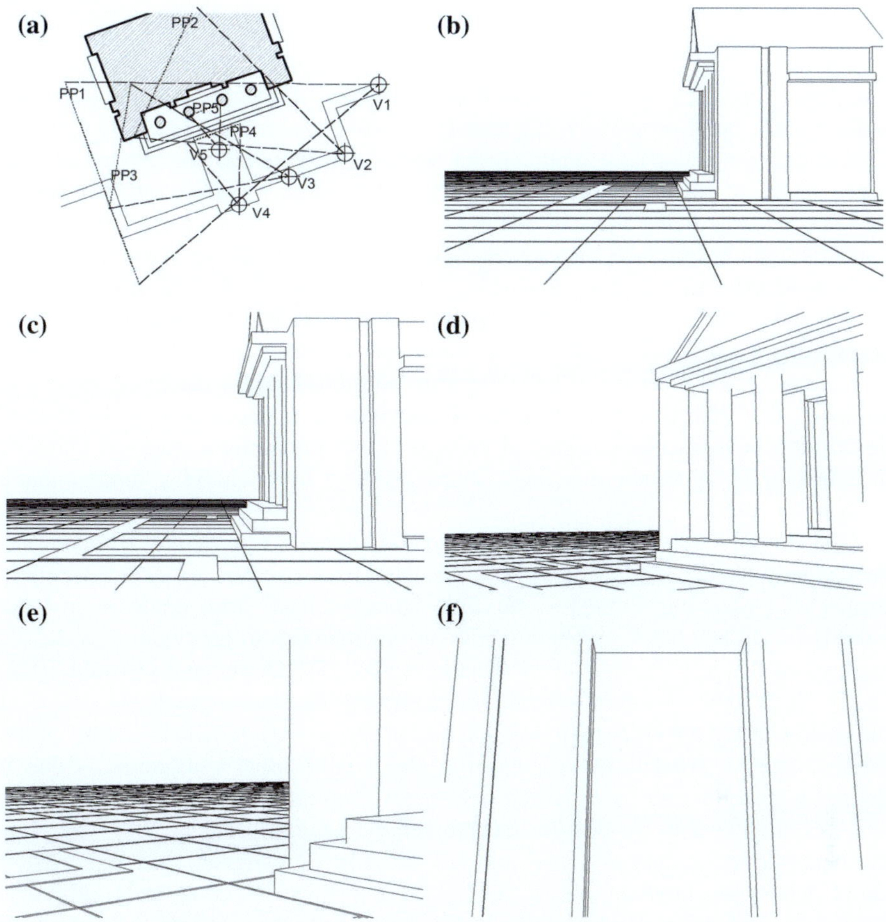

Fig. 8.19 Variable position, multi-point perspective alternative

acknowledges the importance of the limits of human vision. However, for this variation to be useful, there must be a logical rationale for determining the particular path chosen, along with the eye level and the intervals used to generate the perspectives. Without this information, or at least a well-reasoned approach to it, this last method would appear to be the least consistent and useful of the five. Nevertheless, it is actually ideal for testing a common argument about Wright's architecture. Specifically, it has been suggested that the way a person experiences space and form while moving along a defined passage through Wright's architecture, creates a carefully choreographed visual experience. To test this idea the changing fractal dimensions that occur along this path may be measured and compared with the theorised conditions. This approach is demonstrated in the next section.

8.8.2 Method and Results

The idea of undertaking an analysis of the experience of walking into and through a building has been previously suggested by Bovill, when he argued that fractal dimensions should at least maintain the same level of visual complexity as 'one approaches and enters a building' (1996: 3). While there are no examples of fractal analysis actually being used in this way, studies of the visual experience of the approach path both to and through the buildings by Frank Lloyd Wright have also been conducted in the past. Probably the most famous of these studies is by Hildebrand (1991) who, drawing on prospect-refuge theory (Appleton 1975, 1988), argues that a distinctive pattern of spatio-visual experience is found in Wright's architecture when a person follows the path from the entry to the living room. As evidence for this position, Hildebrand provides a phenomenologically-framed description of this route in terms of changing visual conditions, through many of Wright's most important dwellings, along with an accompanying diagrammatic analysis of each.

Hildebrand's theory of the pattern of spatial experience in Wright's houses has been widely quoted and is seemingly an accepted reading of the essence of Wright's formal and planning strategies. Multiple attempts have been made to test the validity of Hildebrand's argument using various methods to provide a quantitative analysis of the paths he identified (Bhatia et al. 2013; Dawes and Ostwald 2014, 2015). Such examples demonstrate that analysing the spatio-visual experience of movement along a path is both possible and useful in Wright's architecture. In the present context, what is most interesting about Hildebrand's argument is that it defines a path through space which is allegedly significant for analysis. The existence of Hildebrand's proposition overcomes the general problem with several of the perspective variants of fractal analysis described previously, that the decision about where to produce views for measuring may be completely arbitrary. Therefore, following Hildebrand's logic, and using the variable or sequential position, multiple-point perspective approach, this section analyses the visual properties of a path through the *Robie House*. The purpose of this application of the fractal analysis method is to test the hypothesis that the degree of visual complexity observed while moving into and through Wright's *Robie House* will, on average, reduce from beginning to end. This hypothesis is broadly in line with one facet of Hildebrand's (1991) argument, and replicates accepted views in environmental preference theory about positive aspects of the experience of the interior.

There are actually multiple potential paths through the ground floor of the *Robie House* that fit Hildebrand's general definition. The *Robie House* path that is analysed hereafter follows the everyday entry route, rather than the formal path that a guest would follow. It proceeds from the street through the forecourt and into the house, through the central hallway, upstairs and around the circulation zone into the living room, where it ends. Along this path perspective images at one metre increments (steps) were generated and analysed using the box-counting method. The perspective eye level for the images was 1.65 m (to match Wright's stature)

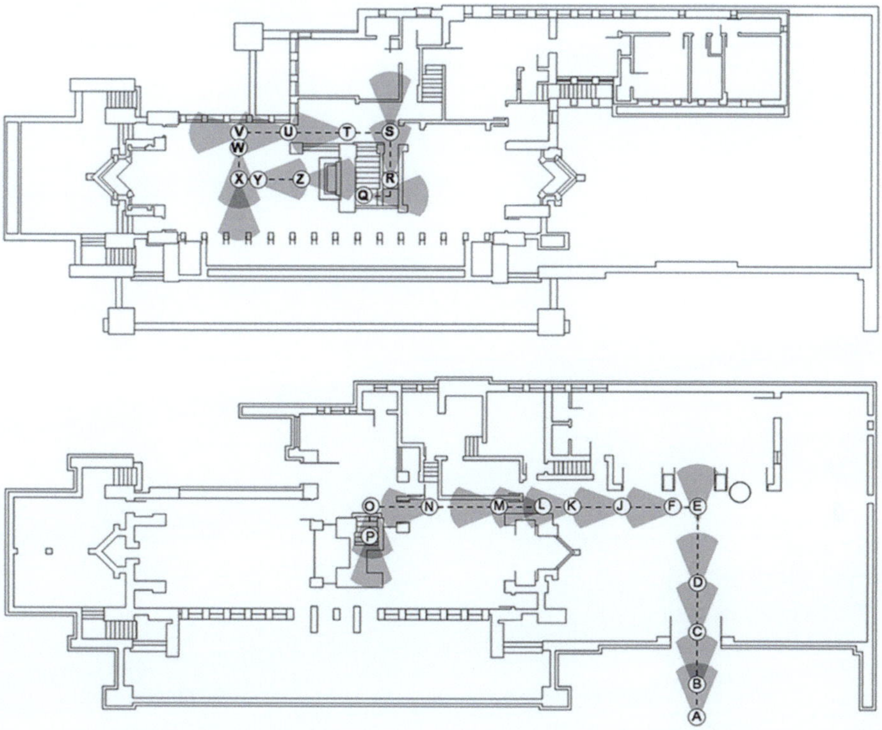

Fig. 8.20 Perspective route through the ground and upper floors of the *Robie House*

and a high-acuity cone of vision of 90° was used. The location of the path through the *Robie House* is recorded in Fig. 8.20. This figure also includes a diagrammatic representation of the direction and location of twenty-four of the fifty-two cones of vision that were used to generate the perspectives. Selected perspective views that are indicative of the more interesting positions along their path and their *D* results are presented in Table 8.10. The complete set of results, which chart the changing visual complexity of passage through the *Robie House*, are given in Fig. 8.21.

While the purpose of the present section is not explicitly to test Hildebrand's argument, it is of interest that one of his suggestions is that these paths commence with a high degree of mystery and visual complexity and that this property reduces towards the end of the path. Importantly, this reduction is not meant to be a linear sequence, but rather a shifting pattern of rising and falling levels that gradually reduce (from left to right in Fig. 8.21). When this property was previously tested using isovists (Ostwald and Dawes 2013b), a marginal fall in mystery and spatial complexity was noted, largely as a result of the spatio-visual geometry of the plan, which is more complex on the lower level and less so in the relatively open-planned upper floor. However that study did not take into account the elaborate decorative modelling and detailing of the roof that was so typical of Wright at this time.

Table 8.10 Selected perspective views, with their fractal dimension results (D)

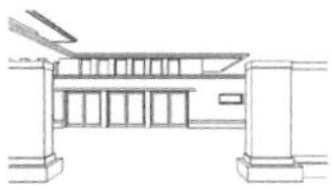	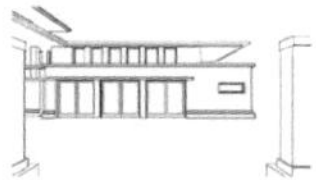	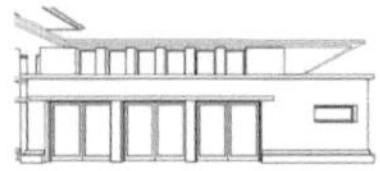	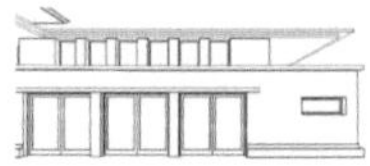
A(1) $D = 1.4658$	B(2) $D = 1.4480$	C(4) $D = 1.5919$	D(5) $D = 1.5640$
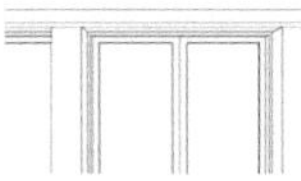			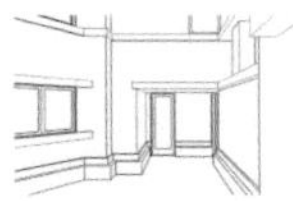
E(11) $D = 1.4098$	F(12) $D = 1.4944$	I(16) $D = 1.5350$	J(18) $D = 1.4218$
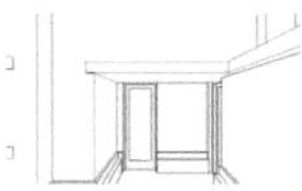	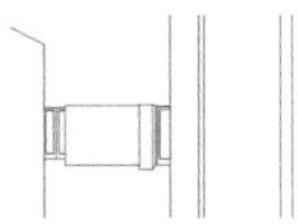	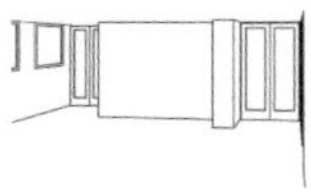	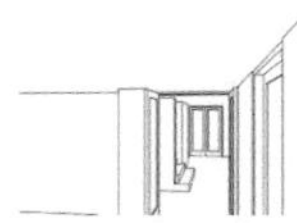
K(20) $D = 1.3901$	L(24) $D = 1.2937$	M(25) $D = 1.4119$	N(28) $D = 1.4801$
	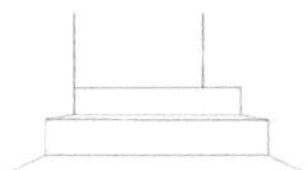		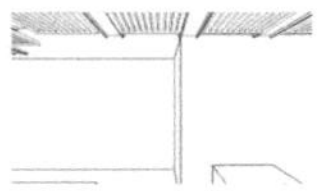
O(33) $D = 1.3530$	P(35) $D = 1.1925$	Q(36) $D = 1.5795$	R(38) $D = 1.4157$
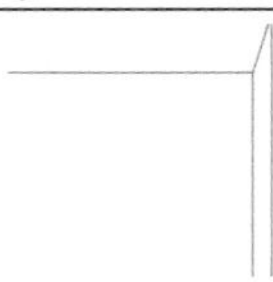	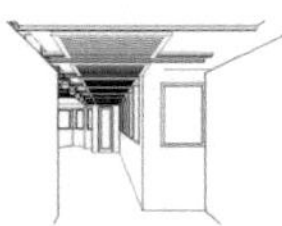		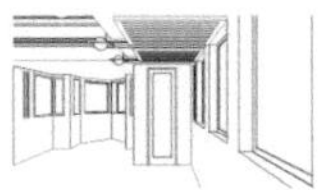
S(39) $D = 1.2348$	T(41) $D = 1.5867$	U(44) $D = 1.6231$	V(48) $D = 1.5659$
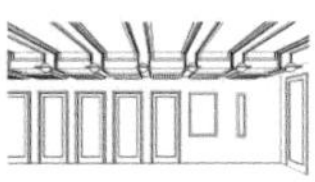	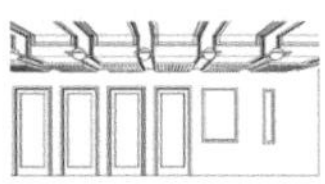	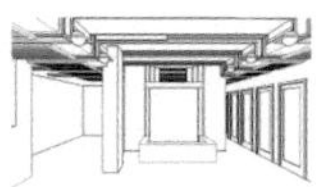	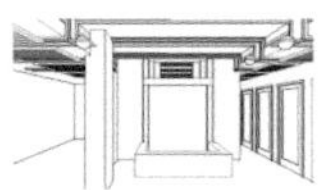
W(49) $D = 1.6068$	X(50) $D = 1.5688$	Y(51) $D = 1.6022$	Z(52) $D = 1.5898$

The route reference is denoted as a letter, the graph reference is the number following in brackets

The fractal analysis results for the three-dimensional visual complexity of the *Robie House* path are variable, but generally rise from left to right. This outcome is heavily influenced by the decorative mouldings in the living space and the window and lighting forms, none of which were considered in the previous analysis. A more comprehensive study of a larger number of works would ultimately be required to test Hildebrand's assumptions, along with a much tighter definition of what visual complexity actually entails in his analysis of Wright's architecture. For example, if Hildebrand's definition of visual complexity is largely spatial, the evidence supports him, but if it includes decorative modelling, the data does not support him.

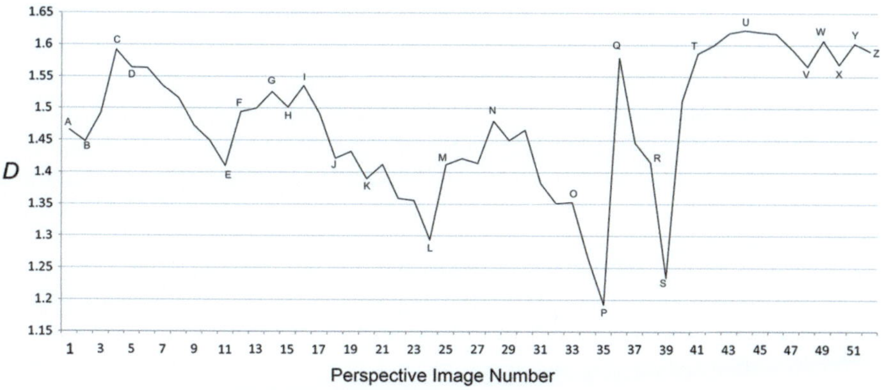

Fig. 8.21 Fractal dimensions of perspective views generated along a path through the *Robie House*

Ultimately, the results for the path analysis demonstrate how visual experience changes as we move through a building. The low points in the graph generally relate to positions on the pathway where the viewer is in very close proximity to the building and thus there is little to see. The higher results are for views that take in more information, or are further from surfaces and other limits caused by physical forms or occlusion. While this result might be obvious, it would be interesting to compare the same method for the *Villa Savoye*, which does not possess such a high degree of detail and might, conceivably, generate a more consistent set of *D* results. Further speculation on this topic is beyond the scope of the present work. Nevertheless, Bovill argues that, '[a]s one approaches and enters a building, there should always be another smaller-scale, interesting detail that expresses the overall intent of the composition' (1996: 3). The *Robie House* displays a consistent level of growth in visual complexity as it is traversed. With an almost 43 % range in the results, the experience of the form of the *Robie House* (rather than its materiality) is clearly one of increasing complexity.

8.9 Conclusion

Frank Lloyd Wright's architecture is conventionally divided into several stylistic periods, some of which, including his Prairie and Usonian works, were refined over more than one hundred constructed works. Historians and critics regard both of these styles as exhibiting a high degree of consistency and, for the five Prairie style works examined in this chapter, the results strongly support the conventional interpretation. While there is very little correspondence between the form of their plans and of their elevations (formal coherence, $R^2 = 0.0849$), viewed separately, the plans and elevations in the set do exhibit a degree of uniformity in their mean, median and range results. Similarly, the Usonian houses have more closely related plans and elevations across the set of five works (formal coherence, $R^2 = 0.6902$), with the difference between the means for plans and elevations at <10 %. The optimal sub-set of Usonian works also displays mean composite results that are <2 %. In contrast, the Textile-block works are more distinctive and diverse, being only five houses from a short-lived, almost experimental style. While the trendlines for plans and elevations in the Prairie and Usonian Styles were almost flat, for the Textile-block houses both rose over time. Thus, Wright's Textile-block works suggest an evolving experiment in design and construction, rather than the steady-state results for the other two styles that had been refined over longer periods or more extensive examples.

Finally, after considering the standard way in which Bovill and many others (including ourselves) have used orthographic images of historic buildings for fractal analysis, an alternative range of perspective variations has been proposed in this chapter. Each of these is, in a sense, more realistic than the orthographic view because they offer a closer simulation of the way in which humans experience architecture. They are superior to the standard method in all but one, important, way. Bovill's method has the advantage that it is a straightforward, repeatable and potentially universal process. It has to be the starting point for any comparative study. Nevertheless, the five variations set out in this chapter, and especially the final one, suggest that there are powerful applications of fractal analysis that have not yet been developed or tested but which will be useful for producing a more nuanced or detailed reading of visual complexity in the built environment.

Chapter 9
The Avant-Garde and Abstraction

The works of three architects who have variously been described as Late Modernists or Neo-Modernists are examined in this chapter. Peter Eisenman, John Hejduk and Richard Meier first came to prominence in the 1970s as members of the 'New York Five' and were later promoted as forming the core of the 'Whites' group of *Avant-Garde* Modernists. Jencks (1977) describes their architecture of this era as involving a high level of formal abstraction that is both a continuation of Modernism's aesthetic agenda and, more controversially, a rejection of its functionalist expression.

The analysis undertaken in this chapter is of the plans and elevations of five early career designs from each of these three architects. For Eisenman, the five houses are from his first numbered sequence of works: *Houses I*, *II*, *III*, *IV* and *VI*. Eisenman's *House V* is not included in the set, as it was not developed to a sufficient level of detail to analyse it. Hejduk's five designs, also part of a numbered series of experimental works called the *Texas House* series, are: *Houses 1*, *4*, *5*, *6* and *7*. Like Eisenman's numbered works, not all of Hejduk's designs in the Texas House series were published in full, and so *Houses 2* and *3* are not included in the set. Meier's five designs are the *Smith House*, *Hoffmann House*, *Saltzman House*, *Douglas House* and the *Shamberg House*. The plans and elevations used for the calculations in this chapter were all adapted from published design or construction drawings (Hejduk 1979, 1985; Eisenman et al. 1975, 1987; Eisenman and Dobney 1995; Meier et al. 1996; Cassarà 1997).

The final section of the chapter is inspired by Kenneth Frampton's critique of the New York Five, in terms of the way their formal compositions can be differentiated using rotational expression. Developing Frampton's theme, the chapter examines one design by each of the three architects, using sub-cardinal orthogonal elevations to compare the visual properties of these three buildings when viewed both frontally and from thirty-six different rotational positions. The three houses investigated in this way are Eisenman's *House I*, Hejduk's *House 7* and Meier's *Hoffman House*.

M.J. Ostwald and J. Vaughan, *The Fractal Dimension of Architecture*,
Mathematics and the Built Environment 1, DOI 10.1007/978-3-319-32426-5_9

9.1 The New York Five

Arthur Drexler was appointed director of the New York Museum of Modern Art's department of Architecture and Design in 1956. Over the following two decades he curated several influential exhibitions and hosted a series of informal meetings inspired by European salons, where designers and theorists could explore new ideas. One such informal collective called themselves the Committee of Architects for the Study of the Environment (CASE). While details about the formation, membership and name of this group have been disputed (Drexler 1975; Crosbie 2011), it is well documented that at one particular meeting in 1969, the works of five architects were informally displayed and discussed, with Drexler arguing that they could be 'said to constitute a New York school' (1975: 1). These five architects were Peter Eisenman, John Hejduk, Richard Meier, Charles Gwathmey and Michael Graves. Drexler proposed that all five shared a fascination with three-dimensional formal composition, producing abstract geometric designs in the spirit of early works by Le Corbusier, Giuseppe Terragni and Louis Kahn. Furthermore, the work of the five also featured a common scale and sense of materiality which, coupled with an often intense intellectualisation of the design process, distinguished their approach from that of other New York practices of the time.

Despite Drexler's observations about their similarities, there were also clear differences. For example, Eisenman and Hejduk were the most exploratory of the five, their work often being framed as a type of abstract, serial experimentation with form. In contrast, Meier's and Gwathmey's designs were more practical and commercial; although both displayed a similar concern with formal manipulation, this was typically undertaken within the constraints of a distinct site, program and budget. The connection between these four architects and the final member of the group, Graves, was even more tenuous, with the latter's historicist, decorative and painterly predisposition visible in even his earliest works. Nevertheless, the book *Five Architects: Eisenman, Graves, Gwathmey, Hejduk, Meier* was published in 1971, including projects by each of the designers along with essays by Colin Rowe and Kenneth Frampton (Eisenman et al. 1975). These essays famously suggest that in these architects' works, the form or shape of a building is more important than its program or site. Thus, while these architects appeared to be producing work in the style of the early Modernists, each, in different ways, rejected the notion that form is necessarily derived from functional expression.

Following the publication of *Five Architects*, the architectural media christened the group the 'Whites', not only as a reference to the dominant colour of their architecture, but also to its formal purity. The five Whites might have been swiftly forgotten by history if it hadn't been for the journal *Architectural Forum*, which published a strong repudiation of their arguments in its May 1973 issue. Under the title 'Five on Five', Romaldo Giurgola, Allan Greenberg, Charles Moore, Jaquelin T. Robertson and Robert Stern used the *Architectural Forum* to admonish the Whites on their lack of social, cultural and historical sensitivity (Giurgola et al.

1973). 'Five on Five' effectively marshalled the polemical theories of Robert Venturi against those of Colin Rowe, who had by then been identified as most clearly articulating the underlying concerns and values of the Whites (Somol 1998). However, rather than undermining the Whites, 'Five on Five' effectively raised the profile of all concerned, pitting them against the newly named 'Grays', and thus allowing the concerns of a revived Modernism to be positioned in opposition to those of a fledgling Post-Modernism. That the architectural press artificially promulgated much of this debate is, in hindsight, obvious. When, in the late 1970s, Graves renounced Neo-Modernism in favour of Post-Modernism it was presented as a momentous event, and when Eisenman and Robertson formed a joint partnership it was similarly regarded as newsworthy. But in reality, both the Whites and the Grays shared a common concern for raising the quality of architectural debate, including testing ideas both in print and in practice, and their similarities were often more marked than their differences (Jencks 1977).

In recent years, the architecture of Meier, Eisenman and Hejduk has been described as Late Modernist, although this category is probably more apt for Meier than the others. Eisenman's and Hejduk's architecture continued to evolve in different ways over the decades that followed, as new ideas and new media offered opportunities for continued experimentation. Therefore, their work remained more in the spirit of the twentieth-century *Avant-Garde* than Modernism, hence the title they are often collectively given, which is also used for this chapter.

One of the reasons the early designs of these three architects are of interest in the present context is because they all feature a distinctive approach to form-making. Whereas historicist, eclectic and humanist concepts shaped the works of the Grays, few of these factors or themes have an innate formal expression. This means that, with some exceptions relating to factors like orientation and address, the values of the Grays cannot be readily equated to levels or types of formal expression. This is also why, in Chap. 10 on Post-Modern architecture in the present book, the issues studied using fractal analysis are those pertaining to permeability as a reflection of function. For the three Whites, whose architecture of that time could be regarded as the result of a series of aesthetic experiments, fractal analysis is ideal for examining and comparing changing formal qualities. Finally, their work is of interest because Salingaros (2004) is critical of the alleged lack of distribution of scale in architecture of this type, an issue which can be tested using fractal analysis.

The origins of Salingaros's (2004) criticism of *Avant-Garde* Modernism can be traced to his view that the early designs of the Deconstructivist movement either lacked human scale or did not provide a sufficient cascade of detail across all levels to accommodate divergent social and phenomenal needs. That Salingaros has such an opinion is intriguing, because he is well known for using fractal geometry analogously, to promote new models of architectural and urban planning (2005, 2007). However, Salingaros's use of fractal *geometry*, rather than fractal *dimensions*, is where the differences of opinions lie. For example, much like Christopher Alexander, Salingaros views the deep, unpredictable order of the vibrant city through the lens of fractal geometry, identifying beneficial aspects and offering strategies to achieve some of these properties. While such a definition is difficult to test in the present context, it is possible to measure the actual fractal dimensions and cascade of detail present in the early works of the Whites to determine if their properties are in accordance with Salingaros's position.

9.2 Peter Eisenman

Peter Eisenman, is an architectural writer, designer and theorist who 'has constantly striven to reveal the critical function of architecture' (Whiting 2004: 394). Eisenman was born in 1932 in New Jersey, completing his Bachelor of Architecture in 1955 at Cornell University and his Master of Science in Architecture at Columbia University in 1960. The following year he moved to England, where he taught at Cambridge University and commenced a Ph.D. under the tutelage of Colin Rowe. Rowe's theories proposed the systematic suppression of 'the semantic dimension of architecture in favour of a more abstract and conceptual analysis of visual form' (Mallgrave and Goodman 2011: 31). After returning to America in 1967, Eisenman formed the Institute for Architecture and Urban Studies (IAUS), described by a former member as a 'think tank, publishing house, proving ground and platform for young architectural talents and provocateurs' (Ockman 2011: ii). Eisenman was both director of IAUS and joint editor of its journal, *Oppositions*, until its final issue in 1982.

Although Eisenman worked on the design of several buildings in the early 1960s, it wasn't until he commenced his *House* series (1967–1980) that his first completed projects were produced. In that series, Eisenman set out to investigate the classic nine-square house, a seemingly universal planning type which was influential in both Renaissance and Modernist architecture. However, whereas, for example, Palladio's nine-square houses explored notions of harmonic proportions and Le Corbusier's examined functional expression, for Eisenman this classic *parti* provided the basis for a series of abstract, but rigorous, form-making exercises. Sanford Kwinter describes Eisenman's approach at this time as being to subject this architectural type to a 'maniacally articulated' series of 'disruptions and deletions' (1995: 10), applying rules seemingly without any regard for the resultant form of the design or its capacity to function as a home. Mario Gandelsonas suggests that in such projects Eisenman was 'drawn to concentrate his attention on the only objective material provided by architecture, that is form itself' (Gandelsonas 1982: 8). Thus, Eisenman's anti-functional and non-representational designs were not simply a means to an end, but rather an end in themselves. C. Ray Smith describes Eisenman's work up to this period as comprising an architectural language that was 'photogenic, cubistic, and volumetric, interlocking and ambiguous, ... It is also purely intellectual, coldly alienating, [and] maddeningly non-functional' (1980: 231).

After completing the *House* series, many of which remained unbuilt, Eisenman entered into an architectural practice with one of the original Grays, Jaquelin T. Robertson, and their firm found international renown in the construction of public buildings and urban projects including the 1981 *Berlin IBA Social Housing* and the 1983 *Wexner Centre for the Visual Arts*. In 1986 Eisenman collaborated with philosopher Jacques Derrida on *Choral Works*, a landscape project for the Parc de la Villette in Paris. After that time Eisenman's designs, in part inspired by Derrida's ideas, were famously presented as part of the 1988 Deconstructivist Architecture exhibition at the Museum of Modern Art, New York, and within two years he began to experiment with digital modelling in an attempt to accommodate a higher level of formal complexity in his work (Galofaro 1999). This period in his career also included several projects inspired by the work of the Post-Structuralist philosophers Gilles Deleuze and Felix Guattari, whose theory of folding was used by Eisenman as a rationale for his designs for the 1991 *Rebstock Master Plan* and the 1996 *Aronoff Centre for Design and Art*. More recently, Eisenman has completed the 2005 *Memorial to the Murdered Jews of Europe* in Berlin and the 2013 *City of Culture of Galicia*, both of which are described by critics as having Piranesi-like qualities, a reference to the multi-layered, evocative and often gloomy delineations of eighteenth-century Italian artist Giovanni Battista Piranesi (Kwinter 1995; Whiting 2004).

Eisenman's ultimate legacy as an architect may well be his demonstration that a range of philosophical and cultural theories—including Post-structuralism and Deconstructivism—can be productively linked to architectural form generation and interpretation. Eisenman's many influential publications include *Notes on Conceptual Architecture* (Eisenman et al. 1970), *In My Father's House Are Many Mansions* (Eisenman 1979), *The Futility of Objects* (Eisenman 1984), *Houses of Cards* (Eisenman et al. 1987) and *Diagram Diaries* (Eisenman 1999). However, as other scholars have noted, the seeds of this later work were planted in the first few of the designs of his *House* series. It was in these small works that Eisenman 'laid down the choreographic lexicon from which his later work would never fully depart' (Kwinter 1995: 10–11).

9.2.1 Five Houses (1968–1975) by Peter Eisenman

Greg Lynn describes Peter Eisenman's early house designs as being wholly concerned with 'layered traces and imprints of orthogonal movement and transformation within a turbulent but nonetheless closed system of nonfigurative cubic grids' (2004: 162). Sanford Kwinter likens Eisenman's forms to ripples in a pool of water, because their 'structure always emanates from an initial pattern that is knocked away from equilibrium. The disturbance then travels, reaches a limit, then turns back toward itself to form a self-interfering wave' (1995: 13). This generative, or iterative, process, where the traces of a series of form-making stages can be seen in the final object, is readily visible in five of Eisenman's earliest works, *Houses I, II, III, IV* and *VI*. In each case, Eisenman conceptualised the house as an abstract series of diagrams that recorded the process wherein the traditional architectural plan (its 'grammar' in the linguistic analogies of the era) could be progressively modified using a set of external rules ('applied syntactical structures'). According to Eisenman, these rules 'differed from the traditional idea of a formal essence in that they had no specific form attached to them, nor did they suggest any specific form, but rather could be considered as unformed possibilities for organization' (Eisenman 1999: 62–63).

The designs that resulted from this process of formal or structural abstraction were described at the time as 'non-representational' architecture, a reference to the artistic Avant-Garde of the mid-twentieth-century in Europe. Whereas representational art depicted something which was, on some level at least, recognisable in its subject matter, non-representational art was sufficiently abstract or hermetic as to deny any clear content other than the work itself. Thus, such works demonstrate that a painting could not simply be assumed to be 'of something,' instead it could just be the end-state of a process or an object in its own right (Greenberg 1961). Therefore, the forms found in Eisenman's houses were not a response to the needs of a client, a particular site, or even cultural expectations of domesticity. They were, instead, an expression of the process which generated them; the fact that a few were built on rural sites, and did function as homes, was largely incidental.

House I, also known as the *Barenholtz Pavilion*, was completed in Princeton, New Jersey, in 1968. Commissioned by the Barenholtz family, Eisenman describes *House I* as an attempt to develop an architecture that is created 'in a logically consistent manner, potentially independent of its function and meaning' (Eisenman 1975: 15). The completed house is timber-framed, with white painted timber panels on the exterior and interior (Fig. 9.1).

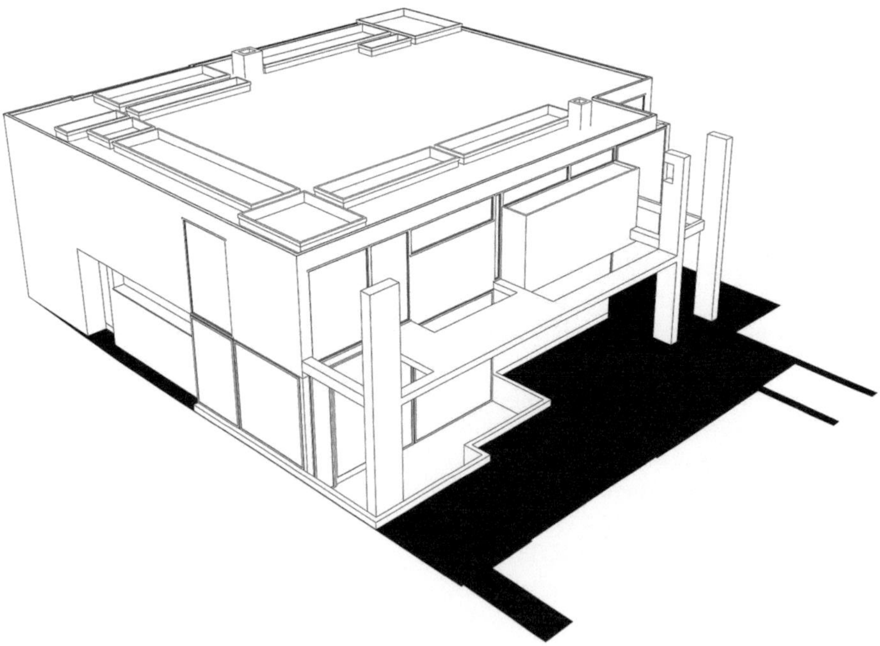

Fig. 9.1 *House I*, perspective view

House II, in Hardwick, Vermont, was designed for the Falk family and completed in 1970. The house is sited on the crest of a low hill with views in three directions. It is timber-framed, clad in painted plywood panels and does not feature the type of 'traditional details associated with conventional houses' of the region (Davidson 2006: 37). Silvio Cassarà describes *House II* as being focused on the architectural expression of two types of volumetric and structural relationships. 'To articulate these ways of conceiving and producing … information in *House II*, certain formal means were chosen, each involving an overloading of the object with formal references' (2006: 82). In *House II*, the layering of voids within the structure produces a series of perforated planes that intersect with each other, leaving the relationship between exterior and interior spaces ambiguous.

House III was designed for the Miller Family in Lakeville, Connecticut, and completed in 1971. Like *Houses I* and *II*, it is timber-framed and clad, with a painted finish. The house has been described as an attempt to 'produce a physical environment which could be generated by a limited set of formational and transformational rules' (Eisenman and Dobney 1995: 34). *House III*'s position in Eisenman's lexicon is associated with the introduction of the 45° angle into an otherwise orthogonal 90° plan.

House IV, while designed around the same time as *House III*, also marks a return to the planning strategies of *Houses I* and *II*. An unrealised project designed for a site in Falls Village, Connecticut, *House IV* is an elaborate investigation of the process of design transformation wherein various structural systems are allowed to trace solids and voids in the overlapping multi-level plan of the house. *House IV* is significant because the formal transformations occur in three dimensions; prior to this, the operations in Eisenman's series of designs were essentially planar in nature.

House VI was built on a rural site in Cornwall, Connecticut, in 1975 as a weekend home for the Frank family. Eisenman perceived this house as the last of the series, and 'the end of a particular process' (Davidson 2006: 80). It is a two-storey timber-framed building with a useable rooftop. The house is constructed from painted timber panels externally and painted wallboard internally. Eisenman explains that the design sets out to invert the 'archetypal spatial relationships' found in canonical design, including questioning the relationships between 'oblique/frontal views, inside/outside, center/periphery [and] beginning/end' (Davidson 2006: 96). In this way it could be seen that all of these spatial relationships might take equal precedence; the formal complexity of the plan and the elevations would thus share similar characteristic qualities.

After *House VI* Eisenman's numbered designs became more theoretical and experimental, with the seventh, eight and ninth works in the series being largely incomplete or unpublished and the concluding trilogy of works, *Houses X*, *11a* and *El Even Odd* charting a new theoretical trajectory. Throughout most of this series, Eisenman sought to develop a consistent rule-based form-making process that not only rejected function, but also gravity and environment. If he was successful in this endeavour, then there should be evidence in his later (Houses *IV* and *VI*) plans and elevations of a high level of formal similarity. This is because without gravity, site or program there are no forces, other than Eisenman's rules, shaping the appearance of each elevation; and without functional spaces nothing, other than Eisenman's rules, are moderating the form of the plan (Table 9.1). Table 9.1 provides example plans and elevations from this set.

9.2.2 Eisenman Houses, Results and Analysis

The fractal dimension results of Eisenman's houses show that the design with the lowest average visual complexity is his first, *House I* ($\mu_E = 1.3035$) and the highest is the diagonally formed *House III* ($\mu_E = 1.4609$), defining a range of $R_{\{E\%\}} = 30.52$ for all elevations of the set. Thus, as a group, the formal compositions of the twenty elevations are visually dissimilar. The median for all elevations is 1.3940 and the standard deviation is 0.0821. The highest individual elevation result is from *House III* ($D_{E1} = 1.4856$) and the highest individual plan is from the first floor of this house ($D_{P1} = 1.4680$). However, the highest plan average is found in *House II* ($\mu_P = 1.3972$) and the lowest plan average is from *House I* ($\mu_P = 1.3309$). Unlike the elevations, the complexity of the plans has some level of correspondence ($R_{\{P\%\}} = 16.57$), with the median for all plans being 1.3538 and the standard deviation 0.0457 (Tables 9.2 and 9.3).

When all of the plan and elevation results are considered together, the aggregate is $\mu_{\{E+P\}} = 1.3707$ and the composite range is $R_{\{\mu E+P\%\}} = 11.28$. When the optimal sub-set is further examined (*Houses II*, *IV* and *VI*), the aggregate result for plans and elevations is only minimally reduced, to $\mu_{[E+P]} = 1.3701$, although the composite range is much reduced, to $R_{[\mu E+P\%]} = 3.09$. This is a notable shift in range, with the visual correspondence between all plans and elevations in the optimal set being very similar.

When graphed (Fig. 9.2), the results for the five Eisenman houses confirm several common interpretations of the development of his work, and also some of its unique properties. In the first instance, over the course of designing his first five houses, Eisenman described a growing fascination with the creation of form at the expense of function or site. Past research has shown that one sign of a lack of connection to site can be that the fractal dimensions of different elevations of a house are very similar (Ostwald and Vaughan 2011). A review of Eisenman's results shows that, while the first two houses have a markedly different look to each of their elevations (*House I* $R_{E\%} = 18.08$, *House II* $R_{E\%} = 26.70$), the ranges for the elevations of the last three houses are substantially reduced (*House III* $R_{E\%} = 4.43$, *House IV* $R_{E\%} = 6.66$, *House VI* $R_{E\%} = 7.72$). Past research has also suggested that there is a practical threshold to plan complexity below which there is little or no functional differentiation of spaces and above which a house would be difficult to inhabit. The range of individual fractal dimensions for all of the plans in this set ($1.3023 < D_P < 1.4680$) supports this notion.

A final observation arising from the data is that if, as Eisenman argues, function is almost completely ignored, then it might be imagined that both plans and

Table 9.1 Eisenman set, example elevations and floor plans (level 4 representation—not shown at uniform scale)

House I

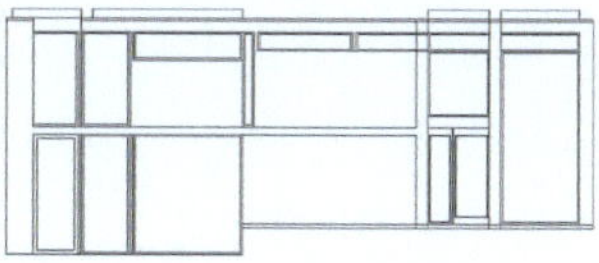

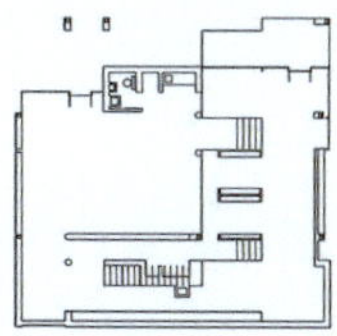

House II

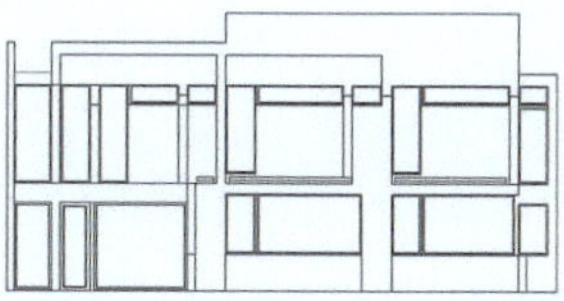

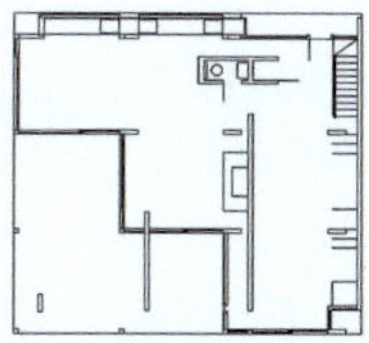

House III

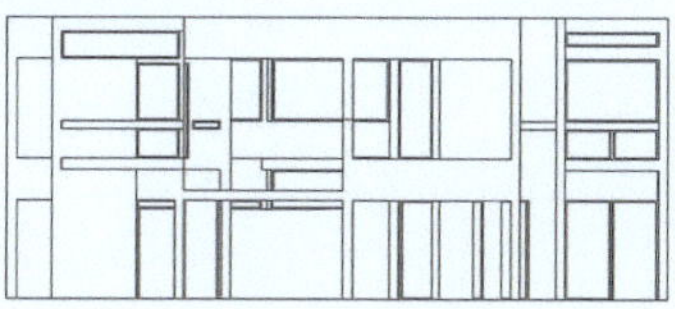

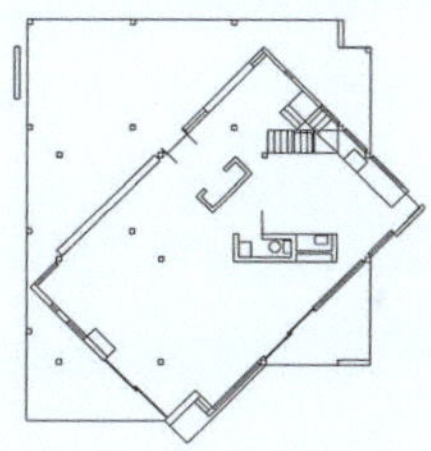

House IV

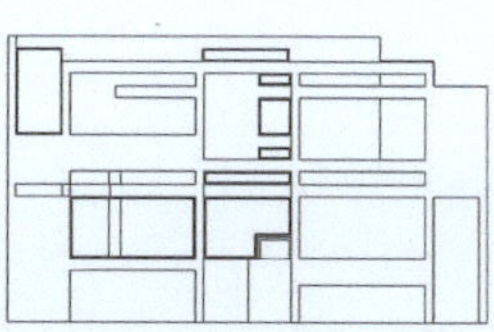

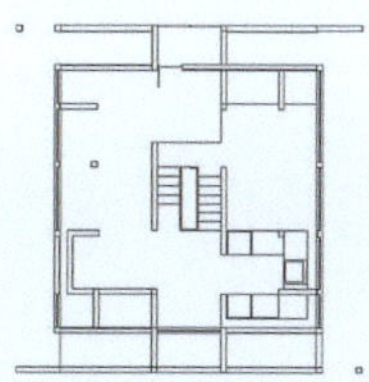

House VI

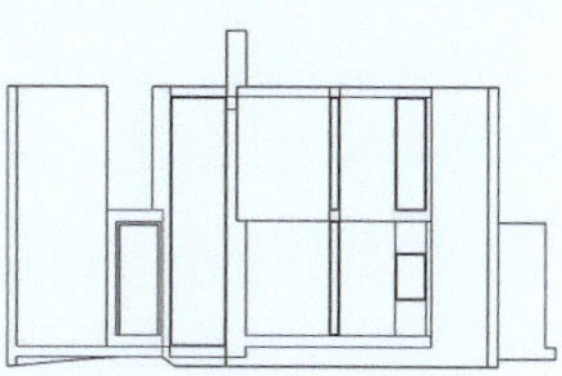

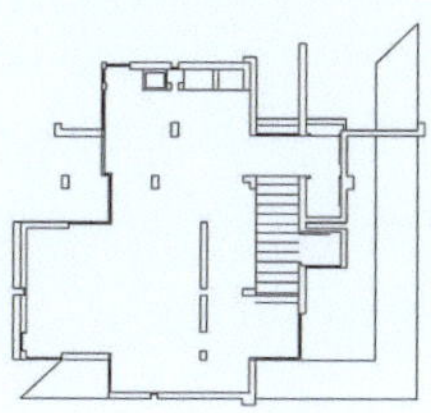

Table 9.2 Eisenman set, results

Houses		House I	House II*	House III	House IV*	House VI*	Opt. [*]	Set {…}
Elevations	D_{E1}	1.2835	1.1804	1.4856	1.3771	1.3251		
	D_{E2}	1.2582	1.4221	1.4413	1.3632	1.3668		
	D_{E3}	1.2458	1.4474	1.4509	1.4298	1.3335		
	D_{E4}	1.4266	1.4238	1.4657	1.3856	1.4023		
	μ_E	1.3035	1.3684	1.4609	1.3889	1.3569		
	$\mu_{[E]}/\mu_{\{E\}}$						1.3714	1.3757
	$M_{[E]}/M_{\{E\}}$						1.3814	1.3940
	$std_{[E]}/std_{\{E\}}$						0.0712	0.0821
Plans	D_{PO}	1.3146	1.3464	1.3103	1.3531	1.3023		
	D_{PI}	1.3387	1.4268	1.4680	1.3872	1.3538		
	D_{PR}	1.3394	1.4184	1.3745	1.3623	1.3655		
	μ_P	1.3309	1.3972	1.3843	1.3675	1.3405		
	$\mu_{[P]}/\mu_{\{P\}}$						1.3684	1.3641
	$M_{[P]}/M_{\{P\}}$						1.3623	1.3538
	$std_{[P]}/std_{\{P\}}$						0.0381	0.0457
Composite	μ_{E+P}	1.3153	1.3808	1.4280	1.3798	1.3499		
Aggregate	$\mu_{[E+P]}/\mu_{\{E+P\}}$						1.3701	1.3707

Table 9.3 Eisenman set, comparative values

Houses		House I	House II*	House III	House IV*	House VI*	Opt. [*]	Set {…}
Elevations	R_{ED}	0.1808	0.2670	0.0443	0.0666	0.0772		
	$R_{E\%}$	18.0800	26.7000	4.43	6.66	7.72		
	$R_{[ED]}/R_{\{ED\}}$						0.2670	0.3052
	$R_{[E\%]}/R_{\{E\%\}}$						26.7000	30.5200
Plans	R_{PD}	0.0248	0.0804	0.1577	0.0341	0.0632		
	$R_{P\%}$	2.4800	8.0400	15.7700	3.4100	6.3200		
	$R_{[PD]}/R_{\{PD\}}$						0.1245	0.1657
	$R_{[P\%]}/R_{\{P\%\}}$						12.4500	16.5700
Composite	$R_{[\mu E+PD]}/R_{\{\mu E+PD\}}$						0.0309	0.1128
	$R_{[\mu E+P\%]}/R_{\{\mu E+P\%\}}$						3.0900	11.2800

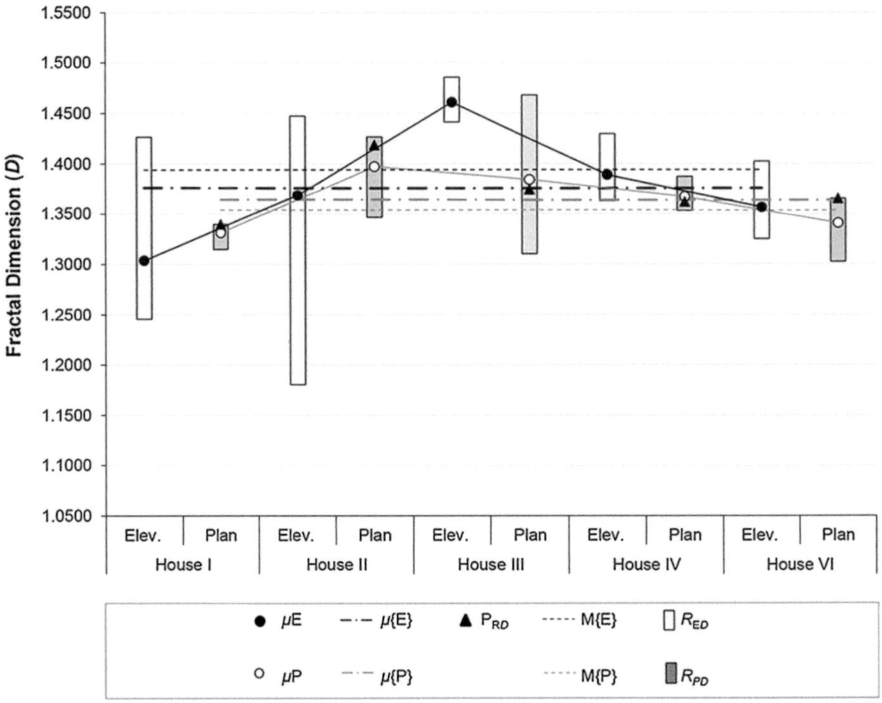

Fig. 9.2 Eisenman set, graphed results

elevations would begin to occupy a tighter range of dimensional results, or possess a more similar range of visual correlations. The results for Eisenman's designs do indeed become increasingly similar in plan and elevation as the series progresses, with the results for *Houses IV* and *VI* describing an object that is more akin to a large sculpture than a house, one that could conceivably be tipped onto its side with no apparent diminution of its functionality or change in its visual expression.

9.3 John Hejduk

John Hejduk was born in New York in 1929 and his professional education commenced in 1947 at the Cooper Union for the Advancement of Science and Art, New York. From 1950 to 1952 he undertook studies towards a Bachelor of Architecture at the University of Cincinnati, Ohio, then relocated to Harvard for his Masters in Architecture. After receiving a Fulbright Scholarship in 1954, Hejduk spent a year in Italy at the University of Rome. Throughout this period he also worked for various architectural firms, later describing the time as one of 'intense excitement, the falling into the void of discovery, the moment when the genetic coding was being formulated' (Hejduk 1985: 26). After completing his studies, Hejduk taught at the University of Texas in Austin before returning to New York and employment with I.M. Pei, after which he became assistant Professor at Cornell University and then moved to Yale. Hejduk returned to the Cooper Union in 1965, where he became Dean, a position he held until his death in 2000.

While he was based in Austin, Hejduk worked closely with Colin Rowe, and began to develop a series of 'speculative designs' that set out to investigate 'the limits and potentials of architectural expression' (Mical 2004: 599). These works, known as the *Texas House series*, included seven designs that sought to 'typify pictorial formalism in architecture' (Linder 2004: 10). Hejduk describes the series as part of his 'monochromatic period' because it 'was purposefully dry, sparse, hard [and] reductive; a search for certain essences' (Hejduk 1985: 34). In 1963 he commenced work on the *Diamond Projects*, 'a series of carefully calibrated and measured formal transformations that owe much to Mondrian, Le Corbusier, and Mies van der Rohe' (Hays 2002: np). By 1965, Hejduk had started his own practice, and his designs for *House 10* and *One-Half House*, both from 1966 and the *Bernstein House* from 1968 were included in *Five Architects*.

After the *Diamond Projects* Hejduk began his third great sequence of experimental works, the *Wall House* series in which 'the walls, often separated from and extending beyond functional volumes, became significant metaphorical representations of passage, boundary, and mass' (Mical 2004: 600). In the early part of the 1980s, with the rise of Post-Modernism, Hejduk set out to reinvent the notion of the architectural masque, producing a series of housing designs and urban installations which 'act as guides through the topology of Dwelling to the experience of the Undwellable' (Libeskind 1985: 12). Both the 1981 *Berlin Masque* and the 1998 *La Máscara de la Medusa* in Buenos Aires were part of this series. Hejduk's final works include individual designs, short series of projects and artworks such as his 1996 *Christ Chapel* and *Cathedral*, the 1998 *Chapel*, *Museum for War and Peace* in 1999 and *Enclosures* in 2000, all of which, according to K. Michael Hays, 'trace a certain failure or loss—the moment in which architecture glimpses its inadequacy —but holds out the possibility of new orders and perceptions' (2002: np).

Throughout his career, Hejduk's interest gradually shifted from the creation of form to the theory and art of architecture. Hejduk's intensively poetic and often cryptic architecture found an enthusiastic audience through its regular exhibition and publication, although only a small number of projects were constructed, including the renovation of the *Cooper Union Building* in 1974, *Kreuzberg Housing* in 1988 and *Tegel Housing* in Berlin in 1988. Hejduk's *Wall House 2*, also known as the *Bye House*, was designed in 1973 and constructed posthumously in 2001. According to Thomas Mical, 'Hejduk decisively influenced many generations of designers with this unique vision of the vocation of architecture' (2004: 599). Some of Hejduk's major publications include *Fabrications* (Hejduk 1974), *Mask of Medusa* (Hejduk 1985), *Soundings* (Hejduk and Shkapich 1993) and *Adjusting Foundations* (Hejduk and Shkapich 1995).

9.3.1 Five Houses (1954–1963) by John Hejduk

Of the multiple series of experimental designs proposed by Hejduk, five of the *Texas House series* are analysed here (*Houses 1*, *4*, *5*, *6* and *7*). Although none of these designs were ever realized, Hejduk had a clear vision of the geography of their siting (all were designed for sites in Texas) and the building materials of each house. Like Eisenman's numbered series of designs, which were also influenced by the ideas of Rowe, all of the *Texas House series* are based on attempts to resolve a design generated from a sixteen-column, nine-square plan. In part, because these designs were produced with only a notional site and program, Hejduk chose a high number of abstract influences and constraints to use for each house (Balfour 1980). For this reason Eisenman explains that these designs are 'concerned with the *process* of reduction, of methodologically cutting down to produce a very complex essence' (1979a: 10).

House 1 was designed for a sloping site and was to be constructed with an exposed, white-painted steel frame and grey-framed, off-white infill panels. The house is one storey in height, raised above its sloping site on the slender steel frame. Eisenman (1979a) describes this house as the progenitor of the entire *Texas Houses series*.

House 4 is a two-storey house which fits neatly in the nine-square framework. It is designed to be constructed using steel-capped masonry wall panels, with grey painted steel frames holding translucent and transparent glass panels. Set on a flat site, with two specific site layouts proposed by Hejduk, the house has regular elevations, with the opposite ones appearing to be nearly the same. Hejduk described this design as 'perhaps the most complete and most resolved' (1985: 42) of the series.

House 5 'confronts the problem of the nine squares as an asymmetrical condition' (Hejduk 1985: 40) and, although asymmetrical in planning, its simple façades are even more similar to each other than those found in *House 4*. The small, open-plan, single-storey house is set on a flat site and is constructed from a white-painted steel frame with glass infill panels. According to Eisenman, *House 5* '[i]s the conclusion of the investigation of the horizontal dimension, while its site plan contains a first signal of the transition to the later houses' (1979b: 12).

House 6 continues to demonstrate Hejduk's shifting concerns, as the designs become more vertical in their spatial and formal explorations. In this three-storey house (plus basement), Hejduk expands the nine-square planning of the house with a stair tower that adjoins a walkway to each floor. The elevation and the plan of the house both contain the nine-square system, causing 'the volumetric nature [to] become obvious …. Now, for the first time, there is a full three-dimensional lattice' (Eisenman 1979b: 15). The proposed construction is similar to *House 4*, with load-bearing masonry walls and grey-painted steel framing holding transparent and translucent glazing.

House 7, the last design of the series, is a seven-storey house, square in plan and almost square in elevation (Fig. 9.3). The materials and finishes were specifically

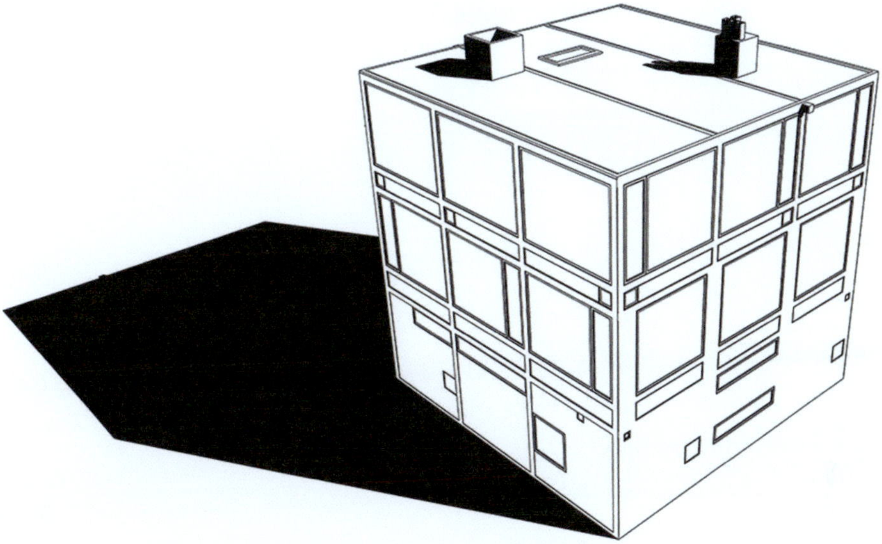

Fig. 9.3 *House 7*, perspective view

not finalized in this design. With this project Hejduk aimed to 'increase the visual scale' (1985: 43) of the design and proposed that 'the program is for a very complex interweaving of spaces' (1985: 43) where multiple families can exist within one house, but are isolated from one another through an interlocking 'system of space enclosures' (43).

Hejduk and Eisenman were well aware that the conventional correlation between form and function produced what was commonly called 'figurative' or 'representational' design; that is, houses with pitched roofs, ceremonial front doors, picture windows, attic dormers and porches. In contrast, and under the influence of Rowe, they separately proposed a 'non-figurative' design process that, like 'non-representational' art, either contained no recognizable features or, if such features existed, deployed them in a subversive manner. However, Hejduk's (1979) approach differed from that of Eisenman and, rather than attempting to produce a consistent three-dimensional form (plans which visually resemble sections and elevations), Hejduk's strategy was to gradually attenuate or stress the relationship between plan and elevation, stretching each façade to accommodate an increasingly complex but still human-scaled set of plans. If Hejduk's architecture actually conforms to his stated position, we should be able to uncover this visual quality. Table 9.4 presents example plans and elevations form this set.

Table 9.4 Hejduk set, example elevations and floor plans (level 4 representation—not shown at uniform scale)

House 1

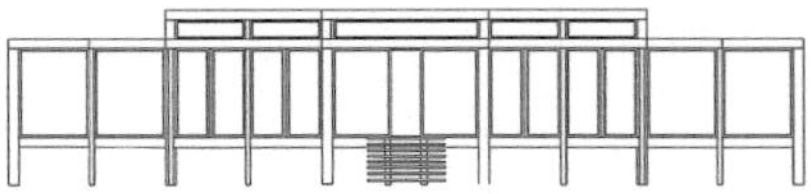

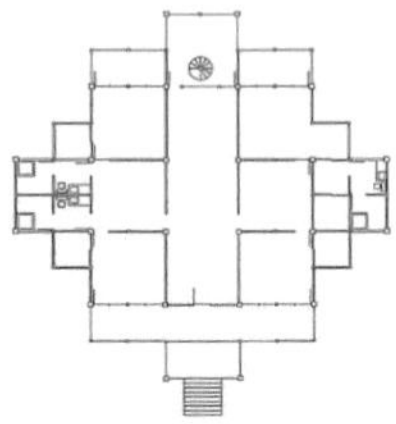

House 4

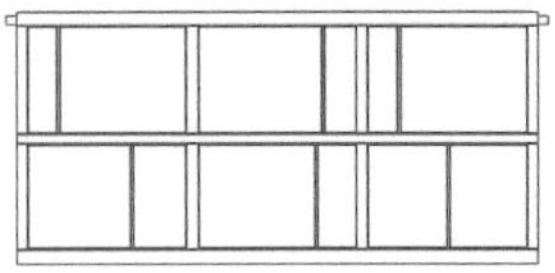

House 5

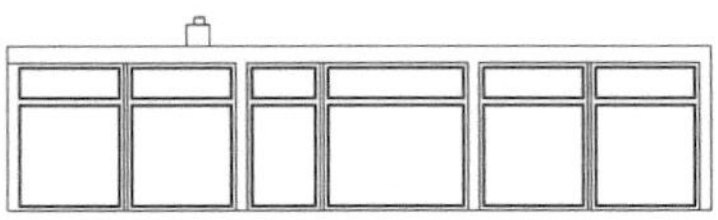

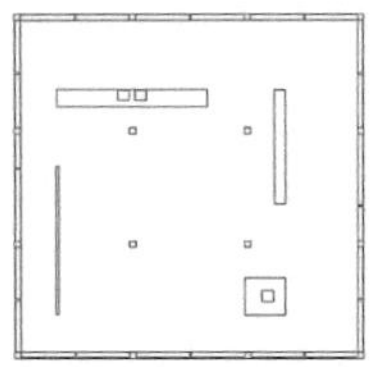

House 6

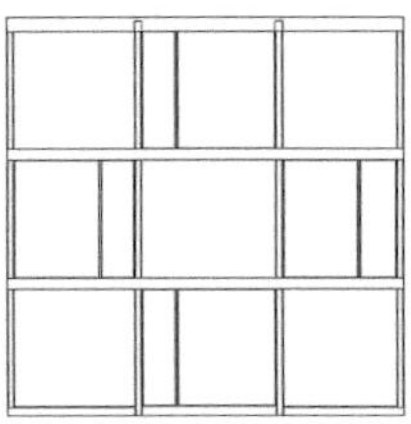

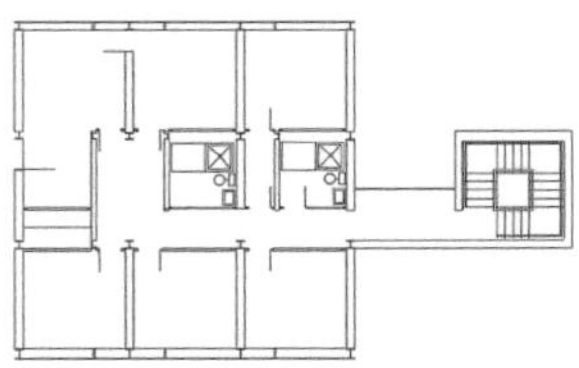

House 7

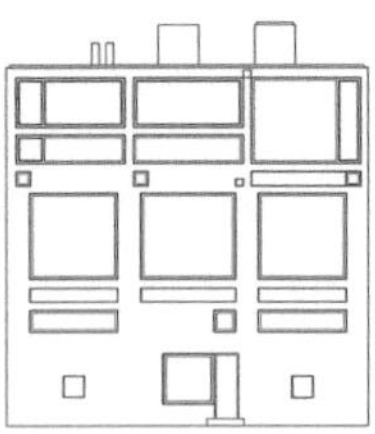

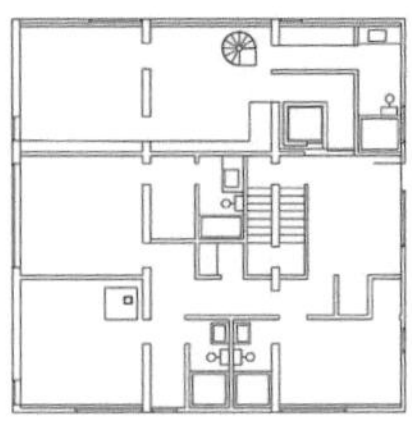

9.3.2 Hejduk Houses, Results and Analysis

Of the five *Texas Houses*, the lowest average elevation result is found in *House 6* (μ_E = 1.3955) and the lowest individual elevation is from the same design (D_{E3} = 1.3587). The highest average elevation is found in *House 1* (μ_E = 1.4830), with a range of $R_{\{E\%\}}$ = 14.75 for all elevations of the set. The median for all elevations is 1.4503 and the standard deviation is 0.0400. The highest plan average is in *House 7* (μ_P = 1.3757), the highest individual plan is the ground floor of the same house (D_{P0} = 1.4187) and the lowest plan average is found in *House 5* (μ_P = 1.1629). The range of all the plans is $R_{\{P\%\}}$ = 33.57, more than double the range of the elevations of this set. The median for all plans is 1.3689 and the standard deviation is 0.0901, once again being much higher than the result for the elevations (Tables 9.5 and 9.6).

When all of the plan and elevation results are considered together, the aggregate is $\mu_{\{E+P\}}$ = 1.3862 and the composite range is $R_{\{\mu E+P\%\}}$ = 4.29, a result which places the five houses within a 'very similar' order of visual complexity. *Houses 1*, *6* and *7* are the optimal sub-set. Considering only these three, the aggregate result for all plans and elevations is increased to $\mu_{[E+P]}$ = 1.3989 and the composite range is reduced to $R_{[\mu E+P\%]}$ = 2.11. This effectively halves the range result, suggesting even further that the visual properties of these houses, in terms of their characteristic distribution of form, are almost indistinguishable (less than ±1.06 % difference).

Three of Hejduk's houses, *1*, *4* and *5*, all have a similar program, with pure, geometric one-or two-storey volumes that feature similar treatments for each elevation. Indeed, it is hard to determine which is the front, back or side of these houses. However, while all of the elevations feature abstract geometric compositions, there is also evidence of the detailed thought that has gone into the placement of their openings and structural features. Overall, the plans for these houses are much simpler than their elevations, with *House 5* so reduced in its internal planning that the entire house has no individual rooms. This feature can be seen in the chart of the results (Fig. 9.4).

House 7 follows a similar approach to that taken in *Houses 1*, *4* and *5*, except that the four largely identical façades of *House 7*, each grossly over-scaled, hide a seven-level diversified environment within. This is precisely the message found in the fractal analysis results, with a tight cluster of high results for the elevations ($R_{E\%}$ = 4.24), meaning that they all look alike, coupled with a wider range of results for the seven plans ($R_{P\%}$ = 11.85).

House 6 is the project by Hejduk that is most akin to Eisenman's work, insofar as form and geometry rather than site or program govern the design. Just as in Eisenman's work, the dominance of form, in both plan and elevation, is represented in the data as a set of plans and elevations with an almost identical fractal range and set of values.

Table 9.5 Hejduk set, results

Houses		House 1*	House 4	House 5	House 6*	House 7*	Opt. [*]	Set {…}
Elevations	D_{E1}	1.5061	1.4089	1.4641	1.4248	1.4679		
	D_{E2}	1.5062	1.4146	1.4720	1.4228	1.4664		
	D_{E3}	1.4599	1.3952	1.4599	1.3587	1.4436		
	D_{E4}	1.4599	1.3949	1.4570	1.3756	1.4255		
	μ_E	1.4830	1.4034	1.4633	1.3955	1.4509		
	$\mu_{[E]}/\mu_{\{E\}}$						1.4431	1.4392
	$M_{[E]}/M_{\{E\}}$						1.4518	1.4503
	$std_{[E]}/std_{\{E\}}$						0.0451	0.0400
Plans	D_{P-1}					1.4127		
	D_{P0}	1.3401	1.3490	1.2427	1.4115	1.4187		
	D_{P1}	–	1.4031	–	1.3878	1.4026		
	D_{P2}	–	–	–	1.3857	1.3766		
	D_{P3}	–	–	–	–	1.3738		
	D_{P4}	–	–	–	–	1.3689		
	D_{P5}	–	–	–	–	1.3524		
	D_{PR}	1.1639	1.2671	1.0830	1.3122	1.3002		
	μ_P	1.2520	1.3397	1.1629	1.3743	1.3757		
	$\mu_{[P]}/\mu_{\{P\}}$						1.3581	1.3343
	$M_{[P]}/M_{\{P\}}$						1.3766	1.3689
	$std_{[P]}/std_{\{P\}}$						0.0691	0.0901
Composite	μ_{E+P}	1.4060	1.3761	1.3631	1.3849	1.4008		
Aggregate	$\mu_{[E+P]}/\mu_{\{E+P\}}$						1.3989	1.3862

Table 9.6 Hejduk set, comparative values

Houses	House 1*	House 4	House 5	House 6*	House 7*	House 1*	Opt. [*]	Set {…}
Elevations	R_{ED}	0.0463	0.0197	0.0150	0.0661	0.0424		
	$R_{E\%}$	4.63	1.9700	1.5000	6.61	4.24		
	$R_{[ED]}/R_{\{ED\}}$						0.1475	0.1475
	$R_{[E\%]}/R_{\{E\%\}}$						14.7500	14.7500
Plans	R_{PD}	0.1762	0.1360	0.1597	0.0993	0.1185		
	$R_{P\%}$	17.6200	13.6000	15.9700	9.9300	11.8500		
	$R_{[PD]}/R_{\{PD\}}$						0.2548	0.3357
	$R_{[P\%]}/R_{\{P\%\}}$						25.4800	33.5700
Composite	$R_{[\mu E+PD]}/R_{\{\mu E+PD\}}$						0.0211	0.0429
	$R_{[\mu E+P\%]}/R_{\{\mu E+P\%\}}$						2.1100	4.2900

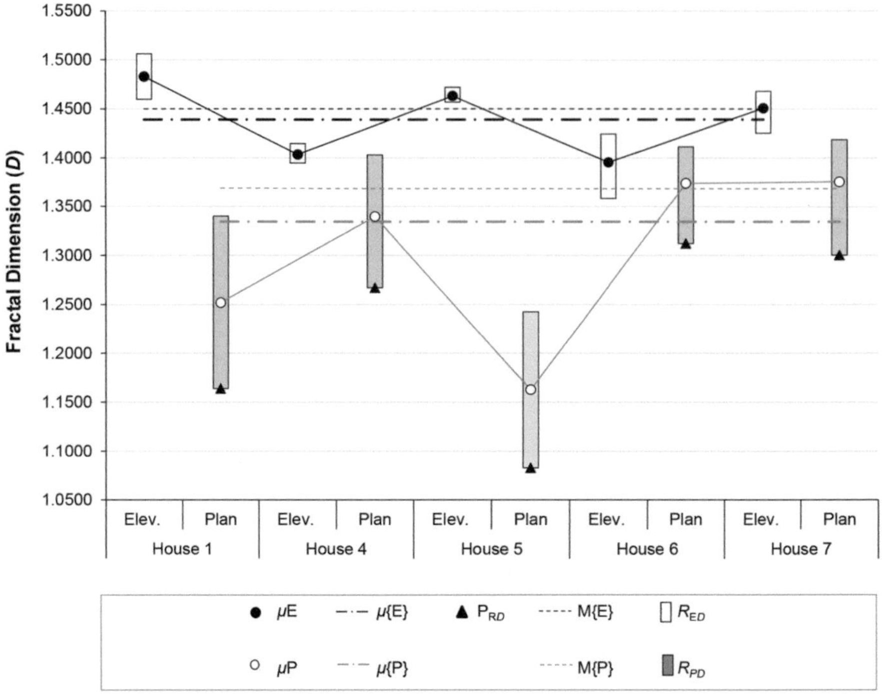

Fig. 9.4 Hejduk set, graphed results

9.4 Richard Meier

Richard Meier was born in 1934 in New Jersey, USA. Described as 'one of the most consistent of contemporary architects' (Jodidio 1995: 6), Meier began studying architecture, painting and art history at Cornell University in 1952, and was enrolled there at the same time as Eisenman. After completing his studies in 1957, Meier was employed by Davis, Brody and Wisniewski, a New York firm specialising in public housing. After a year working there, Meier spent six months travelling in Europe and even sought a position with two of his Modernist heroes, Le Corbusier and Alvar Aalto. However, after failing to find employment in Europe, Meier returned to New York and after a brief period with Skidmore, Owings and Merrill, began working for Marcel Breuer where her remained from 1961 to 1963. During this time he joined a small artists' studio to spend time painting with others, including the Late-Modernist abstract painter Frank Stella.

In 1963 Meier left Breuer's office and began his own practice, initially producing several small renovations. While Le Corbusier was Meier's strongest source of inspiration, his work was also shaped by that of Aalto, Mies and Wright. In 1964 Meier even stayed for a weekend at Wright's *Fallingwater*, an experience which greatly inspired him in the creation of his first true design, a house for his parents in

New Jersey that was completed in 1965. Despite the influences of earlier masters, as Paul Goldberger clarifies, 'Meier's architecture is emphatically post-machine age' (1996: 11).

Built in 1967, Meier's *Smith House* is generally considered to be the first that clearly prefigures the values and concerns that shaped the rest of his career. According to Meier, '[i]t was in this house that I began to work in a more consciously complex way with interpenetrating spaces and transparency, partially influenced by the Rowe and Slutzky distinction between literal and phenomenal transparency' (Meier 1999: 12). After the *Smith House*, Meier continued to produce domestic designs in a similar formal style. Usually painted white and with a flat roof, the houses are unexpectedly scaled, often featuring dominant chimneys, staircases and tubular steel detailing. The flat or regularly curved planar walls in these houses are generally layered with a series of large, deep openings, the shadows caused by this façade configuration contrasting with the flat white surfaces that frame them. The *Smith House* and the *Saltzman House* were prominently featured in *Five Architects*, leading Joseph Rykwert to suggest that 'the persistent use of white in [Meier's] buildings is perhaps the main reason for the group's sobriquet "white architects"' (1997: 214).

While continuing with domestic commissions in the late 1960s, Meier also began to design public buildings and high-density housing. Many of these remained unrealised projects until the *Westbeth Artists' Housing* was completed in New York in 1970. This project was followed by several large-scale commercial and public buildings and, according to Goldberger, Meier 'moved gracefully and easily from the scale of the houses with which his practice began to the larger scale of civic and commercial buildings' (1996: 10). The commission for the most significant of these, the *Getty Center*, was granted after Meier received the Pritzker prize in 1984. Completed in 1997, the *Getty Center* is a sprawling complex covering an entire hillside in Los Angeles, a commission that Meier describes as 'the most important event of my career' (2003: 452).

Like Eisenman and Hejduk, Meier was influenced by the theories of transparency proposed by Colin Rowe and Robert Slutzky. Meier called the effect of transparency on architecture 'dematerialization' (qtd. in Sherman 2011: 202) and his resulting forms were able to 'create a space where light is an omnipresent element' (Jodidio 1995: 8). The architecture of Meier is an interesting subject for fractal analysis because he is very concerned with the form and appearance of his works. Meier's fascination with gridded, geometric, formal systems ultimately signals the significance of the experience of architectural complexity, the distinct visual character of Meier's architecture serving to 'emphasise this quintessential function' (Blaser 1996: 52).

9.4.1 *Five Houses (1967–1974) by Richard Meier*

Four of the five houses by Meier were built on the east coast of the USA, and the fifth on Lake Michigan. Meier has retrospectively divided his domestic architecture into several types, and these five early houses fit into his 'first type', where he set out to 'employ a series of private/public, opaque/transparent juxtapositions, an articulation of structure and circulation, and a taut relationship between all of these dichotomies' (Meier et al. 1996: 8). Meier further elaborates on his process of designing the forms of his houses, stating that in 'every house there is a search for clarity, which in my architecture is a call for basic geometric form' (1996: 8). Just as Eisenman and Hejduk's early houses can be seen as a rigorous series of investigations into the nature of architectural form, so too Meier's works feature a methodical examination of the relationship between solid and void, literal and phenomenal transparency and the curve and the right angle.

The *Smith House*, sited on a large seafront site in Connecticut, was built in 1967 as a weekend retreat for Carole and Fred Smith and their two children. This white, freestanding, three-storey house with a roof terrace is reminiscent of the early Modernist work of Le Corbusier (Fig. 9.5). Meier describes the large areas of glass set in delicate timber frames on the façade of this house as 'a skin tautly stretched over the outside of the building in which the inherent opposition of solid and transparent is brought together in the function of enclosure' (Drexler 1975: 111).

Fig. 9.5 *Smith House*, perspective view

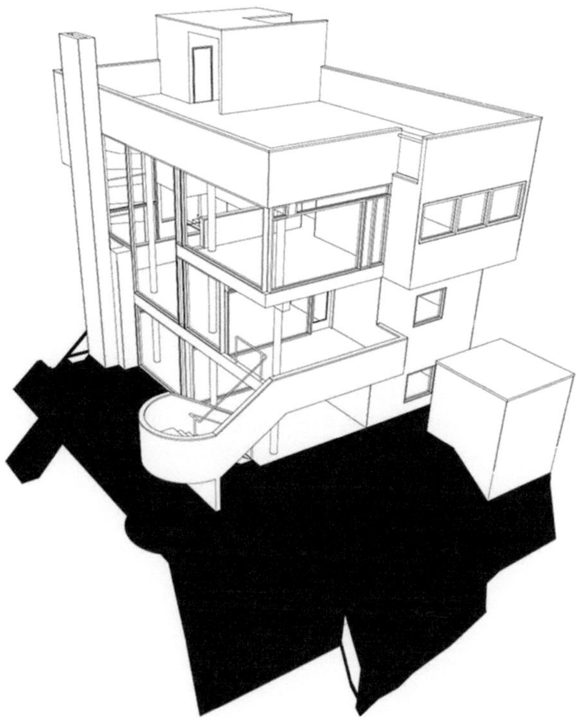

The *Hoffman House*, in East Hampton, was also completed in 1967, a two-storey, white-painted timber family weekend house with a white-painted brick fireplace and chimney. The floor plan has a mix of 90° and 45° angles, creating intersecting forms in the external volume. The architect states that, by 'virtue of the interpenetrating angular geometries, all of the major spaces in the house take on a dynamic quality, providing a complex spatiality within a minimal overall envelope' (Meier 1984: 35).

Also built in East Hampton and completed in 1969, the *Saltzman House* is a family home. Taking advantage of ocean views, the living area is set in the top section of the house and includes double-height windows and a terrace. Constructed of white-painted, vertical timber cladding externally, the house features a large curved wall around one entire corner of the house (Fig. 9.6). With grandly scaled forms and many terraces recessed into the façade, Frampton suggests that this house 'has a Cubistic introversion which, despite its extensive glazing, tends to look into itself rather than out to the site' (1976: 10).

In 1976 Meier won an award from the American Institute of Architects for his *Douglas House* (1973), stunningly sited above the water of Harbour Springs, Michigan. This white building is perched on a steep hillside, surrounded by dense, dark, fir trees. While having the appearance of a cast concrete structure, the house is actually constructed of white-painted timber and steel. With glazing contained in thin frames, tubular steel balustrading, external open tread stairs and tall exterior steel smokestack-like chimneys, the house strongly echoes a machine or nautical aesthetic. Entry to the house is by a bridge to the top of the building, which then descends another three levels. To David Morton this house represents 'an uncommon understanding and synthesis of modern architecture, of where it has been, and of where at least some of it is going' (1975: 44).

The *Shamberg House* was built in 1974 outside the regional town of Chappaqua, in the south of New York State (Fig. 9.7). The flat, regular rectangular form of this white building is only altered by the stairway and entry bridge, and an organically shaped balcony to the south. The south elevation is almost entirely glazed, the glass

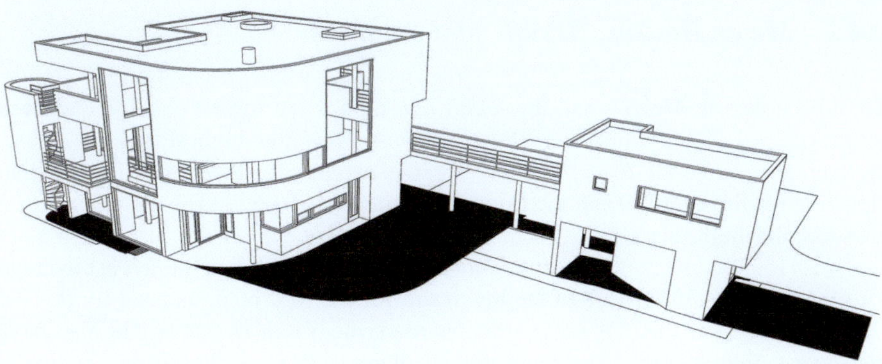

Fig. 9.6 *Saltzman House*, perspective view

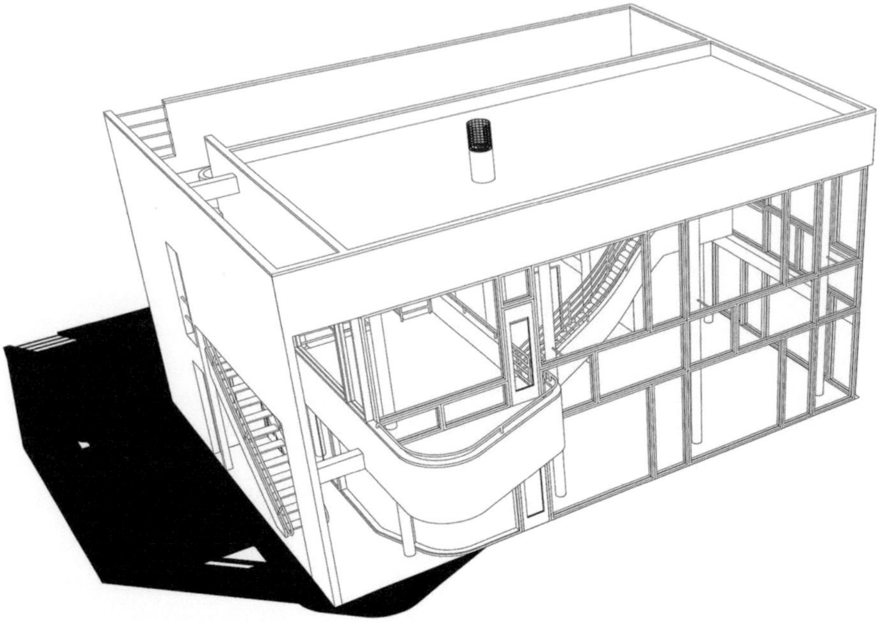

Fig. 9.7 *Shamberg House*, perspective view

façade to the three storey void inside like a white Mondrian painting. The building was designed for an older couple who intended it as a home for themselves as well as an occasional gathering place for their whole family. Cassarà sees this house as a type of Esprit Nouveau pavilion, a 'reverent tribute to the sanctuary of the Modern Movement' (1997: 69), while also noting that '[t]he process of erosion of the pure solid is endowed in this project with new connotations: while the overall unity of the basic cuboid is conserved, it is opened out by the transparent elements' (1997: 68). Table 9.7 provides an example of plans and elevations of this set.

9.4.2 Meier Houses, Results and Analysis

The first house in Meier's set, the *Smith House*, has the lowest average elevation result ($\mu_E = 1.3532$) while the *Douglas House* has the highest ($\mu_E = 1.4164$), producing a range of $R_{\{E\%\}} = 26.27$ for the set of all elevations. The median elevation result is 1.3770 and the standard deviation is 0.0679. The *Douglas House* also has the highest average fractal dimension result for all the plans in the set ($\mu_P = 1.4152$); conversely, it is the *Saltzman House* which has the lowest average for plans ($\mu_P = 1.3681$), due to the low result for the roof plan, unusual for this set ($D_{PR} = 1.2462$). If this anomaly is disregarded, the *Smith House* would also have the lowest plan average. The range for all plans is $R_{\{P\%\}} = 20.07$, the median is 1.3878 and the standard deviation is 0.0470 (Tables 9.8 and 9.9, Fig. 9.8).

Table 9.7 Meier set, example elevations and floor plans (level 4 representation—not shown at uniform scale)

Smith House

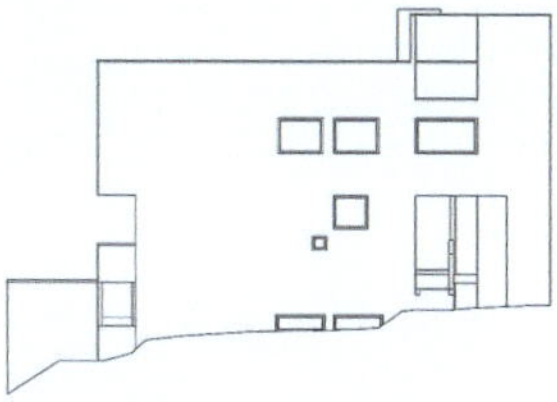

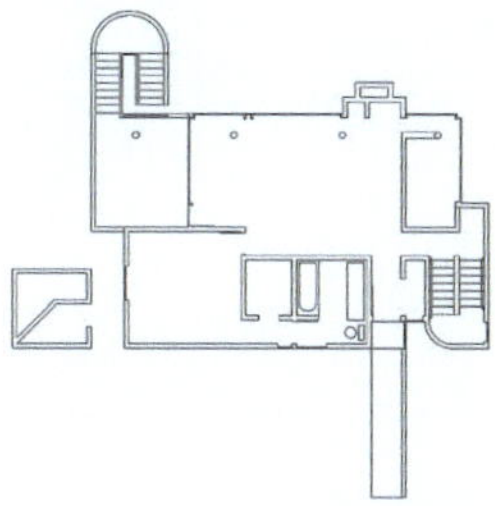

Hoffman House

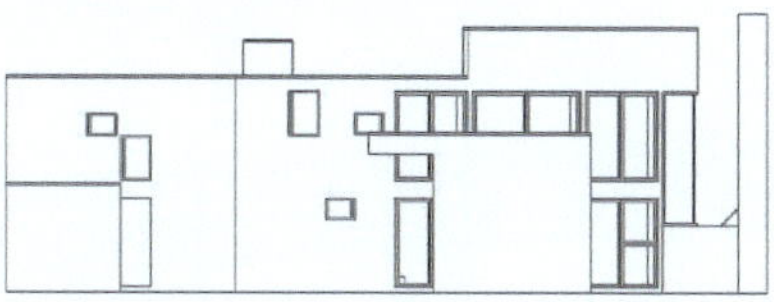

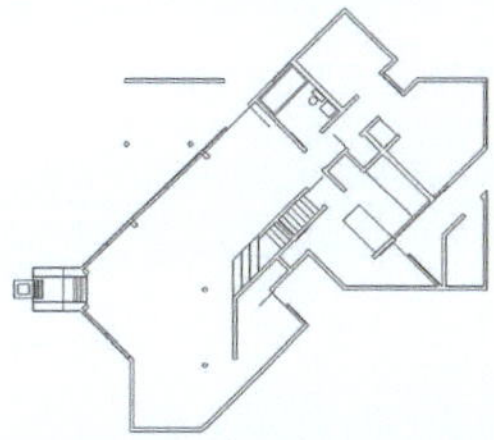

Saltzman House

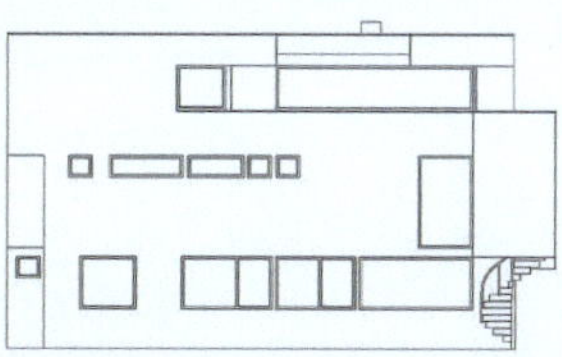

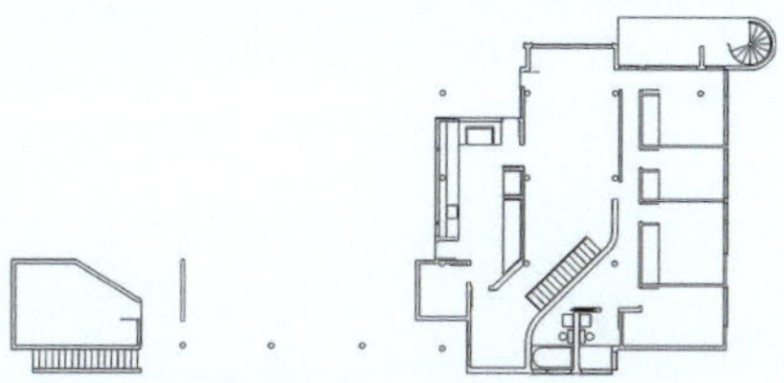

Douglas House

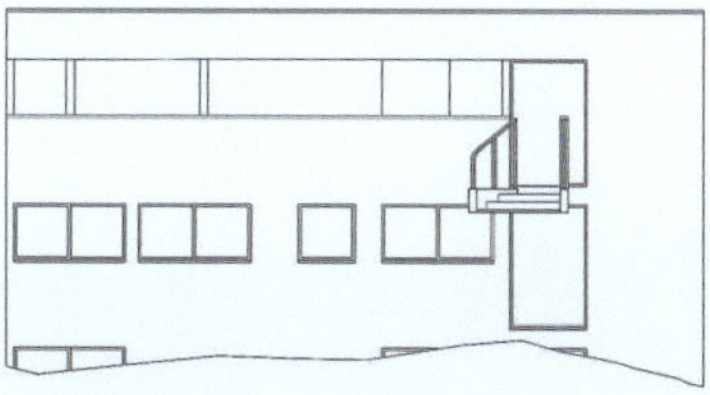

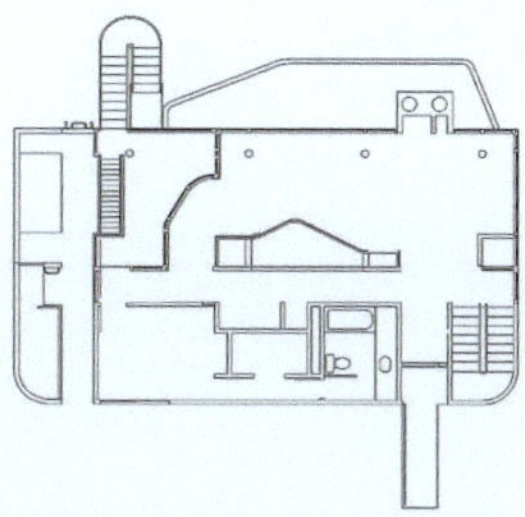

Shamberg House

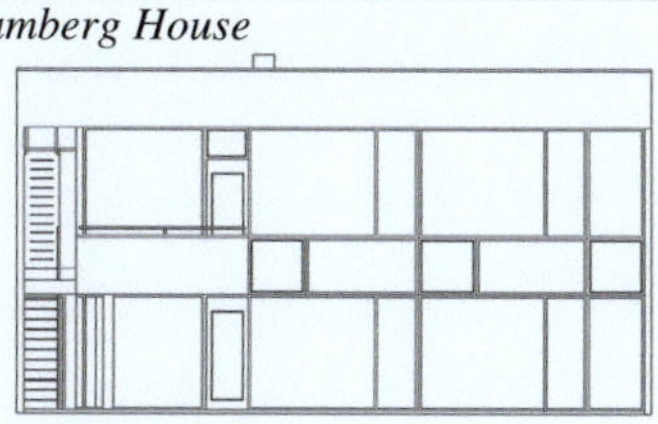

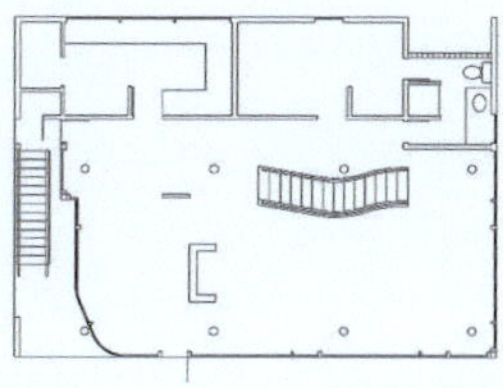

Table 9.8 Meier set, results

Houses		Smith	Hoffman*	Saltzman*	Douglas	Shamberg*	Opt. [*]	Set {…}
Elevations	D_{E1}	1.4032	1.3595	1.2636	1.3745	1.3158		
	D_{E2}	1.2937	1.2777	1.4472	1.4280	1.4555		
	D_{E3}	1.3154	1.3794	1.3851	1.3369	1.3440		
	D_{E4}	1.4005	1.4624	1.3830	1.5263	1.3216		
	μ_E	1.3532	1.3698	1.3697	1.4164	1.3592		
	$\mu_{[E]}/\mu_{\{E\}}$						1.3662	1.3737
	$M_{[E]}/M_{\{E\}}$						1.3695	1.3770
	$std_{[E]}/std_{\{E\}}$						0.0661	0.0679
Plans	D_{P-1}	1.3850	–	–	–	–		
	D_{P0}	1.3878	1.3836	1.3935	1.4159	1.3618		
	D_{P1}	1.3845	1.4174	1.4304	1.4328	1.4371		
	D_{P2}	–	–	1.4021	1.4469	–		
	D_{P3}	–	–	–	1.3948	–		
	D_{PR}	1.3287	1.3406	1.2462	1.3855	1.3467		
	μ_P	1.3715	1.3805	1.3681	1.4152	1.3819		
	$\mu_{[P]}/\mu_{\{P\}}$						1.3759	1.3853
	$M_{[P]}/M_{\{P\}}$						1.3886	1.3878
	$std_{[P]}/std_{\{P\}}$						0.0564	0.0470
Composite	μ_{E+P}	1.3624	1.3744	1.3689	1.4157	1.3689		
Aggregate	$\mu_{[E+P]}/\mu_{\{E+P\}}$						1.3706	1.3781

Table 9.9 Meier set, comparative values

Houses		Smith	Hoffman*	Saltzman*	Douglas	Shamberg*	Opt. [*]	Set {…}
Elevations	R_{ED}	0.1095	0.1847	0.1836	0.1894	0.1397		
	$R_{E\%}$	10.9500	18.4700	18.3600	18.9400	13.9700		
	$R_{[ED]}/R_{\{ED\}}$						0.1988	0.2627
	$R_{[E\%]}/R_{\{E\%\}}$						19.88	26.27
Plans	R_{PD}	0.0591	0.0768	0.1842	0.0614	0.0904		
	$R_{P\%}$	5.9100	7.6800	18.4200	6.1400	9.0400		
	$R_{[PD]}/R_{\{PD\}}$						0.1909	0.2007
	$R_{[P\%]}/R_{\{P\%\}}$						19.0900	20.0700
Composite	$R_{[\mu E+PD]}/R_{\{\mu E+PD\}}$						0.0055	0.0534
	$R_{[\mu E+P\%]}/R_{\{\mu E+P\%\}}$						0.5500	5.3400

Considering the composite results, the trend for the highest result continues with the *Douglas House* (μ_{E+P} = 1.4157) and the lowest with the *Smith House* (μ_{E+P} = 1.3624). Unusually, the *Saltzman* and *Shamberg Houses* have an identical composite result (μ_{E+P} = 1.3689). The optimal sub-set comprises the *Hoffman*, *Saltzman* and *Shamberg* houses. The aggregate result for these three is ($\mu_{[E+P]}$ = 1.3706), only marginally lower than the aggregate for the entire set, ($\mu_{\{E+P\}}$ = 1.3781). The similarity of the two aggregate results shows the remarkable nature of the results for Meier: they are all so similar across the complete set that most results could be considered as part of the optimal sub-set. The only disparity arises from the west elevation of the *Douglas House*. Due to its unusual siting, a different design approach was required for the iconic western façade, which faces down the steep and exposed slope.

Overall, the plan and elevation results have a striking correspondence of range in values between the elevation set ($1.2636 < D_E < 1.5263$) and the plan set ($1.2462 < D_P < 1.4469$). This data supports Richard Rogers' assertion that in Meier's architecture the 'plan always generates the form' while the section creates 'movement and scale'. Rogers concludes that 'you can "read" a Meier façade simply by reading the plan and section' (Rogers qtd in Meier et al. 1996: 233). This effect is further emphasised in the optimal set, where the range is not even a whole percentage, at only $R_{\{\mu E+P\%\}}$ = 0.55, which means that in terms of visual expression, measured using fractal dimensions, the *Hoffman*, *Saltzman* and *Shamburg* houses are virtually indistinguishable from each other.

9.5 Comparison of the Three Whites

When viewed collectively, the results for Eisenman and Meier have strong similarities, confirming their general grouping as designers whose architecture has a distinct Late Modernist aesthetic expression. The unbuilt nature of Hejduk's work, which is also more abstract than that of the other two, may account for the difference in his results. Eisenman's and Meier's works display a similar pattern of fractal dimension measures, with the results for their plans and elevations overlapping, a property which is unusual for most twentieth-century designers, but in accordance with their conceptual ideas about the close relationship between plans, elevations and sections. Eisenman and Meier also share a similar level of complexity in their works, with the average of all of Eisenman's elevations ($M_{\{E\}}$ = 1.3940) being close to Meier's ($M_{\{E\}}$ = 1.3770); and this is also the case for Eisenman's ($M_{\{P\}}$ = 1.3538) and Meier's ($M_{\{P\}}$ = 1.3878) plans. Furthermore, the maximum range between all medians (4 %) for Eisenman and Meier shows the averages to be, once again, very similar. Hejduk's plans and elevations follow a more typical pattern for other twentieth-century architects, with results for elevations and plans only rarely overlapping. While Hejduk's plan average was similar to

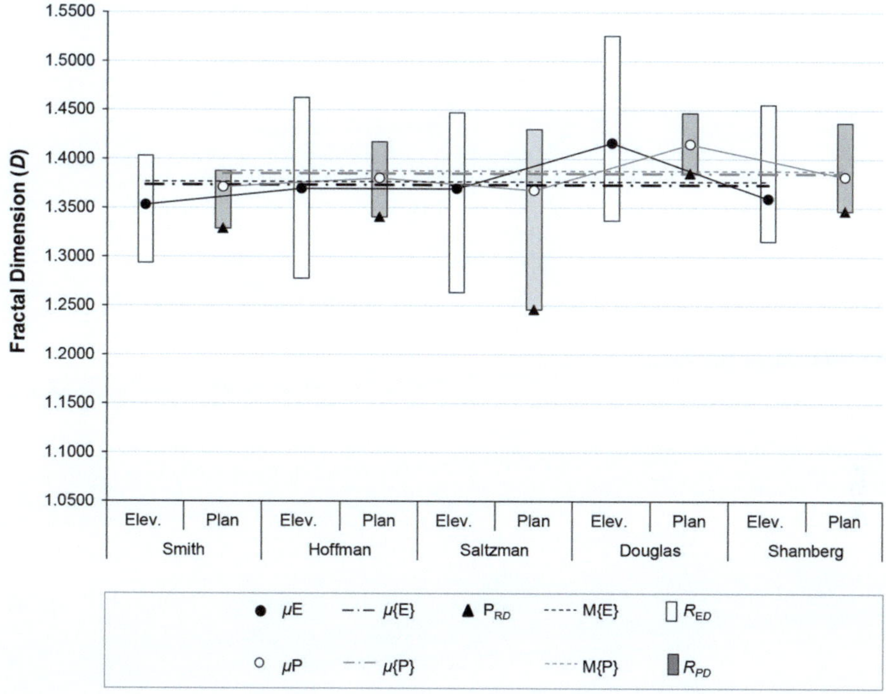

Fig. 9.8 Meier set, graphed results

Eisenman's and Meier's ($M_{\{P\}}$ = 1.3689), Hejduk's elevation average was higher ($M_{\{E\}}$ = 1.4503) and the difference between Hejduk's two medians was greater (8 %), leading to a lower formal coherence outcome (Fig. 9.14). The formal coherence result for Eisenman of R^2 = 0.4597 (Fig. 9.9), is in the mid-range of positive correlations, while the result for Hejduk shows even less correlation, R^2 = 0.427 (Fig. 9.10). In contrast, the close relationships between plans and elevations in Meier's architecture is strongly apparent in his formal coherence, R^2 = 0.8279, the highest result for this indicator recorded in the present book (Fig. 9.11).

Fig. 9.9 Eisenman set, formal coherence graph

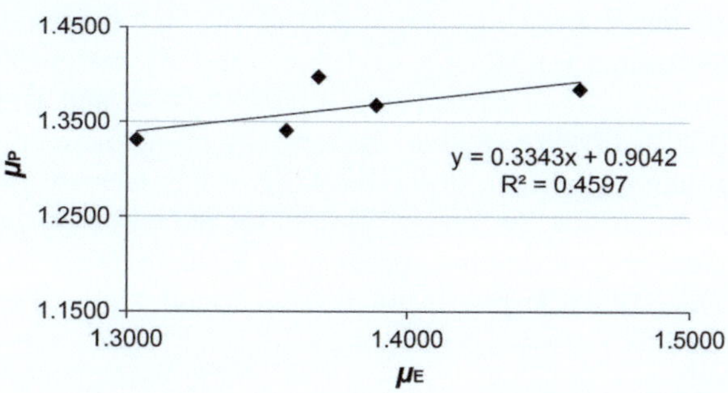

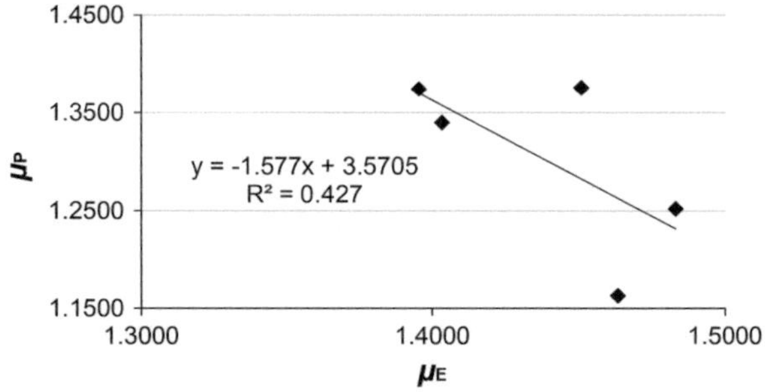

Fig. 9.10 Hejduk set, formal coherence graph

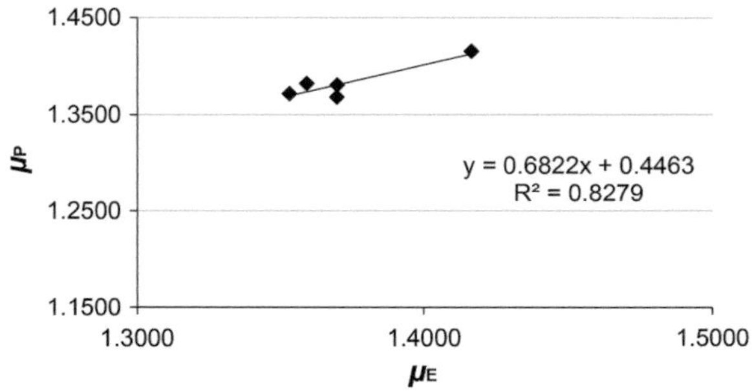

Fig. 9.11 Meier set, formal coherence graph

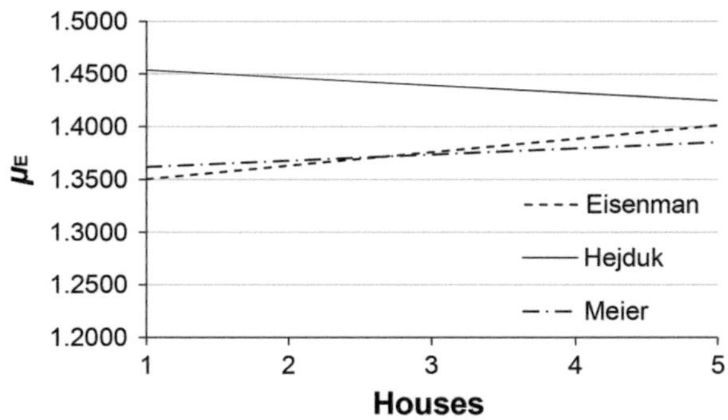

Fig. 9.12 Eisenman, Hejduk and Meier sets, linear trendline data for elevations (μ_E)

The visual complexity of Eisenman's elevations increases over the same time, as does that of Meier's elevations, while Hejduk's elevations become increasingly simplified over time (Fig. 9.12). Comparing the trends in average plan complexity, the results for Meier's work increase only marginally over the course of the five projects, while Eisenman's is almost static and Hejduk's increases dramatically (Fig. 9.13). For Hejduk, the intensity of the increase in complexity in his planning is almost matched in the intensity of the decrease in the complexity of his elevations, so that the average results for the entire set of Hejduk's houses (combined plans and elevations) appear to be stable over the nine years of his work studied. The composite results for Eisenman and Meier show a gentle increase in visual

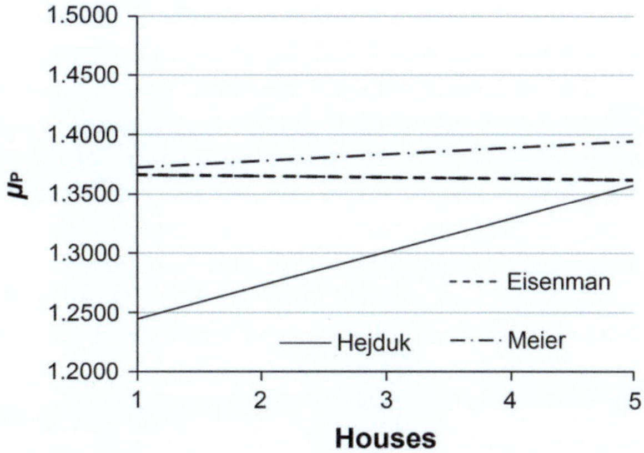

Fig. 9.13 Eisenman, Hejduk and Meier sets, linear trendline data for plans (μ_P)

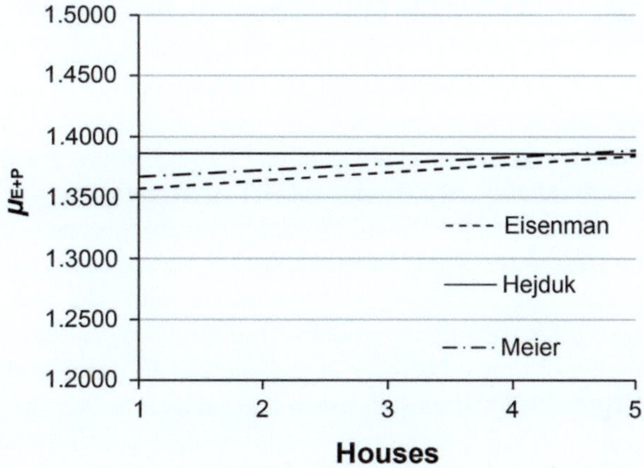

Fig. 9.14 Eisenman, Hejduk and Meier sets, linear trendline data for composite (μ_{E+P}) results

complexity over time, which may reflect their growing confidence in their capacity to experiment with form (Fig. 9.14).

Now consider these houses in the context of the broad criticism of proto-Deconstructivist architecture, as lacking human scale or a sufficient cascade of detail suitable for accommodating a range of social interactions. The mathematical results for the fifteen houses clearly show that each and every house has scale features at human size as well as larger and smaller scale features. Indeed, as all of the designs function (to a greater or lesser extent) as houses—including having kitchens, bathrooms and garages—it is not surprising that these structures have human-scale elements. Moreover, almost all of the houses have a range of scale spaces that can accommodate different functions. Curiously, this is one of the strengths of Hejduk's work, there being a wide range of complexity present in individual houses on different levels of the plan. This implies that in, for example, *House 7*, a greater array of spaces and a heightened capacity to accommodate different social practices, actions and events exists than in the typical family home. It is hard to make such a case for the last two of Eisenman's houses, both of which are admittedly quite small and have a range of complex forms with a high degree of consistency, but which still accommodate most domestic functions.

The question of whether the later Deconstructivist works from the 1980s have similar scale features, or features with a similar range of scales, is beyond the scope of the present work. However, the majority of the fifteen houses considered here have human-scale features and possess a sufficient range of detail across multiple scales to accommodate a range of domestic social and cultural functions. The message of this simple analysis then is clearly that, just because a house does not resemble the conventional image of 'house' (with a pitched roof, picket fence, brick chimney and porch), this does not mean that it lacks equivalent functional capacity or a comparable level of visual interest.

9.6 Frontality, Rotation and the Whites

In the history and theory of art, the concept of 'frontality' refers to the arrangement of elements on planes that are perpendicular to the line of sight and parallel to the picture plane. Frontality can also be used more generally to describe the 'head-on representation of a figure, object or scene' (Mayer 1969: 158). Frontality was a common compositional strategy in Renaissance architecture, where building façades were often intended to be viewed by visitors from a single, dominant perspective. Palladian architecture in particular was criticised as being frontally biased because it constructed a building's façade proportions using harmonic ratios that could only be truly appreciated from a position at right angles to the façade (Kruft 1994). Baroque artists and architects eventually rejected overt frontality in favour of sculptures and designs which were intended to be experienced from a multitude of perspectives and positions (Norberg-Schultz 1971).

In his contribution to *Five Architects*, Kenneth Frampton compares the way in which each of the designers approached the presentation of their buildings as 'objects in space'. Because the architecture of the Whites had such clear sculptural properties, Frampton was able to illuminate their differences using a variation of the concept of frontality and its dissolution, or questioning promoted by the application of rotational strategies and forms. In this way Frampton echoed the by then fashionable condemnation of the 'frontality of Western sculpture … as an expedient, an optical solution arising from a pictorial origin, based solely on the "constitutive activity of the eye of the onlooker"' (Bois et al. 1981: 26). In the field of sculpture, frontality was resisted using 'contrapposto', a technique for 'representing the various parts of the body so that they are obliquely balanced around a central vertical axis: for example, the upper portion of the torso will twist in one direction, while the lower part twists in the opposing direction' (Lucie-Smith 2003: 64). The architectural equivalent of contrapposto was rotation, the natural geometric response to an otherwise orthogonally restricted design practice at the time.

Eisenman described frontality as a 'pictorial convention' based on 'traditional codes that operate in painting' but which are 'almost unknown to most ordinary viewers' (2004: 49–51). This incompatibility between the architect or artist's desire to present their work in an idealised way, and the viewer's incapacity to experience it in this way, was interpreted as representing the 'separation of the self and the world' (Golub 2004: 200). In contrast, it was thought that buildings which exhibit rotation accommodate a 'self' that was part of the 'world', not idealised into a single universal or normative-height, masculine viewpoint. In *Five Architects*, Frampton describes rotation as 'asymmetrical spinning' (1975: 9) akin to Rowe's idea of 'a building as an entity gyrating around horizontal (and vertical) axes' (1996: 192). Rowe defines rotation as 'architectural contrapposto' which 'presumes a pictorial or a sculptural condition of permanent argument' wherein the building 'is simultaneously static and is also set in motion' (1996: 192). Frampton's argument is not that the works of the Whites all embrace the same approach to rotation, but rather that they use different types of contrapposto to resist the frontal proclivities of early Modernism. This could be reframed as the hypothesis that *the designs of Eisenman, Hejduk and Meier show a more consistent formal expression when viewed from multiple perspectives than from a single perspective*. If this hypothesis is true, it should be measurable using a variation of the fractal analysis method which compares frontal and rotational views.

9.6.1 The Analytical Method

This section of the chapter analyses rotational views of Eisenman's *House I*, Hejduk's *House 7* and Meier's *Hoffman House*. While the previous section of the chapter analysed the four cardinal elevations of these houses, this section compares an additional twelve elevations derived from the sub-cardinal points, while continuing to employ an orthogonal viewpoint. Thus, for this purpose, elevations were prepared of each house from the following orientations: north (N), north-northeast (NNE), northeast (NE), east (E), east-northeast (ENE), east-southeast (ESE), southeast (SE), south-southeast (SSE), south (S), south-southwest (SSW), southwest (SW), west-southwest (WSW), west (W), west-northwest (WNW), northwest (NW) and north-northwest (NNW). The fractal dimensions of these forty-eight elevations (thirty-six sub-cardinal and twelve cardinal) are then measured, graphed and compared.

9.6.1.1 Eisenman's *House I* Rotated

Eisenman states that *House I* 'differed from the traditional idea of a formal essence' because it has 'no specific form … but rather could be considered as unformed possibilities for organization' (1999: 62–63). In terms of frontality and rotation, such an ambiguous house might be difficult to analyse, which is probably what Frampton means when he says that in *House I* 'the play between frontalization and rotation amounts to an ever present conflict which at no point is ever allowed to be

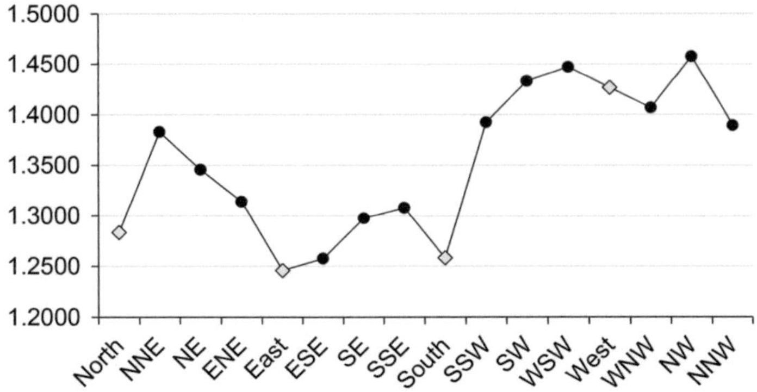

Fig. 9.15 *House I*, results for four cardinal elevations plus twelve sub-cardinal elevations

Fig. 9.16 *House I*, west-southwest elevation (*top*) and northwest elevation (*bottom*)

resolved' (1975: 9). The *House I* results are the least consistent of the three analysed using rotation (Figs. 9.15 and 9.16). The first three cardinal elevations, north, east and south, all provide low results, under $D = 1.2835$, and, apart from the ESE result, all others are higher than these three cardinals. However, what might appear to be a rising and falling pattern as the viewer rotates around the axis of the house—from the north to the south moving in an anti-clockwise direction—then suddenly changes, with further rotation towards the west. Between SSW and NNW the highest set of results $1.3893 < D < 1.4576$ are found. Interestingly, this part of the house is the 'back', in functional terms the private address, while the eastern elevation, with the lowest overall score ($D = 1.2458$), is the public entrance, or 'front' of the house.

9.6.1.2 Hejduk's *House 7* Rotated

According to Stan Allen Hejduk's architecture 'does not fix the gaze of the spectator in a "face-to-face" confrontation, but multiplies and redirects the gaze' (1996a: 94). At the heart of this cryptic commentary on Hejduk's work is the contested relationship between frontality and rotation. Frampton maintained in *Five Architects* that the frontality of the 'total mass' of Hejduk's architecture is 'in contrast to the rotation of its extremities' (1975: 9). Allen rejects this position maintaining that in order for a design to be read, 'the basic orthogonality of the cubic form must be maintained' (1996a: 94). For Allen, each elevation in Hejduk's architecture 'establishes, through the rectilinearity of its geometry, a condition of frontality regardless of the viewer's position' (1996a: 94). This difference of opinion, between Frampton and Allen, can be productively assessed in relation to *House 7*.

The set of sixteen results for *House 7* are all in a tight range ($R_{[E\%]} = 6.87$), with the difference between the cardinal and sub-cardinal elevations being less than ±3.44 % (Figs. 9.17 and 9.18). The four cardinal elevations are marginally less complex than their proximate sub-cardinal neighbours, with indirect angles of viewing adding only slightly to the overall visual complexity. Thus, the fractal dimension results for the rotation of *House 7* describe the design as one which doesn't change markedly as it is viewed from different angles, echoing Allen's proposal that in Hejduk's domestic architecture, every different rotational position constitutes a variation of the same frontal façade.

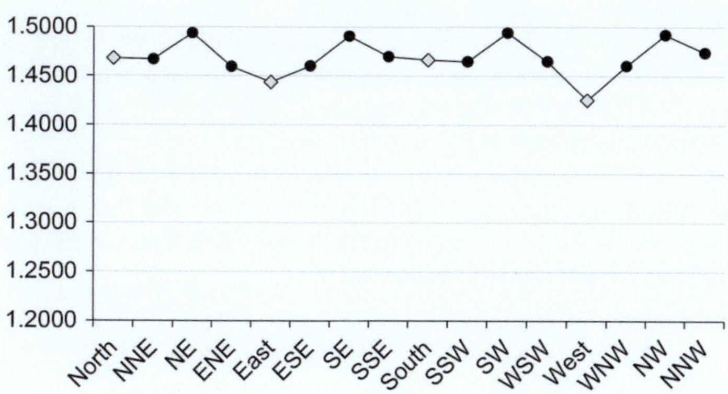

Fig. 9.17 *House 7*, results for four cardinal elevations plus twelve sub-cardinal elevations

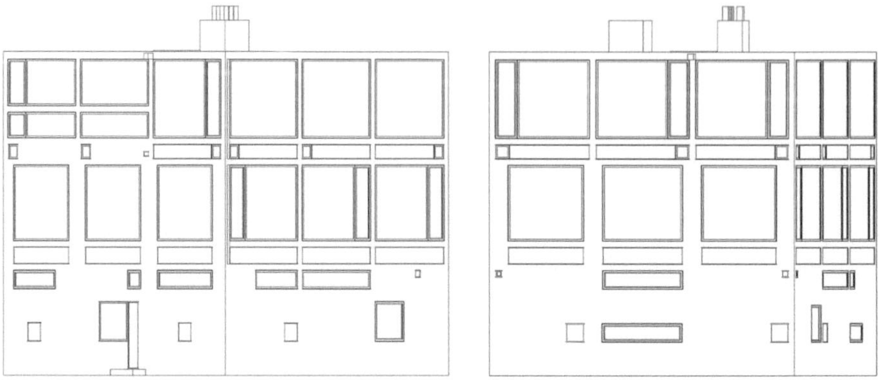

Fig. 9.18 *House 7*, northeast elevation (*left*) and west-southwest elevation (*right*)

9.6.1.3 Meier's *Hoffman House* Rotated

Meier's *Hoffman House* was selected for this analysis because it features his first exploration of the impact of diagonal planning, a development that seems to directly address the ideas of frontality and rotation raised by Frampton. The *Hoffman House* has intersecting 90° and 45° angled plans, (see the plan in Table 9.7) with the siting and landscaping arranged so that the southeast elevation is the most visible from the road, where it 'appears as a three-dimensional abstraction of interlocking geometries' (Meier 1984: 35). The house is then approached by way of a driveway from the south and, as the visitor circles the building, the southeast façade becomes dominant again, before the entrance, in the eastern façade, is finally reached.

The range of rotational results for Meier's *Hoffman House*, while still comparable ($R_{[E\%]} = 15.05$), cannot be said to be in a close range, and they do not form any clear pattern (Figs. 9.19 and 9.20). These fluctuating results may, however, reflect Cassarà's argument that in the *Hoffman House*, 'frontality and rotation confront one another …, simply syllabising the grammar and syntax of the construction' (1997: 30). While not forming a distinct pattern, the results do show a significant lowering in complexity to the southern extremities, where the elevations face the approach path. Viewed in this way, as a visitor would experience the house, its formal complexity increases when seen from the road ($D_E = 1.2759$), along the driveway ($D_E = 1.2777$) and to the front door ($D_E = 1.3794$). Rather than forming a data set that reflects a regular house that has been rotated, this data is perhaps more similar to the experience of the approach path studied in Chap. 8 on Frank Lloyd Wright, but from an orthogonal exterior view.

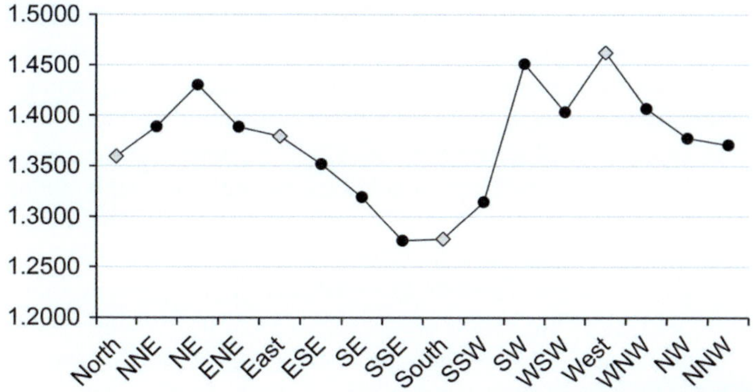

Fig. 9.19 *Hoffman House*, results for four cardinal elevations plus twelve sub-cardinal elevations

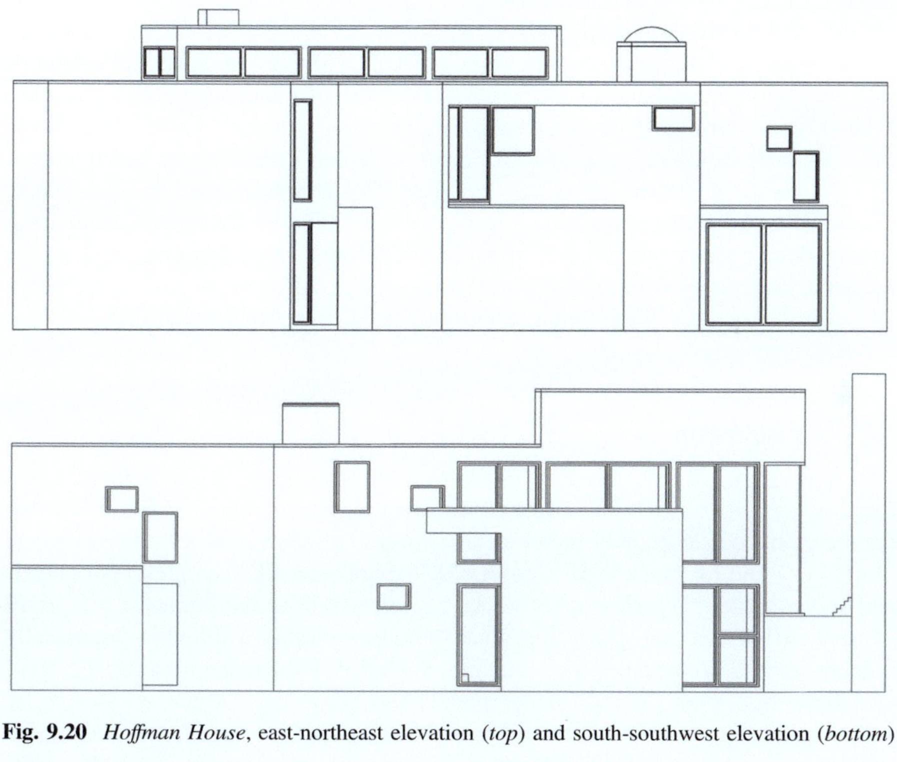

Fig. 9.20 *Hoffman House*, east-northeast elevation (*top*) and south-southwest elevation (*bottom*)

9.6.1.4 Deciphering Frontality Versus Rotation

As Frampton no doubt realised when he wrote his chapter for *Five Architects*, there is no consistency in the way these designers approach the relationship between frontality and rotation. Moreover, as many scholars have observed, the very concepts of frontality and rotation may be more appropriate for the criticism of art or sculpture—objects typically viewed from fixed, external viewpoints—than architecture. As Robert Somol suggests, terms such as frontality and rotation are merely an 'articulation of a series of dialectics', exemplifying the 'logic of contradiction and ambiguity' (Somol 1998: 86–87). Rosemarie Bletter also takes issue with Frampton's argument, declaring that his terms 'are too broadly defined' (1979: 205) to be useful.

The results in this chapter show that the orthogonal views of Hejduk's *House 7* and Eisenman's *House I* are generally less complex than the rotated views, but this could be a by-product of the nature of any cubic building. When a cube is viewed askew, its visual complexity is likely to increase by virtue of the compressed version of the adjacent face now being apparent. In contrast, the results for the irregular form of Meier's *Hoffman House* do not isolate the cardinal views in the same way. Overall, these three sets of results could be interpreted as implying that Hejduk's and Eisenman's houses actually represent special types of frontality, while Meier's demonstrates rotation.

According to Bletter, Frampton argues that in 'Hejduk's projects the notion of frontality and rotation is resolved, while in the work of Eisenman … and Meier the intrinsic conflict of these two systems of organization remains' (1979: 205). This is true only if by 'resolved', Frampton means that Hejduk's architecture, when viewed through rotation, begins to create a single façade experience. In that sense, Eisenman's and Meier's works *are* less resolved, although the latter's work most clearly demonstrates the impact of formal rotation in its elevations, and in the former's may, as Frampton also implies, be most evident in its interior qualities, although this is something which cannot be tested using the present data.

9.7 Conclusion

There are several remarkable features in the data presented in this chapter, including the incredible consistency of Meier's early houses, which feature a composite range of <5.34 % and for the optimal sub-set, <0.54 %. Given different sites, client needs and budgets, Meier's early houses display an unwavering commitment to a clear set of aesthetic values and goals. Eisenman's houses display a different but equally notable set of characteristics. For example, in *House IV* the range between the set of elevations was <0.66 % and for the plans <3.41 %, both of which are striking in isolation, but the difference between the two is also <2.14 %. The equivalent difference between the average plan and elevation results for Eisenman's *House VI* is even less at <1.64 %. If ever a pair of houses could be considered abstract sculptural objects, capable of being turned on their sides while still resembling their original states, then *House IV* and *House VI* are it.

Hejduk's architecture is less consistent across the set of his five works, but within specific houses, and especially when viewed under rotation, they exhibit some equally significant characteristics. For example, the rotational analysis of his

House 7 illuminates a complex argument that has ebbed and flowed around his work for several decades. What is so special about its rotational properties, when his work seems to be the most orthogonal of the group? The answer is that the characteristic complexity of his architecture remains stable as it is viewed under rotation, being within a range of $<5.02\ \%$. We have interpreted this, in the context of past critiques of his architecture, as implying the presence of a single underlying elevation treatment, which is most visible through movement.

Fundamentally, the early houses of Eisenman, Hejduk and Meier privilege form over other factors. These architects demonstrate that the visual appearance of a house is not necessarily a function of its siting (orientation), address (approach) and program (the types of spaces in its interior), moderated in minor and consistent ways by the exigencies of materiality and style. Architectural expression need not be representational or figurative, that is it need not be restricted to the application of a conventional set of architectural elements such as columns and architraves, it can also arise from the manipulation of form, using rigorously applied rules and principles.

Chapter 10
Post-modernism

Ten houses, five by Robert Venturi and Denise Scott Brown and five by Frank Gehry, are the focus of this chapter. Venturi and Scott Brown collaborated on the design of many iconic Post-Modern buildings and in this chapter we measure the geometric properties of the *Beach House*, *Vanna Venturi House*, *House in Vail*, *House in Delaware* and *House on Long Island.* The elevations and plans that we reconstructed for the fractal analysis were derived from published archival drawings and interpreted using photographs of the completed buildings (von Moos 1987; Stadler et al. 2008). The second set of houses featured in this chapter are by architect Frank Gehry, who gained world wide attention in the 1990s for his titanium-clad biomorphic buildings. However, from the late 1970s to the mid 1980s he designed a range of modestly scaled and eclectically styled works, including a series of houses that are often regarded as being the link between Post-Modernism and Deconstructivism. The five houses by Gehry that are examined in this chapter, all designed for or built in California, are the *Familian House*, *Gunther House*, *Wagner House*, *Spiller House* and *Norton House*. Published design and construction drawings were used as the basis for the analysis of Gehry's work, along with, for the unbuilt projects, photographs of presentation models (Friedman 2009; Dal Co et al. 2003; Gehry et al. 1990).

This chapter commences with an overview of the Post-Modern movement before examining the data derived from the ten houses. In the final part of the chapter, a comparison is undertaken of the impact of façade permeability on the visual expression of a set of the Post-Modern works by Venturi and Scott Brown and a set of modernist designs by Le Corbusier. The purpose of this analysis is to look for correlations between formal modelling (the shape of the design) and any openings in that form (typically an expression of function). Because the relationship between form and function was highly contested by both modern and post-modern designers, this secondary approach offers a new way of examining this issue.

M.J. Ostwald and J. Vaughan, *The Fractal Dimension of Architecture*,
Mathematics and the Built Environment 1, DOI 10.1007/978-3-319-32426-5_10

10.1 Post-modernity

For a philosopher or cultural theorist, the era that saw the eclipse of Modernism was simply known as 'post-'—meaning 'after'—Modernism, a description of a time when change began to occur in politics, technology and economics (Harvey 1990; Jameson 1990). In parallel with the ascendance of this political view of the world, architects developed an alternative interpretation of the concept of Post-Modernity which entailed the rejection of functionalism and a return to historicist, social and cultural expressions and values. While the philosophical and the architectural interpretations of post-modernity do have subtle and important similarities, for the purposes of the present chapter, the concept is used in its architectural sense.

The origin of the architectural variant of Post-Modernity is conventionally traced to Hudnut's (1945) article 'The Post Modern House'. However, the Post-Modern movement only began to gain momentum in architecture in the 1960s and 1970s, when criticisms of the modern-day city and Modern architecture were published by, amongst many others, Jacobs (1961), Brolin (1976) and Alexander (1977). Today Post-Modernism is regarded as a movement that dominated architectural production and discourse from the 1960s to the 1980s. Its canonical texts include Robert Venturi's *Complexity and Contradiction in Architecture* (1966) and Venturi and Scott Brown's *Learning from Las Vegas* (1972). Charles Jencks wrote extensively on the movement, defining key moments in its history and highlighting a growing number of examples of its application in design. Jencks's *Adhocism: the Case for Improvisation* (Jencks and Silver 1972), *Modern Movements in Architecture* (Jencks 1973) and *The Language of Post-Modern Architecture* (Jencks 1977) documented and promoted this approach, raising its profile internationally. Along with texts by Stern (1988) and Moore (2004), these works collectively encouraged a generation of architects to reject functionalism and embrace humour, historicism and a newfound respect for human culture.

One of the earliest completed Post-Modern buildings was Venturi's modest design for a house for his mother, the *Vanna Venturi House*. In the years that followed, and in the aftermath of the publication of works by Venturi, Scott Brown and Jencks, several exhibitions promoted the value of historic forms, symbols and social patterns. For example, in 1975 *The Museum of Modern Art* in New York mounted a major exhibition of drawings from the Beaux Arts School; a movement explicitly rejected by Modernist designers for its ornate, impressionistic aesthetic. The display of these nineteenth-century works in a venue that had long been regarded as the bastion of contemporary values, polarised intellectual debate and helped to legitimise a growing volume of Post-Modern discourse on the value of history. In 1980 the Venice Biennale featured works by many Post-Modern architects, further authorising a revival of interest in decoration, colour and whimsy. Marvin Trachtenberg argues that at this time the Post-Modern 'current strengthened into a broad, irresistible movement, with most leading architects converted to its cause' (1986: 553). By the mid-1980s, Michael Graves and Robert Stern, both ardent supporters of Post-Modernism, 'sought to address popular taste in startlingly pastel and boldly historicizing designs' (Ghirardo 2003: 21), their work being emblematic of the later flamboyant era of Post-Modern architecture.

Internationally, Arata Isozaki in Japan and James Stirling in Britain designed major buildings with Post-Modern stylistic leanings and in the USA proponents of the movement included Stanley Tigerman, Frank Gehry, Philip Johnson, Barbara Stauffacher Solomon and the members of SITE (Alison Sky, Michelle Stone and James Wines). In Europe, in a context which was more embedded in the values and experience of historic architecture, a more socially informed and classically derived variation of Post-Modernism was promoted by Leon Krier, Rob Krier, Hans Hollein, Vittorio Gregotti, Giorgio Grassi, Ricardo Bofill, Aldo Rossi and Bruno Reichlin.

Despite manifest differences between the American and European approaches that had been combined under the heading Post-Modernism, the two shared a similarly critical view of the austerity of Modernism and Functionalism. The Post-Modernists argued that 'purism' should be replaced with 'pluralism' and that Mies's maxim 'less is more' should be rejected in favour of Venturi's 'less is a bore'. Post-Modernists sought to improve architecture and urban design by creating forms which embodied meaning, allowed for messiness to thrive and respected cultural diversity. Jencks summarized this position by observing that Post-Modern architecture was pluralist and inclusive, offering 'a resistance to single explanations, a respect for difference and a celebration of the regional, local and particular' (1992: 11). However, rather than denying its Modernist forebears, 'it still carries the burden of a process which is international and in some senses universal. In this sense it has a permanent tension and is always hybrid, mixed, ambiguous, or … "doubly-coded"' (Jencks 1992: 11).

Jencks' concept of 'doubly-coded' relates to the Post-Modern practice of working with, or being aware of, multiple simultaneous levels of meaning. For example, while admiring classical Greek and Roman architecture, most Post-Modernists did not wish to replicate such works; instead, they set out to create an ornamental pastiche of periods and places by applying iconic, identifiable building elements to often blocky, pitch-roofed and otherwise more contemporary building forms. Their historic formal references were also often simplified, lined with large geometric swathes of colour, and containing architectural symbols and signs that were arranged in contradictory combinations. The overall effect was often intended to be ironic, although the satire was inevitably lost on the general public, being reliant on a detailed knowledge of architectural history. This is the type of double-coding that Jencks argues is central to the promise and appeal of Post-Modernism.

In hindsight, historians have described Post-Modernism as both a positive and a negative force in design. For example, Gössel and Leuthäuser describe it as 'superficial' (2012: 485) yet simultaneously 'a model for successful urban repair' (2012: 488). Trachtenberg argues that the movement 'embraced overt historicism, garish symbolism, vivid ornamentation, and humble vernacular models' (1986: 553) but was also 'a dynamic, self-confident, new international "style" of surprising breadth and depth' (1986: 570). For much of the last two decades architecture schools internationally have tended to understate the significance of the Post-Modern movement, deriding its failures while forgetting that beneath its often irreverent surface, it served to reemphasize the importance of embedding social and cultural symbols and values in our buildings and cities.

10.2 Robert Venturi and Denise Scott Brown

Denise Scott Brown met Robert Venturi in 1960 when she was teaching at the University of Pennsylvania. They shared a passion for architectural theory and social discourse and began to collaborate, preparing joint lectures and research. Scott Brown was born in North Rhodesia in 1931, and began her architectural studies in 1948 at the University of Witwatersrand in South Africa before moving to London, where she completed her studies at the Architectural Association in 1952. In London, Scott Brown was influenced by Alison and Peter Smithson and their idea 'that architects should design for the real life of the street and for the way communities actually work' (Venturi et al. 1992: 8). After graduating she travelled and worked throughout Europe, England and South Africa until in 1958 she enrolled in a master's degree at the University of Pennsylvania, where her education was shaped by the teachings of the urban sociologist Herbert Gans.

Venturi was born in 1925 in Philadelphia and studied architecture at Princeton, where under the tutelage of Donald Drew Egbert and Jean Labatut he developed 'an understanding of modernism in the context of history' (Minnite 2001: 245). After completing his master's degree in 1950, Venturi worked for Eero Saarinen and Louis Kahn. In 1954 he won the Rome Prize Fellowship and spent the next two years touring Europe, studying Christian architecture of the fourth and fifth centuries, and ornamental architecture from the fourteenth to the eighteenth centuries. He later stated that 'I have always loved history, and if I hadn't been an architect, I would have been an art historian' (qtd. in Barriere et al. 1997: 129). Venturi became an associate professor at the University of Pennsylvania in 1964 and two years later he published *Complexity and Contradiction in Architecture*, a book which has since been recognised as a key theoretical underpinning for the Post-Modern movement. In that work Venturi argues that a 'valid architecture evokes many levels of meaning and combinations of focus: its space and its elements become readable and workable in several ways at once' (Venturi 1966: 23).

In the years after the publication of *Complexity and Contradiction in Architecture*, Venturi and Scott Brown began to teach at Yale. It was during this period that they took a studio group to Nevada to observe the commercial architecture and signage of Las Vegas, an event that led to the production of their highly influential book, co-written with Steven Izenour, *Learning from Las Vegas* (Venturi et al. 1972). Their book argued for the importance of visual analysis as a tool for developing a sociological understanding of the way space and form operate and communicate. Amongst other provocative ideas contained in the book—like praising car parking lots, signage and luxury casinos—*Learning From Las Vegas* somewhat flippantly proposed that buildings could be classified as either 'ducks' or 'decorated sheds'. A 'duck' is a building that uses formal modelling which exceeds its functional needs, to express its purpose. A 'decorated shed' is a building that relies on applied signs and symbols to represent or communicate its purpose. Using this categorisation, Venturi and Scott Brown argued that duck-buildings (the archetypical approach of a functionally expressive modernism) should be rejected in favour of the more modest, truthful, and historically valid type, the decorated shed.

While Venturi and Scott Brown were writing these two great works, their architectural practice was also flourishing. In 1964 the iconic *Vanna Venturi House* was completed, a design which Peter Eisenman argues was 'the first American building to propose an ideological break with modern abstraction at the same time that it is rooted in this tradition' (2008: 131). In 1969 Scott Brown officially joined Venturi's firm and 'saw to it that their discourse, based as it was on the literary essay, came to be underpinned by the arguments of empirical sociology' (von Moos 1987: 11). The firm was renamed Venturi, Scott Brown and Rauch in 1980, and completed several significant public buildings and urban spaces. In 1989 the firm became Venturi and Scott Brown and Associates, with Scott Brown being primarily responsible for the design of urban projects while Venturi remaining in charge of domestic works. However, as Stern notes, '[t]he nature of the collaboration was so intense on every level' (qtd. in Pogrebin 2013: C1), that both architects are often credited on all designs from the firm. For this reason, in the present book we typically describe both Venturi and Scott Brown as being responsible for the designs in the set even though the authorship of these works is actually more diverse, and the first was substantially designed prior to Venturi meeting Scott Brown.

10.2.1 Five Houses (1959–1990) by Venturi and Scott Brown

The first house in the Venturi and Scott Brown set is the *Beach House*, an unbuilt holiday home designed in 1959 for a site in New Jersey. The pitched roof of this single-storey dwelling appears to have been split open to accommodate an over-sized chimney, a feature that dominates the external appearance of the house, as well as its internal planning. The house has two main elevations, the beachside façade with a deep veranda, and the street façade where the entrance repeats the fireplace and chimney motif. The remaining façades have no openings and are sufficiently narrow that, 'expressively, the house has only two elevations' (Venturi 1966: 106). The *Beach House* was designed to have a roof of timber boards and cedar shingle walls.

With a similar visual character to the *Beach House*, the *Vanna Venturi House* in suburban Pennsylvania was completed in 1964 (Fig. 10.1). Taking five years to design, according to Scott Brown 'Vanna's house contains in embryo almost

Fig. 10.1 *Vanna Venturi House*, perspective view

everything we've done since' (qtd. in Tsukui 2009: 65). The two-storey home is a symbolic meditation on the 'archetypal, Anglo-Saxon images of house—the roof, the chimney, the symmetrical (more-or-less) windows and centrally placed front door' (Venturi et al. 1992: 13).

The *House in Vail*, on a sloping, forested site in Colorado, was built in 1977 as a ski-lodge. Described as 'a variation on both the witch's house in Hansel and Gretel and on the severity of a Polish synagogue' (von Moos 1987: 272), the four-storey structure is somewhat toadstool-shaped and clad externally in cedar boards with a timber shingle roof.

The *House in Delaware*, designed for Peter and Karen Flint, was completed in 1980. Set in a rural area adjoining a forest, the exterior of the three-storey house appears to combine a traditional Delaware 'cornfield farmhouse' with 'cookie cut' (Scully and Mead 1989: 22), oversized and flattened symbols derived from classical architecture. The house is constructed using the same methods and materials found in vernacular buildings in the area, including stone and timber walls and shingle roofing (Fig. 10.2).

Completed in 1990, the *House on Long Island* for the de Havenon family is on a forested site in East Hampton. This three-storey house has an appearance to match the shingle-style resort-type houses in the area, but it is also clearly Post-Modern, with bulbous Doric columns holding up the traditional porch, and over-sized windows which look out over the landscape (Fig. 10.3).

Collectively, and despite more than thirty years separating these five designs, there is a clear aesthetic lineage across the work (Table 10.1). The over-scaled chimney elements, the fairy-tale or vernacular pitched-roof forms, the 'more-or-less' symmetrical façade elements and the overt decoration are all part of the visual character of the architecture. In plan, and despite a general growth in scale across the works, the early houses have many similarities, especially given the way the central chimney element is used to link multiple levels vertically, and then to frame low-pitched habitable spaces within the roof-planes. Certainly there is an

Fig. 10.2 *House in Delaware*, perspective view

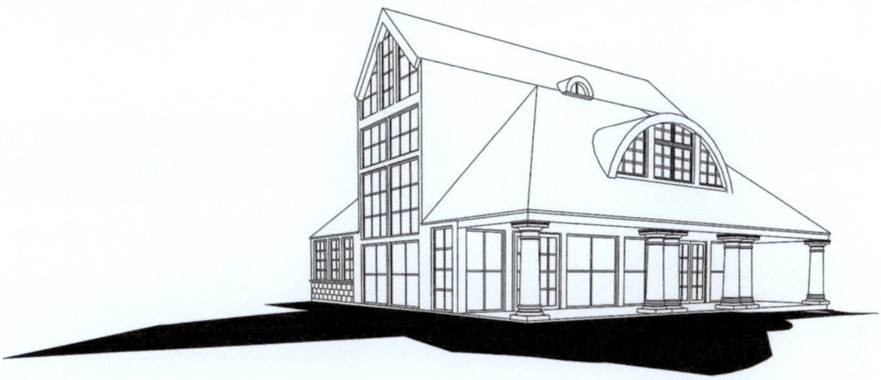

Fig. 10.3 *House in Long Island*, perspective view

increase in the visual complexity of the exterior of these buildings over time, but their planning is seemingly consistent and overall they appear to be part of an aesthetically related sequence of works. Table 10.1 shows example elevations and plans for the Venturi and Scott Brown set.

10.2.2 Venturi and Scott Brown Houses, Results and Analysis

The lowest average fractal dimension for an elevation in the Venturi and Scott Brown set is found in the *Beach House* (μ_E = 1.2940) and the lowest individual elevation is also in this design (D_{E4} = 1.2540). The highest average elevation result is for the *House on Long Island* (μ_E = 1.5019) and the highest individual elevation

Table 10.1 Venturi and Scott Brown set, example elevations and floor plans (level 4 representation—not shown at uniform scale)

Beach House

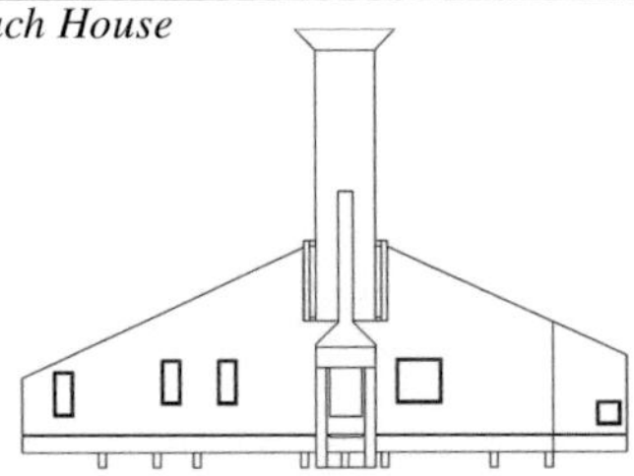

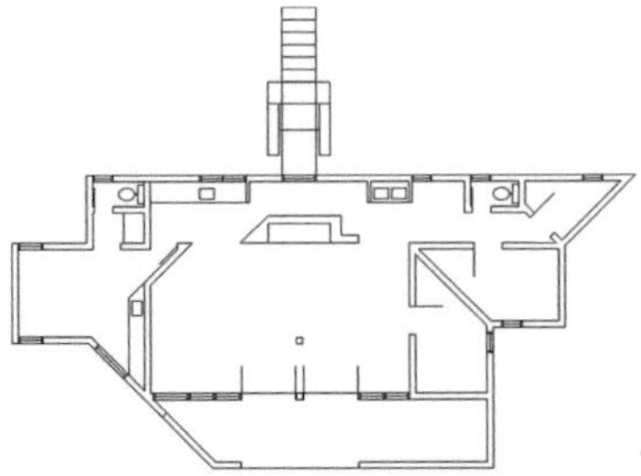

Vanna Venturi House

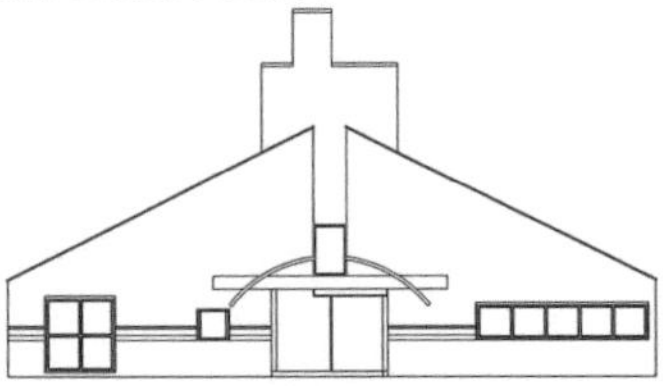

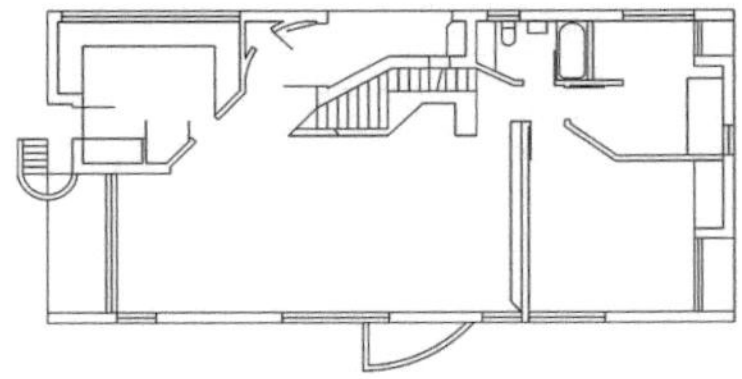

House in Vail

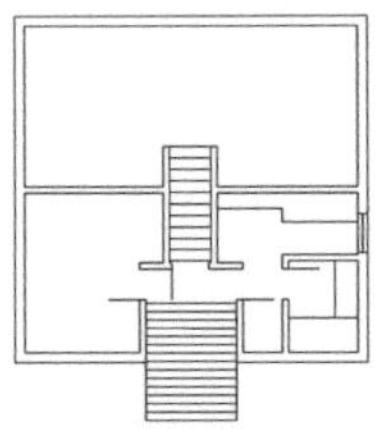

House in Delaware

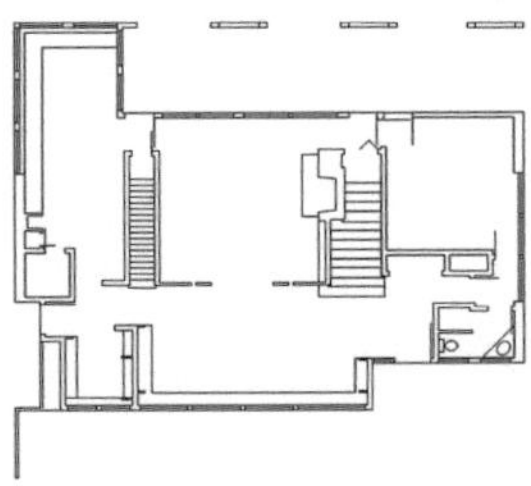

House on Long Island

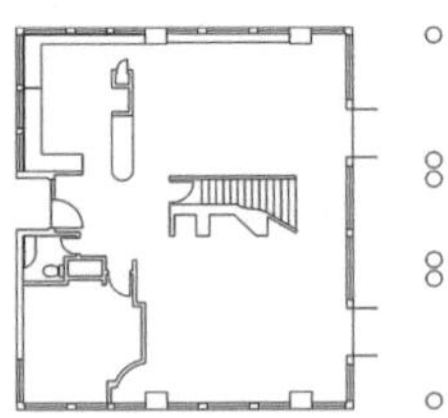

result is also for this house (D_{E1} = 1.5236). The average for all of the elevations in the Venturi and Scott Brown set is 1.4118 and the medium 1.4140. The standard deviation for the elevation data is 0.0812 (Tables 10.2 and 10.3, Fig. 10.4).

While the most complex individual plan is the first floor of the *House on Long Island* (D_{P1} = 1.3989), the highest average plan is not from this house, but the *House in Delaware* (μ_P = 1.3709). Both the lowest individual plan (D_{PR} = 1.2430) and the lowest average plan (μ_P = 1.3105) are in the *Vanna Venturi House*. The median for all plans is 1.3321 and the standard deviation is 0.0487; thus, the plan data is more consistently clustered than the elevation data. In all cases the roof is the least complex of the set of 'plan' views examined. The range of all the plans, ($R_{\{P\%\}}$ = 15.59), suggests there is some visual correspondence between them, whereas the complete range of the elevations, $R_{\{E\%\}}$ = 26.96, is best described as dissimilar.

The aggregate result for all plans and elevations in the Venturi and Scott Brown set is $\mu_{\{E+P\}}$ = 1.3719 and the composite range is $R_{\{\mu E+P\%\}}$ = 13.29. Considering only the three houses of the optimal sub-set, the *Beach House*, the *Vanna Venturi House* and the *House in Vail*, the aggregate result for plans and elevations is only minimally increased to $\mu_{[E+P]}$ = 1.3738 while the composite range is much reduced to $R_{[\mu E+P\%]}$ = 5.78. This means the smaller sub-set is made up of a group of houses which are very similar in appearance, whereas when the entire set are considered as a group, they have a much lower level of visual correspondence. It is the aggregate result for the *House on Long Island* that significantly undermines the consistency of the larger set.

The *Beach House* presents an unusual case because the fractal dimensions for its elevations are lower than, or equal to, that of its plans. Usually, as in the case for most of the results in this book, the elevations have higher *D* results than the plans. However, the result for the *Beach House* may be a result of its unbuilt nature and the possibility that a fuller consideration was given to resolving its interior planning than its elevations. The potential validity of this explanation is reinforced by the fact that the plan range for the *Beach House* is in keeping with that of the other four houses.

The remaining four houses could also be grouped in pairs. The *Vanna Venturi House* and the *House in Vail*, designed within ten years of each other, have similar results in plan and elevation, both lower than the plan and elevation averages for the set of five houses. The other two houses, the *House in Delaware* and the *House on Long Island* also have similar results to each other, being generally higher than the average for the entire set. The *House on Long Island*, which produced some of the highest fractal dimension values of the set, has a very similar range across all elevations ($R_{E\%}$ = 5.85), which fits with Stanislaus von Moos' statement that this house is 'like a pavilion without distinctions between back and front or particular accommodation for a front entrance' (1999: 284).

The most notable feature of this set is related to fact that the interior planning of these houses has a consistent visual expression until the early 1980s, when it was marginally increased in line with the scale of the houses. However, the complexity of the elevations increased more markedly over time from the *Beach House* project, stabilising for the *Vanna Venturi House* and the *House in Vail* and then increasing

Table 10.2 Venturi and Scott Brown set, results

Houses		Beach*	Vanna*	Vail*	Delaware	Long Island	Opt. [*]	Set {…}
Elevations	D_{E1}	1.3326	1.3616	1.4010	1.5094	1.5236		
	D_{E2}	1.3327	1.4078	1.4242	1.4576	1.5104		
	D_{E3}	1.2566	1.4201	1.4030	1.4711	1.5085		
	D_{E4}	1.2540	1.3829	1.3357	1.4787	1.4651		
	μ_E	1.2940	1.3931	1.3910	1.4792	1.5019		
	$\mu_{[E]}/\mu_{\{E\}}$						1.3594	1.4118
	$M_{[E]}/M_{\{E\}}$						1.3723	1.4140
	$std_{[E]}/std_{\{E\}}$						0.0588	0.0812
Plans	D_{P0}	1.3631	1.3760	1.3321	1.3769	1.3306		
	D_{P1}	–	1.3125	1.3644	1.3982	1.3989		
	D_{P2}	–	–	1.3321	–	1.3352		
	D_{P3}	–	–	1.2806	–	–		
	D_{PR}	1.2602	1.2430	1.2544	1.3377	1.2993		
	μ_P	1.3117	1.3105	1.3127	1.3709	1.3410		
	$\mu_{[P]}/\mu_{\{P\}}$						1.3118	1.3291
	$M_{[P]}/M_{\{P\}}$						1.3223	1.3321
	$std_{[P]}/std_{\{P\}}$						0.0495	0.0487
Composite	μ_{E+P}	1.2999	1.3577	1.3475	1.4328	1.4215		
Aggregate	$\mu_{[E+P]}/\mu_{\{E+P\}}$						1.3738	1.3719

Table 10.3 Venturi and Scott Brown set, comparative values

Houses		Beach*	Vanna*	Vail*	Delaware	Long Island	Opt. [*]	Set {…}
Elevations	R_{ED}	0.0787	0.0585	0.0885	0.0518	0.0585		
	$R_{E\%}$	7.8700	5.8500	8.8500	5.1800	5.8500		
	$R_{[ED]}/R_{\{ED\}}$						0.1702	0.26960
	$R_{[E\%]}/R_{\{E\%\}}$						17.0200	26.9600
Plans	R_{PD}	0.1029	0.1330	0.1100	0.0605	0.0996		
	$R_{P\%}$	10.2900	13.3000	11.0000	6.0500	9.9600		
	$R_{\ [PD]}/R_{\{PD\}}$						0.1330	0.1559
	$R_{[P\%]}/R_{\{P\%\}}$						13.3000	15.5900
Composite	$R_{[\mu E+PD]}/R_{\{\mu E+PD\}}$						0.0578	0.1329
	$R_{[\mu E+P\%]}/R_{\{\mu E+P\%\}}$						5.7800	13.2900

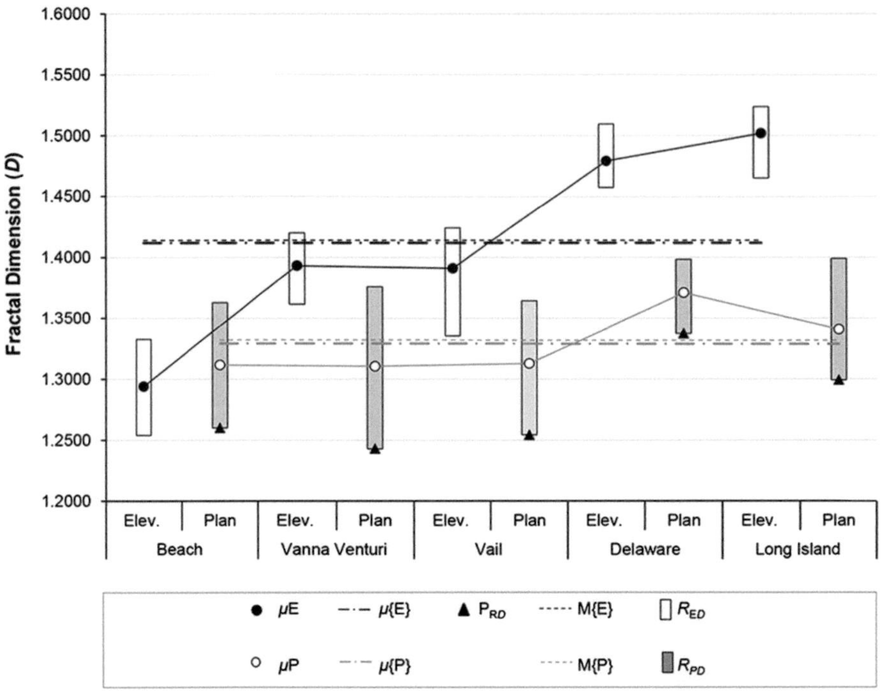

Fig. 10.4 Venturi and Scott Brown set, graphed results

for the remainder of the period being studied. Thus, while the elevation range is not especially close, the progression in growth of complexity over time is potentially a more significant pattern.

10.3 Frank Gehry

In the last few decades Frank Gehry's name has become synonymous with a type of amorphous, sculptural architecture seen in his 1997 *Guggenheim* in Bilbao, the 1996 *Dancing House* in Prague, the 2003 *Walt Disney Concert Hall* in Los Angeles and the 2004 *Jay Pritzker Pavilion* in Chicago. Yet for much of Gehry's career his practice was focussed on small commercial buildings and domestic-scale works. While Gehry's later designs have adopted a relatively consistent formal and material palette of curved titanium, his early works were more eclectic and experimental. Kurt Forster describes Gehry's architecture in appearance as like 'alien intruders into the landscape' (1998: 9). The constant feature Forster identifies across the body of Gehry's work is its use of unexpected forms and materials, each of which are appropriate for their scale and purpose (titanium for civic buildings, aluminium for commercial buildings and wire mesh for houses) but applied in unusual and subversive ways. Thus, Gehry's work, 'transforms the familiar' in such a way that it 'estranges it for the viewer' (Forster 1998: 9).

Born in 1929 in Toronto, in his youth Gehry was introduced to architectural concepts by his maternal grandmother, who encouraged him to create imaginary cities out of irregular blocks. Gehry saw this childhood activity as shaping his design ethos and as granting him a 'licence to play' (Isenberg 2009: 16). In 1947, Gehry moved with his family to Los Angeles, a city which he described as dominated by 'visual chaos' but also as 'a magical place teeming with experimentation and freedom' (qtd. in Isenberg 2009: 14). While studying architecture at the University of Southern California, Gehry took a particular interest in the work of Rudolf Schindler, Richard Neutra and Frank Lloyd Wright. During his time at university Gehry began working for the famous shopping centre designer Victor Gruen, and after completing his degree in 1954, Gehry spent time in the US Army. Returning to civilian life, Gehry enrolled in a degree in city planning at Harvard Graduate School of Design, but withdrew from these studies to work for the architectural firm Perry, Shaw, Hepburn and Dean, and later for landscape architect Sasaki Hideo in Massachusetts. In 1961 Gehry travelled to Paris, where he worked for André Rémondet, and then took time to explore the great works of European Modernism as well as French and German Romanesque architecture. The following year he returned to Los Angeles and began an architectural practice with Gregory Walsh. In 1966 Gehry set up his own firm, Frank O. Gehry and Associates. In 2001 the firm became Gehry Partners, LLP, and still operates from Los Angeles.

Rybczynski (2002) argues that if Venturi, Scott Brown and Izenour's classification system—'ducks' and 'decorated sheds'—had been used to examine Gehry's later career work, they would have been defined as inappropriately placed ducks. However, his early designs were more akin to decorated sheds, being described by Francesco Dal Co as characterised by 'eclectic indecision' (1998: 44). These qualities have often been linked to Gehry's close connection to the art world, a fact which Michael Sorkin argues 'saved Gehry from the seemingly inescapable consequences of universalism' (1999: 36). Sylvia Lavin even suggests that all of Gehry's early house designs could be considered 'artists houses', with 'the ones for artists only more so' (2009: 12).

According to Anette Fierro the singular quality of these early domestic works arises from Gehry's 'manipulation of the stature of the ordinary through overt exploitation of the vernacular' (1997: 19). These houses feature the typical Californian housing construction methods of the era, including a 'balloon-frame' (timber stud framing) clad in a hardware-store palette of corrugated metal, timber sheeting and chain-link fencing. It is the unusual application of these mundane materials and methods that makes Gehry's work unique. For example, where the metal cladding would normally be expected to be on the roof, it is found on the walls. Where the stud framing would usually be hidden in the wall cavity, it is exposed, and where the chain link fencing would be removed after construction, in Gehry's architecture it remains as an integral part of the finished house. The houses of this period celebrated the 'ugly' and 'ordinary' in such a way that Gehry was immediately, and despite his occasional protestations, accepted as a Post-Modernist.

The end of Gehry's Post-Modern period is often traced to the year 1989, when he received the Pritzker Architecture Prize. Prior to that time he had primarily designed buildings for sites in the USA, but throughout the 1990s he began to expand his practice internationally, experimenting with a more expressionist style and buoyed by larger budgets and greater opportunities. Today Gehry's architecture is often linked to applications of computer-aided design (CAD) and manufacturing (CAM), but in his best-known and most successful works, these technical tools were simply devices to achieve a particular vision, rather than being, as they are so often presented, integral to the production of every concept.

10.3.1 Five Houses (1978–1984) by Frank Gehry

The *Familian*, *Gunther* and *Wagner* Houses are all unbuilt projects designed by Gehry in 1978. According to Mildred Friedman, all three appear to be 'under construction' (2009: 143), like 'sketches in wood', they affirm the architect's stated belief that 'a structure in process is always more poetic than the finished work' (Gehry qtd. in Friedman 2009: 143). Like Gehry's 1977 partial remodelling of his own home in Santa Monica—to expose its structure and challenge conventional perceptions of 'home'—the *Familian*, *Gunther* and *Wagner* Houses question the 'distinction between the complete and the unfinished, the stationary and the static' (Dal Co et al. 2003: 178). Indeed the historic importance of these three works is that they demonstrate the growth of an idea, famously developed in a series of details and interventions in Gehry's own home, into a series of full-scale houses.

The first of the three unbuilt designs, the *Familian House* is a three-storey structure designed for a flat beachfront site in Santa Monica. Its white stucco walls appear to have been perforated with a series of rectangular cuts, each exposing the underlying stud framing of the structure and forming window mullions for the many openings in the house. The house comprises two large pavilions, joined by a walkway constructed of timber members set at odd angles, giving the design a sense of visual instability. These awkward elements project from the elevations and plans of the house, sometimes serving functional purposes (skylights and windows) while in other parts they are more decorative or semiotic, communicating that this is a work which is under construction (Fig. 10.5).

The *Gunther House* was designed for a sloping site overlooking the ocean in California. The three-storey structure is based on a rectilinear plan, although its overall form is substantially complicated by the presence of a secondary system of chain-link mesh cubes, mounted on square frames, making the building look as if it may still be under construction. Gehry's use of diagonal and circular elements in the façades, along with the additional 'ghost structure' of mesh, undermines the otherwise rectilinear form of this house. The house was designed to be clad in stucco, redwood timber panelling, concrete block and chain-link mesh (Fig. 10.6). Forster argues that the *Gunther House* 'fractured the very notion of the house as a unifying shell' taking its form from 'the shape of loosely assembled and barely

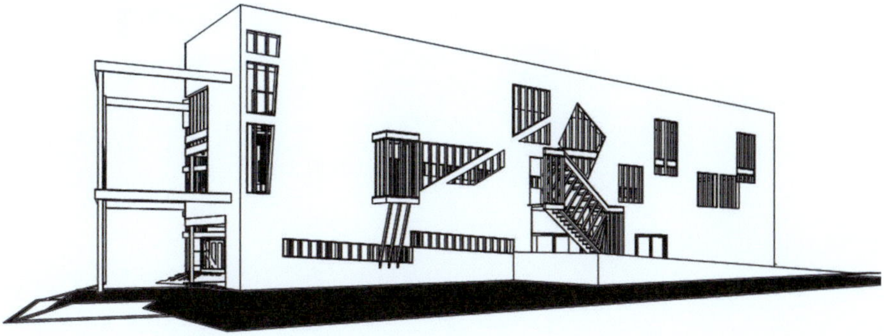

Fig. 10.5 *Familian House*, perspective view

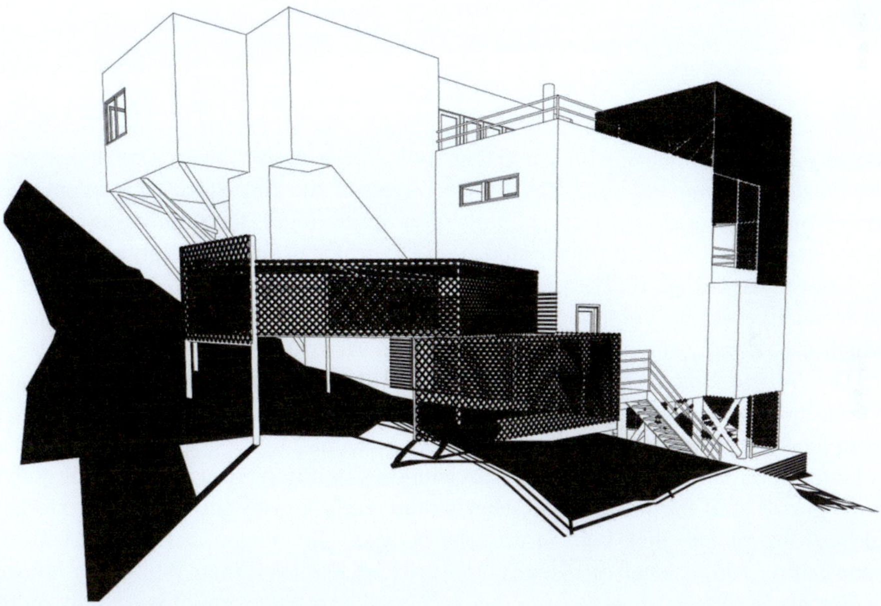

Fig. 10.6 *Gunther House*, perspective view

concatenated volumes, each resting on its own footing and tending in a different direction' (Forster 1998: 25).

The *Wagner House* was designed for a steeply sloping site in California. The building, which has a parallelogram-shaped plan, is 'designed to appear like a stucco shoebox about to slide down a fragile hillside site in Malibu' (Friedman 2009: 143). A tall, separate chain-link fence at the lower section of the site seems ready to catch the building if it moves, enhancing the precarious appearance of the house. Externally the house is clad in corrugated metal sheeting on the walls and roof.

Fig. 10.7 *Spiller House*, perspective view

The *Spiller House*, built in 1980 on a small block close to Venice Beach, is actually two apartments which together look like a large house. The house was designed for filmmaker Jane Spiller, who lived in the four-storey apartment and rented out the two-storey section. Built on a tight budget, the building uses traditional materials including timber stud frame, corrugated metal and plywood, albeit in a novel way (Fig. 10.7). Aesthetically and tectonically, the *Spiller House* is a close relative to the *Gunther* and *Wagner* houses and features several elements which are strongly reminiscent of the *Familian House.*

The *Norton House* was built for Lyn and Bill Norton in 1984 on a narrow site bordering the sands of Venice beach. The bulk of the three-storey house is set back from the street front while a 'lifeguard tower'—a small home-office perched atop a column, dominates the beachfront. The windows on the *Norton House* employ the same glazed stud frames used in Gehry's other early houses and while the external surfaces are planar, they vary in material between tile, stucco and painted panels, contributing to the level of visual complexity of its elevations. While this house, like others in this set, departs from the traditional domestic form, it is actually more in keeping with the eclectic character of the neighbouring houses along the Venice beach boardwalk. Indeed, the *Norton House*, despite some similarities to the previous four in this set, marks a turning point in Gehry's oeuvre, as it is the last of his exposed-frame designs, and its lacks the appearance of being perpetually 'under construction' that characterised the earlier works. From this moment in time, Gehry began to experiment with more eclectic geometric formal compositions, his decorated and unusual shed-constructions becoming increasingly feathered and web-footed. Table 10.4 provides example elevations and plans for the Gehry set.

Table 10.4 Gehry set, example elevations and floor plans (level 4 representation—not shown at uniform scale)

Familian House

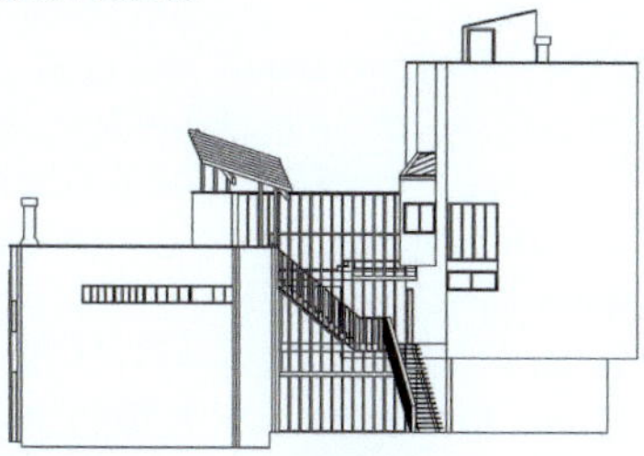

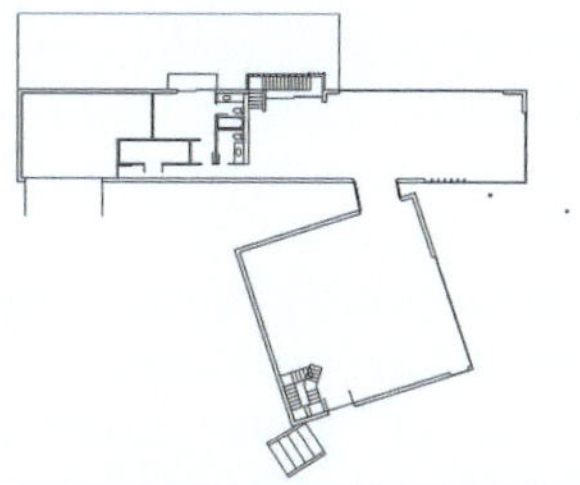

Spiller House

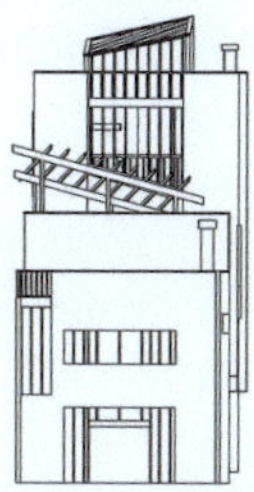

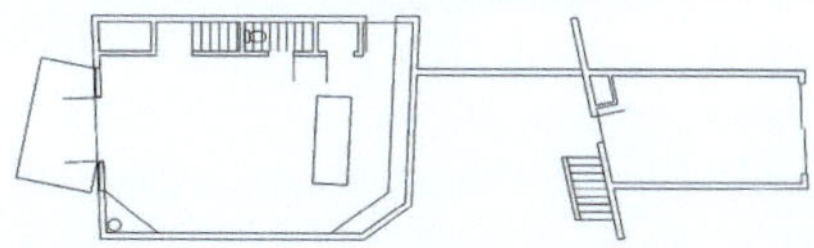

Wagner House

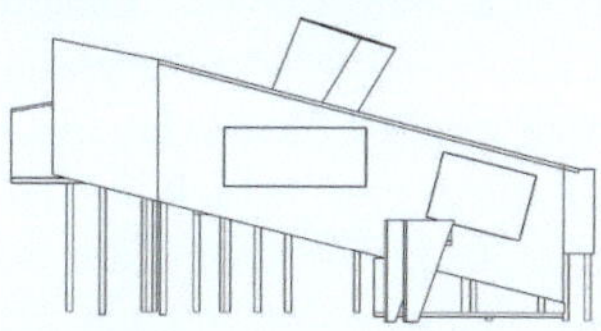

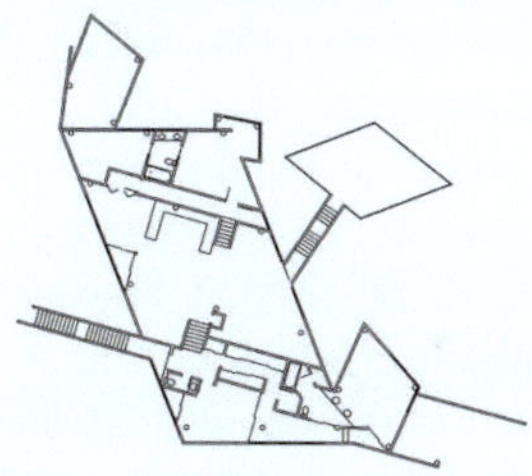

Gunther House

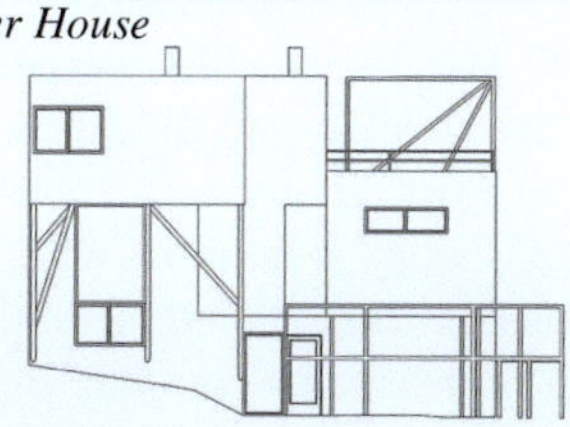

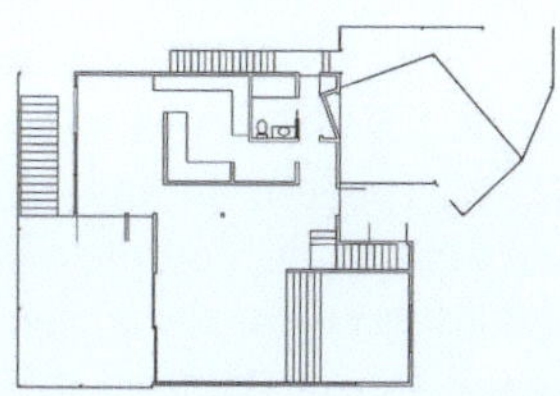

Norton House

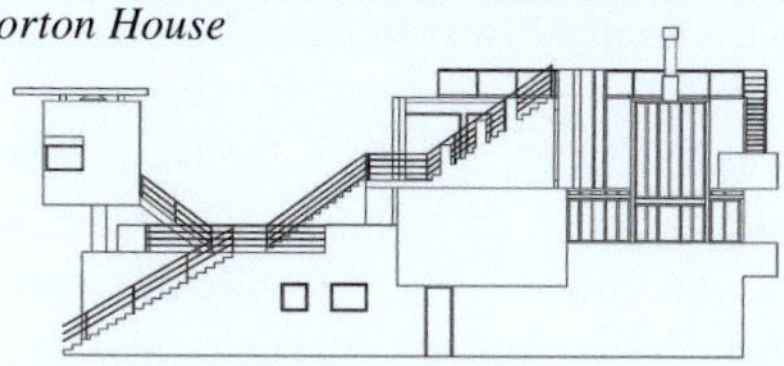

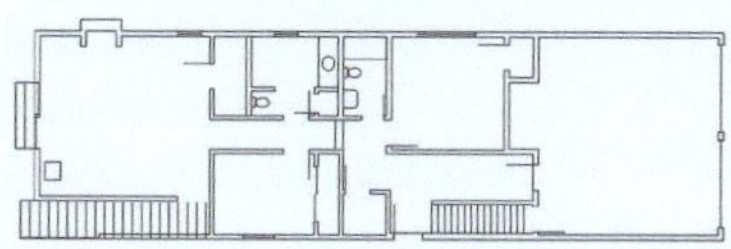

10.3.2 Frank Gehry's Houses (1978–1984), Results and Analysis

In the set of Gehry's houses, the lowest average elevation result is found in the *Wagner House* ($\mu_E = 1.3761$) and the highest is in the *Familian House* ($\mu_E = 1.4897$), leading to a range for the set of $R_{\{E\%\}} = 26.23$. The median elevation result is 1.4468, the average is 1.4377 and the standard deviation is 0.0757. Like the elevations, results for the plans show the lowest average is also the *Wagner House* ($\mu_P = 1.3169$) whereas the highest is the *Spiller House* ($\mu_P = 1.4199$). The range for the set of plan results is $R_{\{P\%\}} = 24.68$, the median is 1.3694 and the standard deviation is 0.0641, meaning that the plan data is slightly less distributed from the average than it is for the elevation data (Tables 10.5 and 10.6, Fig. 10.8).

The optimal sub-set comprises the *Familian House*, *Gunther House* and *Spiller House*. Because these are also the three most complex houses of the set, the mean result for the sub-set is $\mu_{[E+P]} = 1.4224$, a figure which is a slight increase from the overall aggregate result $\mu_{\{E+P\}} = 1.4049$. However, the composite range is dramatically lowered from a set of houses which are already visually similar ($R_{\{\mu E+P\%\}} = 7.89$) to an optimal set which are virtually indistinguishable ($R_{[\mu E+P\%]} = 1.85$).

The fractal dimension results for the roof plans of the houses in this set reflect Gehry's design strategy, which departs from a typical house layout. Generally, for most architects the roof plan of a house is the least complex of its plans; this is certainly the case for the Venturi and Scott Brown houses. However, in the set of Gehry's houses, it is only the *Gunther House* roof plan which is the lowest dimension of all of that house's plans ($D_{PR} = 1.3202$). The roof plans for every other house by Gehry are the second highest fractal dimension in each plan set.

The fractal dimensions of the elevations and plans of the *Familian House* are less similar to each other than any others in the set, with a difference of 4.01 % between the average values of each ($\mu_E = 1.4897$ and $\mu_P = 1.3496$ respectively). While the *Spiller House* has a very similar average in plan and elevation, with a difference of 1.5 % ($\mu_E = 1.4353$ and $\mu_P = 1.4199$ respectively).

There is a pattern in the results for four of Gehry's houses, with a clear downward trend in the complexity of the elevations over time; conversely, there is a slight upward trend for the plans, as they become increasingly complex over time. The *Wagner House* disrupts this trend, possibly because the design was not as developed as the other houses in the set; it is distinctly lower in fractal dimension in both plan and elevation than the others.

Frank Gehry's mid-career architecture was often classified as Post-Modern by virtue of its eclectic aesthetic expression, along with its occasionally subversive intent. The five houses in this chapter are the last of his tectonically inspired assemblages, and the first of his more sculptural compositions. As such, they are not expected to show a high degree of consistency. However, what is visible across the set of fractal analysis results is a series of designs wherein the relationship between the character of the plans and elevations gradually became aligned as his work

Table 10.5 Gehry set, results

Houses		Familian*	Gunther*	Wagner	Spiller*	Norton	Opt. [*]	Set {…}
Elevations	D_{E1}	1.4958	1.4779	1.4433	1.3517	1.3297		
	D_{E2}	1.5002	1.4502	1.3198	1.5195	1.4770		
	D_{E3}	1.5821	1.4130	1.4107	1.5121	1.4076		
	D_{E4}	1.3805	1.5273	1.3305	1.3579	1.4675		
	μ_E	1.4897	1.4671	1.3761	1.4353	1.4205		
	$\mu_{[E]}/\mu_{\{E\}}$						1.4640	1.4377
	$M_{[E]}/M_{\{E\}}$						1.4869	1.4468
	$std_{[E]}/std_{\{E\}}$						0.0735	0.0757
Plans	D_{P0}	1.3061	1.3546	1.2432	1.2902	1.3694		
	D_{P1}	1.3816	1.3669	1.4139	1.4189	1.3603		
	D_{P2}	–	1.3794	–	1.4194	1.4283		
	D_{P3}	–	–	–	1.4900	–		
	D_{PR}	1.3612	1.3202	1.2935	1.4811	1.4117		
	μ_P	1.3496	1.3553	1.3169	1.4199	1.3924		
	$\mu_{[P]}/\mu_{\{P\}}$						1.3808	1.3732
	$M_{[P]}/M_{\{P\}}$						1.3732	1.3694
	$std_{[P]}/std_{\{P\}}$						0.0630	0.0641
Composite	μ_{E+P}	1.4296	1.4112	1.3507	1.4268	1.4064		
Aggregate	$\mu_{[E+P]}/\mu_{\{E+P\}}$						1.4224	1.4049

Table 10.6 Gehry set, comparative values

Houses		Familian*	Gunther*	Wagner	Spiller*	Norton	Opt. [*]	Set {…}
Elevations	R_{ED}	0.2016	0.1143	0.1235	0.1678	0.1473		
	$R_{E\%}$	20.1600	11.4300	12.3500	16.7800	14.7300		
	$R_{[ED]}/R_{\{ED\}}$						0.2304	0.2623
	$R_{[E\%]}/R_{\{E\%\}}$						23.0400	26.2300
Plans	R_{PD}	0.0755	0.0592	0.1707	0.1998	0.0680		
	$R_{P\%}$	7.5500	5.9200	17.0700	19.9800	6.8000		
	$R_{[PD]}/R_{\{PD\}}$						0.1998	0.2468
	$R_{[P\%]}/R_{\{P\%\}}$						19.9800	24.6800
Composite	$R_{[\mu E+PD]}/R_{\{\mu E+PD\}}$						0.0185	0.0789
	$R_{[\mu E+P\%]}/R_{\{\mu E+P\%\}}$						1.8500	7.8900

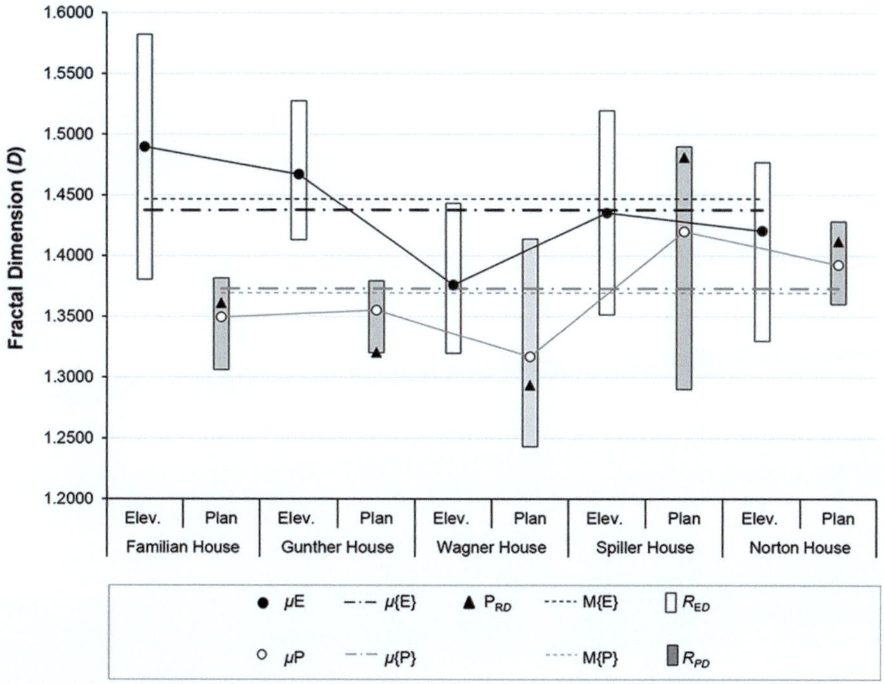

Fig. 10.8 Gehry set, graphed results

became less reliant on a specific material palette and construction method and sought more abstract and yet three-dimensionally consistent formal properties.

10.4 Comparison of the Post-Modernist Works

Comparing the aggregate results of the two Post-Modern sets, the most obvious difference is that Gehry's buildings are more visually complex ($\mu_{\{E+P\}} = 1.4049$) than those of Venturi and Scott Brown ($\mu_{\{E+P\}} = 1.3719$). The formal coherence graphs for the two sets of houses show that the correlation between the formal properties of the plans and elevations is stronger in the work of Venturi and Scott Brown ($R^2 = 0.5736$) than Gehry ($R^2 = 0.0332$) (Figs. 10.9 and 10.10). This means that while the two sets of houses are somewhat similar in their overall average visual complexity, the manner in which the elevations and plans relate to each other varies markedly. Venturi and Scott Brown's houses, in their planning, are somewhat related to their three-dimensional exterior forms, but the formal coherence graph suggests that Gehry's are not at all related.

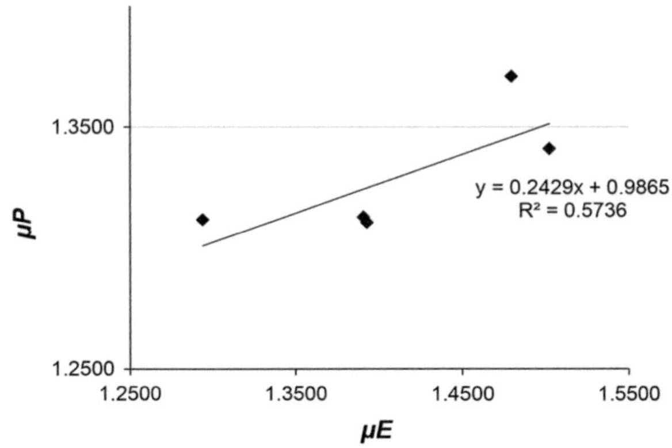

Fig. 10.9 Venturi and Scott Brown set, formal coherence graph

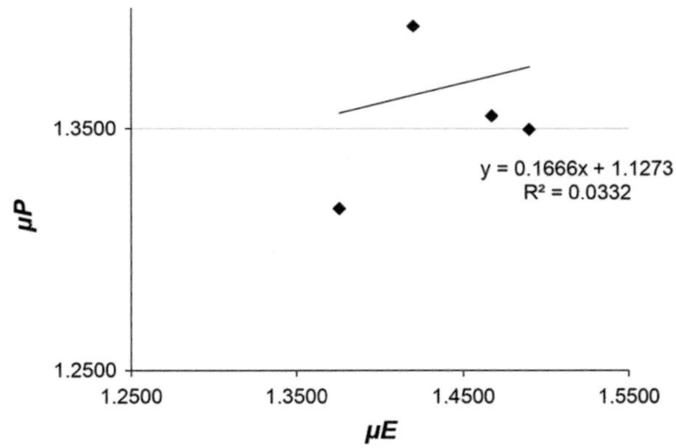

Fig. 10.10 Gehry set, formal coherence graph

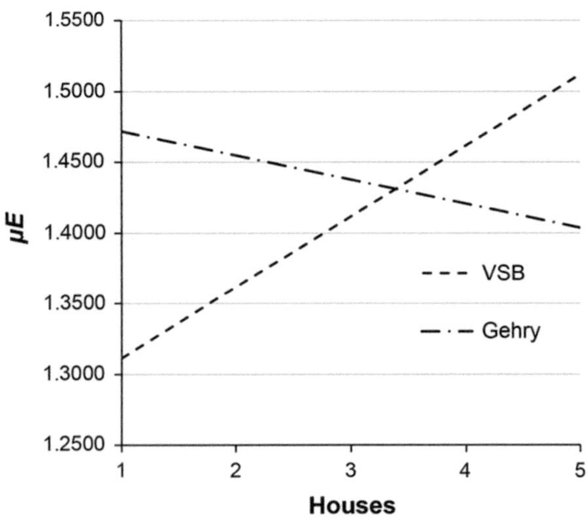

Fig. 10.11 Venturi and Scott Brown (VSB) and Gehry sets, linear trendline data for elevations (μ_E)

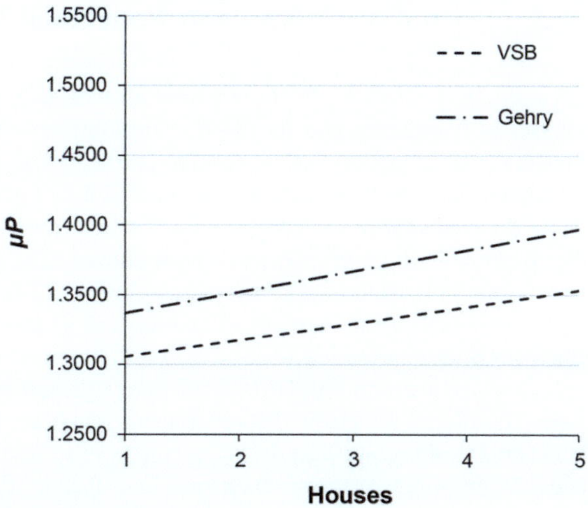

Fig. 10.12 Venturi and Scott Brown (VSB) and Gehry sets, linear trendline data for plans (μ_P)

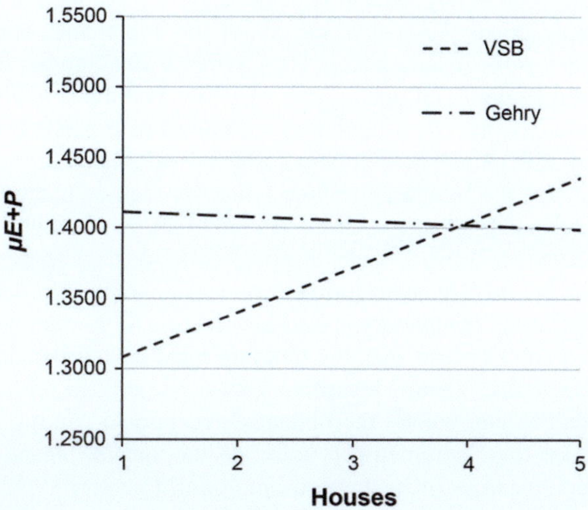

Fig. 10.13 Venturi and Scott Brown (VSB) and Gehry sets, linear trendline data for composite (μ_{E+P}) results

Comparing the linear trends in average complexity for the sets of houses over time, the fractal dimensions of the elevations in Venturi and Scott Brown's designs typically increase, while those for Gehry decrease (Fig. 10.11). The plan complexity of both sets increases (Fig. 10.12) and the composite results, combining plans and elevations, more closely correspond to the elevation trends (Fig. 10.13).

10.5 Formal Modelling and Functional Permeability

An opening in a façade, be it a window or door, is a type of mediation between the interior of a building and its exterior environment. The number of openings in a façade is also a measure of its relative permeability. Façade permeability is both a parameter that can be readily measured, and a factor that communicates something about the character of the building. For example, Moore, Allen and Lyndon suggest that windows 'do more than let in light and air. The way they are placed in a wall affects our understanding of the whole house' (1974: 211). The relative permeability of a façade (the number of openings in it) and the density of its fenestration (the pattern of distribution of its windows) can both be determined. The former can be calculated using a simple tally of openings and the latter using fractal dimensions. Furthermore, given the relationship between the two, a partial correlation between *D* values and number of openings in a façade might be expected. This is because the more openings there are in a façade, the more visual complexity it might be expected to have.

While this assumption about the relationship between openings and visual expression might seem reasonable, it is possible that a highly sculptural built form may not express anything about the functional, environmental or programmatic factors it accommodates. Thus, rather than assuming the two are necessarily related, if the number of openings is very low, and the *D* value is relatively high, the façade must be heavily articulated, sculptured or modelled. Venturi describes an example of this situation in Edwin Lutyens' original design for the *Liverpool Cathedral*, where the 'scattered minute windows seen as black dots impose themselves in an independent pattern on the symmetrical, monumental forms of the whole building' (1966: 65–6).

An inverse correlation between the number of openings in a façade and its level of formal complexity is also informative. If the number of openings is high, and the complexity low, then the building must have little additional detail other than these openings. Venturi identifies such a property in the *Low House* by McKim, Mead and White, where 'the blatantly exceptional window positions … contradicted the consistent symmetrical order of the outside shape to admit the circumstantial complexities of its domestic program' (Venturi 1966: 55). In this way, the number of openings in a façade and its fractal dimension can be used in combination to differentiate between four different responses to function and form in architecture (Table 10.7).

10.5.1 The Analytical Method

This analysis undertaken in this section tests the hypothesis that *a comparison between the number of openings in a façade, and the geometric modelling of that façade, will reveal the extent to which a building is dominated by form or function.*

Table 10.7 Interpreting the relationship between form and permeability

D_E	# Windows	Interpretation of results
High	High	Building has a complex form with many openings Building form may be a reflection of functional permeability Likely to be a design which is dominated by openings
High	Low	Building has a complex form with few openings Building form is not a reflection of functional permeability Likely to be a design which is highly sculptural or modelled
Low	High	Building has a simple form with many openings Building form is not a reflection of functional permeability Possible in situations where small, repetitive windows are placed in an otherwise blank wall
Low	Low	Building has a simple form with few openings Building form may be a reflection of functional permeability Likely to be a minimalist design

To test this hypothesis, fractal dimension data derived from two sets of houses is augmented with permeability information. The two sets are Le Corbusier's Modern architecture and Venturi and Scott Brown's Post-Modern architecture. This pairing is interesting in light of Venturi's criticism of Modernism's fixation on form and its concomitant rejection of decoration. However, before undertaking the analysis, it is notable that Venturi and Scott Brown's architecture is actually slightly *less* visually complex ($\mu_{\{E+P\}} = 1.3719$) than Le Corbusier's ($\mu_{\{E+P\}} = 1.3825$). This result might be unexpected, given Venturi's admonishment of Modernism's 'less is more' attitude, although it could also be explained, in Venturi's terms, as a product of Modernism's duck-like fixation on form.

The approach used for comparing the two sets involves calculating the number of permeable openings in each elevation for the two sets of five houses. The D_E data for each individual elevation is then combined with data derived from the number of openings. The augmented data for each set of houses is then charted to seek patterns in the relationship between façade complexity and permeability. For the purposes of this analysis, permeable 'openings' include conventional doors and garage doors as well as windows. While this definition is adequate for the majority of houses, some architectural styles pose particular challenges to this variation of the fractal analysis method. The following guidelines inform the process used in this chapter for counting openings.

First, as the analytical method is fundamentally concerned with design intent, the focus of this approach is on counting the major openings (voids surrounded by solid surfaces) that penetrate a façade or, more correctly, are visible in an elevation. This implies that the count is not to be dominated by the number of panels or panes within the window or door opening unless these are clearly deliberate parts of the aesthetic expression of a façade. This is because divisions within openings are often a function of the exigencies of materials and structural systems. For example, consider an opening in a façade that is fitted with a single aluminium-framed unit which has one sliding glass panel and one fixed glass panel; this is effectively one window and it is counted as such. Similarly, a window in an Arts and Crafts style house that is made up of thirty or more diamond-shaped panes and associated lead mullions is still counted as just one window. The bigger difficulty with deciding how to count windows is associated with curtain-wall structures and expansive glass walls. For example, several of Le Corbusier's Modernist villas have extensive horizontal bands of glass in their façades that are broken by the deliberate placement of vertical steel mullions. At first this would seem to be a case of a single window, but if we look more closely the determination changes. For example, in most cases the mullions are not structural and not simply determined by the availability of glass in certain sizes. The placement of such mullions often corresponds to a change in the interior function of the house. Under such circumstances, the opening is classified as multiple.

To broadly interpret the data against the four possible categories of results (Table 10.7) two midpoints on the scales had to be selected. For the number of windows in an elevation, a typical upper limit of 11 was postulated based on the set of designs examined in this book, leading to a value of 5.5 being used as the midpoint to divide the data into 'high' and 'low' numbers of openings. For fractal dimension, a typical mid-point result of 1.40 was used to differentiate 'high' and 'low' complexity in these designs.

10.5.2 Results and Discussion

The permeability and complexity chart for the five Modernist houses Le Corbusier shows that ten of his seventeen elevations are in the high D_E range and high number of openings category (top-right sector), four are in the low D_E and high openings sector (bottom-right), one is in the low D_E and low openings sector (bottom-left), and two are in the high D_E and low openings sector (top-left) (Fig. 10.14). This means that, based on the logical assumptions about the relationship between visual complexity and practical permeability, eleven of the seventeen elevations (64.7 %) adhere to a pattern wherein form broadly follows function (top-right and bottom-left

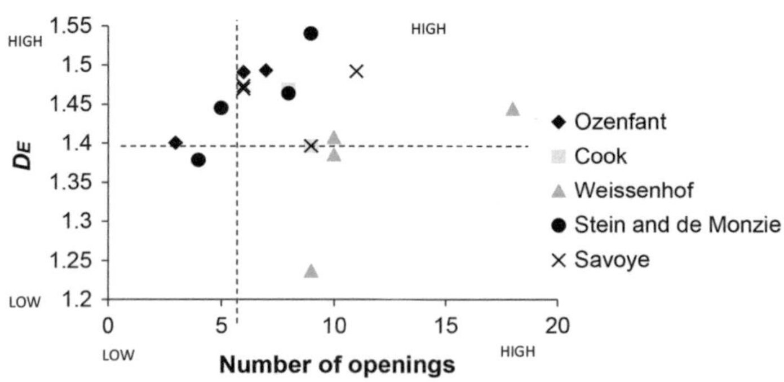

Fig. 10.14 Correlation of elevation complexity and permeability results for Le Corbusier

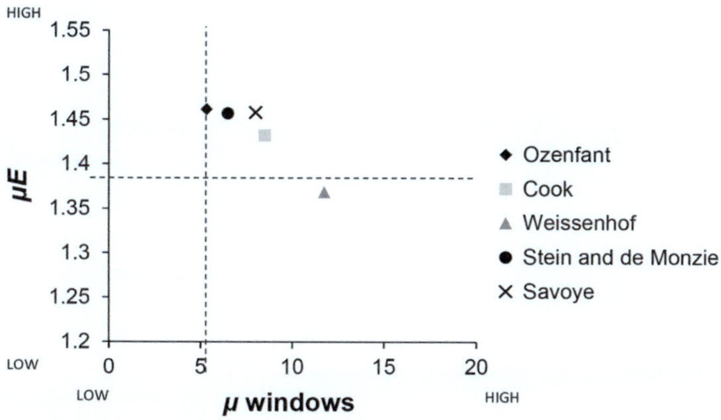

Fig. 10.15 Correlation of average elevation complexity and average permeability results for Le Corbusier

sectors). Significantly, five of the six that do not fall into these sectors are actually very close to them, and only one elevation, for the *Weissenhof-Siedlung Villa 13*, is a clear outlier. Indeed, the *Weissenhof-Siedlung Villa 13* appears unique in the set with two of its elevations containing a high number of windows and a low fractal dimension, suggesting that the windows are the only factor contributing to the complexity of its exterior. The *Villa Savoye* results are predominantly clustered in the top-right sector, corroborating the view that it is a functional building insofar as it has a clear relationship between permeability and expression. The *Maison-Atelier Ozenfant* results for permeability vary; the two main elevations have high window counts and high fractal dimensions, akin to the *Villa Savoye*, while the southern, or 'back' elevation is more restrained in its form and has lower permeability.

While examining the data elevation-by-elevation is informative, some of these results may be shaped by outliers, or elevations that for practical reasons are not viable locations for windows and openings. Thus, an alternative way of looking at this data is to compare average results for D_E and for the number of openings for each of Le Corbusier's houses (Fig. 10.15). Of the five average results, three of the houses (*Savoye*, *Cook* and *Stein-de Monzie*) are in the high complexity, high number of openings sector, and a further house (*Ozenfant*) is very close to that sector, having an average number of windows just less than 5.5. The *Weissenhof-Siedlung Villa 13* is the only outlier. Furthermore, on balance, there is a slight dominance of complexity over permeability in the final data, confirming that while openings are definitely not the only factor shaping the formal expression of Le Corbusier's architecture, their presence is nevertheless a critical part of the visual character of these works.

In most cases the elevation permeability and expression chart for the Post-Modern houses of Venturi and Scott Brown shows a tighter clustering of elevations for each building than for those found in the works of Le Corbusier (Fig. 10.16). The fact that all of Venturi and Scott Brown's designs are

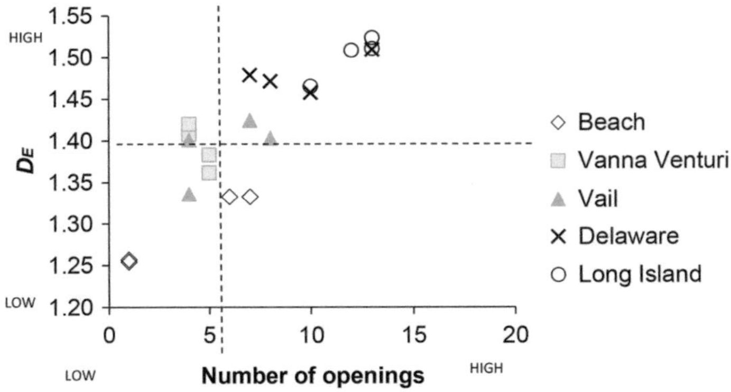

Fig. 10.16 Correlation of elevation complexity and permeability results for Venturi and Scott Brown

free-standing, suburban or rural structures, whereas several of Le Corbusier's are on tight urban sites, may partially explain this difference. The most dominant group in the chart of Venturi and Scott Brown results is, like Le Corbusier's, the high D_E and high number of openings sector of the graph (top-right), with ten elevations in this quadrant. This sector includes all of the elevations of the *House in Delaware* and the *House on Long Island* along with two of the elevations from the *House in Vail*. The next most common sector, with five elevations, is the low D_E and a low number of openings quadrant (bottom-left). Two of the elevations each from the *Beach House* (which are so close as to appear identical in the chart) and the *Vanna Venturi House*, along with one façade from the *House in Vail* are in this quadrant. There are three elevations in the top-left sector, (two from the *Vanna Venturi House* and one from the *House in Vail*) and only two elevations in the bottom-right—low D_E and high number of openings—both from the *Beach House*. Thus, fifteen of Venturi and Scott Brown's elevations fall into the duck-like 'form follows function' sectors, the top-right and bottom-left of the chart.

While this sector-result may appear to have some similarities to that of Le Corbusier's houses, the spread of the data is very different; the results for Venturi and Scott Brown are more evenly scattered along an almost 45° line rising across the chart, while Le Corbusier's are more clustered in the upper range. This trend is even more notable when the average results for each house (D_E and number of openings) are charted (Fig. 10.17). While four of Le Corbusier's houses are neatly grouped around the upper-middle section of the graph, Venturi and Scott Brown's tend to be more evenly distributed in the central part of each sector of the graph.

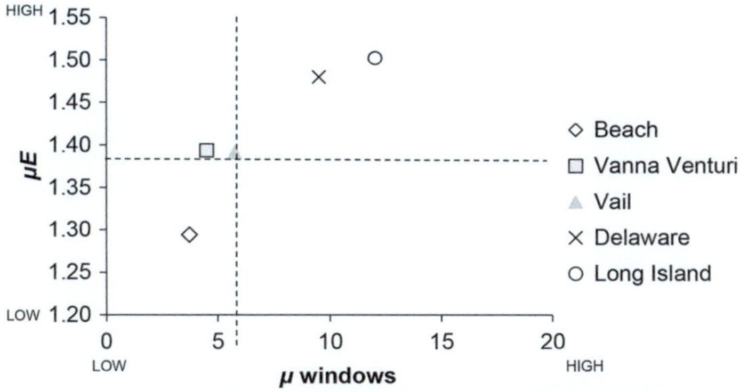

Fig. 10.17 Correlation of mean elevation complexity and mean permeability results for Venturi and Scott Brown

10.6 Conclusion

Robert Venturi was highly critical of the artificial simplicity and apparent formal purity found in many Modernist works, arguing that 'complexity must be constant in architecture. It must correspond in form and function' (1966: 19). In the five Venturi and Scott Brown houses examined in this chapter there is a correlation between the typical level of visual complexity found in an elevation, and the number of openings in that elevation. While there are many other factors in addition to permeability that shape this result, the results do imply that for Venturi and Scott Brown the interior room arrangement does determine, by way of the number and type of windows and openings in a façade, its exterior expression. The somewhat mannered (and even contrived) window forms notwithstanding, complexity is a constant in architecture, in the precise sense that Venturi originally described in 1966.

For Le Corbusier it is apparent in the correlation analysis of elevation complexity and permeability that the relationship between form and openings is less direct than it is for Venturi and Scott Brown. While there is a clear tendency in Le Corbusier's houses for more openings to be in the more complex elevations, there are evidently other factors, some of them quite substantial, driving higher formal complexity results. Le Corbusier's desire for a functional or machine aesthetic, his use of metal railings, exposed structure and roof garden screens, all add to the visual complexity, but are not directly associated with permeability.

Both the results for Le Corbusier and Venturi and Scott Brown confirm that a comparison between the number of openings in an elevation, and the geometric modelling of that elevation, does indeed begin to reveal something about the extent to which a building's expression is dominated by form or function.

Chapter 11
Minimalism and Regionalism

Minimalism is a design approach that relies on simple and unadorned shapes, or the repetition of a limited formal vocabulary and palette of materials and colours, to emphasise perceptions of lightness, spaciousness, order or purity. Regionalism is an architectural movement that advocates using local materials and tectonic practices in combination with building forms that are attuned to immediate environmental conditions, to create a contemporary, but still recognisably vernacular architecture. While there are clear differences between Minimalism and Regionalism, the two design strategies often result in the production of visually restrained and formally refined buildings that evoke poetic or phenomenal responses to materials or settings.

11.1 Introduction

This chapter examines the work of Japanese Minimalist architects Kazuyo Sejima and Atelier Bow-Wow, and Australian Regionalist architects Peter Stutchbury and Glenn Murcutt. While the Minimalist and the Regionalist approaches are considered separately in this chapter, they are often closely related. For example, both Stutchbury and Murcutt are regularly described as producing minimalist designs that emphasise lightness and order, while Sejima and Atelier Bow-Wow have been praised for the way they respond to their urban contexts and rely on Japanese tectonic practices and materials. Atelier Bow-Wow in particular are repeatedly connected to attempts to shape architecture in response to regional and local conditions. Thus, the simple distinction that is so often made between Regionalist and

M.J. Ostwald and J. Vaughan, *The Fractal Dimension of Architecture*,
Mathematics and the Built Environment 1, DOI 10.1007/978-3-319-32426-5_11

Minimalist approaches is one that this chapter will engage with, as all four architects have been connected, in different ways, to both of these design movements. Furthermore, the relationship between Regionalism and Minimalism is also one that can be directly tested using fractal analysis. This is because Minimalism, by its very nature, suggests the presence of a level and type of visual expression that should be lower than the majority of other works in this book. Conversely, Regionalist buildings are often described as being minimal in their formal modelling, but this visual expression is not an explicit characteristic of the movement. Thus, a comparison of the measures derived from these two approaches can be used to illuminate their similarities and differences.

The first of the four architects whose works are examined in this chapter, Kazuyo Sejima, has been described as 'the most radical exemplar of present-day minimalism' (Allen 1996b, p 103). Five of her houses are analysed in this chapter—*Y House*, *S House*, *M House*, *Small House* and *House in a Plum Grove*—and as there are no consistent design drawings available for Sejima's work, a number of sources were used to produce new and consistent reconstructions of her architecture for the present research. For the *Y House*, the majority of the drawings were published in *El Croquis* and missing views were recreated using the images and photographs in the same issue (Sejima 1996). Jun Aoki's adapted drawings of the *S House* and *M House* and Yuko Hasagawa's plans for the *Small House* and the *House in a Plum Grove*—the latter of which was also published in *El Croquis*—were used to reconstruct these designs (Aoki 2003; Sejima 2004a; Hasegawa 2006).

Atelier Bow-Wow is the name chosen by Momoyo Kaijima and Yoshiharu Tsukamoto for their architectural partnership. Their firm's idiosyncratic title, like their approach to design, was inspired by both the constraints and potentialities of domestic existence in the dense urban fabric of Tokyo. The five designs by Bow-Wow analysed in this chapter are the *Ani*, *Mini*, *Shallow*, *Gae* and *Juicy* houses. The digital reconstructions of these designs were created using documentation in *Graphic Anatomy* (Bow-Wow 2007) and *Behaviorology* (Bow-Wow 2010).

The first of the Regional sets comprises five houses by Peter Stutchbury. Throughout his career, Stutchbury's houses have tended to conform to one of two distinct siting and associated formal strategies; either tall, raised houses within the treetops, or linear types on more horizontal sites. The set of five in this chapter are examples of the latter strategy, all being built in suburban bushland or rural sites. They are *Verandah House*, *Beach House*, *Paddock House*, *Invisible House* and *Billabong House,* and drawings for each were provided by the architect.

The final two sets of designs analysed in this book are by Glenn Murcutt. The first of these sets includes his famous 'early' rural works: the *Marie Short, Nicholas, Carruthers, Fredericks* and *Ball-Eastaway* houses. These are all elevated buildings that 'touch the earth lightly', raised above the ground-plane on narrow posts, and have visually enclosed roof types, each with a central ridge running the length of their pavilions. The second set of 'later-career' houses by Murcutt—the *Magney, Simpson-Lee, Fletcher-Page, Southern Highlands* and *Walsh* houses—all sit directly on, or in, the ground, and are generally characterised by having skillion or butterfly roof forms. Nearly all of these houses have been altered or extended since being completed and a few have been published under different names to reflect changes in ownership. In all cases, the version of the house analysed here is the original, and the original naming of each has also been retained. For the purposes of this research, all of these designs were digitally reconstructed using published measured and design drawings (Murcutt 2008, 2012; Beck and Cooper 2002).

In the final part of this chapter, a new variation of the fractal analysis method is used to examine a recurring theme in Regionalist architecture, the visual connection between a building's interior and its surrounding landscape. This connection is allegedly enabled through the use of transparent and layered elements in a building's façade. The final part of this chapter tests this claim by comparing the fractal dimension of two sets of views of Murcutt's houses. The first set treats the building façades as opaque while the second set includes views through open doors and transparent windows or screens. This testing is first undertaken using standard orthogonal viewpoints for all ten houses and then using three-dimensional perspectives for the *Marie Short House*. The results are then compared to determine if transparency has, as Regionalist architects maintain, a significant impact on the visual expression of their designs.

The chapter concludes with a brief examination of the relationship between the Minimalists, at least one of which (Atelier Bow-Wow) is closely related to local contextual issues, and the Regionalists, at least one of which (Murcutt) has been repeatedly described as a minimalist designer. Through this process the chapter comments on the legitimacy of differentiating between these two approaches on the basis of visual expression.

11.2 Minimalism

In many parts of the world, the 1980s was an era synonymous with excess, as widespread economic prosperity gave architects and their clients hitherto unavailable opportunities to construct complex, extravagant and even gaudy buildings to cater for an increasingly sybaritic lifestyle. Some of these designs were produced under the auspices of the Post-Modern movement while, by the end of the 1980s, a different but equally theatrical and elitist style was being promoted under the umbrella of the Deconstructivist movement (Frampton 1992). As these approaches gained momentum in the first half of the 1990s, a counter-response was slowly being formulated and presented in the architectural media. This rejoinder came from architects who rejected both the compositional excesses and the philosophical posturings of their Post-Modernist and Deconstructivist counterparts. In place of this, they called for a return to the pure design aesthetic of early Modernists like Mies van der Rohe and to the unadorned and distilled practices of literalist and abstract artists. It was the latter connection to the visual arts that gave this counter-movement its most common name, Minimalism.

In the art world, Minimalism developed in the early 1960s with 'the purpose of obtaining maximum tension with minimal means' (Rosell 2005, p. 6). Architectural examples of Minimalism were produced in the UK by John Pawson, David Chipperfield and Tony Fretton; the last of these three created the *Lisson Galleries* in 1992, with its abstract and timeless façade of white walls and frosted glass surfaces. In Switzerland, Herzog and de Meuron's 1995 *Signal Building in Auf dem Wolf* is an enigmatic six-storey structure with a façade made of twisted strips of copper. In France, Bernard Tschumi and Jean Nouvel each produced iconic minimalist works while buildings such as Dominique Perrault's *APLIX Factory* in 1999, a mirror-clad industrial complex, used 'visual deception and revelation' (Ursprung 2003, p. 102) to challenge viewer perceptions.

In Japan, minimalist architecture was not only an antidote to the brashness of the 1980s but also to the overpowering presence of traffic, advertising, jumbled building scales and imposing roadways. The disordered pandemonium of the Japanese city was a result of the century-long cycle of demolition and rebuilding caused by urban fires, earthquakes and the devastation wrought by the second world war. This process of constant transformation, combined with a preponderance of small irregularly shaped sites and a lack of consistent building controls, resulted in a heterogeneous environment which afforded architects a high level of freedom of expression but resulted in a low level of visual coherence (Hein 2004).

By the 1970s, in the search for sanctuary from the urban chaos, Japanese houses were often designed to be inward facing so that they could remove any sense of their surroundings (Zukowsky 1998). Minimalism, as a panacea for this chaos, did not come to Japan by way of Europe, but was already an incipient movement inspired by Zen philosophy and the Japanese ideal of *wabi-sabi.* Apart from expressing 'rusticity and minimalism', *wabi-sabi* is an ethical proposition which promotes 'harmony with nature, and the rejection of the ostentatious, the gaudy, and the wilful' (Mehta and MacDonald 2011, p. 14). For this reason, Minimalism in Japan was not only an aesthetic response, but was also part of a nationalistic agenda to invent a more authentic and yet still contemporary Japanese architecture.

In the 1990s, in the aftermath of the global financial crisis, Japanese architects began to reject Western aesthetic influences, calling for a return to historic or traditional Japanese values of simplicity and clarity. One of the people at the forefront of this change was Tadao Ando, an architect who was already practicing in an 'uncompromisingly reduced architectural language' (Sachs 2003, p. 34). Ando's designs—such as the 2001 *Pulitzer Foundation for the Arts* and the 2008 *Fukutake Hall*—feature smooth concrete façades, their iconic formwork tie-holes providing the only hint of decoration on otherwise blank surfaces. It is primarily through the play of sunlight on and through these structures that Ando's work is viewed as responding to the natural world. A parallel approach is found in the architecture of Kuma, whose design strategy throughout the 1990s was to 'erase architecture and confront materials' (Kengo Kuma and Associates 2010, p. 30), that is to reduce the reliance on form and to celebrate architecture's textural and tectonic expression. In a similar way, albeit for a different purpose, Toyo Ito employed simple forms and advanced technology in designs like the *Sendai Mediatheque*, a building in which an irregular forest of thin steel columns supports a monolithic, translucent structure.

Probably the most famous Minimalist designer practicing in Japan today is Kazuyo Sejima. According to Luigi Prestinenza Puglisi Sejima's work commences with a consideration of 'very complicated things that gradually become simple' (2008, p. 215), her designs employing transparent or translucent elements combined with slender structures to create predominantly white, visually lightweight structures. An even more extreme case of the minimal surface is found in the work of Shigeru Ban, an architect whose designs often don't feature walls at all. His *Curtain Wall House*, *Wall-Less House* and *Naked House*, respectively from 1995, 1997 and 2000, have few enclosed elements and are often, with the exception of temporary screens, open to the elements (Ruby and Ruby 2003).

While the early Minimalist houses of Ando, Ito, Kuma and Sejima were often inward-facing courtyard homes, the more recent trend, which is also seen in the work of Ban, is to embrace connections to the urban landscape. As Bognár (2008) observes, the more recent trend is to 'seek to discover value in the given, less-than-perfect conditions of the city or the environment in general, and benefit from them, regardless of how demanding or problematic they may be' (63). The architectural firm that best exemplifies this site-based approach to Minimalism is Atelier Bow-Wow.

It is this last trend in Japanese Minimalism that most closely connects it to Regionalism. Quim Rosell defines Minimalism as being 'based on a reduction of architecture down to its essential concepts of space, light and form, rather than on mechanics of subtraction, negation or absence of ornament' (Rosell 2005, p. 6). However, there is also an increasing tendency for Minimalism to choreograph 'an intense dialogue with the site and its surroundings to the extent of transforming them and endowing them with a new identity' (Rosell 2005, p. 6). This connection to site, coupled with the use of a distilled combination of forms and materials, begins to blur the boundaries between Minimalism and Regionalism that existed in the early 1990s.

11.3 Kazuyo Sejima

Born in 1956 in the Ibaraki Prefecture in Japan, Kazuyo Sejima worked for Toyo Ito for six years after completing her Masters degree in architecture at the Japan Women's University in 1981. In 1987 Sejima formed her own practice, Kazuyo Sejima and Associates, and soon produced the 'Platform' series of designs, which comprises two built works (*Platform I, Vacation House* and *Platform II, Studio*) and an unrealised project (*Platform III, House*). These early independent designs display the genesis of Sejima's design method, an approach that is concerned with understanding and illuminating people's movements within and through the functional zones of a building. To achieve this aim, Sejima was forced to create architectural forms that would not impede the continuity and visibility of human motion. After a series of increasingly complex attempts to solve this problem, Sejima changed her approach and began to consider 'how ideas of simple volume and mixture of use could interact' (Sejima 1999, p. 118). In 1991 her designs for the *Castelbajac Sport Store* and *Saishunkan Seiyaku Women's Dormitory* convinced her that this approach was viable. This revelation—that extreme simplicity of form and materiality still allowed for a wide range of architectural and functional effects —is regarded as a turning point in Sejima's design methodology (Ito 1996; Hasegawa 2006).

In 1995, while maintaining her own practice, Sejima began collaborating with Japanese architect Ryue Nishizawa, forming SANAA. Over the next decade Sejima and Nishizawa became known for their ultra-Minimalist style, with its 'ordered aesthetic, which is aseptic, abstract, immaterial, anti-hierarchical, monochrome and inflexible' (Puglisi 2008, p. 214). While such a description seeks to capture the rational characteristics of their architecture, the experiential qualities found in buildings like the *House in a Plum Grove* (2003) and the *21st Century Museum of Contemporary Art* (2004), are reliant on the layering of light between translucent building skins, creating intense, delicate structures (Fernandez-Galiano 2007).

The qualities of the site also influence Sejima's design strategy for each building. For example, in each design Sejima seeks out the potential for the building's users to connect to the outside world, either by means of a visual connection, such as through the glazed walls that are found in the *Park Café* in 1998 and the *Onishi Civic Centre* in 2005, or via a direct physical connection though openings to the outside or courtyards, such as are found in the 1997 *M House*. According to Yūko Hasegawa, Sejima achieves this external connection by way of a 'temporal sequence' of 'actions and events caused by living in that building. When the movements of the people inside the building are visible from without, the sequence of events becomes a part of its external appearance' (2006, p. 9). Examples of this external representation are particularly evident in the *Small House* and *Shibaura House*.

The work of Sejima on her own, and the collaborative work of Sejima and Nishizawa, has been exhibited and published internationally in books and journals. They have won many national and international awards and honours: SANAA was awarded the Pritzker in 2010, the same year in which Sejima was the director of the Venice Architecture Biennale.

11.3.1 Five Houses (1994–2003) by Kazuyo Sejima

Hasegawa argues that Sejima and Nishizawa's designs for domestic architecture accept that '[c]onventional prototypes of human relationships have begun to fall apart' and that we must now question the 'typical human relations upon which modern architectural formulas are based' (2006, p. 16). Five of the houses that explore this new way of living are the focus of this section. They are: *Y House*, *S*

House, *M House*, *Small House* and *House in a Plum Grove*. All five were built in dense, residential areas in Japan, and designed to accommodate up to three generations of a family. With integral courtyard spaces that are accessible from most areas of the home, these designs demonstrate Sejima's use of circulation as a design strategy. They conform to Sejima's simple, geometrical approach, featuring seemingly thin, transparent walls, monochromatic finishes and flat roofs. These features are typical of what Luis Fernandez-Galiano sees as 'architecture in the negative, achieved through a stripping-down' process wherein Sejima's 'buildings strive to divest themselves of thickness, dispense with inertia [and] rid themselves of density' (2007, p. 175). Level 4 representations of the set of five houses are shown in Table 11.1.

The 1994 *Y House* in Katsuura is located in the Chiba prefecture, in the Greater Tokyo Area of Japan. Katsuura's housing stock was traditionally two storeys high with white walls and dark, sloping, tiled roofs. In contrast, the *Y House* is a three-storey, flat-roofed structure with two walls that are almost fully glazed and a tall, green marble-tiled façade to the street (Fig. 11.1). A private residence for a couple and their two children, the rooms in the *Y House* are all connected by way of various circulation routes to one or another of the external courtyards on the long sides of the house. According to Sejima (1999), the *Y House* offered her a chance to explore ideas about the equality of circulatory and non-circulatory spaces and the relationship between the interior and the exterior.

Designed in conjunction with Ryue Nishizawa, the *S House* in Okayama is a home for an extended family including two children and grandparents. This social structure shaped both the program and the form of the *S House,* as the client 'requested that this cohabitation of two families be reflected in the design' (Chermayeff et al. 2007, p. 90). The *S House*, a small two-storey cubic volume, has an external skin of clear corrugated polycarbonate sheeting on a timber frame creating a double-storey void that acts as circulation space on the ground floor and a connection to the top floor. As a result of this planning strategy, all of the rooms of the *S House* open into this transition zone between exterior and interior. The *S House* requires few additional openings in its façade as it draws light and air through the external skin. Thus, the elevations have no ornamentation, each appearing as a corrugated plane punctuated by small windows. The few windows that do exist are unassumingly framed with minimal steel flashings, and the roof is not visible from the exterior.

The *M House* was constructed in 1997 in Shibuya, an area of Tokyo which was once dominated by large residences and it now increasing in density due to the subdivision of many lots. One characteristic of this neighbourhood is that the southern façades of many residences face the street, offering a street wall 'wherein permanently drawn curtains and high fences … hide large windows' (Nicolin 1998, p. 20). In response to this setting, the *M House* utilises a mixture of corrugated metal cladding and transparent sheeting to suggest an internal space behind the bare walls. Appearing from the street as a single storey, the house has a basement level which contains the living room, dining room, study and studio, and a series of double-storey courtyards flood this lower level with light. With an external

Table 11.1 Sejima set, example elevations and floor plans (level 4 representation— not shown at uniform scale)

Y House	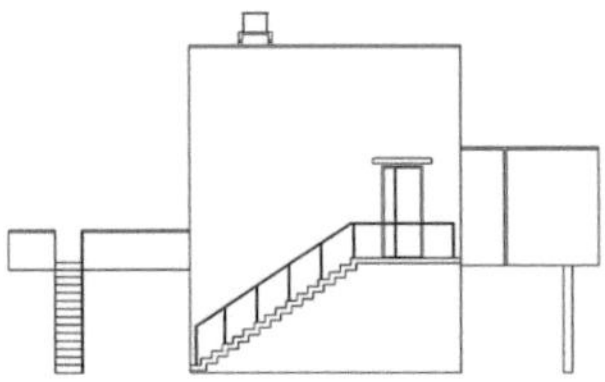	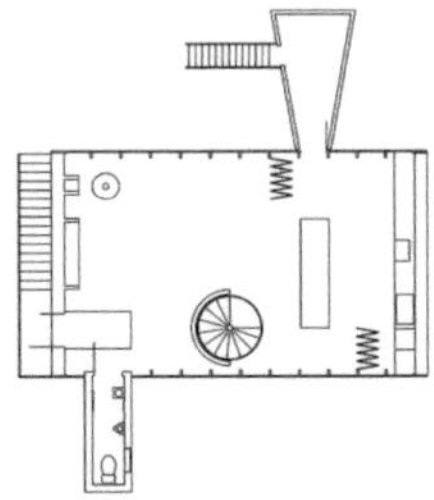
S House	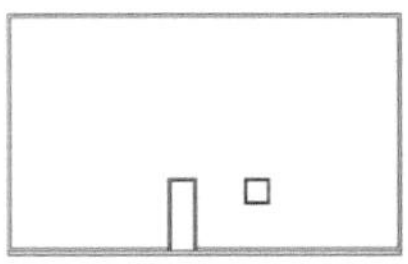	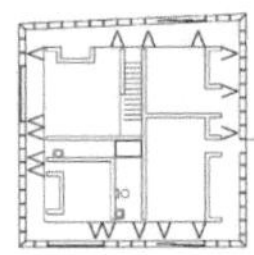
M House	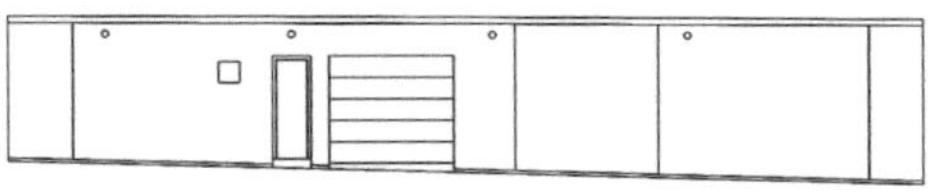	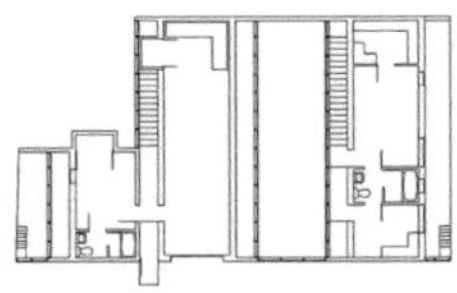
Small House	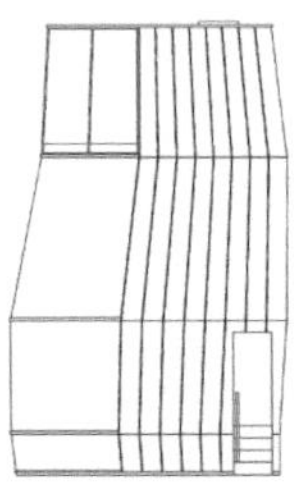	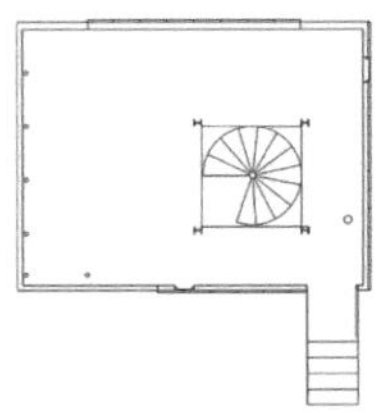
House in a Plum Grove	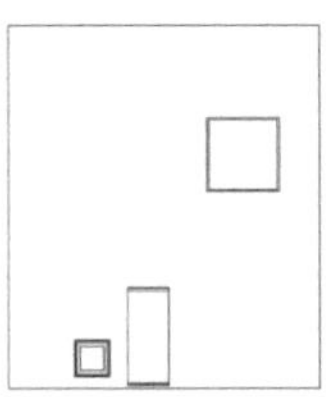	

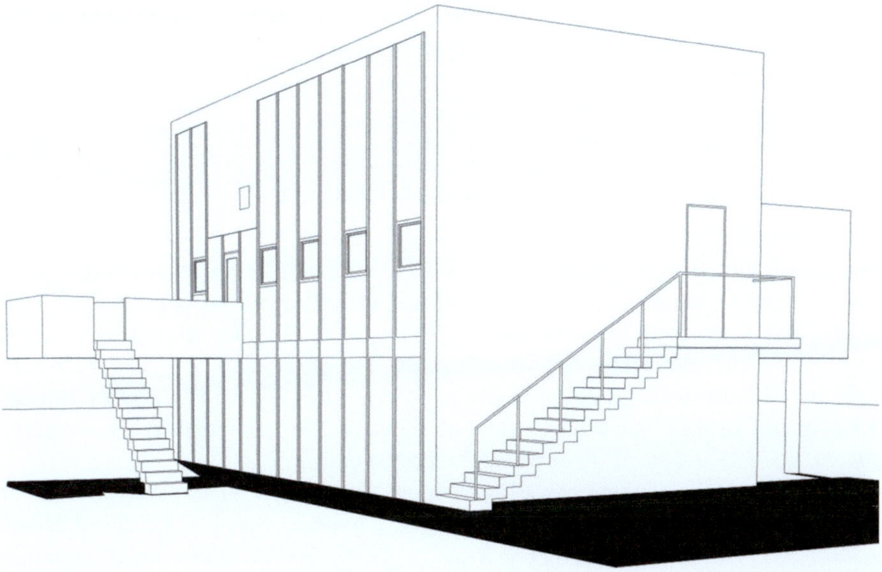

Fig. 11.1 *Y House* perspective view

appearance and materiality similar to the *S House*, featuring uninterrupted corrugated surfaces and recessed doorways framed with a minimum of expression, the *M House* appears to have no windows to the outside at all.

Set on a tiny 60 m^2 site, the *Small House* is located in leafy Aoyama, a place Sejima describes as 'one of the most attractive areas of Tokyo' (2004b, p. 242). With a compact total floor area of 77 m^2, this house rises from its 34 m^2 footprint in an undulating four-storey volume of steel and glass. The design is for a family who had a clear idea of the program they required for their home, the resulting form assuming the shape of these functional areas. According to Ito, Sejima's architecture 'is ultimately the equivalent of the diagram of the space used to abstractly describe the mundane activities presupposed by the structure' (1996, p. 18). The *Small House* is an ideal embodiment of this concept, where each of its concrete floors serves a different purpose. Sloping lightweight walls then connect the shifted slabs of each level, forming the gentle, swaying façade of the building.

Built in Tokyo on a small site bordered with established trees, Sejima's *House in a Plum Grove* meets the requirements of an extended family who wanted a home which 'felt like a connected space' and which would 'save the landscape' so that they would eventually be able to 'enjoy its characteristic plum trees' (Chermayeff et al. 2007, p. 278). This three-storey house has an external skin of steel panels with insulation and gypsum board over structural steel walls. With planning similar to that of the *S House*, the *House in a Plum Grove* differs dramatically in its appearance. With white painted steel walls, the house is covered on all elevations by thinly framed openings of seemingly random locations and sizes. These shapes puncture the interior of the building where '[e]very room has windows that can be viewed from every other space, making the inside of the house one large space' (Hasegawa 2007, p. 186).

11.3.2 Sejima Houses, Results and Analysis

The fractal dimension measurements derived from Sejima's houses indicate that the lowest average elevation result is in the *House in a Plum Grove* ($\mu_E = 1.2183$) and the highest is in the *Small House* ($\mu_E = 1.\ 4417$), leading to a range of $R_{\{E\%\}} = 36.43$. For the entire set of elevations the average is $\mu_{\{E\}} = 1.3256$, the median is 1.2850 and the standard deviation is 0.1108. There is a notable upward skew in the results, caused by just one elevation in each of the *Small House* ($D_{E4} = 1.5353$) and the *Y House* ($D_{E4} = 1.5004$).

While it had the highest elevation average, the *Small House* plans possess the lowest average ($\mu_P = 1.2699$); and the highest average result is for the *Y House* ($\mu_P = 1.3677$) with the range being $R_{\{P\%\}} = 19.97$. For the entire set of plans the mean is $\mu_{\{P\}} = 1.3221$, the median is 1.3308 and the standard deviation is 0.0596. This data is much more tightly clustered than for the set of Sejima's elevations. When all of the plans and elevation results are considered together, the aggregate is $\mu_{\{E+P\}} = 1.3227$. Overall, the visual complexity of the entire set of images is similar, as demonstrated by the composite range, $R_{\{\mu E+P\%\}} = 10.87$ (Tables 11.2 and 11.3, Fig. 11.2).

The optimal sub-set comprises the *S House*, *M House* and *Small House*. The aggregate result for the sub-set is $\mu_{[E+P]} = 1.3328$, a figure which is slightly higher than the overall value. However the composite range result is lowered by more than half, $R_{[\mu E+P\%]} = 4.07$, confirming that the images in the optimal set are very similar in their level of detail.

The results derived from Sejima's houses match the intuitive response to the elevations, confirming that the designs lack a consistent progression of visual detail and that D results would be in the low range. However, the aggregate results for Sejima were not as consistently low as might be anticipated, with the data for the *Small House* boosting the results to a higher than expected level in this context. It is also interesting to note that the D values of the plans of the *Small House* are dramatically lower than its elevations and, compared with the rest of the set, the high fractal dimensions of the elevations for the *Small House* are a clear anomaly. So, while the *Small House* does feature mathematically in the optimum set as a result of its average qualities, it actually features more diverse results than the other houses.

The results for the *House in a Plum Grove* reflect the incredible simplicity of this design, which is, in effect, a flat steel cube with holes cut in it for windows. The windows are placed irregularly upon the wall surface in a random-looking arrangement, giving the building some minor visual interest, but this type of placement does not affect its measured complexity. Without window framing or

Table 11.2 Sejima set, results

Houses		*Y House*	*S House**	*M House **	*Small House**	*House in a Plum Grove*	Opt. [*]	Set {…}
Elevations	D_{E1}	1.4830	1.2613	1.2445	1.4402	1.2021	–	–
	D_{E2}	1.1710	1.4190	1.2854	1.4245	1.1802	–	–
	D_{E3}	1.2984	1.2685	1.2680	1.3668	1.2785	–	–
	D_{E4}	1.5004	1.2845	1.3882	1.5353	1.2123	–	–
	μ_E	1.3632	1.3083	1.2965	1.4417	1.2183	–	–
	$\mu_{[E]}/\mu_{\{E\}}$	–	–	–	–	–	1.3489	1.3256
	$M_{[E]}/M_{\{E\}}$	–	–	–	–	–	1.3261	1.2850
	$std_{[E]}/std_{\{E\}}$	–	–	–	–	–	0.0931	0.1108
Plans	$D_{P\text{-}1}$	–	–	1.3505	1.3176	–	–	–
	D_{P0}	1.3308	1.3925	1.3954	1.3028	1.3013	–	–
	D_{P1}	1.3961	1.3768	–	1.2886	1.3147	–	–
	D_{P2}	1.3875	–	–	1.3455	1.3313	–	–
	D_{PR}	1.3564	1.2151	1.2881	1.1964	1.2334	–	–
	μ_P	1.3677	1.3281	1.3447	1.2902	1.2952	–	–
	$\mu_{[P]}/\mu_{\{P\}}$	–	–	–	–	–	1.3154	1.3221
	$M_{[P]}/M_{\{P\}}$	–	–	–	–	–	1.3176	1.3308
	$std_{[P]}/std_{\{P\}}$	–	–	–	–	–	0.0664	0.0596
Composite	μ_{E+P}	1.3655	1.3168	1.3172	1.3575	1.2567	–	–
Aggregate	$\mu_{[E+P]}/\mu_{\{E+P\}}$	–	–	–	–	–	1.3328	1.3227

Table 11.3 Sejima set, comparative values

Houses		*Y House*	*S House**	*M House* *	*Small House**	*House in a Plum Grove*	Opt. [*]	Set {…}
Elevations	R_{ED}	0.3294	0.1577	0.1437	0.1685	0.0983	–	–
	$R_{E\%}$	0.0032	0.0015	0.0014	0.0016	0.0009	–	–
	$R_{[ED]}/R_{\{ED\}}$	–	–	–	–	–	0.2908	0.3643
	$R_{[E\%]}/R_{\{E\%\}}$	–	–	–	–	–	29.08	36.43
Plans	R_{PD}	0.0653	0.1774	0.1073	0.1491	0.0979	–	–
	$R_{P\%}$	6.53	17.74	10.73	14.91	9.79–	–	–
	$R_{[PD]}/R_{\{PD\}}$	–	–	–	–	–	0.1990	0.1997
	$R_{[P\%]}/R_{\{P\%\}}$	–	–	–	–	–	19.90	19.97
Composite	$R_{[\mu E+PD]}/R_{\{\mu E+PD\}}$	–	–	–	–	–	0.0407	0.1087
	$R_{[\mu E+P\%]}/R_{\{\mu E+P\%\}}$	–	–	–	–	–	4.07	10.87

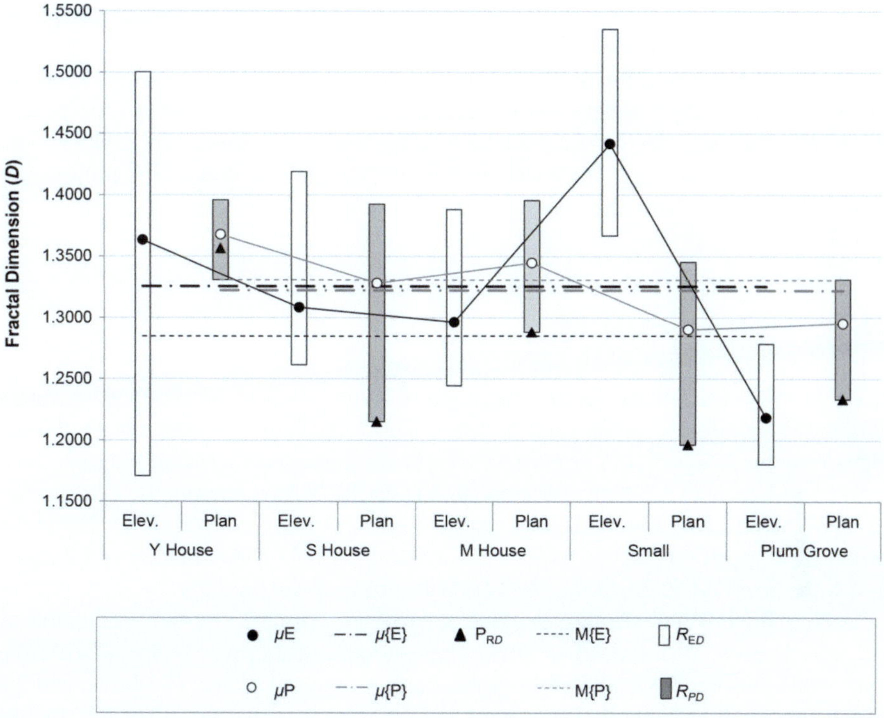

Fig. 11.2 Sejima set, graphed results

mullions, these walls are very minimal, as shown by the results for this house, $1.1802 < D_E < 1.2785$, which are amongst the lowest of any houses analysed in this book.

The design strategy for the *Y House* resulted in a structure with two solid marble bookend façades, with minimal constructed detailing, each producing low dimensions $D_{E2} = 1.170$ and $D_{E3} = 1.2984$. The other two façades feature long stretches of framed glazing with alternating window detailing. Although the visual effect is typical of Sejima's transparency, the framing detail is picked up in the analysis process, bringing these other façade dimensions up to $D_{E1} = 1.4830$ and $D_{E4} = 1.5004$. Framed glazing that makes up an entire wall also causes a rise in fractal dimension for the *M House* and the *S House*. One glazed wall of the *M House* ($D_{E4} = 1.3882$) creates a visual difference in this design which is otherwise very minimal externally, with the other elevations producing results between $D = 1.2445$ and $D = 1.2854$. Sejima described the *S House* as having an 'extremely abstract exterior' (Sejima 1999, p. 119); a description which is consistent with three of the D values for its elevations being between $D = 1.2613$ and 1.2845, while the south elevation is much higher ($D_{E2} = 1.4190$). This south wall is composed of corrugated sheeting which is only interrupted by two simple windows, a combination that would normally produce a very low result. However, as previously explained, in this case the sheeting is almost completely transparent, and the timber framing within the wall is deliberately exposed, raising the visual complexity of this view. Translucent or semi-transparent façades of this type pose a particular challenge for the fractal analysis method in this book. To exclude such readily visible details will artificially reduce key design qualities of the project, but to include them inflates the result higher than might be expected.

11.4 Atelier Bow-Wow

Born in 1969 in Tokyo, Japan, Momoyo Kaijima studied architecture at Japan Women's University, graduating in 1991. The following year she set up Atelier Bow-Wow in Tokyo with her partner, Yoshiharu Tsukamoto. Tsukamoto was born in 1965 in Kanagawa, Japan and graduated from the Tokyo Institute of Technology in 1987. Both Kaijima and Tsukamoto furthered their education at European schools—respectively at the ETH in Zurich and L'Ecole d'Architecture in Paris—and were awarded Ph.D. from the Tokyo Institute of Technology.

Atelier Bow-Wow's design practice is focussed primarily on the production of tall, thin houses on dense urban sites. These houses often possess irregular forms, with walls and windows angled in vertical and horizontal planes to take advantage of the surrounding context. The ceiling heights also vary but are often tall, giving these otherwise diminutive residences a heightened verticality. Bognár argues that these 'tiny shacks, defy existing architectural conventions, and often open up fresh approaches to spatial configuration, use and design' (2008, p. 63), In particular, Atelier Bow-Wow use open vertical internal circulation zones to provide space, light and communication within each house. This property leads Terunobu Fujimori to observe that they effectively use stair landings 'to fulfil the primary functions in their buildings' (2010, p. 128).

The decision to name their firm an *atelier*, a French name for a creative studio, is one of the first clues to the exploratory nature of their work. Kaijima and Tsukamoto not only design, they also dedicate time to research and teaching, proceses through which they 'draw upon various activities of the city's many actors as their wellspring of creativity' (Nango 2010, p. 322). Like the practices of Regionalist architects, Atelier Bow-Wow are directly influenced by the local vernacular. However, for Atelier Bow-Wow it is not the traditional Japanese vernacular with its sliding screens and tatami mats but rather a contemporary vernacular that is of interest; in the same way that Denise Scott Brown and Robert Venturi drew inspiration from their analysis of car parking and casino signage in *Learning from Las Vegas*, Kaijima and Tsukamoto analysed seventy ordinary, ad hoc and hybrid buildings—including an expressway patrol building and a distribution complex—for their book, *Made in Tokyo* (Bow-Wow 2001a). In the same year they also published *Pet Architecture* (Bow-Wow 2001b), a whimsical study of buildings that have been constructed in left-over urban spaces.

Through their published research and design, Atelier Bow-Wow have developed a rich, if constrained view of domestic architecture. For example, their publication *Graphic Anatomy* (Bow-Wow 2007) offers an extremely refined and detailed set of sectional perspectives and programmatic descriptions of twenty-four of their houses, while their follow-up work, *Echo of Space, Space of Echo* (Bow-Wow 2007) documents the relationship between human experience and the urban environment. Their next book, *Behaviorology* (Bow-Wow 2010), charted their shifting design agenda as they moved away from 'the formation of disjunctions' and the presentation of 'anti-normative values' to focus instead 'on integrating those disjunctions as organic relationships' (Nango 2010, p. 334).

Throughout this period, and in parallel with their writings, the designs of Atelier Bow-Wow continued to evolve in distinct ways. For example, for the *Ani*, *Mini* and *Gae House*s Atelier Bow-Wow used a common design approach to 'first establish a distance from the surrounding environment, then remake the relationships in every direction' (Bow-Wow 2007, p. 111). Because of this approach, the three houses all have different visual expressions, as each has been designed in response to its particular urban environment, but share similar organisational strategies. In later designs, they sought to create an urban architecture which is integrated into its context, reaching 'well beyond a building's basic enclosure' to create 'a sense of spatial expansiveness that reaches out to the city' (Bow-Wow 2010, p. 10). More recently, Atelier Bow-Wow have produced several books including *In Praise of Mud* (Bow-Wow 2012), a study of vernacular earth construction in Japan, and *Graphic Anatomy 2* (Bow-Wow 2014), an update on their latest designs. Both of these more recent works continue to strengthen Bow-Wow's growing Regionalist credentials.

11.4.1 Five Houses (1998–2004) by Atelier Bow-Wow

All five of Atelier Bow-Wow's houses studied in the present chapter—*Ani House*, *Mini House*, *Shallow House*, *Gae House* and *Juicy House*—are set on tiny blocks in Tokyo. Furthermore, Atelier Bow-Wow often describe the *Ani*, *Mini* and *Gae* houses as a distinct and consistent set of works. Level 4 representations of the set are given in Table 11.4.

The *Ani House* was built in the Kanagawa Prefecture, a waterfront urban location near Tokyo. After analysing the urban fabric of the region Atelier Bow-Wow decided that 'the residence should not adhere to the beach house type; instead, it required a more communicative approach toward the pressures of real conditions' (Kira 2001, p. 119). With a steel structure, silver-grey corrugated metal cladding and flat steel detailing to the windows and balustrades, the house has an industrial appearance. This three-level building (one level of which is a habitable, windowed basement) is predominantly clad internally in unpainted timber sheeting, which ameliorates the industrial look, and while the interior is sparse, it has a warm appearance (Fig. 11.3).

Table 11.4 Atelier Bow-Wow set, example elevations and floor plans (level 4 representation—not shown at uniform scale)

Ani House

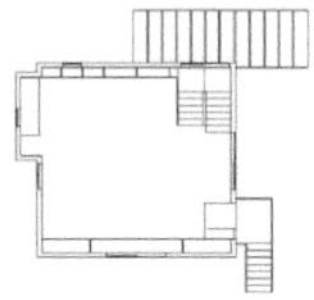

Mini House

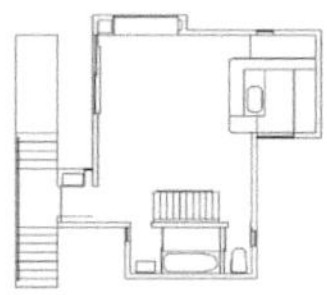

Shallow House

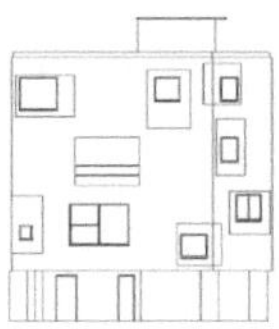

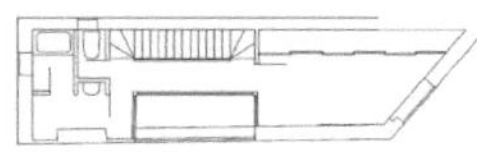

Gae House

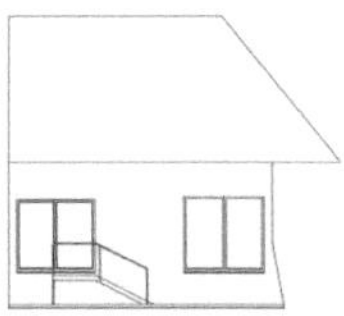

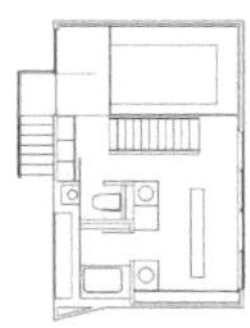

Juicy House

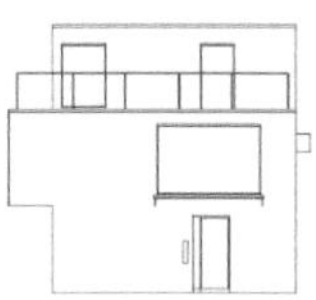

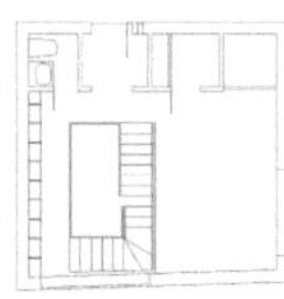

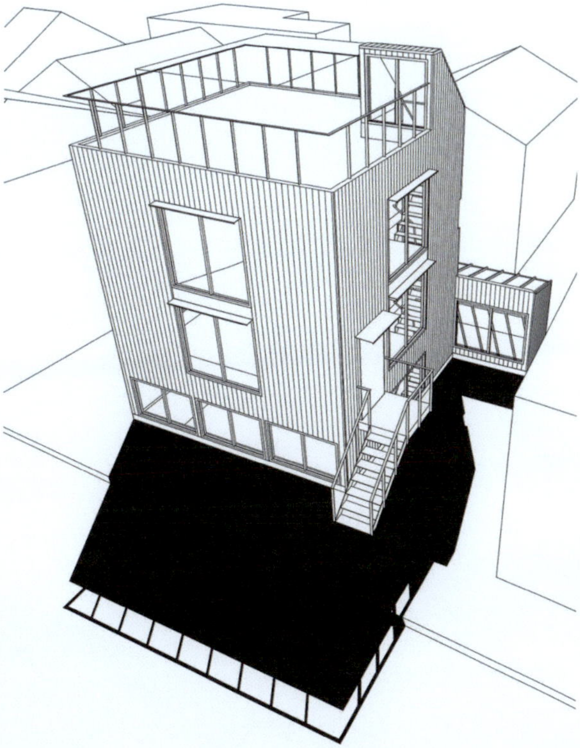

Fig. 11.3 *Ani House*, perspective view

The *Mini House*, completed the following year, has a similar program to the *Ani House*, with two floors above ground and a liveable basement below. The *Mini House* is also made of steel, but the flat profiled steel sheeting is arranged more like a planar sculpture. The street presence of the house is a square silver plane of steel with a recessed layer of red steel in the background, in an almost inverted cross. While from the outside the building looks windowless, inside it is open-planned in section as well as in plan, with white painted tubular steel balustrades and cork flooring throughout (Fig. 11.4).

With an internal width of only 2.5 m, the *Shallow House* is a narrow, tall residence in Shinjuku, a very densely populated area of Tokyo. Four storeys high with an additional rooftop terrace, this house does not have a living space in the basement like other houses in this set. Instead it appears to integrate several of Corbusier's five points of architecture. For example, the building is set on pilotis, with car parking and entry on the ground level; it also features free internal planning and a rooftop garden (Fig. 11.5). The house is constructed in steel and concrete construction and is clad with yellow concrete tiles externally, creating an experience that is 'shallow' while bringing 'a sense of openness in spite of the limited space' (Bow-Wow 2006, p. 69).

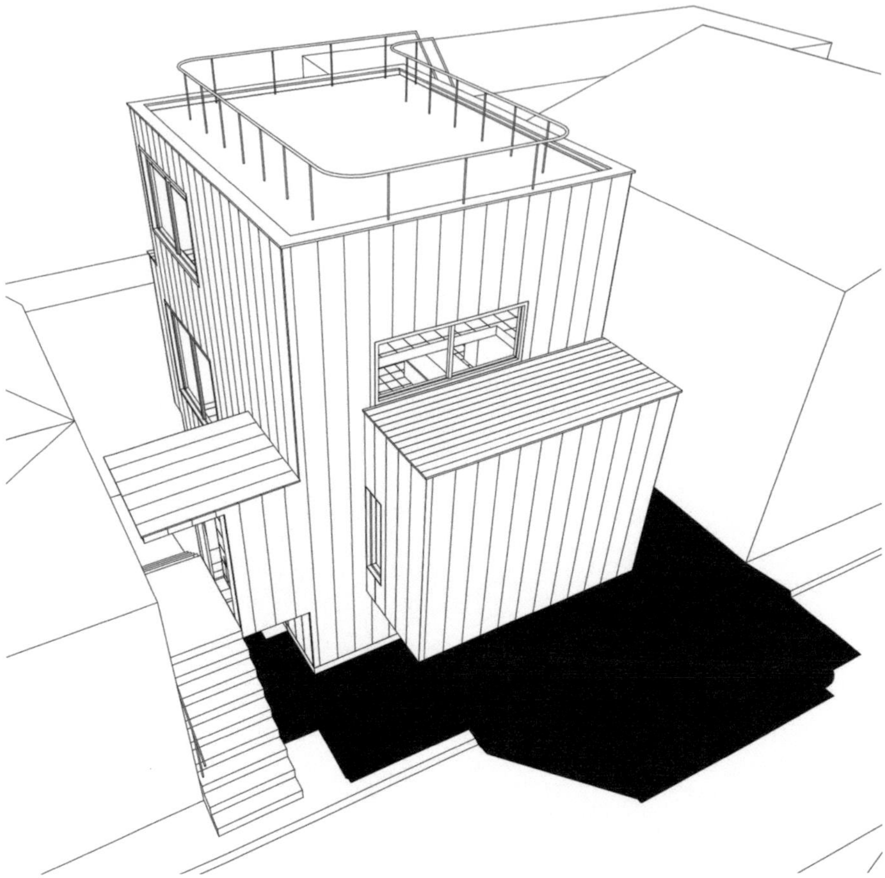

Fig. 11.4 *Mini House*, perspective view

The *Gae House*, built in Tokyo in 2003, echoes the design strategies employed in the *Ani* and *Mini* houses with a similar program: a semi-interred basement (the bedroom) and two floors above (amenities and living/dining/kitchen). The design is visually reminiscent of a child's drawing of a house: a square with a triangle roof on top. This effect is exaggerated by the use of weatherboard-like cladding to walls and roof, which is actually over-scaled, horizontally ridged steel. Developing opportunities to include windows where possible, the triangular roof space is actually part of the top floor, its wide eaves made of glass creating an unusual downward facing window ribbon, invisible on the elevation.

Completed in Tokyo in the following year, the *Juicy House* features a taller variation of the programmatic structure used for the *Mini*, *Ani* and *Gae* houses, with bedrooms in the basement area and a further three stories above. Externally, the *Juicy House* differs from the other three with a form reminiscent of extruded cubes which are steel-framed and rendered with a resin mortar. Internally, the house is

Fig. 11.5 *Shallow House*, perspective view

lined with bright, 'juicy' colours, with perforations that 'appear to be carved out of the orange-tinted space, revealing fragments of the typical residential landscape', allowing for 'impressions of extreme depth (… to the kitchen) and extreme shallowness (… to the white bathroom)' (Bow-Wow 2010, p. 66).

11.4.2 Atelier Bow-Wow Houses, Results and Analysis

The results for Atelier Bow-Wow show that the *Gae House* has the lowest elevation average of the five designs (μ_E = 1.2473), as well as the least complex façade overall (μ_E = 1.1846). The highest elevation average is for the *Ani House* (μ_E = 1. 3789), leading to an overall range of $R_{\{E\%\}}$ = 24.15. For the entire set of elevations the average is $\mu_{\{E\}}$ = 1.2977, the median is 1.3033 and the standard deviation is 0.0747. The *Juicy House* plans possess the lowest average (μ_P = 1.3118) and the highest is for the *Shallow House* (μ_P = 1.3670), with the range for the plans being $R_{\{P\%\}}$ = 35.93. For the entire set of plans the mean is $\mu_{\{P\}}$ = 1.3426, the median is 1.3445 and the standard deviation is 0.0705. When all of the plans and elevation results are considered together, the aggregate is $\mu_{\{E+P\}}$ = 1.3218 (Tables 11.5 and 11.6, Fig. 11.6).

Table 11.5 Atelier Bow-Wow set, results

Houses		*Ani*	*Mini*	*Shallow**	*Gae**	*Juicy**	Opt. [*]	Set {…}
Elevations	D_{E1}	1.3322	1.2744	1.2028	1.1846	1.3275	–	–
	D_{E2}	1.4261	1.2994	1.3787	1.2111	1.2789	–	–
	D_{E3}	1.4231	1.3072	1.1875	1.3089	1.2663	–	–
	D_{E4}	1.3341	1.3700	1.3621	1.2847	1.1936	–	–
	μ_E	1.3789	1.3128	1.2828	1.2473	1.2666	–	–
	$\mu_{[E]}/\mu_{\{E\}}$	–	–	–	–	–	1.2656	1.2977
	$M_{[E]}/M_{\{E\}}$	–	–	–	–	–	1.2726	1.3033
	$std_{[E]}/std_{\{E\}}$	–	–	–	–	–	0.0695	0.0747
Plans	$D_{P\text{-}1}$	1.3425	1.3733		1.3485	1.2907	–	–
	D_{P0}	1.3525	1.3507	1.3457	1.4311	1.3413	–	–
	D_{P1}	1.3433	1.3349	1.3381	1.4459	1.2997	–	–
	D_{P2}	–	–	1.3912	–	1.3349	–	–
	D_{P3}	–	–	1.3270	–	–	–	–
	D_{PR}	1.3595	1.3757	1.4328	1.0866	1.2923	–	–
	μ_P	1.3495	1.3587	1.3670	1.3280	1.3118	–	–
	$\mu_{[P]}/\mu_{\{P\}}$	–	–	–	–	–	1.3361	1.3426
	$M_{[P]}/M_{\{P\}}$	–	–	–	–	–	1.3397	1.3445
	$std_{[P]}/std_{\{P\}}$	–	–	–	–	–	0.0883	0.0705
Composite	μ_{E+P}	1.3642	1.3357	1.3295	1.2877	1.2917	–	–
Aggregate	$\mu_{[E+P]}/\mu_{\{E+P\}}$	–	–	–	–	–	1.3036	1.3218

Table 11.6 Atelier Bow-Wow set, comparative values

Houses		*Ani*	*Mini*	*Shallow**	*Gae**	*Juicy**	Opt. [*]	Set {…}
Elevations	R_{ED}	0.0939	0.0956	0.1912	0.1243	0.1339	–	–
	$R_{E\%}$	9.39	9.56	19.12	12.43	13.39	–	–
	$R_{[ED]}/R_{\{ED\}}$	–	–	–	–	–	0.1941	0.2415
	$R_{[E\%]}/R_{\{E\%\}}$	–	–	–	–	–	19.41	24.15
Plans	R_{PD}	0.0170	0.0408	0.1058	0.3593	0.0506	–	–
	$R_{P\%}$	1.70	4.08	10.58	35.93	5.06	–	–
	$R_{[PD]}/R_{\{PD\}}$	–	–	–	–	–	0.3593	0.3593
	$R_{[P\%]}/R_{\{P\%\}}$	–	–	–	–	–	35.93	35.93
Composite	$R_{[\mu E+PD]}/R_{\{\mu E+PD\}}$	–	–	–	–	–	0.0419	0.0765
	$R_{[\mu E+P\%]}/R_{\{\mu E+P\%\}}$	–	–	–	–	–	4.19	7.65

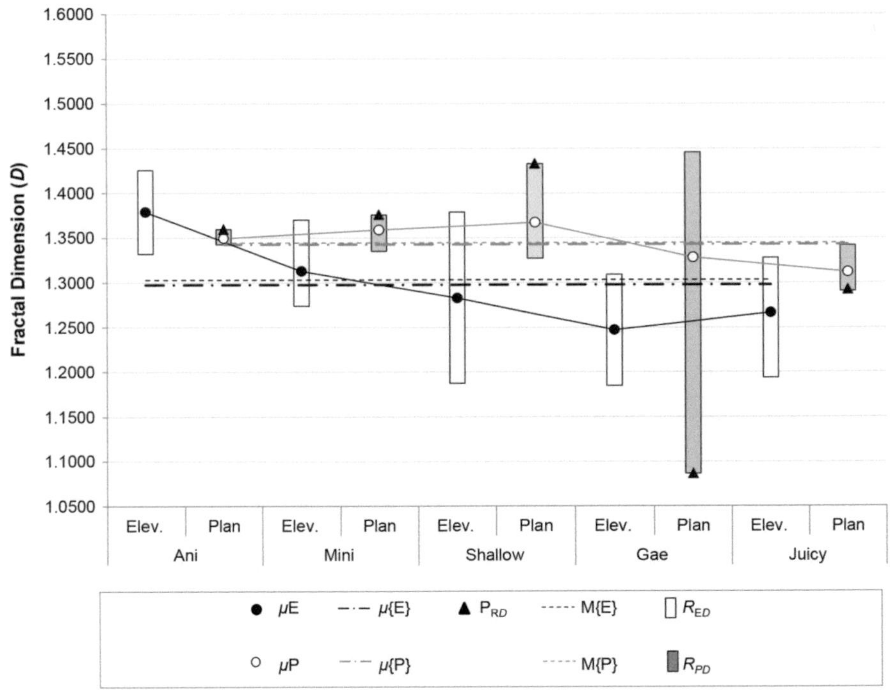

Fig. 11.6 Atelier Bow-Wow set, graphed results

When consolidated, the composite range ($R_{\{\mu E+P\%\}}$ = 7.65) is notably tighter than that of the elevations or plans separately. This is because, unlike the majority of designs considered in the present book, Bow-Wow's architecture is both vertically stacked and horizontally layered, with a consistent overall compositional strategy at work, but where individual plans or elevations in each design can be quite different because of the way they respond to their immediate contexts. Thus, elevations which have no direct view do not need modulation or windows, leading to some very low results (D_{E1} = 1.1846 for the *Gae House*), while some others (D_{E2} = 1.4261 for the *Ani House*) have more potential to respond to contextual challenges and opportunities, leading to higher results.

The optimal sub-set comprises the *Shallow House, Gae House* and *Juicy House.* The aggregate result for the optimal sub-set is $\mu_{[E+P]}$ = 1.3036, a figure which is slightly lower than the overall value. The composite range result is also lowered to $R_{[\mu E+P\%]}$ = 4.19, meaning that, in terms of complexity, this optimal set is very similar in their appearance.

The design similarities between the *Ani* and *Mini* houses are reflected in the elevation results, with the range for the elevations being almost the same for both houses ($R_{E\%}$ = 9.39, $R_{E\%}$ = 9.56, respectively). These two houses also have a similar pattern when the results are graphed, with the tight cluster of visual complexity of the elevations covering the even tighter cluster of plan dimensions. This means that the floor plans and the elevations of these two houses all have a similar level of visual complexity. This trend is partially repeated for the other three houses, with plan and elevation results often producing similar figures. The *Juicy House* has the least visual complexity overall of the set, its highest dimension, for its ground floor plan (D_{PO} = 1.3413), is well below most of the other plans of the Atelier Bow-Wow set, and its elevations fall generally below the average.

The houses of Atelier Bow-Wow are often designed to have more than just four elevations. With such limited space, the roof can be as important as the walls, often providing openings for the house, or acting as a sloping wall or an outdoor space. In this analysis of Atelier Bow-Wow, the roof plan is a fundamental part of the data and, unlike most other houses analysed in this book, its visual complexity is of a level similar to the elevations. In particular, for the *Ani*, *Mini* and *Shallow* houses, the roof has the highest fractal dimension of the set of plans. In this regard, the *Gae House* is the only anomaly, because while the roof of the *Gae House* is clearly an important part of the design, containing the entire living space within it as well as the downward facing windows, these are not revealed in the roof plan, which comprises only two simple planes (D_{PR} = 1.0866).

11.5 Regionalism

Histories of the rise of the Modern Movement in architecture typically focus on examples drawn from Europe and North America in the first half of the twentieth-century. In such works the European cases are often used to emphasise the role of functionalism in design along with architectural propositions for the creation of healthy and egalitarian cities. However, in North America, Henry-Russell Hitchcock and Philip Johnson celebrated the formalist and tectonic tendencies of European and American Modernism, emphasising the movement's attempt to articulate a universal language of aesthetic expression. This focus on one particular facet of the larger project of Modernity allowed Hitchcock and Johnson to selectively rename it the 'International Style' (Hitchcock and Johnson 1995).

It would be an over-simplification of the International Style to say that it consistently venerated the placeless qualities of architecture over the local and the regional, but this tendency is notable in many works. Even the title, International Style, seemed to affirm its applicability to any geographic location and to suggest that its omniscience was somehow inescapable. Modern architects did indeed employ 'a similar palette of industrial forms and materials for buildings in widely diverse geographical areas, be-cause they wanted to express a universal view of modern life and industrialization that transcended individual places' (Gelernter and Dubrucq 2004, p. 1090). With Le Corbusier's 1922 proposal for a *Ville Contemporaine* as a precedent, Modern architects advocated an approach to urban projects that commenced with the demolition of all previous buildings and infrastructure along with the social and cultural systems supported by this urban fabric. Thereafter, on this *tabula rasa*, they proposed the construction of a highly repetitive urban infrastructure. In the case of the *Ville Contemporaine*, this consisted of a series of sixty-storey, cruciform-plan skyscrapers, deployed in a rigid grid across the landscape. The apparent ubiquity and inevitability of such urban projects effectively promulgated a version of Modernism that decried the value of tradition and rejected regional differences. However, Modernism was neither so universal in its expression nor so wilfully ignorant of tradition as the early histories and manifestos suggest.

Across Eastern Europe, South America and the Asia-Pacific region, local issues —including political imperatives, climatic conditions and cultural values—were often the catalyst for regional variations of Modernism. For example, in the 1930s several influential European architects moved to Australia, bringing with them first-hand experience of Modernist concepts, themes and works. Amongst these émigrés were Hugh and Eva Burich, Harry Seidler and Frederick Romberg, each of whom continued to foster an interest in Modernism through their designs, publications and exhibitions (Goad 2012). However, the Modernist inclination of these émigrés was tempered by a combination of the Australian climate, the availability of materials and local construction techniques. Some of these influences were resisted, in the desire to achieve a pure language of functional expression, while others were embraced, for often pragmatic reasons, to create a regional interpretation of Modernism. By the 1950s and 1960s, against a backdrop of growing criticisms of Modernism in Europe and America, such regional variations began to appear as ingenious attempts to reconcile functionalism and tradition. These designs combined the open-planned, clean-lined, design aesthetic of Modernism with an appreciation for vernacular expressions and environmental conditions. Over the years that followed, this particular approach to design began to be called Regionalism.

Regionalist buildings are usually constructed from local materials and using methods that are familiar to local tradespeople. They are also designed to be responsive to local climatic and environmental conditions. Unlike pop-art inspired, Post-Modern interpretations of vernacular architecture, Regionalist buildings celebrate the physical experience of living in particular spaces (architecture) in particular locations (a specific region and landscape). According to Mallgrave and Goodman, Regionalism is 'at heart humanistic and is opposed [to] the trendy acceptance of historical forms' (2011, p. 99).

In Australia, as elsewhere in the world, the spread of the Regionalist design ethos was supported by the existence of the poetic, yet still emphatically Modern, designs of the American Frank Lloyd Wright and the Finnish Alvar Aalto. The works of these architects, both of whom demonstrated a profound respect for site, became well known at a time when the Australian landscape was being appreciated for its raw beauty and celebrated in the drawings of many great artists, including Wright's protégée, Marion Mahoney Griffin.

Some of the earliest Regionalist designs in Australia were completed by the so-called 'Sydney School'. These works were designed to take advantage of a local climate that suited outdoor living for much of the year. As such, they were focused on both internal and external space and sought to accommodate the landscape up to and even within their interiors. Early designs of the Sydney School from the 1950s had the clean lines of European Modernism, with predominantly orthogonal planes and flat roofs. However, they were constructed of local, unrendered 'clinker' bricks, or of locally quarried sandstone, and their horizontal lines and structures were often emphasised in bold, dark timberwork.

Respecting the Modernist tradition, Sydney School buildings often had large expanses of glazing. However, these were now carefully shaded against the harsh Australian sun and placed to capture views of the surrounding landscape. Furthermore, where modernist houses of the Northern hemisphere were often flat in section, architects of the Sydney School were forced to accommodate steep blocks of waterfront land in densely treed enclaves. For this reason, these architects commenced their design process with a carefully considered response to the site, its vegetation, watercourses and views. Some architects reacted to these sites with dramatically projecting homes—such as Bill and Ruth Lucas's *Lucas House* and Neville Gruzman's *Holland House*—while others, like Ken Woolley, reciprocated in the design of his *Woolley House* in 1962, which followed the fall of the land, stepping down towards the water.

Achieving these carefully composed relationships between the building and the land, while still maintaining Modernist open-planning principles, meant that 'the houses were spatially complex with interpenetrating volumes, often with split levels, raked ceilings and exposed structure' (Urford 2012, p. 675). Julie Willis and Philip Goad argue that the central achievement of the Sydney School was to create 'locally developed versions of contemporary architecture [that] drew on international architectural currents' to create works that 'were distinctly Australian by their engagement with the immediate place … for which they had been designed' (2012, p. 56).

By the 1970s, the rustic, masonry expression of the Sydney School had given way to more 'lightweight structures of timber and steel' (Taylor 1990, p. 164). It was at this time that Glenn Murcutt began to produce buildings that used vernacular materials, like corrugated steel, to create simple, modern interpretations of local farmhouse structures. According to Jennifer Taylor, it was Murcutt's *Marie Short House* 'that clearly set forth the differences between the new works and that of the previous decades' (1990, p. 168). Murcutt's early architecture, which was strongly reminiscent of Mies van der Rohe's pavilion forms, used local materials and construction techniques to create flexible, light-filled spaces with only a minimal presence in their rural landscapes.

In the two decades that followed, a growing number of Australian architects began to adopt similar, regionally-informed approaches to contemporary design. In the early part of this period Morrice Shaw, Richard Leplastrier and John Andrews joined Murcutt in producing Regionalist Australian designs. More recently, Alex Popov, Peter Stutchbury, Paul Pholeros and Neil Durbach have continued to develop and expand the Regionalist tradition, producing contemporary works that engage with material expression and the poetry of their sites (Jahn and Bingham-Hall 1997; MacMahon 2001; Urford 2012; McGillick 2005).

11.6 Peter Stutchbury

Rick Joy describes the character of Peter Stutchbury's architecture as arising from 'a rigorous investigation' of the properties of a site, to express 'nature's potential mysteries and atmospheres' (2010, p. 5). Joy is not alone in tracing the essence of Stutchbury's architecture to the poetry of its landscape setting, with most descriptions referring to its relationship with both a specific site and a broader regional context. Stutchbury explains the origins of this approach to design as arising from his childhood visits to his mother's rural property in Western New South Wales. It was his time at this farm, coupled with his regular forays into Lane Cove National Park, which fostered his appreciation of the Australian landscape. Julie Oliver notes that it was from his mother that Stutchbury learnt a 'fundamental philosoph[y] … that there was education to be found in the landscape' (Oliver 2001, p. 95). From the teachings of his father, an engineer, Stutchbury developed an enthusiasm for assembling and constructing projects.

Stutchbury was born in Sydney in 1954 and in 1972 began studying architecture at the University of Newcastle. After graduating in 1978, Stutchbury spent the next three years working for Quay Partnership in Sydney. In 1982 he went to Papua New Guinea, where his uncle was a missionary. It was in Papua that Stutchbury completed his first building, a church in Port Moresby with a simple but powerful roofline whose pitch varied across its length. On his return to Australia, Stutchbury established his own practice in Sydney and during the 1980s he designed several houses which were raised above their bushland sites and had their highest levels amongst the tree canopies. Importantly, Stutchbury not only designed these houses, but in many cases he also assisted in the construction process. These early works often feature a series of experimental roof forms that rely on clerestory sections to accommodate sunlight or views. Key designs from this period include the 1983 *Primmer Residence*, 1984 *Mary Gilbert House* and *The Castle* from 1987.

In 1990 Stutchbury travelled extensively around the world; after returning to Sydney he was joined by landscape architect Phoebe Pape in a new partnership, Stutchbury and Pape, a collaboration that further emphasized the importance of landscape in Stutchbury's architecture (Drew 2000). Working in this partnership, Stutchbury's designs began to question the lifestyle fostered and accommodated within a typical Australian home. Anna Johnson argues that this led Stutchbury to propose that 'the notion of dwelling manifests itself as a "glorified" and more permanent form of camping' (2008, p. 155). This concept of the house as a type of temporary shelter was 'achieved through flexibility of planning, and—crucially—by the establishment of a reciprocal relationship whereby the landscape and the house are equals, and in a state of continual exchanges' (Johnson 2008, p. 155). Some of the first designs to achieve this quality include the *Israel House* in 1992, a tall structure with a tiny footprint and a curved roof. The slender, Japanese-influenced aesthetic developed for this design is repeated and refined in several other houses of this decade, such as the *Kangaroo Valley Pavilion* in 1998 and the 1999 *Reeves House*. These designs minimize and thereby challenge the conventional distinction between 'indoor' and 'outdoor', by using large sliding sections of walls and the continuation of internal materials into external living spaces and into the landscape. The manipulation of the roofline of these houses is also a 'persistent theme in Stutchbury's work' whereby 'the roof literally becomes an umbrella, a technical bird's wing to form a warped or folding surface, which both shades and admits light' (Goad 2012, p. 662).

Throughout the 1990s Stutchbury's practice took on a growing number of non-domestic works, producing five award-winning designs for the University of Newcastle, including *Birabahn* (with Richard Leplastrier and Sue Harper in 2002 and the *Design Faculty* (with EJE architects) in 1994. The latter was described by MacMahon as 'confident in its simple, passive approach to modifying the micro-climate' (2001, p. 117). Other notable non-domestic works from this era include the *Archery Pavilion* for the Sydney Olympic Games in 1998 and the *Deepwater Woolshed* in 2003.

In 2007 Pape left the partnership and the firm was renamed Peter Stutchbury Architecture. By this time, due to growing international recognition of his work, Stutchbury's houses began to enjoy more generous budgets, which allowed for a more extensive exploration of his ideas about the nature of living in Australia. The work produced in this most recent period typically features a highly refined palette of materials and tectonic expressions, often employing unpainted timber alongside steel, concrete and glass. However, these potentially industrial-looking materials are applied with a delicate sophistication so that they have a warm presence. This apparent simplicity of the resultant architecture is achieved through the use of bespoke detailing and assembly techniques that have been intricately realised with the assistance of both the builder and manufacturer. The end result of this process is the production of a series of buildings which are 'clear, direct and readable at first glance' (Joy 2010, p. 4), and which also 'celebrate their own engineering' (McGillick 2005, p. 138). This melding of the poetic and the natural, with the manufactured and artificial, is what Stutchbury is best known for.

11.6.1 Five Houses (2004–2011) by Stutchbury

Peter Stutchbury's domestic works generally conform to one of two types: tall, thin 'tree houses' in forests or long, narrow projections in the landscape. The five houses selected for analysis in the present chapter are all of the second type. Four of these are on flat sites in New South Wales, while the fifth is on a mountainside in the same state (Table 11.7). Table 11.7 shows level 4 representation of this set.

The *Verandah House* (2004) is on a flat site in Bayview, a leafy, harbour-front suburb to the north of Sydney. A 'verandah' is a semi-enclosed but still outdoor room, that is an integral part of the house as a whole, providing 'a more appropriate dwelling space for most Australian buildings than the traditional enclosed and mechanically cooled "room" spaces that now predominate' in society (McEoin 2011, p. 151). The house has a long, narrow, horizontal form constructed from unpainted

Table 11.7 Stutchbury set, example elevations and floor plans (level 4 representation— not shown at uniform scale)

Verandah House

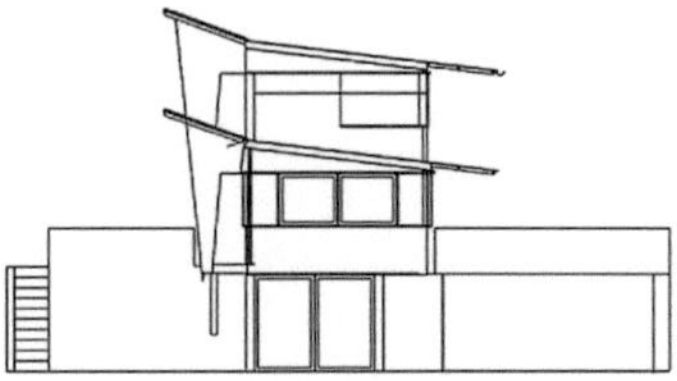

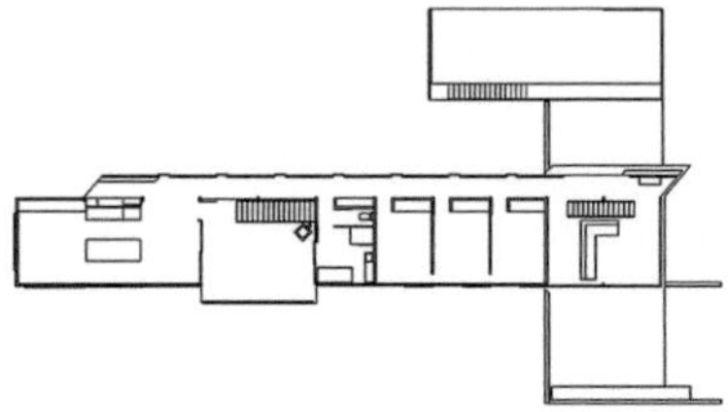

Beach House

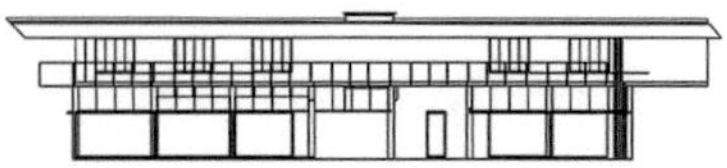

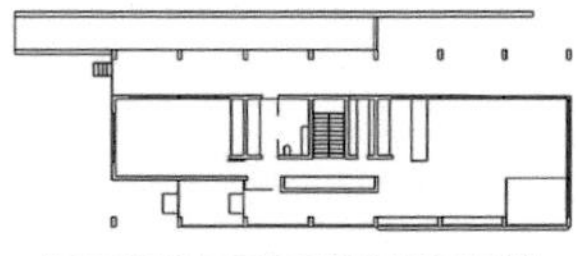

Paddock House

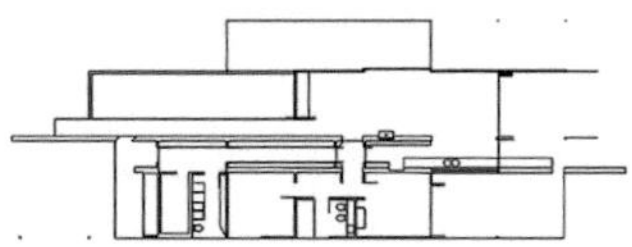

Invisible House

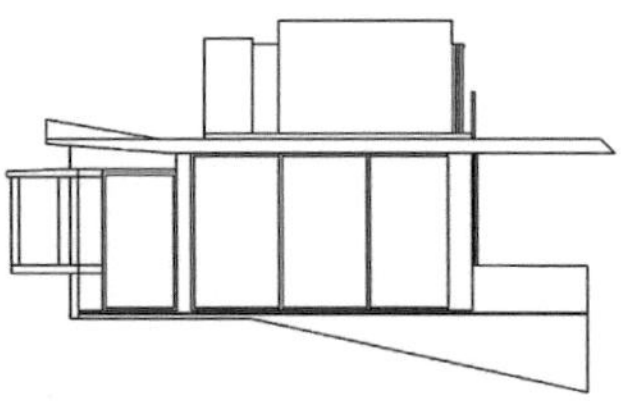

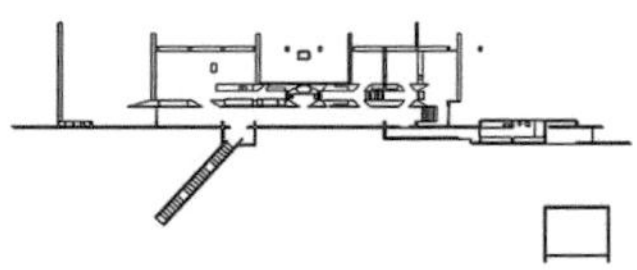

Billabong House

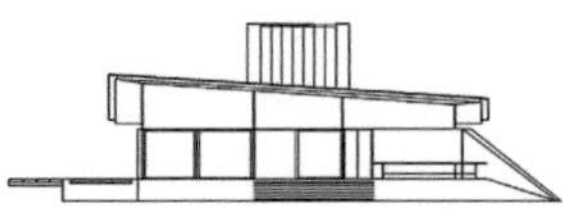

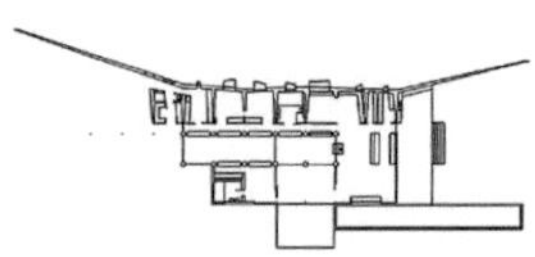

structural plywood, metal sheeting and timber. The main living area is elevated above the ground, creating a usable outdoor space underneath. The name of the house is deceptive, as it does not feature a traditional Australian verandah or projecting balcony, rather the house functions in its entirety as a new type of verandah.

The *Beach House* (2006) is built on a flat, oceanfront site in Newport, a suburb to the north of Sydney. This long, two-storey structure has an open-plan living area on the ground level with walls that peel back to allow access to the site and the beach. The structure of the lower level comprises a series of orthogonal, concrete portal frames that support a metal-clad storey above. The second level contains bedrooms lined with wide, louvre panels.

The *Paddock House* (2007) is a modern reinvention of the traditional farmhouse. The design is a long, rectangular, single-storey home set on a flat rural site in Tarago, a regional farming community in the Southern Tablelands of New South Wales. The design is notable for its large flat roof that seems to hover above the house, a feeling which is accentuated by a line of glazing that visually separates the house and the roof. According to Stutchbury, the *Paddock House* 'sits like an insect in the landscape, it does not demand high energy to maintain and includes the landscape as an active aesthetic partner' (2011a, p. 70).

The *Invisible House* (2010) is sited in the Blue Mountains west of Sydney. Unlike the more temperate climates associated with the other houses in this set, the Blue Mountains region is cold in winter and can be extremely hot in summer. To accommodate these challenges, the *Invisible House* uses stone walls and a concrete structure and slab to provide thermal mass, while creating a flat roof that has a permanent layer of water on it, like a rooftop pond, for summer cooling. The house dips down behind its mountain ridge, with only north facing sun-scoop skylights visible from the street. For Stutchbury, the *Invisible House* questions the presence of architecture in the landscape, it 'can be there or cannot .... If the roof, with water, reflects the sky this building will never be found' (2010, p. 134).

The *Billabong House* (2011) is set in a large, flat paddock in Holbrook, a rural area in southern New South Wales. The form of Stutchbury's design responds to both the extreme horizontality of the landscape as well as the thin, vertical geometry of the sparse vegetation. The resulting house, made from concrete, glass, stone, steel and hardwood, is a single-storey structure that wraps around a central courtyard, with large concrete culverts reflecting the massive trees in the surrounding landscape (Stutchbury 2011b).

11.6.2 Stutchbury Houses, Results and Analysis

The results for Peter Stutchbury's houses (Tables 11.8 and 11.9, Fig. 11.7) show that the least complex elevation average is in the *Invisible House* ($\mu_E = 1.3981$) and the highest is in the *Beach House* ($\mu_E = 1.\ 5038$). The range for the entire set of the elevation results is $R_{\{E\%\}} = 21.05$, the average is $\mu_{\{E\}} = 1.4422$, the median is 1.4527 and the standard deviation is 0.0552. While it had the lowest elevation

Table 11.8 Stutchbury set, results

Houses		*Verandah**	*Billabong*	*Beach*	*Invisible**	*Paddock**	Opt. [*]	Set {…}
Elevations	D_{E1}	1.3408	0.4627	1.5356	1.3251	1.4454	–	–
	D_{E2}	1.3786	1.4249	1.4945	1.3690	1.4677	–	–
	D_{E3}	1.4921	1.4600	1.4644	1.4761	1.4600	–	–
	D_{E4}	1.4404	1.4411	1.5208	1.4222	1.4227	–	–
	μ_E	1.4130	1.4472	1.5038	1.3981	1.4490	–	–
	$\mu_{[E]}/\mu_{\{E\}}$	–	–	–	–	–	1.4200	1.4422
	$M_{[E]}/M_{\{E\}}$	–	–	–	–	–	1.4316	1.4527
	$std_{[E]}/std_{\{E\}}$	–	–	–	–	–	0.0547	0.0552
Plans	D_{P0}	1.3206	1.3734	1.3630	1.4026	1.3183	–	–
	D_{P1}	1.4096	–	1.3700	1.3907	–	–	–
	D_{P2}	1.3434	–	–	–	–	–	–
	D_{PR}	1.2929	1.2554	1.2877	1.3383	1.1006	–	–
	μ_P	1.3416	1.3144	1.3402	1.3772	1.2095	–	–
	$\mu_{[P]}/\mu_{\{P\}}$	–	–	–	–	–	1.3241	1.3262
	$M_{[P]}/M_{\{P\}}$	–	–	–	–	–	1.3383	1.3409
	$std_{[P]}/std_{\{P\}}$	–	–	–	–	–	0.0932	0.0792
Composite	μ_{E+P}	1.3773	1.4029	1.4337	1.3891	1.3691	–	–
Aggregate	$\mu_{[E+P]}/\mu_{\{E+P\}}$	–	–	–		–	1.3789	1.3944

Table 11.9 Stutchbury set, comparative values

Houses		*Verandah**	*Billabong*	*Beach*	*Invisible**	*Paddock**	Opt. [*]	Set {...}
Elevations	R_{ED}	0.1513	0.0378	0.0712	0.1510	0.0450	–	–
	$R_{E\%}$	15.13	3.7800	7.12	15.10	4.50	–	–
	$R_{[ED]}/R_{\{ED\}}$	–	–	–	–	–	0.1670	0.2105
	$R_{[E\%]}/R_{\{E\%\}}$	–	–	–	–	–	16.70	21.05
Plans	R_{PD}	0.1167	0.1180	0.0823	0.0643	0.2177	–	–
	$R_{P\%}$	11.67	11.80	8.23	6.43	21.77	–	–
	$R_{[PD]}/R_{\{PD\}}$	–	–	–	–	–	0.3090	0.3090
	$R_{[P\%]}/R_{\{P\%\}}$	–	–	–	–	–	30.90	30.90
Composite	$R_{[\mu E+PD]}/R_{\{\mu E+PD\}}$	–	–	–	–	–	0.0200	0.0646
	$R_{[\mu E+P\%]}/R_{\{\mu E+P\%\}}$	–	–	–	–	–	2.00	6.46

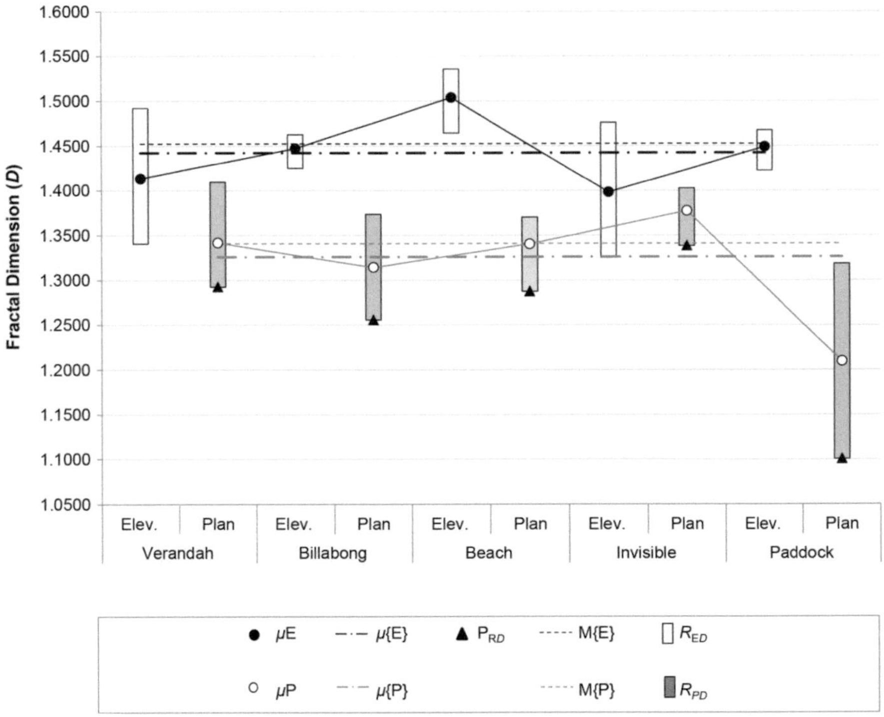

Fig. 11.7 Stutchbury set, results graphed

average, the *Invisible House* plans feature the highest average (μ_P = 1.3772) while the lowest average plan result is for *Paddock House* (μ_P = 1.2095). For the entire set of plans, the range is $R_{\{P\%\}}$ = 30.90, the mean is $\mu_{\{P\}}$ = 1.3262, the median is 1.3409 and the standard deviation is 0.0792. When all of the plans and elevation results are considered together, the aggregate is $\mu_{\{E+P\}}$ = 1.3944. Overall, the visual complexity of the entire set of images is similar, as demonstrated by the tight composite range; $R_{\{\mu E+P\%\}}$ = 6.46.

The optimal sub-set comprises the *Verandah House, Invisible House* and *Paddock House.* The aggregate result for the optimal sub-set is $\mu_{[E+P]}$ = 1.3789, a figure which is slightly lower than the overall value. However the composite range result is significantly lowered to $R_{[\mu E+P\%]}$ = 2.00, confirming that the images in the optimal set are very similar in their level of detail.

Overall the results for Stutchbury show a fairly consistent trend. The individual elevations typically have a higher level of detail ($1.3251 < D_E < 1.5356$) and the plan dimensions tend toward the lower spectrum of the range of elevational complexity or below, while also remaining in a fairly consistent cluster. The floor plan of the *Paddock House* fits well inside the typical result pattern; however, the roof plan for this house is essentially made of two flat planes, constructed without ridge capping or flashing and with only a small chimney projecting out, leading to the very low result of D_{PR} = 1.1006.

11.7 Glenn Murcutt

Glenn Murcutt's architecture is often positioned in architectural histories as an exemplar of Regionalist design, his works being both emphatically Modern and yet site specific. He rejects superficial historicist tendencies and kitsch cultural representation in architecture in favour of simple forms and materials that evoke a poetic or timeless connection to nature and a practical response to the environment.

Born in 1936, Murcutt's early childhood was spent with his Australian parents in Papua New Guinea. In 1941, the family returned to Sydney, where Murcutt's father worked in the building industry and shared his construction skills and interest in architecture with his son (Fromonot 1995). This led Murcutt to enrol in the architecture program at the University of New South Wales in 1956. He entered university at a time when the Sydney School was becoming active and his formal architectural education included an extensive introduction to the Modern movement in Europe and America. However, according to Graham Jahn and Patrick Bingham-Hall, the academics of this era 'trained a new generation to throw out conservative values and think again about the image of architecture as understood through space, technology and engineering imperatives' (1997, p. 146). Murcutt continued his association with the Sydney School after completing his degree in 1961, first working for Allen and Jack, and later Ancher, Mortlock, Murray and Woolley, both prominent firms in this movement (Taylor 1990). Indeed, later in his career, when he was called upon to identify architects who had inspired his own style, Murcutt listed the partner of one of these firms, Sydney Ancher, alongside Mies van der Rohe and Pierre Chareau as his major influences. Murcutt remained with the last of these Sydney School firms until 1969 when he created his own architectural practice.

Philip Drew describes the first few years of Murcutt's solo work as an 'initial exploratory phase … during which he worked through a Miesian derived idiom in a series of accomplished glass pavilions' (Drew 1994, p. 685). At this time Murcutt 'moved away from the somewhat rigid Modernism of his earlier houses and renewed contact with the tradition of Australian architecture' (Fromonot 1995, p. 21). His newly-evolved style was first expressed in 1975 in the *Marie Short House* and it continued to be repeated in several prominent projects until the completion of the *Ball-Eastaway House* in 1983. The works of this period are typically long, thin, pavilion-style houses with permeable façades, elevated above the site on delicate posts. Reminiscent of rural sheep-shearing sheds, these houses were often constructed from corrugated metal cladding.

After the *Ball-Eastaway House,* Murcutt's architecture underwent a second stage of evolution, during which he developed a series of asymmetrical roof forms to accommodate different environmental conditions and also placed his designs firmly on or even into the earth. Murcutt argues that his work of this era seeks 'to convey something of the discrete character of elements in the Australian landscape [and] to offer [his] interpretations in built form' (1991, p. 7). However, while these changes occurred in his architectural expression in the 1990s, Murcutt's overall philosophy remains true to his Regionalist ideals. Murcutt was awarded the Alvar Aalto Medal in 1992, the Richard Neutra Award for Architecture and Teaching in 1998, the Thomas Jefferson Medal for Architecture in 2001, the Pritzker prize in 2002 and the American Institute of Architects Gold Medal in 2009.

Such is the consistency of Murcutt's work that, like Frank Lloyd Wright, his designs have been the subject of a number of computational studies using shape grammar (Hansen and Radford 1986a, b) and Space Syntax analysis (Ostwald 2011a, b). All ten houses included in the present chapter are characteristic of Murcutt's oeuvre, featuring long narrow forms, often separated into pavilions, and all are located in rural settings in New South Wales. However, these ten are also divided into two groups that correspond to the subtle shift in character that occurred after the completion of the *Ball-Eastaway House.*

11.7.1 Five Early Houses (1975–1982) by Glenn Murcutt

The first group of five houses comprises the *Marie Short*, *Nicholas*, *Carruthers*, *Fredericks* and *Ball-Eastaway* houses (Table 11.10). Despite completing several urban houses prior to 1975, these five are widely regarded as the first of Murcutt's characteristic works. Philip Drew describes the first four of these houses as 'really members of a series, [because …] taken together, they represent a progressive development and refinement of the longitudinal house type' (1985, p. 92). These four also directly prefigure a fifth house—an intermediate work in Murcutt's oeuvre —the *Ball-Eastaway House*. Together these five are all elevated pavilion buildings united by verandahs. The buildings 'touch the earth lightly' on posts or piers, and have enclosing, symmetrical roofs with a central ridge (Farrelly 1993; Fromonot 1995). Table 11.10 shows level 4 representation of this set.

The first of the set, the *Marie Short House*, is sited on a raised floodplain, in the bend of a river near Kempsey in New South Wales. This is the first of Murcutt's famous regional houses and it was credited as heralding both a new Australian style as well as being a key Critical Regionalist work (MacMahon 2001; Frampton 2006). The house consists of two similarly sized pavilions that are axially displaced (Fig. 11.8).

Located in the Blue Mountains, the *Nicholas House* and the *Carruthers House* were built in 1980 on adjacent sites as country retreats for two families. While the *Nicholas House*, like the *Marie Short House*, has a two-pavilion *parti*, it is the first of Murcutt's houses where the pavilions are unequally sized to accommodate living spaces in the larger one and services in the smaller. The main pavilion of the *Nicholas House* is clad in timber boards and lined with glass louvres internally and cedar blinds externally. In contrast, the south edge of the house has a distinctive solid wall clad in corrugated iron with a curved roof above. The *Carruthers House* (1980) is, at first glance, even more straightforward in its form and design. Françoise Fromonot describes it as a 'simple timber barn roofed with corrugated iron' (1995, p. 112). With the exception of the chimney, the single pavilion sits

Table 11.10 Murcutt, early set, example elevations and floor plans (level 4 representation— not at uniform scale)

Marie Short House

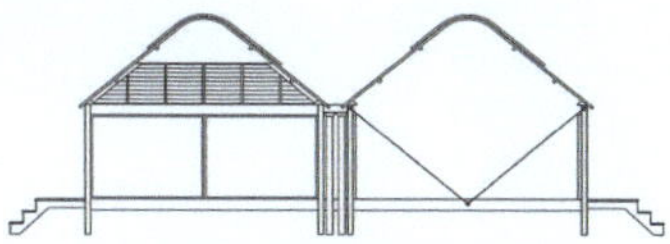
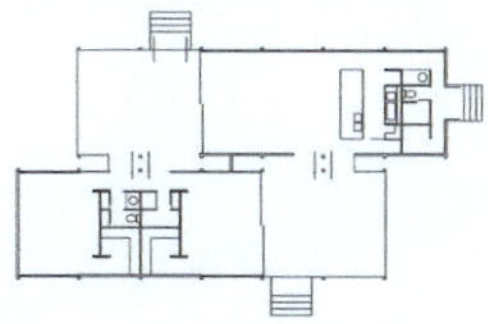

Nicholas House

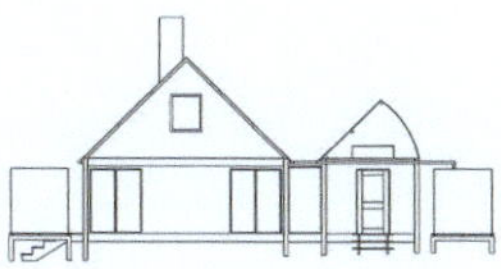
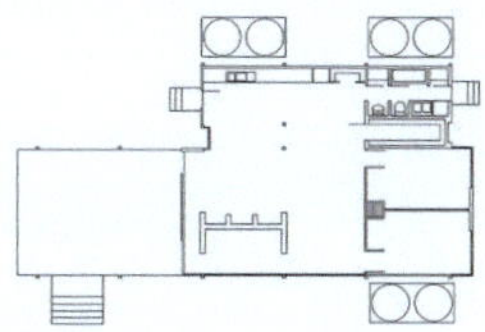

Carruthers House

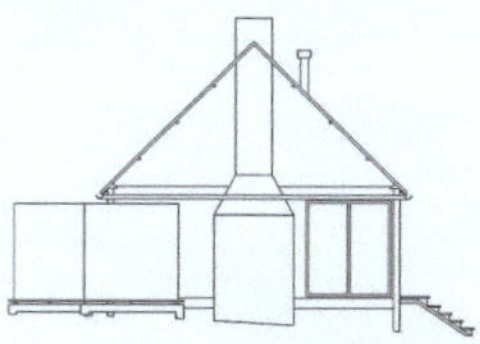
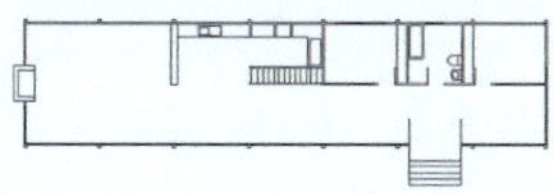

Fredericks House

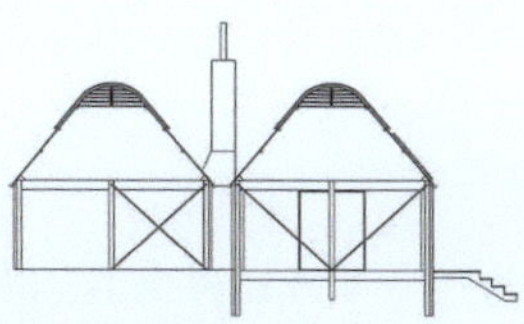
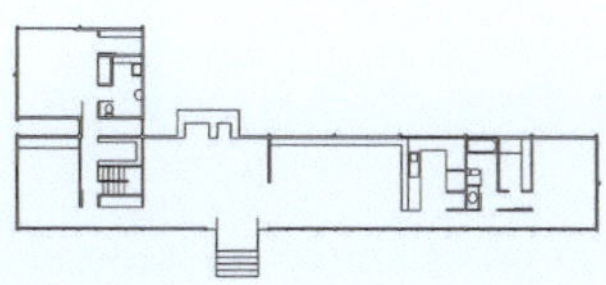

Ball-Eastaway House

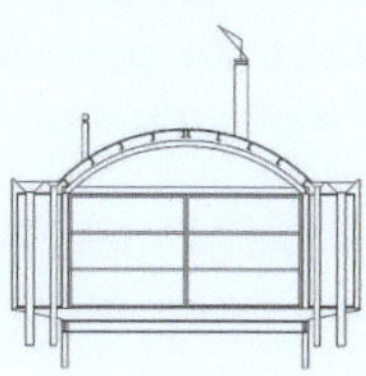
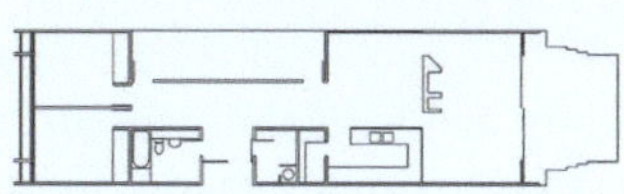

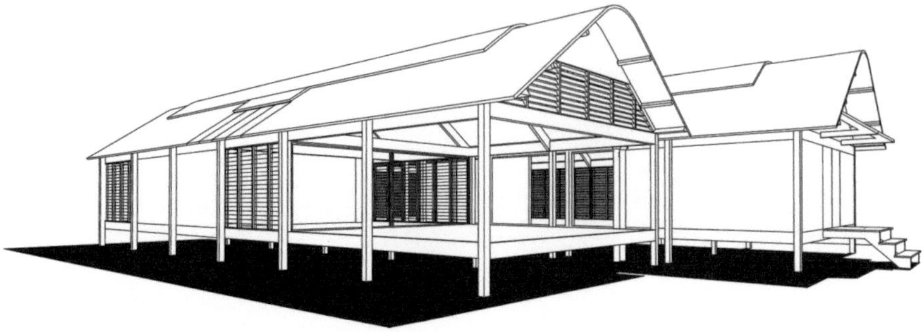

Fig. 11.8 *Marie Short House*, perspective view

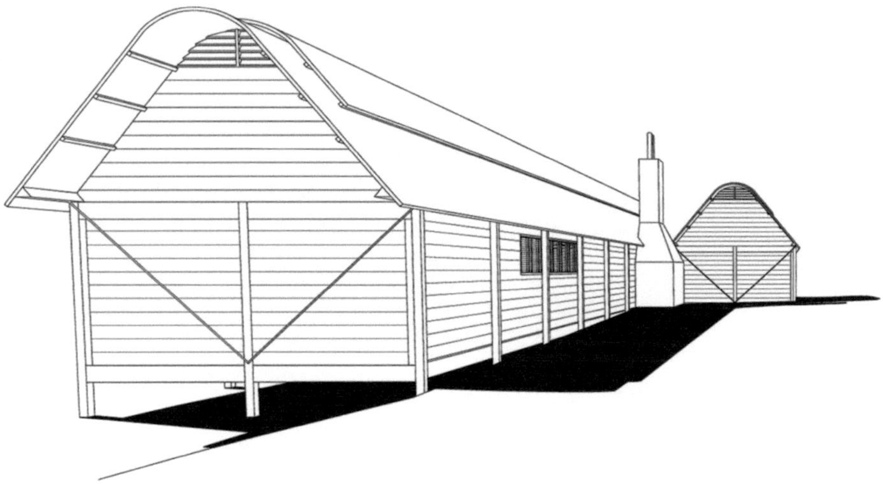

Fig. 11.9 *Fredericks House*, perspective view

lightly on posts above the ground plane. Externally, the south wall is almost fully enclosed protecting the inhabitants from winter winds.

The *Fredericks House*, located in Jamberoo, a regional area south of Sydney, is described by Drew as 'the finest of Murcutt's series of long houses' (1985, p. 121). For Drew, this house achieves a relationship between the landscape and the form of the building that is reminiscent of a temple: '[c]lassical without sacrificing any of its richness to oversimplification, light in appearance, it is the best kind of essentialist minimalist architecture, every bit as impressive as the landscape' (1985, p. 121). The house features two pavilions, of timber post and beam construction, with external western red cedar cladding (Fig. 11.9).

Designed as a house and private gallery for the artists Syd Ball and Lyn Eastaway, the *Ball-Eastaway House* is sited on top of a series of sandstone ledges near a wooded reserve northwest of Sydney. The house has a 'train carriage' plan,

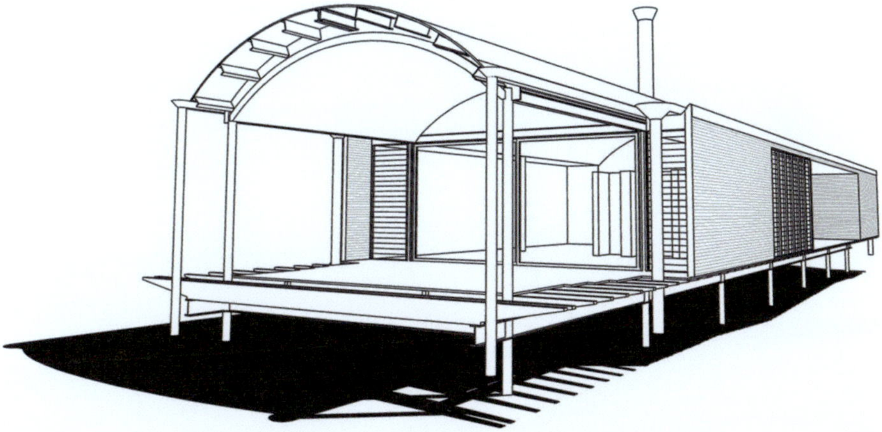

Fig. 11.10 *Ball-Eastaway House*, perspective view

an impression which is exaggerated externally with the building sitting above the ground, as if raised on wheels, and being clad in corrugated steel, with exposed downpipes and vents (Fig. 11.10).

11.7.2 Murcutt, Early Houses, Results and Analysis

The fractal dimension measures for the five houses identify the lowest elevation average in the results is for the *Carruthers House* (μ_E = 1.3780) and the highest is for the *Ball-Eastaway House* (μ_E = 1. 4397). For the entire set of elevations the range is $R_{\{E\%\}}$ = 15.97, the average is $\mu_{\{E\}}$ = 1.4128, the median is 1.4078 and the standard deviation is 0.0389 (Tables 11.11 and 11.12, Fig. 11.11). Thus, this is a consistent set, with a clear pattern of results. While it had the highest elevation average, the *Ball-Eastaway House* features the lowest plan average (μ_P = 1.2616) and the highest plan average is for the *Nicholas House* (μ_P = 1.3518). For the entire set of plans the range is $R_{\{P\%\}}$ = 18.47, the mean is $\mu_{\{P\}}$ = 1.3195, the median is 1.3398 and the standard deviation is 0.0536. When all of the plans and elevation results are considered together, the aggregate is $\mu_{\{E+P\}}$ = 1.3762. The overall composite range is $R_{\{\mu E+P\%\}}$ = 3.15, confirming, as repeatedly suggested by architectural critics, that Murcutt's early houses are very similar in appearance (less than 3.2 % difference).

The optimal sub-set comprises the *Marie Short House,* the *Fredericks House* and, somewhat unexpectedly as it is regarded as an outlier, the *Ball-Eastaway House*. The aggregate result for the optimal sub-set is $\mu_{[E+P]}$ = 1.3751, a figure which is only slightly lower than the overall value. The composite range result is also lowered, to $R_{[\mu E+P\%]}$ = 1.05, confirming that the level of visual complexity is effectively indistinguishable across these three houses.

Table 11.11 Murcutt, early set, results

Houses		*Marie Short**	*Nicholas*	*Carruthers*	*Fredericks**	*Ball Eastaway**	Opt. [*]	Set {…}
Elevations	D_{E1}	1.4346	1.3963	1.4375	1.4096	1.4382	–	–
	D_{E2}	1.3656	1.4060	1.3786	1.3820	1.4375	–	–
	D_{E3}	1.4570	1.5010	1.3431	1.3951	1.4452	–	–
	D_{E4}	1.4440	1.3925	1.3528	1.4013	1.4379	–	–
	μ_E	1.4253	1.4240	1.3780	1.3970	1.4397	–	–
	$\mu_{[E]}/\mu_{\{E\}}$	–	–	–	–	–	1.4207	1.4128
	$M_{[E]}/M_{\{E\}}$	–	–	–	–	–	1.4361	1.4078
	$std_{[E]}/std_{\{E\}}$	–	–	–	–	–	0.0290	0.0389
Plans	D_{P0}	1.3280	1.3411	1.3571	1.3474	1.3396	–	–
	D_{P1}	–	1.3461	1.3221	1.3464	–	–	–
	D_{PR}	1.2278	1.3682	1.3398	1.3067	1.1835	–	–
	μ_P	1.2779	1.3518	1.3397	1.3335	1.2616	–	–
	$\mu_{[P]}/\mu_{\{P\}}$	–	–	–	–	–	1.2971	1.3195
	$M_{[P]}/M_{\{P\}}$	–	–	–	–	–	1.3280	1.3398
	$std_{[P]}/std_{\{P\}}$	–	–	–	–	–	0.0652	0.0536
Composite	μ_{E+P}	1.3762	1.3930	1.3616	1.3698	1.3803	–	–
Aggregate	$\mu_{[E+P]}/\mu_{\{E+P\}}$	–	–	–	–	–	1.3751	1.3762

Table 11.12 Murcutt, early set, comparative values

Houses		*Marie Short**	*Nicholas*	*Carruthers*	*Fredericks**	*Ball Eastaway**	Opt. [*]	Set {…}
Elevations	R_{ED}	0.0914	0.1085	0.0944	0.0276	0.0077	–	–
	$R_{E\%}$	9.14	10.85	9.44	2.76	0.77	–	–
	$R_{[ED]}/R_{\{ED\}}$	–	–	–	–	–	0.0914	0.1579
	$R_{[E\%]}/R_{\{E\%\}}$	–	—	–	–	–	9.14	15.79
Plans	R_{PD}	0.1002	0.0271	0.0350	0.0407	0.1561	–	–
	$R_{P\%}$	10.02	2.71	3.50	4.07	15.61	–	–
	$R_{[PD]}/R_{\{PD\}}$	–	–	–	–	–	0.1639	0.1847
	$R_{[P\%]}/R_{\{P\%\}}$	–	–	–	–	–	16.39	18.47
Composite	$R_{[\mu E+PD]}/R_{\{\mu E+PD\}}$	–	–	–	–	–	0.0105	0.0315
	$R_{[\mu E+P\%]}/R_{\{\mu E+P\%\}}$	–	–	–	–	–	1.05	3.15

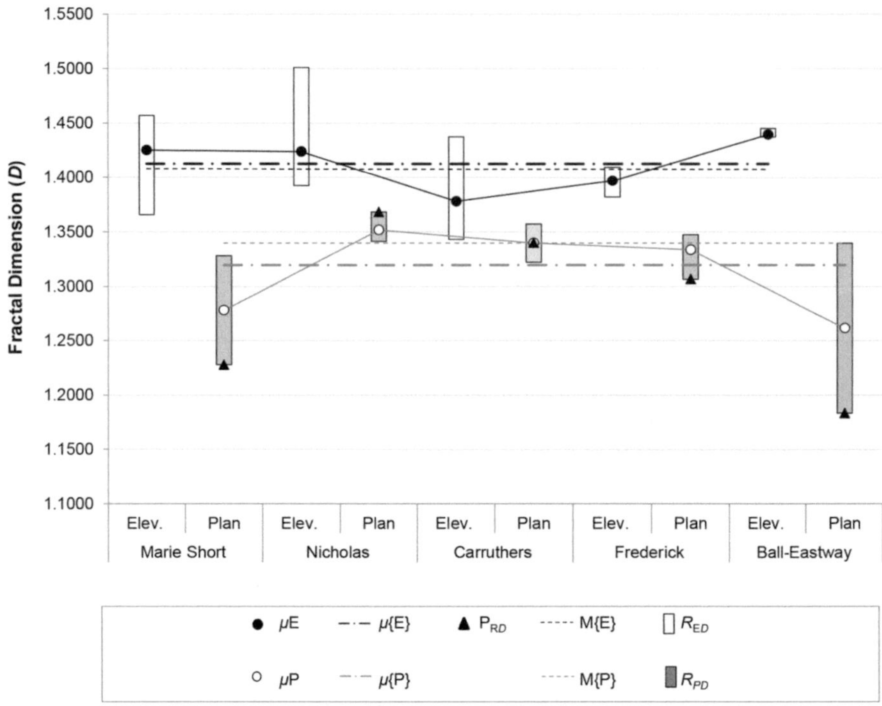

Fig. 11.11 Murcutt, early set, graphed results

The individual elevation results for the early houses fall between $1.3431 < D_E < 1.5010$, with a range between $0.77 < R_{E\%} < 10.85$ for each house, and an overall range of $R_{\{E\}\%} = 15.79$ for the complete set. These results show that Murcutt produces a comparable level of characteristic complexity for all elevations of each of the early houses, while individually the houses range in terms of similarity across their elevations from the strikingly identical *Ball-Eastaway House*, to the tight range present in the *Nicholas House*. The elevations of Murcutt's early houses generally have a higher level of visual complexity than the plans. The *Marie Short House*, the *Fredericks House* and the *Ball-Eastaway House* there is a distinctly higher level of complexity in the elevations than the plans, while in the houses built on adjoining sites, the *Nicholas* and *Carruthers* houses, this separation is less apparent.

For this set of five long, narrow buildings, the elevation images analysed for each house can typically be divided into a pair of square-proportioned façades, (the short ones, with lengths of around 10 m), and a pair of rectangularly-proportioned façades, (the elongated ones, with lengths of up to 40 m). Despite the difference in the shape of the façades, the fractal dimension results are in a noticeably similar range. This demonstrates two things; first, it provides an example of how the fractal analysis method will work despite the scale or proportions of the buildings being compared. Second, it verifies Murcutt's consistency in the level of detail he uses in his buildings and that this was neither dispersed nor concentrated as a function of the façade proportions or purpose. Thus, like a classical temple or Palladian villa, each façade has a similar level of care and attention, as if they have been designed to achieve some higher rationalist purpose, or to be seen and appreciated from all angles.

11.7.3 Five Later Houses (1984–2005) by Glenn Murcutt

After the completion of the *Ball-Eastaway House*, Murcutt's approach to domestic, rural architecture gradually evolved, retaining his linear planning style but with more elaborate building sections, often featuring curvilinear steel structures. These later designs, most of which are larger in scale and are more firmly sited on their ground planes, are characterised by skillion or butterfly roof forms. Table 11.13 provides level 4 representation of this set.

The first of this set of later designs, the *Magney House,* was constructed in 1984 and it is located on an isolated and windswept site on the coastal plains of Bingie Bingie, in southern New South Wales. The house is a single, linear pavilion on an open, gently sloping site. Constructed primarily of steel, Kenneth Frampton describes the house as having 'a peculiarly aeronautic character' (2006, p. 68), a quality that is emphasised by its narrow side profile displaying an undulating, partial butterfly roof form. This roof form, an inversion of Murcutt's prior roof profiles, allows the building's section to reflect the topography of its site.

The 1994 *Simpson-Lee House*, is a double pavilion residence sited on a rock ledge overlooking an escarpment in the Blue Mountains. Both pavilions are linear and have mono-pitch roofs which lift up toward the views, illustrating what Drew describes as 'Murcutt's development of an essentially linear typology where the chief variable is the section and roof profile' (1995, p. 150). Furthermore, Frampton claims that the corrugated sheets laid horizontally on the frame of the house enhance this linear, visual quality giving it a strong 'machinist character' (2006, p. 84). A timber footbridge connects the two pavilions with a water reservoir in between (Fig. 11.12).

The 1998 *Fletcher-Page House* is another narrow, linear building sited on sloping grassland in the hills of the Kangaroo Valley. The roof of this house is tilted upwards to the north and parallel to the hillside, partially hiding the northern elevation. In response to this, on the southern side of the house, large openings are positioned toward the primary views. This opposition between views and sunlight gives this house an unusual spatial quality in comparison to other examples of Murcutt's rural works. The house is constructed with a steel frame that rests on a concrete slab and clad externally in cedar weatherboard.

The *Southern Highlands House* is located in the farming country of Kangaloon, south of Sydney. At almost 600 m^2, this design marks a significant change in the scale of Murcutt's residential work. To buffer against severe winter winds, the house has been designed with a protective shield, in the form of a curved metal plane along the entire length of the southern façade, and wrapping over the roof

Table 11.13 Murcutt, later set, example elevations and floor plans (level 4 representation—not shown at uniform scale)

Magney House

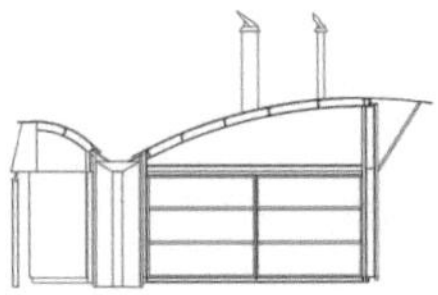

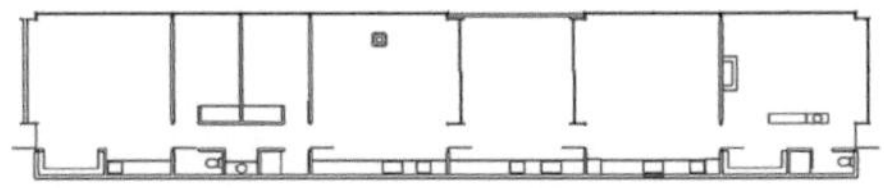

Simpson-Lee House

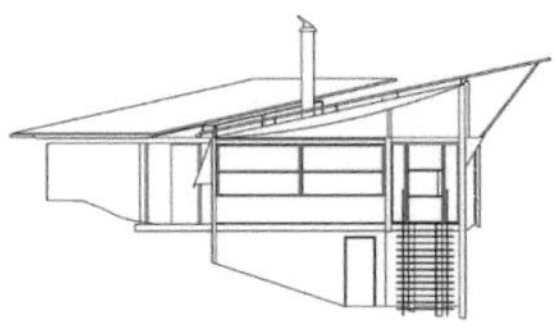

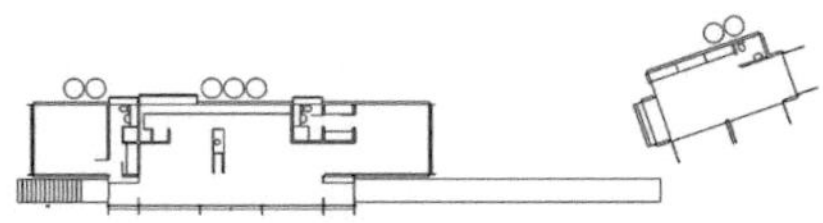

Fletcher-Page House

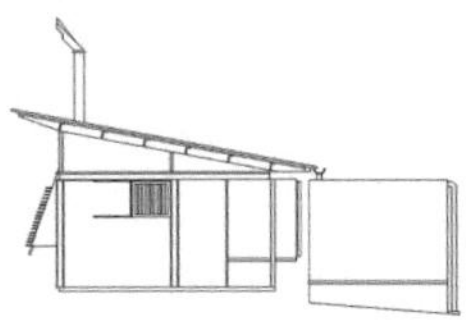

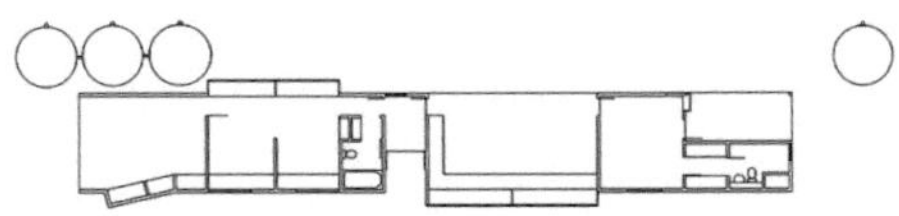

Southern Highlands House

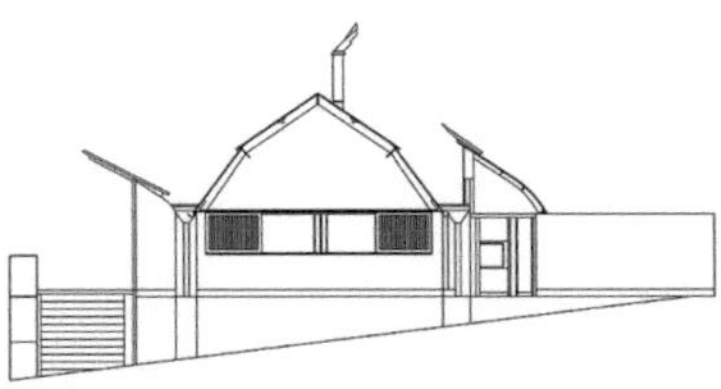

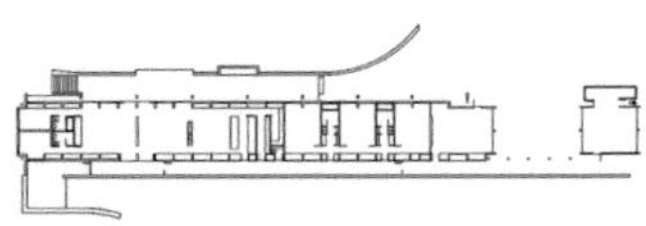

Walsh House

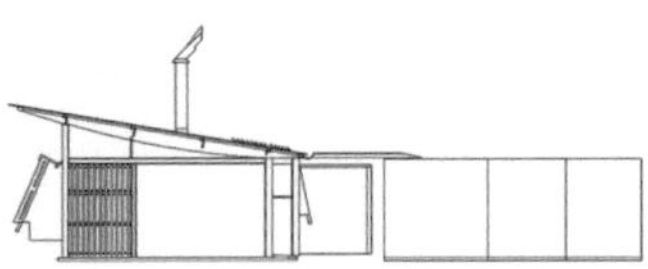

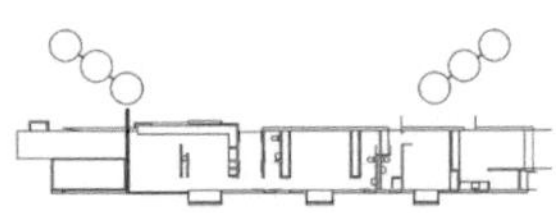

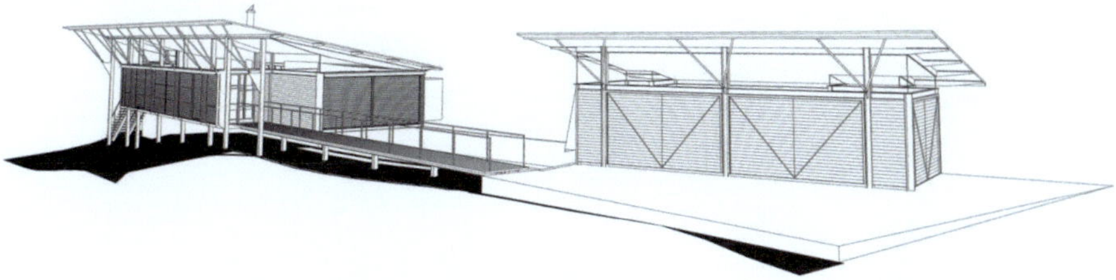

Fig. 11.12 *Simpson-lee House*, perspective view

edge. Maryam Gusheh contrasts the 'gleaming, weight-less, abstractness of the curved southern façade' with the north elevation that is 'highly articulated with adjustable hardwood window shutters and screens which modify sun-penetration and views' (2008, p. 250).

The 2005 *Walsh House* is orientated to the north and sited on a gently sloping, grassy site in the Kangaroo Valley. The detailing and materials used to construct the north and east facades display a higher level of articulation compared to those of the windward facing, south and west facades. Tilted up to the north, the roofline extends past the high northern windows in order to protect them from the summer sun. Designed with minimal openings and constructed from rustic materials, the south and west façades are reminiscent of a more conventional rural dwelling, with its services (water and gas tanks) exposed to view.

11.7.4 Murcutt Later Houses, Results and Analysis

For the set of Murcutt's later houses, the results indicate that the lowest elevation average is found in the *Walsh House* ($\mu_E = 1.4639$) and the highest is for the *Simpson-Lee House* ($\mu_E = 1.4793$). For the entire set of elevations, the range is $R_{\{E\%\}} = 20.24$, the average is $\mu_{\{E\}} = 1.4701$, the median is 1.4543 and the standard deviation is 0.0669. The *Fletcher-Page House* plans feature the lowest average result ($\mu_P = 1.2533$) while the highest is for the *Southern Highland House* ($\mu_P = 1.4575$). For the entire set of plans the range is $R_{\{P\%\}} = 33.05$, the mean is $\mu_{\{P\}} = 1.3206$, the median is 1.3484 and the standard deviation is 0.1047. When all of the plan and elevation results are considered together, the aggregate is $\mu_{\{E+P\}} = 1.4203$ and the overall composite range is $R_{\{\mu E+P\%\}} = 6.32$, suggesting that while Murcutt's later houses are similar, they are not as emphatically consistent as his earlier works (Tables 11.14 and 11.15, Fig. 11.13).

Together, the *Simpson-Lee*, *Magney* and *Walsh* houses form the optimal set, with the complexity of their average plans ($\mu_{[P]} = 1.2975$) slightly lower and average elevations ($\mu_{[E]} = 1.4711$) slightly higher than the full set. The range for the optimal set is almost negligible at $R_{[\mu E+P\%]} = 0.65$, which means these three houses are effectively identical in their level of detail (0.65 % difference).

Table 11.14 Murcutt, later set, results

Houses		*Simpson-Lee**	*Fletcher-Page*	*Magney**	*Southern Highlands*	*Walsh**	Opt. [*]	Set {…}
Elevations	D_{E1}	1.4947	1.5512	1.5756	1.5492	1.5694	–	–
	D_{E2}	1.4651	1.5701	1.4268	1.4250	1.4662	–	–
	D_{E3}	1.4114	1.3946	1.4221	1.4328	1.3876	–	–
	D_{E4}	1.5460	1.3732	1.4554	1.4532	1.4323	–	–
	μ_E	1.4793	1.4723	1.4700	1.4651	1.4639	–	–
	$\mu_{[E]}/\mu_{\{E\}}$	–	–	–	–	–	1.4711	1.4701
	$M_{[E]}/M_{\{E\}}$	–	–	–	–	–	1.4603	1.4543
	$std_{[E]}/std_{\{E\}}$	–	–	–	–	–	0.0629	0.0669
Plans	D_{PO}	1.3635	1.3440	1.3527	1.4218	1.3708	1.3635	1.3440
	D_{PR}	1.1782	1.1626	1.2650	1.4931	1.2546	–	–
	μ_P	1.2709	1.2533	1.3089	1.4575	1.3127	–	–
	$\mu_{[P]}/\mu_{\{P\}}$	–	–	–	–	–	1.2975	1.3206
	$M_{[P]}/M_{\{P\}}$	–	–	–	–	–	1.3089	1.3484
	$std_{[P]}/std_{\{P\}}$	–	–	–	–	–	0.0773	0.1047
Composite	μ_{E+P}	1.4098	1.3993	1.4163	1.4625	1.4135	–	–
Aggregate	$\mu_{[E+P]}/\mu_{\{E+P\}}$	–	–	–	–	–	1.4132	1.4203

Table 11.15 Murcutt, later set, comparative values

Houses		*Simpson-Lee**	*Fletcher-Page*	*Magney**	*Southern Highlands*	*Walsh**	Opt. [*]	Set {…}
Elevations	R_{ED}	0.1346	0.1969	0.1535	0.1242	0.1818	–	–
	$R_{E\%}$	13.46	19.69	15.35	12.42	18.18	–	–
	$R_{[ED]}/R_{\{ED\}}$	–	–	–	–	–	0.1880	0.2024
	$R_{[E\%]}/R_{\{E\%\}}$	–	–	–	–	–	18.80	20.24
Plans	R_{PD}	0.1853	0.1814	0.0877	0.0713	0.1162	–	–
	$R_{P\%}$	18.53	18.14	8.77	7.13	11.62	–	–
	$R_{[PD]}/R_{\{PD\}}$	–	–	–	–	–	0.1926	0.3305
	$R_{[P\%]}/R_{\{P\%\}}$	–	–	–	–	–	19.26	33.05
Composite	$R_{[\mu E+PD]}/R_{\{\mu E+PD\}}$	–	–	–	–	–	0.0065	0.0632
	$R_{[\mu E+P\%]}/R_{\{\mu E+P\%\}}$	–	–	–	–	–	0.65	6.32

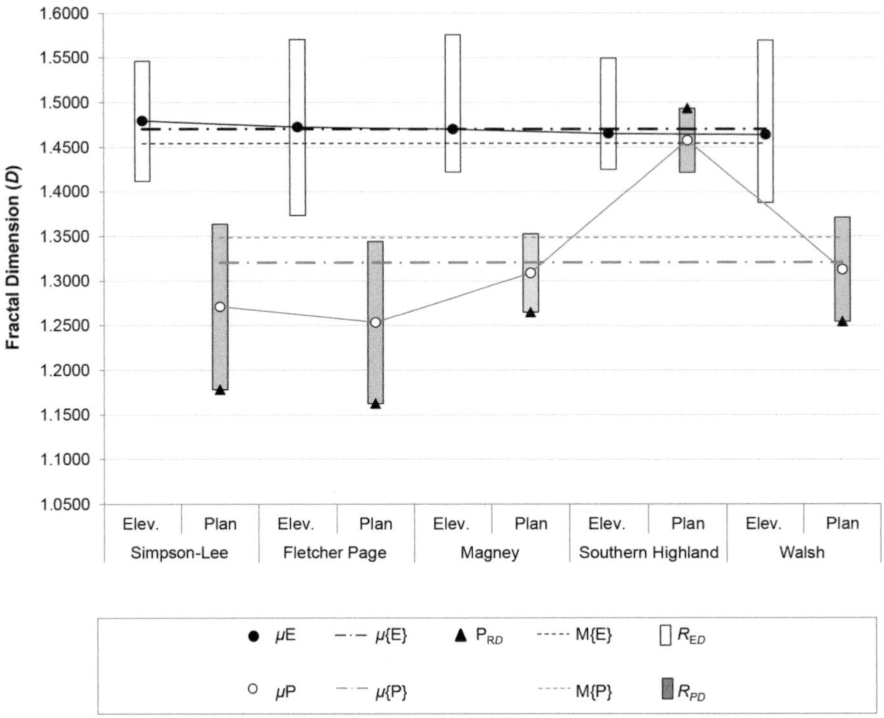

Fig. 11.13 Murcutt, later set, graphed results

The individual results for the elevations of the later houses fall in the range $1.3732 < D_E < 1.5756$, that is they show generally higher fractal dimensions than the earlier set from Murcutt. The later five houses also have a wider range of results ($12.42 < R_{E\%} < 19.69$), confirming that in these works Murcutt began to treat each elevation differently. Not dramatic, but still significant when compared to the earlier set, is an increase in the individual results $1.1782 < D_P < 1.4931$ and their diversity $7.13 < R_{P\%} < 18.53$ in the plans for the later set.

With few exceptions, Murcutt's rural domestic architecture has been described by critics as providing an exemplar of Arcadian minimalism—a rigorous modern evocation of the form and tectonics of the primitive hut. For example, Drew proposes that Murcutt's talent lies in his capacity to shape 'a minimalism that is austere and tough so that all that remains is an irreducible core' (1986, p. 60). Rory Spence describes Murcutt's early houses as constituting a clear formal type: 'the long thin open pavilion' (1986, p. 72). Fromonot argues that Murcutt's houses are all 'variations on the same theme' and that these design 'prototypes' represent a 'relatively homogenous body of work' (1995).

The results in the present chapter confirm these views, despite some diversity in the later set, as Murcutt's formal pattern, or signature, found in the earlier work is still evident in his later houses. The pattern in the later houses shows in the average elevation for each house, these being so similar they almost form a horizontal line on the graph of results. The plans for this set are also almost consistently below the elevations, with a typical maximum plan dimension being in the close vicinity of the median for the plans (M_P = 1.3484). The *Southern Highland House* reverses this trend by having all the fractal dimensions of its plans in the same sector as its elevations, which is unusual among the examples of Murcutt's work studied here.

Descriptions of Murcutt's architecture as minimalist would suggest that a fractal analysis of elevations should reveal a low level of visual complexity; meaning that D will be closer to 1.0 than 2.0. However, Murcutt himself notes that a simple form does not necessarily imply the presence of a simple interior or experience; '[t]he house [may be] very simple. But remember simplicity is the other face of complexity' (Murcutt 2007, p. 26). In this statement Murcutt suggests that the visual properties of his architecture might change depending on the perspective of the viewer. This is certainly reflected in past Space Syntax analysis of the interiors of Murcutt's houses which have indicated that, with few exceptions, the internal configurations were both more complex and less predictable than the canonical literature suggests (Ostwald 2011a, b).

The claims about visual simplicity are also only partially supported by the present research. While the results for the ten houses are generally in the lower half of the fractal dimension spectrum (that is, where $D_E < 1.5$), they do not confirm the suggestion that Murcutt's work has a very low level of visual complexity akin to that of other Minimalist architects (where $D_E < 1.3$). With both the elevation median results in the range $1.4078 < M_{\{E\}} < 1.4543$, and plan median results $1.3398 < M_{\{P\}} < 1.3484$, Murcutt's architecture is, in terms of fractal dimensions, not dissimilar to many Modernist works.

The real explanation for the 'lightness' or 'simplicity' so often remarked upon in Murcutt's architecture may be tied to the fact that Murcutt's houses are often pictured as free-standing objects in the landscape, artificially exaggerating their Acadian or even neo-Platonic properties. Whatever the underlying power of Murcutt's architecture is, while it does possess some transparent façades, and its forms are relatively simple, the real explanation for its character is clearly not just a factor of the form, space and geometry within his work. Several of these factors are tested in the next section.

11.8 Testing Visual Lightness and Transparency

Descriptions of Murcutt's architecture typically emphasise its 'visual lightness' and 'transparency' and argue that these are particular traits of both Australian vernacular architecture and of the Australian landscape. In architectural criticism, visual lightness is a property associated with the degree to which a building reveals its interior through transparent and layered elements in the façade. Yet, there are differences between the properties of lightness and transparency. In 1963 Colin Rowe and Robert Slutzky provided a clarification of these two properties, differentiating them as 'literal' and 'phenomenal' transparency. Literal transparency occurs where a physical form is 'pervious to light and air' (Rowe and Slutzky 1963, p. 45), while phenomenal transparency occurs when a building exhibits 'stratifications' which, whether implied or real, are 'devices by means of which space becomes constructed, substantial, and articulate' (Rowe and Slutzky 1963, p. 53).

Murcutt's houses possess façade elements that fulfil both parts of this definition. For example, his façades often function as diaphragms that control the degree to which sunlight and air are allowed to enter the interior; this is a type of literal transparency. Murcutt's façades are also structured in layers that can be experienced in different ways, a property which is exaggerated by the narrowness of his plans, and can sometimes be both seen through and directly walked through. Of these two types of transparency, the literal and the phenomenal, it is the first that is seemingly strongest in Murcutt's rural houses. This is because, while the façade layers and pavilion forms might be phenomenally transparent, there is still a thinness about them that does not fulfil the spirit of Rowe and Slutzky's definition. Nevertheless, fractal analysis does offer ways of testing the significance of literal and phenomenal transparency in Murcutt's architecture. Using these methods it is possible to determine first, whether it is actually reliant on transparency for its architectural character, and second, which type of transparency is more significant. This two-part test of Murcutt's architecture is the subject of the final section in this chapter.

The first test is a fractal analysis of two variations of each façade of ten of Murcutt's rural domestic designs. The first façade variation is the one recorded in the previous section, where each elevation is measured with all elements represented as opaque and non-reflective and all permeable features as closed. The second elevation variation uses the same set of façades, but with the visual impact of transparency depicted. That is, these elevations treat glass, screens and louvres as open and transparent. The eighty results produced through this analysis are then compared to quantify the difference and determine if it is significant in terms of visual perception. This first test is effectively focussed on the visual impact of *literal* or actual transparency.

The second test presented in this section considers the argument that the particular properties of transparency that are so important in Murcutt's architecture might not be apparent in orthographic projections, and might instead be a result of the suggestion of depth in a façade. To test this idea, a set of twelve perspectival images of the *Marie Short House* are analysed to determine if such views alter the fractal dimension result in both opaque and transparent modes. The twenty-four results of this analysis are then compared to provide a measure of the implied *phenomenal* transparency of this design.

11.8.1 Method and Results for Test 1

For the first test, for each of the ten houses by Murcutt all four elevations were drawn twice. The first representation is a typical elevation, with all doors closed and

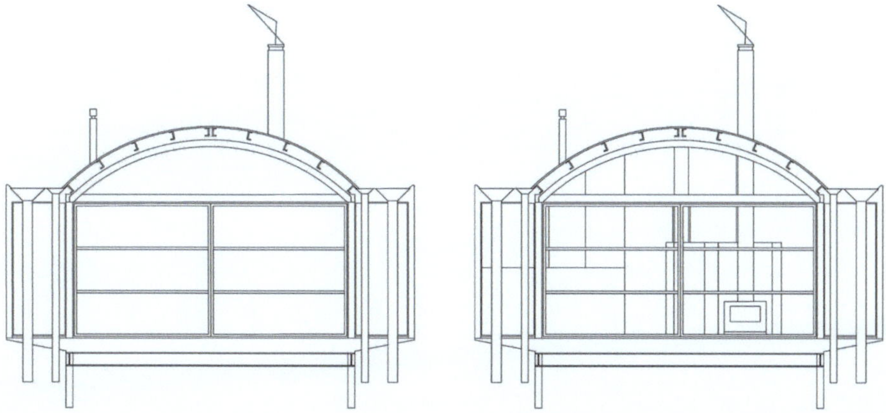

Fig. 11.14 *Ball-Eastaway House*, east elevation; opaque and transparent variations

all windows opaque. The second version of each elevation shows the building with the internal walls and fixed furniture visible through open doors, screens and windows (Fig. 11.14). The procedure used for this first test, where the impact of literal transparency is measured using orthographic projections, is as follows.

i. The 40 (ten houses each with four elevations) 'opaque' views of each individual house are identified as Set 1.
ii. Each elevation of each house is analysed producing a D_E outcome.
iii. The D_E results for the elevations of each house are averaged to produce a separate $D_{(\mu E)}$ result.
iv. The range ($R_{E\%}$) between the highest and lowest D_E results in a single house is calculated and expressed as a percentage.
v. The median ($M_{\{E\}}$) is calculated for the combined D_E results for all elevations.
vi. The 40 'transparent' views of each individual house are identified as Set 2. Steps (ii) to (v) are repeated for Set 2. The results are both tabulated (Table 11.16) and charted (Fig. 11.15).

Table 11.16 Murcutt houses, comparison between transparent and opaque elevation values

House	Opaque mean (μ_E)	Transparent mean (μ_E)	Difference ($R_{E\%}$)
Marie Short	1.4253	1.43265	0.74
Nicholas	1.42395	1.436925	1.29
Carruthers	1.390875	1.40115	1.02
Fredericks	1.397	1.40235	0.53
Ball-Eastaway	1.4397	1.455725	1.6
Simpson-Lee	1.4793	1.4963	1.7
Fletcher-Page	1.472275	1.479575	0.73
Magney	1.469975	1.4863	1.63
Southern Highlands	1.461675	1.463875	0.21
Walsh	1.463875	1.472975	0.91
Average	1.442393	1.452783	1.04

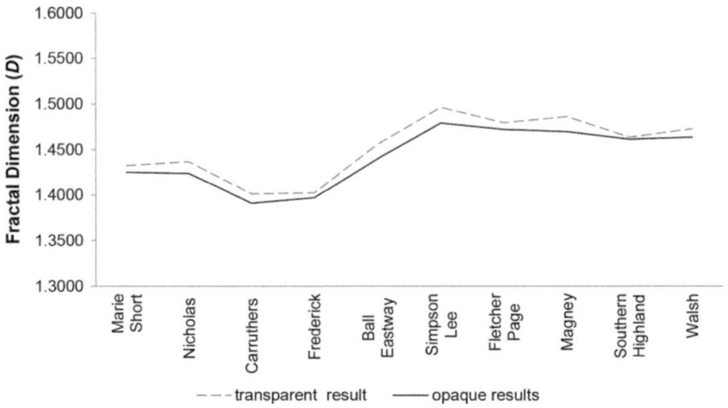

Fig. 11.15 Murcutt houses, comparison between transparent and opaque elevation values

When considering the results of this method, for the opaque elevations, the range is 1.390875 < D_E < 1.472275 or $R_{\{E\}\%}$ = 8.14 % while for the transparent elevations, the range is 1.40115 < D_E < 1.4863 or $R_{\{E\}\%}$ = 8.51 %. The difference between the opaque and transparent results for each house is from a low of 0.21 % for the *Southern Highlands House*, to 1.7 % for the *Simpson-Lee House*. The mean difference, transparent to opaque, is 1.04 %.

11.8.2 Method and Results for Test 2

While the first test produces a straightforward set of results using a repeatable method, in this particular inquiry into the transparency of buildings, the fact that many interior elements in Murcutt's architecture are precisely aligned to exterior columns, walls and window openings, potentially makes them invisible in a transparent elevation drawing. To take this unique effect into account, the second approach utilises a set of perspective line drawings of the *Marie Short House* (Fig. 11.16). The procedure used for this second test, where the impact of phenomenal transparency is measured using perspectival projections, is as follows.

i. The viewing distance used to generate the perspective views is first determined by finding the perpendicular distance from the longest façade of the house that allows for the high acuity zone of a human cone of vision to view the entire façade. This distance is used to define a circle around the plan, from which all perspective viewing positions are generated.
ii. The viewing locations must then be identified along this circle. Taking a line perpendicular to the longest elevation as a starting point, this is defined as 0°. Thereafter, the circle is divided into twelve 30° arcs, creating a set of cardinal and sub-cardinal viewing positions around this compass.

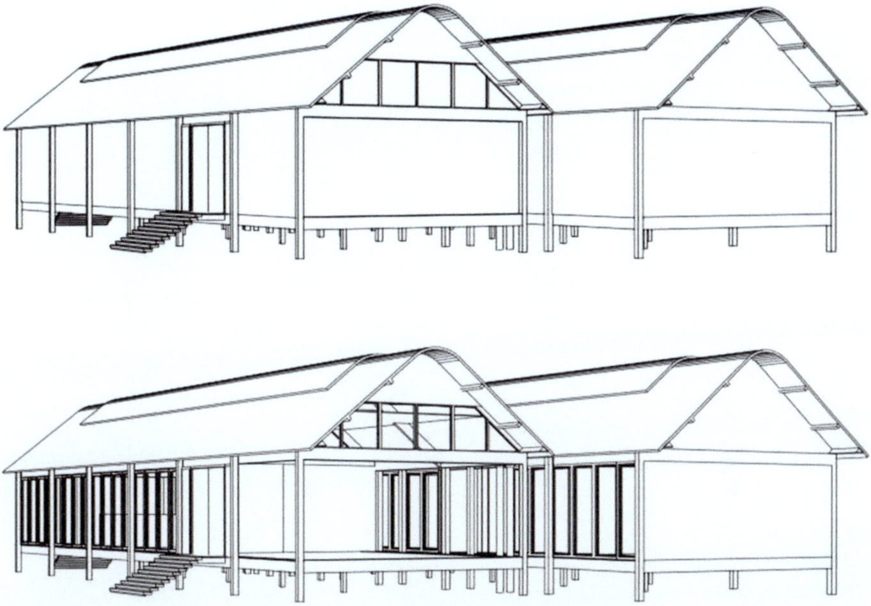

Fig. 11.16 *Marie Short House*, opaque (*upper image*) and transparent (*lower image*) perspective variations

iii. The twelve (sub-cardinal viewpoint) 'opaque' views of the *Marie Short House* are identified as Set 3 and perspective views are generated from these positions.
iv. Each view is analysed producing a D_{PerO} measure.
v. The D_{PerO} results for the views of the house are averaged to produce a separate D (μ_{PerO}) result.
vi. The twelve 'transparent' views of the *Marie Short House* are identified as Set 4. Steps (iii) to (v) are repeated for Set 4 producing D_{PerT} and μ_{PerT} results. The results are both tabulated (Table 11.17) and charted (Fig. 11.17).

From the data produced using this method, the lowest level of difference between the opaque and transparent views of the *Marie Short House* is found at the 180° location, with a 0.3 % change in detail. The highest difference is at the 30° location, with a 14.3 % rise in the level of detail. The mean difference for the complete set of data is 5.58 %, while the difference between μ_{PO} and μ_{PT} is 5.64 %.

11.8.3 Discussion

The higher fractal dimension results for the transparent views are precisely as expected, because the elevations depicting the interior through the façade naturally

Table 11.17 *Marie Short House*, comparison between transparent and opaque perspective values

View location (degrees)	Opaque (D_{PerO})	Transparent (D_{PerT})	Difference (%)
30	1.4416	1.5855	14.3
60	1.4973	1.5950	9.7
90	1.5380	1.6113	7.3
120	1.5474	1.5567	0.9
150	1.6026	1.6084	0.5
180	1.5477	1.5512	0.3
210	1.5034	1.5240	2.0
240	1.5462	1.5708	2.4
270	1.5334	1.5501	1.6
300	1.4597	1.5094	4.9
330	1.5294	1.6353	10.5
360	1.5022	1.6276	12.5
Average	1.5207 (μ_{PO})	1.5771 (μ_{PT})	5.58

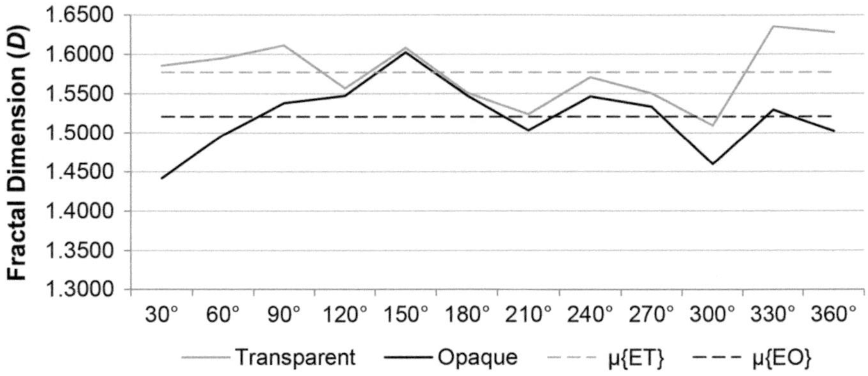

Fig. 11.17 *Marie Short House,* results

contain at least as much detail as the opaque versions in all cases. What was unexpected is the low level of difference between the results of the two tests. In the first instance, testing for literal transparency, the mean difference is 1.04 % and for phenomenal transparency for the *Marie Short House,* 5.58 %. The most obvious interpretation of the results is that literal transparency, contrary to popular opinion, is not a strong determinant of visual complexity in Murcutt's architecture. Phenomenal transparency, which is potentially closer to the way an architectural critic would describe visual lightness, is marginally higher, but still not especially significant.

11.9 Comparative Results

Comparing the aggregate results of the five sets of Minimalist and Regionalist designs, the most obvious difference is that the Japanese buildings are less visually complex (Sejima $\mu_{\{E+P\}} = 1.3227$, Bow-Wow $\mu_{\{E+P\}} = 1.3218$) than the Australian buildings (Stutchbury $\mu_{\{E+P\}} = 1.3944$, Early Murcutt $\mu_{\{E+P\}} = 1.3762$, Late Murcutt $\mu_{\{E+P\}} = 1.4203$). Thus, despite the Regionalists being described as presenting minimal or simple formal compositions, they are not as visually restrained as the Japanese cases examined here.

The formal coherence graphs for both Sejima and Atelier Bow-Wow present a variable level of elevational complexity in opposition to a more consistent level of plan detail (Fig. 11.18, 11.19). In these graphs there is no clear correlation between plan and elevation results in the Sejima graph ($R^2 = 0.00016$) and only slightly stronger correlation for Atelier Bow-Wow ($R^2 = 0.20843$). For Stutchbury, the correlation is also low ($R^2 = 0.066$) (Fig. 11.20) while for Murcutt's later works, the formal coherence pattern is almost in opposition to that of Sejima. Sejima's plans were largely consistent, while her elevations varied, whereas Murcutt's elevations are consistent, while his plans vary in their level of visual complexity (Figs. 11.21, 11.22).

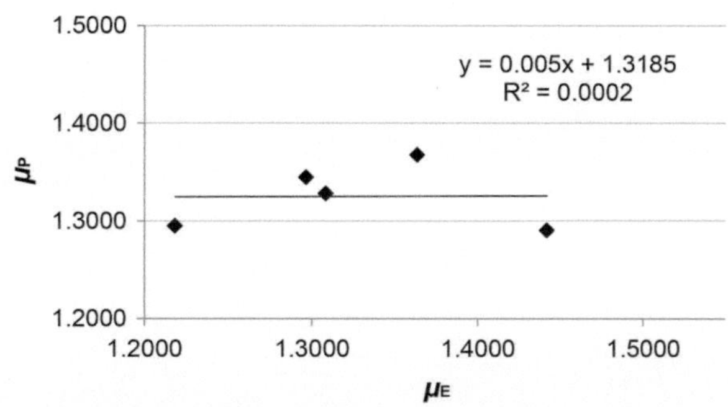

Fig. 11.18 Sejima set, formal coherence graph

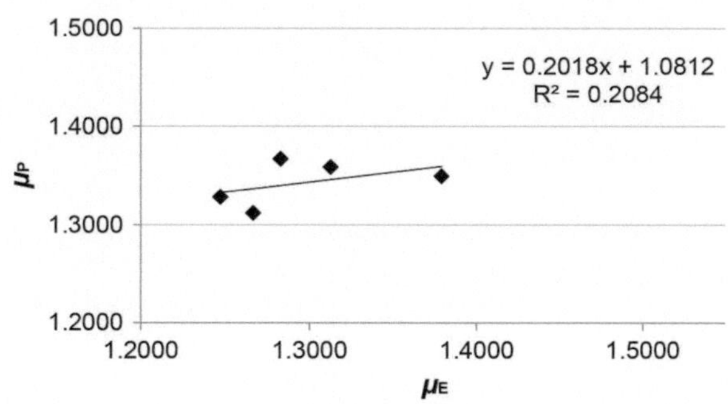

Fig. 11.19 Atelier Bow-Wow set, formal coherence graph

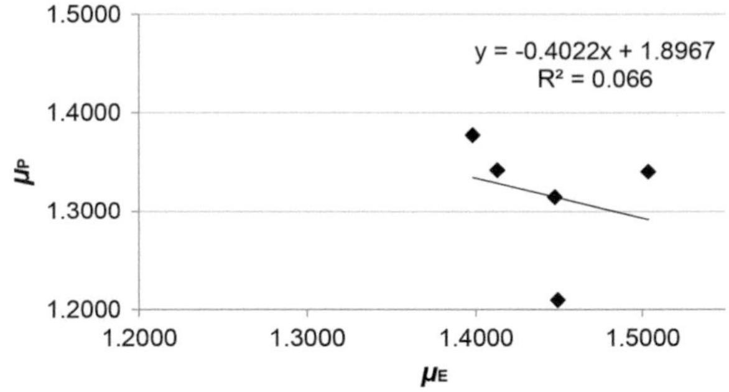

Fig. 11.20 Stutchbury set, formal coherence graph

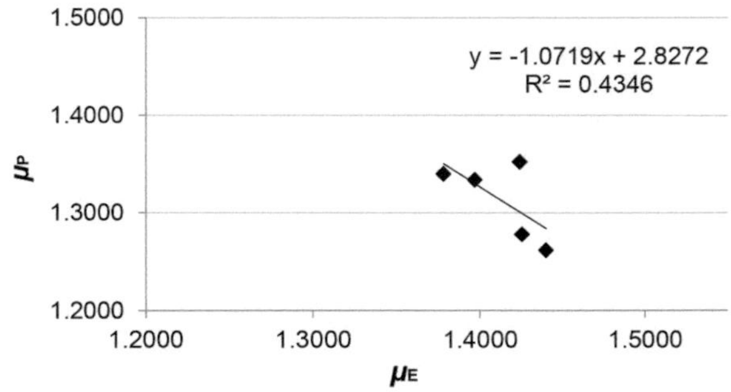

Fig. 11.21 Murcutt, early set, formal coherence graph

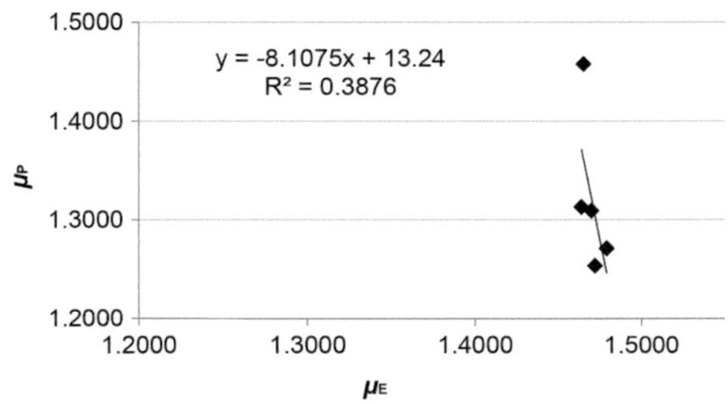

Fig. 11.22 Murcutt, later set, formal coherence graph

If the houses in each set, by each architect, are sequenced in the order in which their designs were produced, and then trendlines are extrapolated from this data, a series of broad indicators about these architects' changing approaches can be seen. For example, when considering elevations for all four architects in isolation from plans, both the designs of Sejima and Bow-Wow display a growing level of simplicity, or increasingly minimal expression, while Murcutt and Stutchbury are more consistent or unchanging (Fig. 11.23). Considering plans in isolation, the only increasing trendline is associated with Murcutt's later designs, wherein the scale of the buildings increased and the number of rooms grew. In all other cases the gross

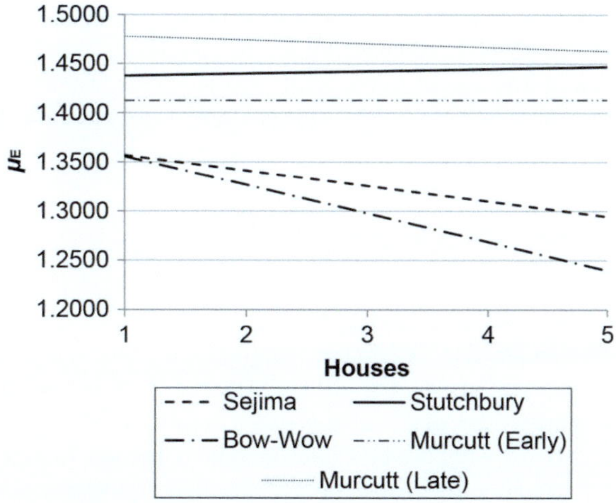

Fig. 11.23 Sejima, Atelier Bow-Wow, Murcutt and Stutchbury sets, linear trendline data for elevations (μ_E)

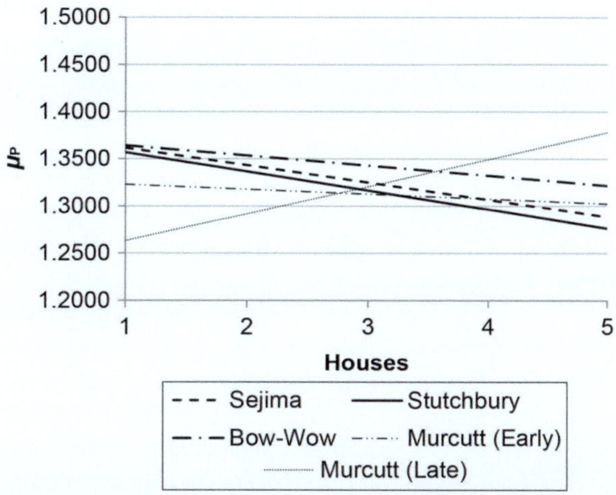

Fig. 11.24 Sejima, Atelier Bow-Wow, Murcutt and Stutchbury sets linear trendline data for plans (μ_P)

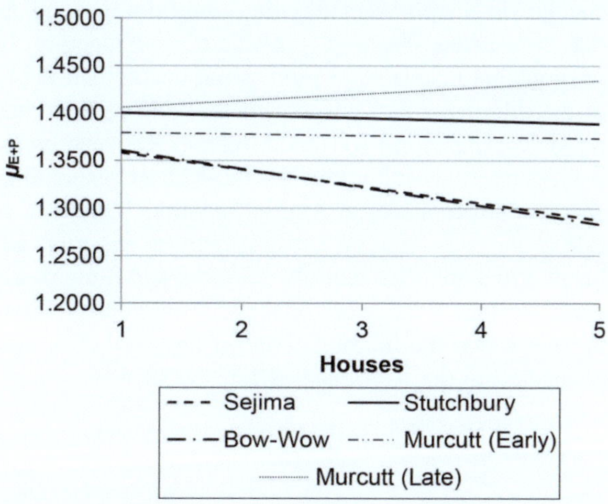

Fig. 11.25 Sejima, Atelier Bow-Wow, Murcutt and Stutchbury sets linear trendline data for composite (μ_{E+P}) results

trend is towards an increase in simplicity (Fig. 11.24). When plans and elevations are combined for each house, Atelier Bow-Wow's and Sejima's works feature a strong downward trend, while the Regionalist architects show a relatively flat line for Stutchbury and early Murcutt, and a marginal rise for later Murcutt (Fig. 11.25).

11.10 Conclusion

Despite obvious connections between the design philosophies of the Japanese Minimalist and Australian Regionalist architects, the two approaches ultimately present different, yet still distinctive patterns in their fractal dimensions. For example, like many of the other architectural movements analysed previously in this book, Stutchbury's and Murcutt's houses feature elevation dimensions which are typically in a higher range than their equivalent plan dimensions. In contrast, the Japanese minimalist designers produced plans and elevations that fall broadly into the same numerical range. The only other example of such a pattern in the present book is found in the works of the Avant-Garde architects, who deliberately set out to create objects which were consistent in all three dimensions. However, this pattern in the Minimalist data is more a result of two disconnected factors. The first of these arises from the aesthetic character of Minimalist designs, which deliberately minimises the impact of formal and material modelling, resulting in a low set of fractal dimensions for the elevations. The second factor is that the tight, vertical planning of these houses—a direct result of the context and scale of their sites—forces a large number of programmatic zones to be accommodated in a series of overlapping spaces connected by circulation areas. This approach to functional planning produces relatively high fractal dimensions for plans. In combination, these two factors create the overlapping pattern of dimensions. This is a very different explanation from that previously uncovered for Eisenman, Hejduk and Meier, whose works superficially feature a similar mathematical relationship.

The final section in this chapter examined the importance of transparency in the architecture of Glenn Murcutt. The results of the two studies clearly indicate that changes in form and materiality could not account for the quality of transparency that is so often observed in Murcutt's architecture. In the tests provided here, the phenomenal results were more pronounced than the literal ones, but the magnitude of the difference between transparent and opaque measures remained limited. One possible explanation for this difference between measured and observed properties is that historians and critics are most likely responding to more than just the formal and spatial properties of Murcutt's work. Materials, colours and textures all play a role in the way Murcutt's architecture is experienced and portrayed. In particular, his interiors are often warmly coloured and illuminated, in stark contrast to the often grey, weathered exteriors of these same houses. Such factors no doubt exaggerate the sense that the interior is on display to the outside world, and that it is closely connected to the landscape as a result of this.

Chapter 12
Conclusion

Throughout Part II of this book, the fractal dimensions of eighty-five buildings designed by fifteen prominent architects or practices were measured. The time period covered by these works spans from Wright's 1901 *Henderson House* to Stutchbury's 2007 *Invisible House*. For these eighty-five designs, all visible elevations, all habitable plans and the roof plan were recreated in a consistent representational format, producing 335 elevations and 290 plans. The fractal dimensions of these 625 orthographic drawings, along with those of more than 200 additional views of these buildings, were measured using over 9000 grid comparisons to count in excess of five million pieces of data. Whereas results for specific architect's works, or stylistic sets, were presented in previous chapters, the purpose of this concluding chapter is to draw together the complete set of data in order to consider larger questions about the formal and visual properties of buildings. However, before doing this, it is important to remember that the houses used to generate these results possess a particular set of characteristics.

The houses analysed in this book are *not* representative of the properties of the larger population of housing produced during the twentieth-century. These works have also *not* been chosen to provide an indicative coverage of design in every year, every country or culture. They are also, despite our best intentions, not a balanced selection taking into account gender and race. Instead, the set of works chosen are ones that historians and critics have identified as embodying the visual characteristics of significant styles, approaches and eras. Therefore, these are not just houses by famous architects, but designs which have been praised, analysed and widely disseminated by the architectural media. These designs are justly regarded as canonical buildings in the history of architecture and they are important because they represent major theories or approaches that have shaped design culture over the last century. For all of these reasons, a comparison of the complete set of results can only reveal something about the visual properties of a diverse group of historically significant and critically lauded designs. Arguably then, the most significant findings of this book are contained in each of the major analytical chapters and especially in the additional applications of the method to testing specific arguments

M.J. Ostwald and J. Vaughan, *The Fractal Dimension of Architecture*,
Mathematics and the Built Environment 1, DOI 10.1007/978-3-319-32426-5_12

about these movements. However, an analysis of the complete set of results offers the potential to revisit the three overarching hypotheses that were introduced in Chap. 1.

12.1 Presentation of Results

In the present chapter, the complete set of fractal dimension data is analysed using four approaches, the first three of which correspond to the hypotheses framed in Chap. 1. First, the data is examined chronologically, in terms of the date when each design was constructed or, if unbuilt, finalised. Second, the data is clustered into stylistically similar or connected sets. Third, a rank order, based on formal coherence, is established to compare the fifteen architects. Finally, the data is sorted using both complexity and consistency criteria for each set and each movement.

For the first of these approaches, the eighty-five houses are ordered chronologically, and the average fractal dimensions for the elevations (μ_E), plans with roofs (μ_P), plans without roofs ($\mu_{P\text{-}R}$) and plans and elevations combined (μ_{E+P}) are graphed. The results are also graphed according to the range of their average elevations (R_E), plans with roofs (R_P) and plans without roofs ($R_{P\text{-}R}$). All of these results are tabulated, divided in upper and lower quartiles and the interquartile range, allowing for specific architect's works, and even individual houses, to be differentiated in terms of their position in the data. Importantly, for each of these seven sets of data linear trendlines are developed. These trendlines can be used to test the first of the three overarching hypotheses identified in Chap. 1. The first hypothesis states that, *as the complexity of social groupings and functions contained within the home has reduced over time, the fractal dimensions of plans and elevations should decrease to reflect this change*. An examination of these trendlines should provide evidence for or against this proposition. In all cases, if the hypothesis is true, the trend-line should fall from left to right. The more pronounced the angle of this fall, the truer the hypothesis is, based on this data.

The second analytical approach sorts the data by stylistic periods. In addition to calculating the mean and range of the average elevations (μ_E), plans (μ_P) and the two combined (μ_{E+P}), in this section the standard deviation is also graphed and the results tabled by quartiles. This data is used to investigate the second hypothesis, which holds that, in architecture *each stylistic genre or movement possesses a distinct visual character that is measurable using fractal dimensions*. If this hypothesis is true, when the data is sorted by stylistic periods there should be a clear clustering apparent which readily differentiates each style.

In the third approach to the complete data set, the formal coherence results developed for each architect or practice are ranked. These results compare an architect's plans and elevations in a stylistic set, and measure the relationship between the two. For example, for an architect like Richard Meier, whose houses often have similar levels of complexity in plan and elevation, higher formal coherence results are to be anticipated, but for many of the others, like Wright's

Prairie Style houses, the elevations and plans have different visual characters and thus much lower levels of formal coherence. This test reveals something of the essence of each architect's approach to making three-dimensional form. As such, it can be used to investigate the third hypothesis, which states that *individual architects will present distinctive patterns of three-dimensional formal and spatial measures across multiple designs*.

In the fourth and final approach to the data, a series of matrix charts are prepared comparing complexity (D) and consistency (R), divided into bands using quartiles and the interquartile range. By creating a separate matrix map for both plan and elevation data, this last method demonstrates how multiple criteria might begin to be used to more comprehensively differentiate and classify designs or styles using fractal dimensions.

12.2 Chronological Analysis

For this first approach to the data, it is significant that the chronological spread of houses is fairly consistent from 1901 to 2007, with the only notable gap from 1939 to 1945, coinciding with the second world war. This relative evenness in the spread of data is of interest because the plans and elevations of a house are thought to reflect the social patterns and values of the era in which it was produced. Thus, one hundred years ago architect-designed houses catered to large extended families, with multiple servants and guest quarters, leading to complex separations of functional zones in a plan. However, today much has changed and open-plan living, with smaller family units, no servants, and less differentiation and control, has become the standard for much of the developed world. Similarly, in terms of the visual expression of elevations, one hundred years ago only small planes of glass were readily available to most people, meaning that façades with multiple window perforations were common. Structural spans were also limited, leading to the clear visual expression of columns and beams, and roof forms were often complex to take into account the limits of materials and the problems of waterproofing. However, more recently entire glass walls have become commonplace in housing, structural steel has reduced the visual presence of columns and beams and new materials and construction methods have limited the need for many complex changes in form to accommodate waterproofing.

For these reasons, in a chronological sorting of the data, the plans and elevations, even of larger architect-designed homes, would be expected to display a marginal reduction in visual complexity over time. To further test this idea a new data variation is introduced, the plan average *excluding* the roof plan ($\mu_{\text{P-R}}, R_{\text{P-R}}$). In all of the previous chapters, the 'roof elevation' was treated as a special type of plan, and while that data variant is still presented here for consistency and completeness, this new version methodically separates roof plans from floor plans allowing for an assessment of changing social complexity of the interior. It might also be anticipated that the visual complexity of elevations will reduce over time, however the fact that a high level of complexity is no longer required does not necessarily mean that it will reduce. This is because stylistic and cultural factors also come into play when considering the exterior expression of a building.

12.2.1 Average Elevations

The overall average of the elevations is $D = 1.4148$ and the interquartile range is $1.36975 < \mu_E < 1.51035$. The chronological graph of the average visual complexity of each house, grouped by architect, displays several important features in the complete data set (Fig. 12.1). First, there is a clear downward trendline generated for complexity over the century, with the most complex houses ($D > 1.5$) all prior to the 1930s, and the least complex ($D < 1.25$) after the year 2000. As some of the latter houses are part of the Minimalist movement and would be expected to have lower D values, this may not be as striking as it first appears, but over the complete set of eighty-five results, the trendline is still notable.

A second feature of the graph is that some architects clearly produced a range of different aesthetic expressions in a relatively short timeframe. Gray (1920s to 1930s), Eisenman (1960s to 1970s), Sejima and Atelier Bow-Wow (both 1990s to 2010) all show this tendency. However, despite such variation, if the results are considered in quartiles, many of these differences are diminished. For example, Wright produced five of the six houses which fall into the upper quartile range

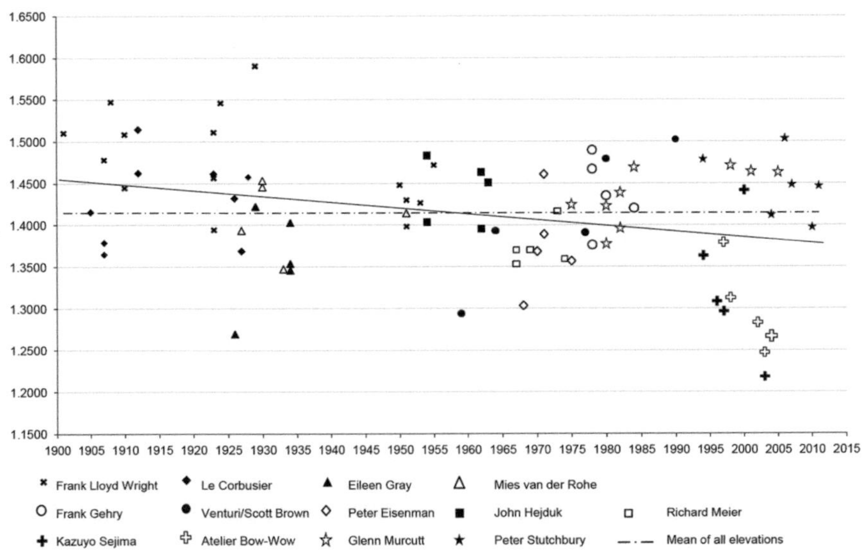

Fig. 12.1 Graph of the average elevation of each house over time

(most visually complex), and Wright's results for the Prairie Style and Textile-block houses skew the quartiles for the entire set, being consistently amongst the most complex buildings analysed in this book. The highest (μ_E) result in the entire set is for Wright's *Lloyd Jones House* ($\mu_E = 1.5906$), and Corbusier's Arts and Craft style *Villa Favre* ($\mu_E = 1.5143$), is the only design in the top quartile not by Wright. At the lower end of the spectrum, Gray, Eisenman, Meier, Sejima and Atelier Bow-Wow share eighteen of the twenty two results in the lowest quartile or least visually complex (Table 12.1).

Sejima's *House in a Plum Grove* ($\mu_E = 1.2183$) has the lowest average elevation dimension and her houses are also consistently amongst the lowest results in the complete set ($1.2183 < \mu_E < 1.3632$). Four of Atelier Bow-Wow's minimalist urban forms are in a similar range ($1.2473 < \mu_E < 1.3789$) and four of Meier's refined Avant-Garde houses are also in the lowest quartile, although more visually complex than the remainder of this category ($1.3532 < \mu_E < 1.4165$).

Table 12.1 Interquartile results, average complexity of elevations

		Lowest quartile	Interquartile range	Upper quartile
		$D < 1.36975$	$1.36975 < D < 1.51035$	$D > 1.51035$
Pre-modernism	Le Corbusier	1	3	1
Organic Modernism	Wright: Prairie Style	–	3	2
	Wright: Textile-block	–	2	3
	Wright: Usonian	–	5	–
Functional Modernism	Le Corbusier	1	4	–
	Gray	3	2	–
	Mies van der Rohe	1	4	–
Post-modernism	Venturi Scott Brown	1	4	–
	Gehry		5	–
Avant-garde and Abstraction	Hejduk		5	–
	Meier	4	1	–
	Eisenman	3	2	–
Minimalism and Regionalism	Murcutt: Early	–	5	–
	Murcutt: Late	–	5	–
	Sejima	4	1	–
	Atelier Bow-Wow	4	1	–
	Stutchbury	–	5	–
TOTALS		22	57	6
Proportion %		26 % Low visual complexity	67 % Average visual complexity	7 % High visual complexity

12.2.2 Average Plans (Including Roofs)

The average plan result (including roofs) is $D = 1.3380$, and the interquartile range is $1.31165 < \mu_P < 1.36805$. Whereas several outliers skew the quartiles in the elevation results, for plan results the data is closer to a 'standard normal' distribution (Table 12.2) with a marginally decreasing linear trendline (Fig. 12.2). The highest fractal dimension for plans is again for a house by Wright, this time the *Ennis House* ($\mu_P = 1.4810$), and again Wright's houses feature in the highest quartile (seven in total), although this time they are not alone, with another fifteen houses in that range, the most significant being the complete set of Meier's works ($1.3861 < \mu_P < 1.4152$). Of less consistency, in terms of plan data, are the designs of Hejduk, which include two houses in the top quartile, two in the lowest and only one, *House 4* ($\mu_P = 1.3397$), in the interquartile range. Hejduk's *House 5* ($\mu_P = 1.1629$) is also the design with the lowest average plan result in the entire set of eighty-five works. Another curiosity is that the plans of the Japanese Minimalists, which dominated the low complexity

Table 12.2 Interquartile results, average complexity of plans, including roofs

		Lowest quartile	Interquartile range	Upper quartile
		$D < 1.31165$	$1.31165 < D < 1.36805$	$D > 1.36805$
Pre-modernism	Le Corbusier	4	1	–
Organic Modernism	Wright: Prairie Style	–	3	2
	Wright: Textile-block	–	1	4
	Wright: Usonian	1	3	1
Functional Modernism	Le Corbusier	1	3	1
	Gray	3	2	–
	Mies van der Rohe	2	3	–
Post-modernism	Venturi Scott Brown	1	3	1
	Gehry	–	3	2
Avant-garde and Abstraction	Hejduk	2	1	2
	Meier			5
	Eisenman		3	2
Minimalism and Regionalism	Murcutt: Early	2	3	–
	Murcutt: Late	2	2	1
	Sejima	2	3	–
	Atelier Bow-Wow	–	5	–
	Stutchbury	–	4	1
TOTALS		20	43	22
Proportion %		24 % Low visual complexity	50 % Average visual complexity	26 % High visual complexity

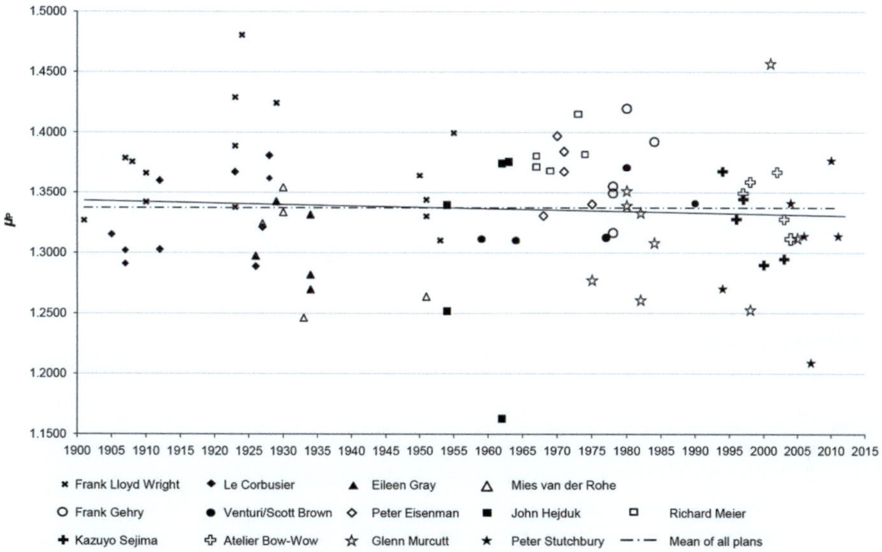

Fig. 12.2 Graph of the average plan of each house over time

quartile for the elevations, do not stand out in a similar way in the plan results. All of Atelier Bow-Wow's houses are in the interquartile range, as are three of Sejima's houses, while the other two have a lower level of complexity. It is Le Corbusier's Pre-Modern works that are predominantly in the lower band for plan complexity.

12.2.3 Average Plans (Excluding Roofs)

The set of data in this section does not include measures derived from the roof plan as part of the calculation of averages. Unlike the previous results (μ_P), those in this section ($\mu_{P\text{-}R}$) are solely concerned with the formal modelling of inhabitable spaces. Whereas the previous results included a roof plan that was often visually simple, lowering the average, in this section these are excluded, although this decision also means that single-storey houses now only possess an average of one result.

The average for all plans, excluding the roof, is $\mu_{P\text{-}R} = 1.3645$, which is slightly higher than the average including roofs. Likewise, the interquartile range is higher than for the plans with roofs, being $1.3396 < D < 1.3875$. Despite this, there is still a relatively normal distribution of the results, with the frequency across quartiles being similar. When comparing both plan type graphs, the graph without roofs included (Fig. 12.3) shows a more distinct clustering of data, with only two obvious outliers, the highest being Wright's *Ennis House* ($\mu_{P\text{-}R} = 1.4955$) and the lowest Hejduk's *House 5* ($\mu_{P\text{-}R} = 1.2427$). Interestingly, by excluding the roof plan, three of Hejduk's houses fall within the interquartile range; only *House 1* is in the lower quartile and

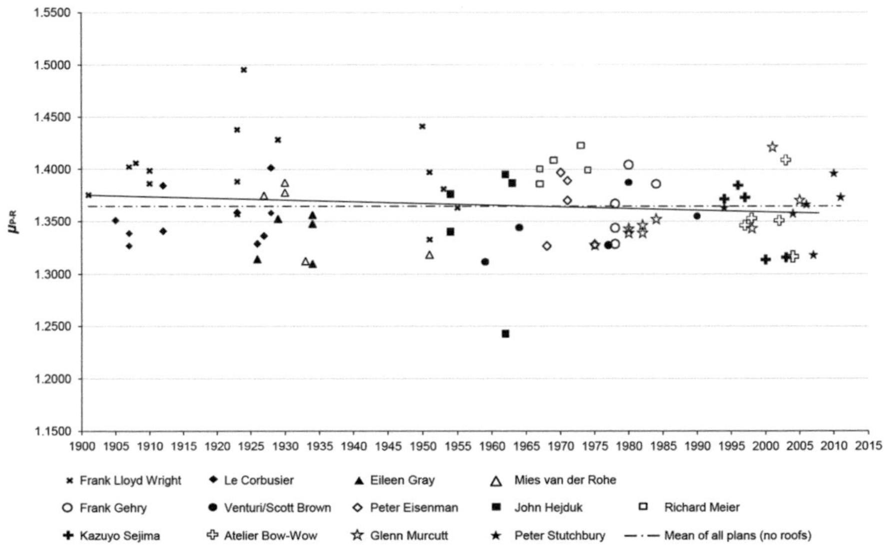

Fig. 12.3 Graph of the average plan of each house over time, excluding roof data

House 6 in the upper. Thus, Hejduk's plans are more consistent when the roof is excluded. Apart from Wright, Meier, and Murcutt's later work, all of the architects examined have at least one house in the lower quartile for $\mu_{P\text{-}R}$ data (Table 12.3).

12.2.4 Elevations and Plans Combined

Because the results in this section combine both plans and elevations, they could be thought of as either being a better reflection of the total character of the buildings they represent or as hiding a large number of subtleties and differences in the data. Because of this, these results are indeed robust in one sense, but they can be misleading to interpret without the aid of additional information.

The mean of all of the averages of the elevations plus plans (roofs included) is $D = 1.3821$, and the interquartile range is $1.3575 < D < 1.4145$ (Fig. 12.4). The spread of the data across quartiles is, once again, close to a standard distribution of results, and is also superficially closer to that found in the plans, than in the elevations (Table 12.4). Significantly, the Hejduk and early Murcutt sets of designs are all in the interquartile range; Stutchbury and Meier have all but one of their houses in the interquartile range, with the others, in both cases, in the upper quartile. All of the houses in the Mies set are in the interquartile range except for the *Lemke House,* which is less complex than the others. Wright's *Ennis House* ($\mu_{E+P} = 1.5243$) is the most complex again, and four out of five of his Textile-block houses and of his Prairie Style houses are all in the upper quartile, as are two of his Usonian houses.

Table 12.3 Interquartile results, average complexity of plans, excluding roofs

		Lowest quartile	Interquartile range	Upper quartile
		$D < 1.3396$	$1.3396 < D < 1.3875$	$D > 1.38755$
Pre-modernism	Le Corbusier	2	3	
Organic Modernism	Wright: Prairie Style	–	2	3
	Wright: Textile-block	–	1	4
	Wright: Usonian	1	2	2
Functional Modernism	Le Corbusier	2	2	1
	Gray	2	3	–
	Mies van der Rohe	2	2	1
Post-modernism	Venturi Scott Brown	2	2	1
	Gehry	1	3	1
Avant-garde and Abstraction	Hejduk	1	3	1
	Meier	–	1	4
	Eisenman	2	1	2
Minimalism and Regionalism	Murcutt: Early	1	4	–
	Murcutt: Late	–	4	1
	Sejima	2	3	–
	Atelier Bow-Wow	1	3	1
	Stutchbury	1	3	1
TOTALS		20	42	23
Proportion %		24 % Low visual complexity	49 % Average visual complexity	27 % High visual complexity

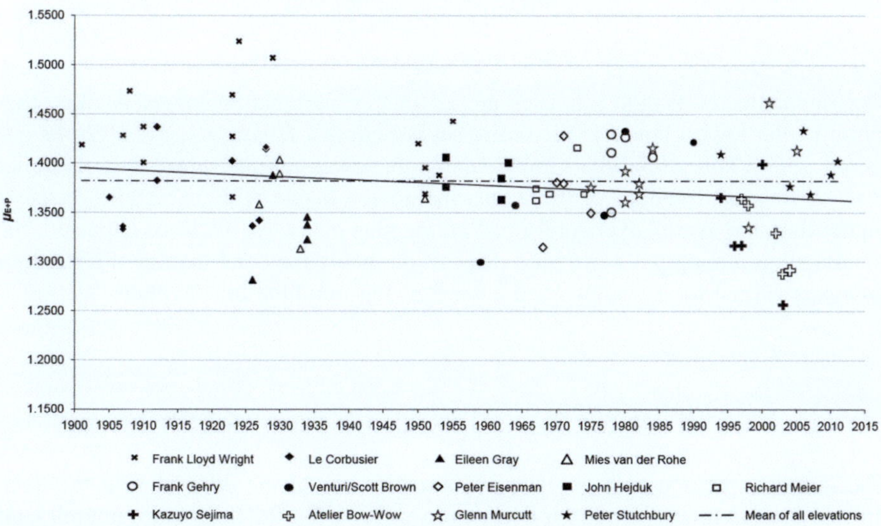

Fig. 12.4 Graph of the average plan + elevation of each house over time

Table 12.4 Interquartile results, average complexity of plans and elevations combined

		Lowest quartile	Interquartile range	Upper quartile
		$D < 1.3575$	$1.3575 < D < 1.4145$	$D > 1.4145$
Pre-modernism	Le Corbusier	2	2	1
Organic Modernism	Wright: Prairie Style	–	1	4
	Wright: Textile-block	–	1	4
	Wright: Usonian	–	3	2
Functional Modernism	Le Corbusier	2	2	1
	Gray	4	1	–
	Mies van der Rohe	1	4	–
Post-modernism	Venturi Scott Brown	2	1	2
	Gehry	1	2	2
Avant-garde and Abstraction	Hejduk	–	5	–
	Meier	–	4	1
	Eisenman	2	2	1
Minimalism and Regionalism	Murcutt: Early		5	–
	Murcutt: Late		3	2
	Sejima	4	1	–
	Atelier Bow-Wow	4	1	–
	Stutchbury	–	4	1
TOTALS		22	42	21
Proportion %		26 % Low visual complexity	49 % Average visual complexity	25 % High visual complexity

The remainder of Wright's houses are all in the interquartile range. At the other extreme, the lowest house in the entire set is Sejima's *House in a Plum Grove* ($\mu_{E+P} = 1.2567$). Four of Sejima's houses are in the lower quartile, as are four of Atelier Bow-Wow's, and neither of these sets includes a house with a fractal dimension higher than the overall average for elevations plus plans. As might be expected, the trendline has a marginal downward disposition, being the combination of a stronger downward trend for elevations and a weaker, flat trendline for the plans.

12.2.5 Elevation Ranges

The range, expressed as a percentage, is a measure of the difference in the data, from highest to lowest result, for every house. However, the range does not tell us if a building is visually complex or not; it only indicates how diverse or inconsistent the results are. A low R value implies that each of the elevations or plans in a building have a high degree of formal similarity or consistency. Conversely, a high R value confirms that at least one of the elevations or plans, in terms of formal similarity, diverges from the others. Whereas range is a measure of the degree of difference in a set, the standard deviation reflects the clustering of data within the range. In combination, the range and standard deviation can be used to interpret the degree to which there is variation in the results for specific houses or architect's works.

The average range for all elevations is 11.40 % and the interquartile range is $6.54 < R_{E\%} < 15.22$, meaning that if a house has a range below $R_{E\%} = 6.54$ its elevations are more consistent (visually similar to each other), while if the range is above $R_{E\%} = 15.22$, its elevations are inconsistent (visually dissimilar to each other). The individual design with the most visually similar elevations is the *Ball-Eastaway House* in the set of Glenn Murcutt's early designs, with a remarkable range of 0.77 % across its four elevations. However, it is the set of Hejduk's houses which have the overall highest level of consistency, with four of his houses falling in the lowest quartile (most similar). Thus, despite varying levels of absolute elevation complexity (as seen in the previous section), across his complete set, within each and every house Hejduk's elevations are remarkably consistent. In contrast, the least consistent house, in terms of elevation expression, is Sejima's *Y-House* ($R_{E\%}$ = 32.94 %); three of her houses are in the upper (inconsistent) quartile, and two in the interquartile range, meaning that none of her houses are strikingly consistent across all elevations. The sets for later-career Murcutt, Meier, and Wright's Textile-block houses are the only others with more than half of their members in the inconsistent quartile. Conversely, the sets of Wright's Prairie Style houses, Venturi and Scott Brown, and Hejduk are the only ones with a majority of houses in the consistent quartile (Fig. 12.5, Table 12.5).

12.2.6 Plan Ranges (Including Roofs)

The average range for all plans including roofs is 11.78 % (Fig. 12.6). Houses of high visual consistency in plan expression have a range below 6.8 %, while the less consistent have a range over 15.77 %, with the interquartile band falling between these values (Table 12.6). These figures are all similar to the range and quartile limits for elevations.

In these results Wright's work stands out for the evenness or regularity of his planning, with four of the five Textile-block plans in the lower quartile, indicating greater consistency. All of the Usonians are in the interquartile range and only the Prairie Style houses had two representatives in the upper quartile range (the

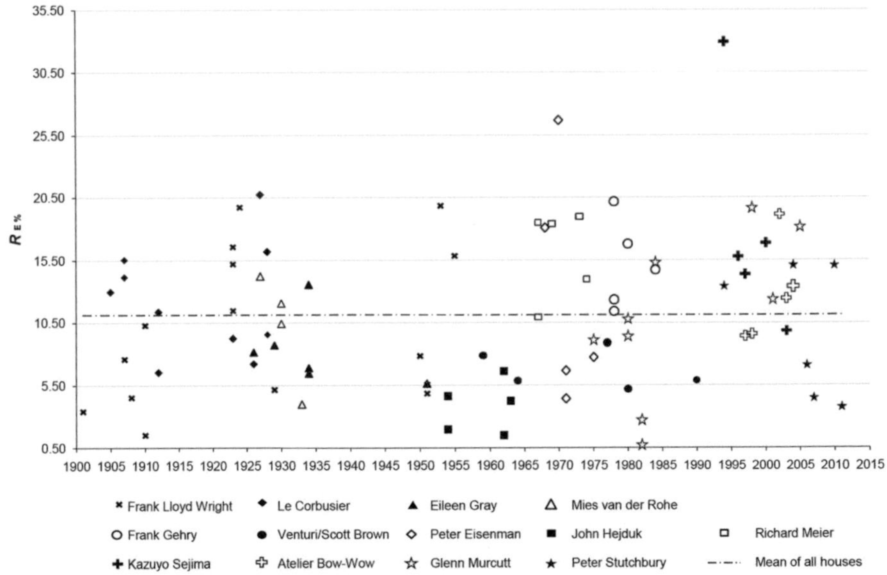

Fig. 12.5 Graph of the ranges of average elevations of each house over time

Henderson and *Zeigler* houses). Only Le Corbusier's Pre-Modern set of houses are contained entirely in the upper quartile, showing these houses possess a high degree of diversity in their planar visual expressions.

12.2.7 Plan Ranges (Excluding Roof)

The plan range in this section is a measure of the difference between at least two plans (not including roofs) in a design. However, because some houses only have one floor level, there may be only one image to analyse in the plan set, and therefore we have treated this as a 0 result, rather than excluding the house from the set. The inclusion of the 0 results changes the overall trends, reducing the average range to 4.72 %, and the interquartile to $0.45 < R_{E\%} < 7.27$ (Fig. 12.7). Because of the impact of this effect, all of Murcutt's late houses and four of Wright's Usonian houses, which are predominantly single-storey, cannot really be considered in this data. For the other sets, there are some changes from the data with the roof included, such as the Japanese Minimalists having all of their houses in the interquartile range and with no architect having a set that falls strongly in the upper or lower quartiles, except for the single level residences (Table 12.7).

Table 12.5 Interquartile results, average elevation range

		Lowest quartile	Interquartile range	Upper quartile
		$R_{E\%} < 6.54$	$6.54 < R_{E\%} < 15.22$	$R_{E\%} > 15.22$
Pre-modernism	Le Corbusier	–	4	1
Organic Modernism	Wright: Prairie Style	3	2	–
	Wright: Textile-block	1	1	3
	Wright: Usonian	2	1	2
Functional Modernism	Le Corbusier	–	3	2
	Gray	1	4	–
	Mies van der Rohe	2	3	–
Post-modernism	Venturi Scott Brown	3	2	–
	Gehry	–	3	2
Avant-garde and Abstraction	Hejduk	4	1	–
	Meier	–	2	3
	Eisenman	1	2	2
Minimalism and Regionalism	Murcutt: Early	2	3	–
	Murcutt: Late	–	2	3
	Sejima	–	2	3
	Atelier Bow-Wow	–	4	1
	Stutchbury	2	3	–
TOTALS		21	42	22
Proportion %		24 % High consistency results	49 % Average consistency results	27 % Low consistency results

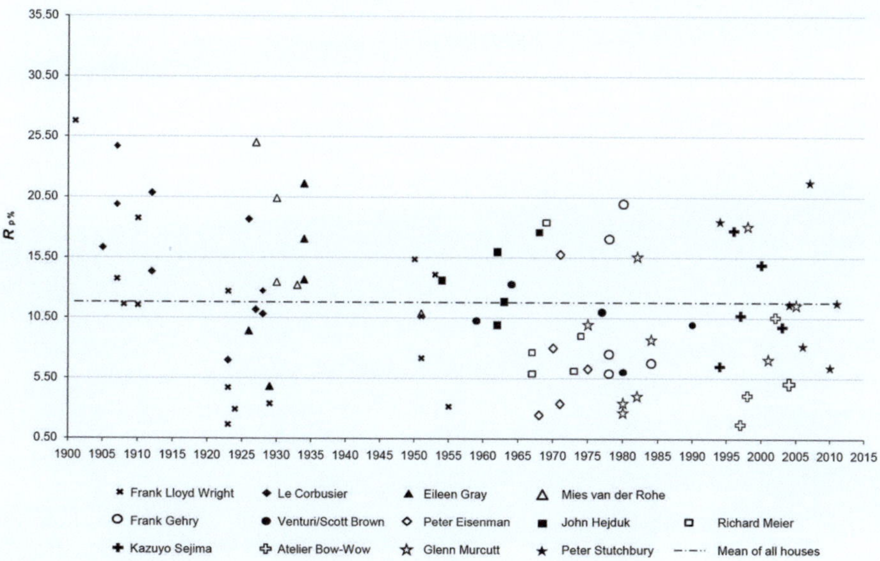

Fig. 12.6 Graph of the ranges of plan elevations of each house over time

Table 12.6 Interquartile results, average plan range including roofs

		Lowest quartile	Interquartile range	Upper quartile
		$R_{P\%} < 6.80$	$6.80 < R_{P\%} < 15.77$	$R_{P\%} > 15.77$
Pre-modernism	Le Corbusier	–	–	5
Organic Modernism	Wright: Prairie Style	–	3	2
	Wright: Textile-block	4	1	–
	Wright: Usonian	–	5	–
Functional Modernism	Le Corbusier	–	4	1
	Gray	1	2	2
	Mies van der Rohe	–	3	2
Post-modernism	Venturi Scott Brown	1	4	–
	Gehry	3		2
Avant-garde and Abstraction	Hejduk	–	3	2
	Meier	3	1	1
	Eisenman	3	2	–
Minimalism and Regionalism	Murcutt: Early	3	2	–
	Murcutt: Late	–	3	2
	Sejima	1	3	1
	Atelier Bow-Wow	3	1	1
	Stutchbury	1	3	1
TOTALS		23	40	22
Proportion %		27 % High consistency results	47 % Average consistency results	26 % Low consistency results

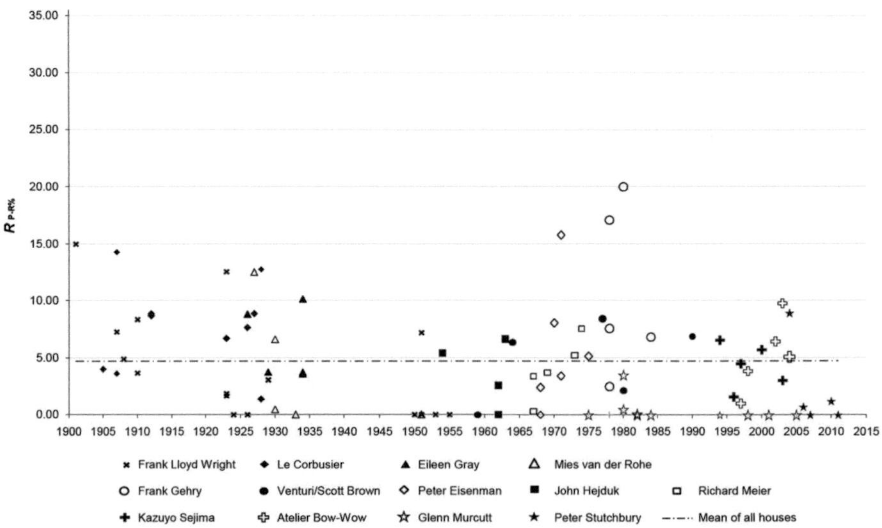

Fig. 12.7 Graph of the ranges of plans of each house over time, excluding roofs

Table 12.7 Interquartile results, average plan range, excluding roofs

		Lowest quartile	Interquartile range	Upper quartile
		$R_{P\text{-}R\%} < 0.45$	$0.45 < R_{P\text{-}R\%} < 7.27$	$R_{P\text{-}R\%} > 7.27$
Pre-modernism	Le Corbusier	–	2	3
Organic Modernism	Wright: Prairie Style	–	3	2
	Wright: Textile-block	1	3	1
	Wright: Usonian	4	1	–
Functional Modernism	Le Corbusier	–	2	3
	Gray	–	3	2
	Mies van der Rohe	3	1	1
Post-modernism	Venturi Scott Brown	–	3	2
	Gehry	–	2	3
Avant-garde and Abstraction	Hejduk	2	3	–
	Meier	1	3	1
	Eisenman		3	2
Minimalism and Regionalism	Murcutt: Early	3	2	–
	Murcutt: Late	5	–	–
	Sejima	–	5	–
	Atelier Bow-Wow	–	5	–
	Stutchbury	2	2	1
TOTALS		21	43	21
Proportion %		25 % High consistency results	50 % Average consistency results	25 % Low consistency results

12.3 Stylistic Period

The approach taken to the data in the preceding section was predicated on developing a chronological analysis of every house, by every architect or practice, sorted by the seven major fractal measures. In the present section, the fractal measures for a set of five houses by an architect (or a set of houses from specific periods in the case of Wright, Le Corbusier and Murcutt) are combined to present a more holistic result for the geometric properties of an architectural movement or style. In this section data is graphed for both plans and elevations on the same chart, starting with a consideration of average dimensions, then average ranges and finally standard deviations. For all three of these charts, the results are also tabulated in quartiles by plan and elevation for each stylistic set.

In this section the *x*-axes in the charts are broadly arrayed chronologically with the earliest to the left of the chart and the more recent to the right, but there is a substantial overlap in timing between several styles. For example, the organic and

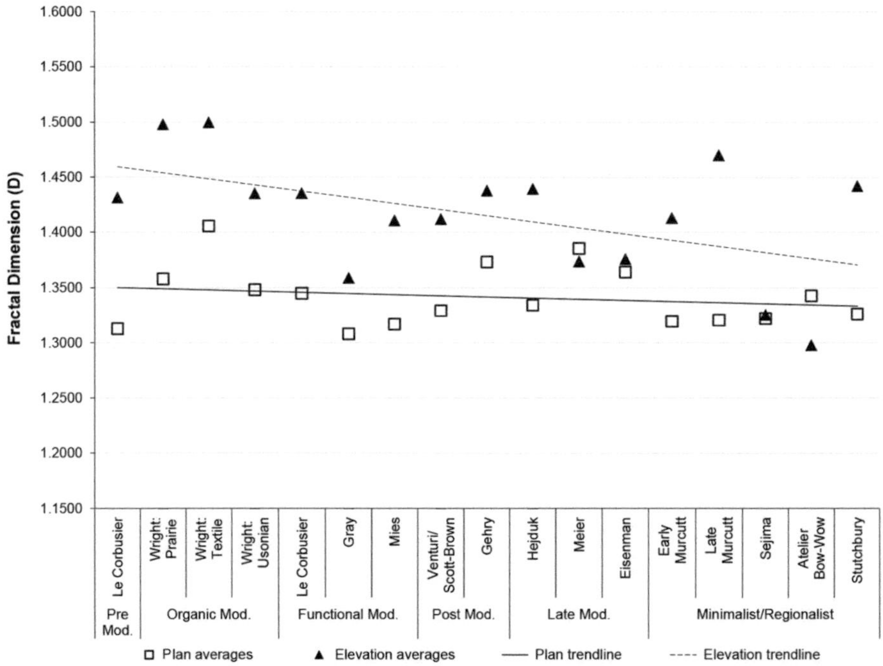

Fig. 12.8 Graph of the average plans and elevations for each architect's set

functionalist variations of Modernity occurred simultaneously in different parts of the word, and there are other overlaps between the Avant-Garde and Post-Modernism and early Murcutt and Post-Modernism. The order chosen here is the one conventionally adopted in architectural histories to describe the sequence of development of these different movements and approaches. Thus, the longitudinal trends in the graphs are not so much historical as historiographical, with the data sequence reflecting the post-rationalised constructions of scholars. This way of presenting the data is most useful for analysing claims about the degree to which fractal dimensions can be used to differentiate stylistic movements or periods.

12.3.1 Averages

The separate average results for elevations and plans show that the visual complexity of the elevations is typically 7.9 % higher than that of plans (Fig. 12.8). The exceptions to this general rule, that elevations are more visually complex than plans, are found in the works of the Avant-Garde and the Minimalists. In particular, the mean elevation results for Meier's houses are similar to that of his plans, being only 1.16 % ($D_{RE} = 0.0116$) less complex and, coincidentally, Eisenman's elevations are only incrementally higher than his plans by exactly the same amount (1.16 %). The results for Sejima's elevations are barely higher than her plans (a difference of +0.0034 or 0.3 %) and the elevations of Atelier Bow-Wow are the most divergent in the set, falling below their plan complexity by 0.0449 (4.49 %). However, as the analysis of formal cohesion later in this chapter reveals, just because Meier and Sejima, for example, have similar average plan and elevation results across their sets of designs, this does not mean that each is consistent in their approach to design. This feature starts to become apparent in the arrangement of these results in the comparative quartiles, where only Wright's and Gray's works have both plans and elevations together in the same quartile, respectively the upper and lower bands (Table 12.8).

What is most striking in this set of results is what is missing: a clear mathematical difference between the stylistic sets. For example, Murcutt's early and late architecture, both of which are regarded as emphatically Regionalist, vary greatly in their levels of visual expression. Similarly, Hejduk and Meier's architecture, both famous examples of Abstraction and Avant-Garde, also differ in this way. Le Corbusier and Mies vary from each other in terms of both plan and elevation results

Table 12.8 Interquartile results, average plans and elevations for each architect's set

		Lowest quartile	Interquartile range	Upper quartile
		$\mu_{[E]} < 1.3757$	$1.3757 < \mu_{[E]} < 1.4392$	$\mu_{[E]} > 1.4392$
		$\mu_{[P]} < 1.3206$	$1.3206 < \mu_{[P]} < 1.3578$	$\mu_{[P]} > 1.3578$
Pre-modernism	Le Corbusier	P	E	–
Organic Modernism	Wright: Prairie Style	–	–	E, P
	Wright: Textile-block	–	–	E, P
	Wright: Usonian	–	E, P	–
Functional Modernism	Le Corbusier	–	E, P	–
	Gray	E, P	–	–
	Mies van der Rohe	P	E	–
Post-modernism	Venturi Scott Brown	P	E	–
	Gehry	–	E	P
Avant-garde and Abstraction	Hejduk	–	P	E
	Meier	E	–	P
	Eisenman	E	–	P
Minimalism and Regionalism	Murcutt: Early	P	E	–
	Murcutt: Late	–	P	E
	Sejima	E	P	–
	Atelier Bow-Wow	E	P	–
	Stutchbury	–	P	E
Elevations		5	7	5
Plans		5	7	5
TOTAL		10 Low visual complexity	14 Average visual complexity	10 High visual complexity

and both are considered ideal examples of Functionalist Modernity. Most dramatically, there is great diversity across the three sets of Frank Lloyd Wright's designs. The most we can observe from these average results is that the Minimalists are, indeed, much lower in their visual expression than the average, and that Wright's two most decorative and visually rich periods, the Prairie Style and Textile-block houses, are at the higher range of complexity. Beyond that, the mean results alone are not useful for differentiating one architectural movement from another. There is simply too much variation in many individual houses (with often different client budgets, needs and siting challenges) to be completely consistent. However, the range and standard deviation results might be useful when interpreting the mean data.

12.3.2 Ranges

The range results for these stylistic sets could be considered a measure of inconsistency. On this basis, the inconsistency in the elevations and plans does tend to increase across the final styles (Fig. 12.9). When we break this result down into interquartile ranges, it can be seen that most ranges of results are in the interquartile range, with many of the earlier styles having both the elevation ranges and plan

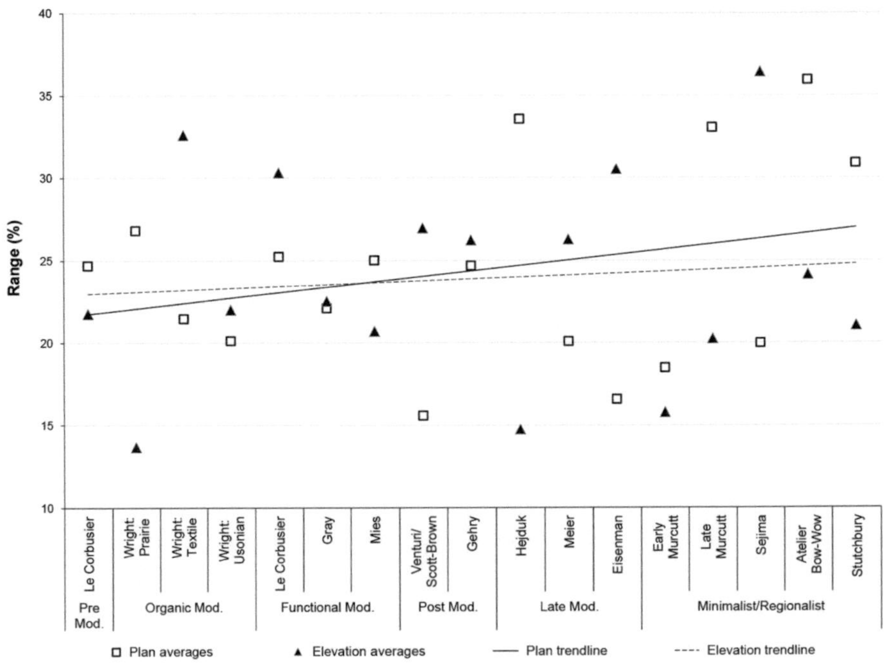

Fig. 12.9 Graph of the range of plans and elevations for each architect's set

Table 12.9 Interquartile results range of plans (P) and elevations (E) for each architect's set

		Lowest quartile	Interquartile range	Upper quartile
		$R_{\{E\%\}} < 20.47$	$20.47 < R_{\{E\%\}} < 28.63$	$R_{\{E\%\}} > 28.63$
		$R_{\{P\%\}} < 20.02$	$20.02 < R_{\{P\%\}]} < 28.86$	$R_{\{P\%\}} > 28.86$
Pre-modernism	Le Corbusier	–	E, P	–
Organic Modernism	Wright: Prairie Style	E	P	–
	Wright: Textile-block	–	P	E
	Wright: Usonian	–	E, P	–
Functional Modernism	Le Corbusier	–	P	E
	Gray	–	E, P	–
	Mies van der Rohe	–	E, P	–
Post-modernism	Venturi Scott Brown	P	E	–
	Gehry	–	E, P	–
Avant-garde and Abstraction	Hejduk	E	–	P
	Meier	–	E, P	–
	Eisenman	P	–	E
Minimalism and Regionalism	Murcutt: Early	E, P	–	–
	Murcutt: Late	–	E	P
	Sejima	–	P	E
	Atelier Bow-Wow	–	E	P
	Stutchbury	–	E	P
Elevations		3	10	4
Plans		3	10	4
TOTAL		6 High consistency results	20 Average consistency results	8 Low consistency results

ranges in the central band (Table 12.9). Only Murcutt's early work is extremely consistent, with the rest of the Minimalist, Regionalist work in the interquartile and upper quartile range, showing less consistency between the houses in their sets.

12.3.3 Standard Deviation

Whereas the range is the difference between the highest and lowest (in this case, average) result in a set, the standard deviation is a measure of the amount of variation in the complete set of data, not just a consideration of the top and bottom. Like the range chart (Fig. 12.9), the standard deviation chart (Fig. 12.10) shows a gradual rise in the results over time, signalling a general increase in variation in the

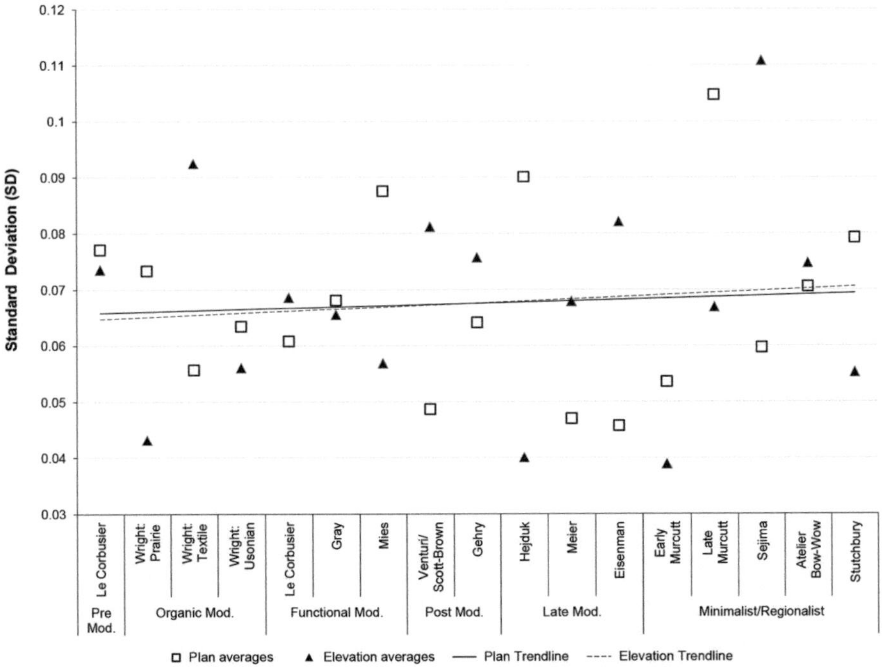

Fig. 12.10 Graph of the standard deviation of plans and elevations for each architect's set

data. When tabulated in quartiles (Table 12.10) no individual architect has both plans and elevations in the most variable (highest) quartile, but the plans and elevations of Murcutt's early work fall within the lowest quartile for least variable results. Several architects' works are in the interquartile range for both their plans and elevations including Le Corbusier's modern houses, and the works of Gray, Gehry and Atelier Bow-Wow. These sets have a consistent degree of variation in both plan and elevation results. In contrast, Stutchbury's elevations are in the least variable quartile and his plans are in the most, whereas Venturi and Scott Brown's results are the reverse of this situation.

12.4 Formal Coherence

The process of comparing the average results for plans and elevations for a set of designs produces only a limited reading of the relationship between the two in an architect's work. For example, it was previously observed that Sejima has average elevation and plan results for her complete set of works which are of a similar *D* value (Fig. 12.8). Meier's results have a related pattern, albeit with a higher *D* value, as both plans and elevations have similar averages. Does this imply that for

Table 12.10 Interquartile results, standard deviation of plans and elevations for each architect's set

		Lowest quartile	Interquartile range	Upper quartile
		$std_{\{E\}} < 0.0556$	$0.0556 < std_{\{E\}} < 0.0784$	$std_{\{E\}} > 0.0784$
		$std_{\{P\}} < 0.0546$	$0.0546 < std_{\{P\}} < 0.0781$	$std_{\{P\}} > 0.0781$
Pre-modernism	Le Corbusier	–	E, P	–
Organic Modernism	Wright: Prairie Style	E	P	–
	Wright: Textile-block	–	P	E
	Wright: Usonian	–	E, P	–
Functional Modernism	Le Corbusier	–	E, P	–
	Gray	–	E, P	–
	Mies van der Rohe	–	E	P
Post-modernism	Venturi Scott Brown	P	–	E
	Gehry	–	E, P	–
Avant-garde and Abstraction	Hejduk	E	–	P
	Meier	P	E	–
	Eisenman	P	–	E
Minimalism and Regionalism	Murcutt: Early	E, P	–	–
	Murcutt: Late	–	E	P
	Sejima	–	P	E
	Atelier Bow-Wow	–	E, P	–
	Stutchbury	E	–	P
Elevations		4	9	4
Plans		4	9	4
TOTAL		8 Low variability of data	14 Average variability of data	10 High variability of data

both of these architects, the elevations of their designs and their plans typically resemble one another? As we will see in this section, this is often true of Meier, but not Sejima. The extreme ranges in many of Sejima's houses are, to a large extent, balanced out when they are averaged together. In contrast, Meier's works are consistent in the relationships he created between elevations and plans. These properties can be uncovered by examining formal coherence across each set of works.

Formal coherence is a measure of the degree of correlation (R^2) between the elevations (x-axis) and plans (y-axis) in each set of designs. A high R^2 value indicates that there is a greater degree of correlation between the visual properties of a set of plans and elevations, a low result means that there is little or no correlation, and a negative result means an inverse correlation, where plans rather than elevations, consistently dominate the relationship (Table 12.11).

In the category, where there is a clear positive correlation between elevations and plans, nine sets of designs are featured. Richard Meier's works display the highest

Table 12.11 Formal coherence, rank order correlation of elevation to plan by set

Correlation	Set	R^2
Positive	Meier	0.828
	Wright: Usonian	0.690
	Wright: Textile-block	0.673
	Mies van der Rohe	0.618
	Venturi Scott Brown	0.574
	Eisenman	0.460
	Murcutt: Early	0.435
	Hejduk	0.427
	Murcutt: Late	0.388
No	Atelier Bow-Wow	0.208
	Le Corbusier: Modern	0.103
	Wright: Prairie Style	0.085
	Stutchbury	0.066
	Gehry	0.033
	Sejima	0.000
Inverse	Le Corbusier: Pre-modern	−0.512
	Gray	−2.666

results (R^2 = 0.8279), followed by Wright's Usonian and Textile-block houses and then the works of Mies van der Rohe. What this means in a descriptive sense is that the designs in these four sets all feature relatively high *D* results for both elevations and plans, and in individual houses these are often relatively close together. This property is readily apparent in Meier's house designs and in Wright's Textile-block works, both of which feature plans that loosely resemble their elevations.

In contrast, the works of Atelier Bow-Wow, Le Corbusier (Modern), Wright (Prairie Style), Stutchbury, Gehry and Sejima, show no strong correlation between plans and elevations. This in itself can be part of the architect's personal signature approach to form. For example, Le Corbusier's Modernist elevations are often heavily modelled while his plans are relatively open and less intricately defined. This result (R^2 = 0.1028) isolates a particular characteristic of Le Corbusier's design signature, with the lack of correlation being for him a recurring feature. Similarly, the result for Sejima is almost 0 (R^2 = 0.00016), even more so than for Le Corbusier, demonstrating a lack of connection between the character of elevations and plans, despite having similar average results across the entire set for both.

There are two architects whose works demonstrate the strength and consistency of their plans, rather than their elevations. Gray's architecture displays a very strong inverse correlation (R^2 = −2.666), meaning that her plans are consistently more detailed and formally complex than her elevations. This can be noticed qualitatively from a review of her collection of drawings, which contain many detailed and annotated plans, and it could be considered that she designed predominantly in plan form. Le Corbusier's Pre-Modern works display a similar tendency, although not as strongly as those of Gray.

12.5 Complexity and Consistency

The further way of analysing this data is to chart it simultaneously using two different characteristics, in this case average fractal dimensions (separately for plans and elevations) and range (also separately for plans and elevations). Using three bands it is possible to create a matrix chart comparing visual complexity with visual consistency for each set. For such an approach, complexity is measured using fractal dimensions (D) and consistency using range (R). The three bands are differentiated as the lowest quartile ($<$IQR), the interquartile range (IQR) and the highest quartile ($>$IQR). Four examples of how the location of a set of works in such a matrix is interpreted are as follows (Table 12.12).

First, the set of Sejima's elevations have low visual complexity, but high variability or inconsistency, placing them into the $D < \text{IQR}$ *and* $R > \text{IQR}$ sector of the matrix (the bottom-right in Fig. 12.11). Second, Hejduk's elevations have a typical or mid-level of visual complexity and his elevation averages fall into a tight range (the centre-left cell in Fig. 12.11). The set of Murcutt's early works feature plans with a low level of overall visual complexity, but where all plans have a highly consistent character. Therefore, Murcutt's early plans fall into the $D < \text{IQR}$ *and* $R < \text{IQR}$ sector of the matrix (bottom-left cell in Fig. 12.12). Finally, the set of Eisenman's plans have high overall levels of visual complexity and high levels of consistency across these plans (that is, the range is low) which falls into the $D > \text{IQR}$ *and* $R < \text{IQR}$ sector of the matrix (top-left cell in Fig. 12.12).

What is interesting about these results is that, in terms of complexity and consistency, many of the architects can be isolated from each other and differentiated mathematically in a useful way. Even the middle cell of the matrix in the plan chart (meaning typical complexity and typical consistency) has only two sets within its bands, and the elevations in the same mid-zone have five sets. Notably, there are several sectors in the chart where none of the sets are located. For example, in the elevation chart, none of the sets have low overall complexity and high consistency (Fig. 12.11, bottom-left cell), and in the plan results, there are two cells in the matrix where no data has been mapped, both of which are for inconsistent visual expressions (high R results) and either high or low visual complexity (respectively, top-right and bottom-right cells in Fig. 12.12). That is, the only inconsistent plan treatments are in the centre category for complexity.

Like the results for formal coherence, these combined results for complexity and consistency offer a better means of classifying and differentiating architectural approaches than using raw or mean D results. They are also seemingly informative when considering stylistic movements, although a larger body of data is required to fully test this idea (Tables 12.13 and 12.14). For example, when considering plans alone, the Regionalist works are clustered completely within two cells in the matrix ($D < \text{IQR}$ and $R < \text{IQR}$, or bottom-left cell: $D = \text{IQR}$ and $R > \text{IQR}$, or centre-right), both of which they dominate, with 100 % in the former, and 50 % in

Table 12.12 Interpreting the matrix chart of combined complexity and consistency bands

D	*R%*	Location	Interpretation of Results	Elevations	Plans
>IQR	<IQR	Top-left	Buildings with high overall visual complexity and where the plans/elevations have a consistent visual character	Wright: Prairie Murcutt: Late	Eisenman
>IQR	IQR	Top-centre	Buildings with high overall visual complexity and where the plans/elevations have a typical range of visual character	Stutchbury	Wright: Textile-block Gehry Meier
>IQR	>IQR	Top-right	Buildings with high overall visual complexity and where the plans/elevations have an inconsistent visual character	Wright: Textile-block	–
IQR	<IQR	Centre-left	Buildings with typical visual complexity and where the plans/elevations have a consistent visual character	Hejduk Murcutt: Early	Venturi Scott Brown Sejima
IQR	IQR	Centre-centre	Buildings with typical visual complexity and where the plans/elevations have a typical range of visual character	Le Corbusier: Pre Modern; Mies Wright: Usonian Venturi Scott Brown Gehry	Wight: Prairie Le Corbusier: Modern
IQR	>IQR	Centre-right	Buildings with typical visual complexity and where the plans/elevations have an inconsistent visual character	Le Corbusier: Modern Eisenman	Hejduk Stutchbury Murcutt: Late Bow-Wow
<IQR	<IQR	Bottom-left	Buildings with low overall visual complexity and where the plans/elevations have a consistent visual character	–	Murcutt: Early
<IQR	IQR	Bottom-centre	Buildings with low overall visual complexity and where the plans/elevations have a typical range of visual character	Gray Meier Bow-Wow	Le Corbusier: Pre-Modern; Mies Gray
<IQR	>IQR	Bottom-right	Buildings with low overall visual complexity and where the plans/elevations have an inconsistent visual character	Sejima	–

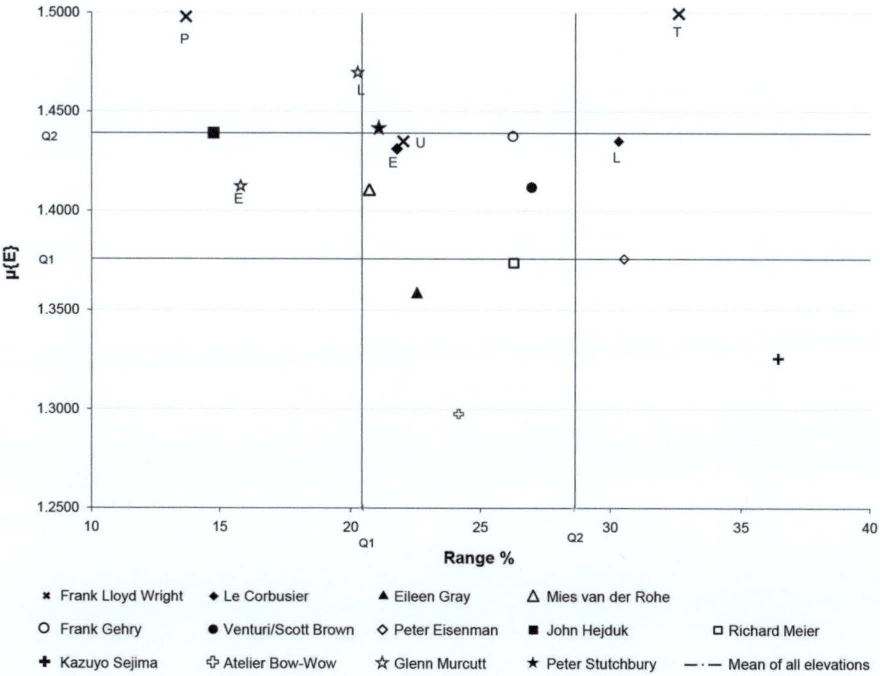

Fig. 12.11 Matrix chart of combined complexity (*D*) and consistency (*R*) bands for average elevations results

the latter. The Modernist movement, in both its Functionalist and Organic guises, also dominates the centre position (D = IQR and R = IQR, or centre-centre) in both plan and elevation results, with 100 % of the former and 60 % of the latter results. The reminder of the centre cell in the elevation results (where D = IQR and R = IQR, or centre-centre) is filled with 40 % Post-Modern designs, the only location in this matrix where they occur at all! While not a perfect set of results for differentiating style, this method suggests the first rigorous approach to the mathematical characterisation of style using fractal dimensions that begins to make sense, given an understanding of the designs themselves. The system of using average dimensions alone is clearly not, as previous sections demonstrate, sufficient to differentiate architectural styles. But this alternative system, combining R and D, for both plans and elevations, offers much greater potential.

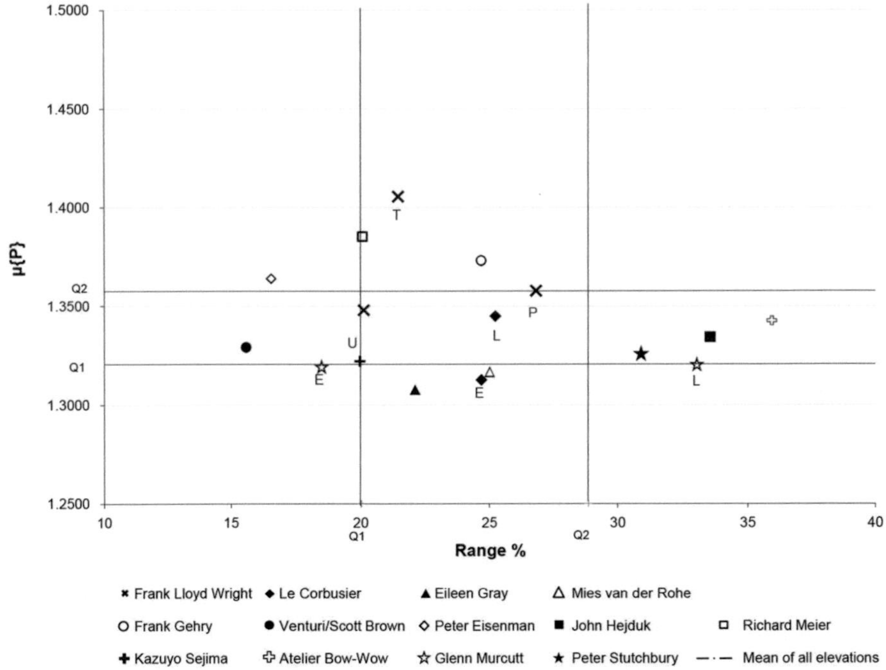

Fig. 12.12 Matrix chart of combined complexity (*D*) and consistency (*R*) bands for average plan results

Table 12.13 Combined complexity and consistency bands, elevation data sorted by style

Elevations	*R* < IQR	*R* = IQR	*R* > IQR
D > IQR	50 % Organicism	100 % Regionalism	100 % Organicism
	50 % Regionalism		
D = IQR	50 % Avant-garde	40 % Functionalism	50 % Functionalism
	50 % Regionalism	40 % Post-modernism	50 % Avant-garde
		20 % Organicism	
D < IQR	–	33 % Functionalism	100 % Minimalism
		33 % Avant-garde	
		33 % Minimalism	

Table 12.14 Combined complexity and consistency bands, plan data sorted by style

Plans	*R* < IQR	*R* = IQR	*R* > IQR
D > IQR	100 % Avant-garde	33 % Organicism	–
		33 % Post-modernism	
		33 % Avant-garde	
D = IQR	50 % Post-modernism	50 % Organicism	50 % Regionalism
	50 % Minimalism	50 % Functionalism	25 % Avant-garde
			25 % Minimalism
D < IQR	100 % Regionalism	66 % Functionalism	–
		33 % Pre-modernism	

12.6 Conclusion

At the start of this chapter we outlined three hypotheses that had previously been identified as points of contention or debate amongst architectural researchers.

The first hypothesis states that, *as the complexity of social groupings and functions contained within the home has reduced over time, the fractal dimensions of plans and elevations should decrease to reflect this change*. In the chronological analysis of the results some evidence to support this position was presented, with trendlines recording a gradual reduction in complexity of both plans and elevations over time. However, the least emphatic of these results was the one for plans without roofs, the very data set which might have been expected to reveal this answer most clearly. There are several ways of interpreting these results in the context of this hypothesis. The most obvious negative reading is that these results have arisen as a by-product of the designs chosen. For example, by choosing Regionalism and Minimalism as the final or most recent movements to examine, it might be argued that this decision had a direct influence on the chronological trendlines. However, Regionalism was not as lacking in visual complexity as first assumed and the Minimalist results had higher levels of plan complexity than anticipated. Conversely, most of the complex designs selected date to the first half of the twentieth-century, even though these were not designs overtly associated with a highly complex façade or plan expressions. For both of these reasons, the impact of the sample selection may not have been so significant as to undermine the validity of the chronological trendlines within the general limits of the sample. Thus, the data provides some evidence to support the first hypothesis, even though it is not completely satisfying.

The second hypothesis proposed that in architecture *each stylistic genre or movement possesses a distinct visual character that is measurable using fractal dimensions*. Past research, using only a small number of designs, has suggested that this case is either wholly true (Wen and Kao 2005), might be partially true (Bovill 1996) or seems unlikely (Ostwald et al. 2008). The average fractal dimension data, sorted by stylistic movements in this chapter, appears to completely undermine the argument that fractal dimensions, in isolation, allow a direct means of differentiating architectural styles. If one simple message has come out of the detailed analysis in this book, it is that the visual expression of style is not a straightforward mathematical measure. However, the *character* of a set of buildings does appear to be at least partially differentiable using a combination of average measures for complexity and either range or standard deviation. While not perfect, the combined complexity and consistency results do differentiate many of the architectural characteristics presented in this book, and in conjunction might take us a step closer to proving this hypothesis at some later stage. Thus, this second hypothesis is clearly rejected as it stands, but an alternative, more viable variant can be proposed which states that *combinations of measures derived using fractal analysis are useful for differentiating architectural character*.

The third hypothesis holds that *individual architects will present distinctive patterns of three-dimensional formal and spatial measures across multiple designs.* As with the second hypothesis, just using mean data, even for large sets of buildings, does not provide any direct method for differentiating the architectural approaches taken across the complete data set. Nevertheless, the formal coherence results do begin to single out several distinct approaches to three-dimensional form that are more nuanced and revealing than the raw data alone suggests. Once again, this is not a proof of the hypothesis, but the beginning of a new line of enquiry about the mathematical-visual properties of architecture, and the idea that a distinct geometric-formal signature might exist for an architect.

Finally, in this chapter, for the first time, we have compiled a series of ranges within which domestic architecture can be classified as being 'low', 'high' or 'average' complexity, and 'consistent', 'typical' or 'inconsistent' in terms of ranges. For future researchers measuring architecture using fractal dimensions, these results are useful for considering both individual buildings (Tables 12.15 and 12.16) and sets of buildings (Tables 12.17 and 12.18).

Table 12.15 Complexity bands for classifying a single building

	<IQR Low visual complexity	IQR Average visual complexity	>IQR High visual complexity
Average elevations	$\mu_E < 1.36975$	$1.36975 < \mu_E < 1.51035$	$\mu_E > 1.51035$
Average plans (incl. roof)	$\mu_P < 1.31165$	$1.31165 < \mu_P < 1.36805$	$\mu_P > 1.36805$
Average plans (excl. roof)	$\mu_{P\text{-}R} < 1.3396$	$1.3396 < \mu_{P\text{-}R} < 1.3875$	$\mu_{P\text{-}R} > 1.38755$
Average elevations and plans	$\mu_{E+P} < 1.3575$	$1.3575 < \mu_{E+P} < 1.4145$	$\mu_{E+P} > 1.4145$

Table 12.16 Consistency bands for classifying a single building

	<IQR Consistent results	IQR Typical level of consistency	>IQR Inconsistent results
Range % elevation	$R_{E\%} < 6.54$	$6.54 < R_{E\%} < 15.22$	$R_{E\%} > 15.22$
Range % plan (incl. roof)	$R_{P\%} < 6.80$	$6.80 < R_{P\%} < 15.77$	$R_{P\%} > 15.77$
Range % plan (excl. roof)	$R_{P\text{-}R\%} < 0.45$	$0.45 < R_{P\text{-}R\%} < 7.27$	$R_{P\text{-}R\%} > 7.27$

Table 12.17 Complexity bands for classifying a set of five or more buildings

	<IQR Low visual complexity	IQR Average visual complexity	>IQR High visual complexity
Mean elevations	$\mu_{[E]} < 1.3757$	$1.3757 < \mu_{[E]} < 1.4392$	$\mu_{[E]} > 1.4392$
Mean plans	$\mu_{[P]} < 1.3206$	$1.3206 < \mu_{[P]} < 1.3578$	$\mu_{[P]} > 1.3578$

Table 12.18 Consistency bands for classifying a set of five or more buildings

	<IQR Consistent results	IQR Typical level of consistency	>IQR Inconsistent results
Standard deviation Elev.	$std_{\{E\}} < 0.0556$	$0.0556 < std_{\{E\}} < 0.0784$	$std_{\{E\}} > 0.0784$
Standard deviation Plans	$std_{\{P\}} < 0.0546$	$0.0546 < std_{\{P\}} < 0.0781$	$std_{\{P\}} > 0.0781$
Range % elevations	$R_{\{E\%\}} < 20.47$	$20.47 < R_{\{E\%\}} < 28.63$	$R_{\{E\%\}} > 28.63$
Range % plans	$R_{\{E\%\}} < 20.02$	$20.02 < R_{\{E\%\}]} < 28.86$	$R_{\{E\%\}} > 28.86$

It is now up to future researchers to examine and extend the body of work that was tested in this book, to use this method to examine new ideas and to develop new applications. Future researchers will also add to this body of knowledge by challenging the ideas contained herein, by further refining the method and productively disagreeing with aspects of its premise, application or interpretation. That is the nature of research, and in the field of architecture, where cultural, social and philosophical issues shape design, the use of a mathematical system of analysis will always require rigour (in its application) and sensitivity (to interpret the results) first, before it can be useful for scholars and practitioners.

References

Akleman, Ergun, Paul Edmundson, and Ozan Ozener. 2005. A generalized vertex truncation scheme to construct intriguing polyhedral shapes. *Visual Mathematics 7*. http://www.mi.sanu.ac.rs/vismath/akleman. Accessed 8 October 2015.

Alber, Markus, and Joachim Peinke. 1998. Improved multi-fractal box-counting algorithm, virtual phase transitions, and negative dimensions. *Physical Review E* 57: 5489–5493. doi:http://dx.doi.org/10.1103/PhysRevE.57.5489.

Alexander, Christopher, Sara Ishikawa, and Murray Silverstein. 1977. *A pattern language: towns, buildings, construction*. New York: Oxford University Press.

Allen, Stan. 1996. Nothing but architecture. In *Hejduk's chronotope*, ed. K. Michael Hays, 79–96. New York: Princeton Architectural Press.

Allen, Stan. 1996. Recent projects: Kazuyo Sejima. *Assemblage* 30: 74–107.

Alofsin, Anthony. 1994. Frank Lloyd Wright and Modernism. In *Frank Lloyd Wright, architect*, ed. Terence Riley, Anthony Alofsin, Museum of Modern Art New York, and Frank Lloyd Wright Foundation, 32–57. New York: Museum of Modern Art.

Ando, Tadao. 2001. *Le Corbusier: Houses*. Tokyo: TOTO Shuppan.

Aoki, Jun. 2003. Kazuyo Sejima+ Ryue Nishizawa: A portrait. *GA Houses* 73: 158–64.

Appleton, Jay. 1975. *The experience of landscape*. London: Wiley.

Appleton, Jay. 1988. Prospects and refuges revisited. In *Environmental aesthetics: theory, research, and applications*, ed. Jack L. Nasar, 27–44. Cambridge, MA: Cambridge University Press.

Asvestas, Pantelis, George Matsopoulos, and Konstantina Nikita. 2000. Applications of fractal theory on medical data processing. In *Advanced Infrastructures for Future Healthcare*, ed. A. Marsh, L. Grandinetti, and T. Kauranne, 425–411. Amsterdam: IOS Press.

Atelier Bow-Wow. 2001a. *Made in Tokyo*. Tokyo: Kajima Instiute Publishing.

Atelier Bow-Wow. 2001b. *Pet architecture guide book*. Tokyo: World photo press.

Atelier Bow-Wow. 2006. *Bow Wow from post bubble city*. Tokyo: Inax.

Atelier Bow-Wow. 2007. *Graphic anatomy*. Tokyo: TOTO Publishing.

Atelier Bow-Wow. 2009. *Atelier Bow-Wow: Echo of space, space of echo*. Tokyo: Izumi Akiyama.

Atelier Bow-Wow. 2010. *Behaviorology*. New York: Rizzoli.

Atelier Bow-Wow. 2012. In praise of mud: A guide to the earth walls of Kyoto. Kyoto: Gakugei Shuppansha.

Atelier Bow-Wow. 2014. *Graphic anatomy 2*. Tokyo: TOTO Publishing.

Baker, Geoffery. 1995. *Design Strategies in Architecture*: *An Approach to the Analysis of Form*. New York: Van Nostrand Reinhold.

Baker, Geoffrey H. 1996. *Le Corbusier, the creative search: The formative years of Charles-Edouard Jeanneret*. New York: Van Nostrand Reinhold.

Balfour, Alan. 1980. John Hejduk. In *Contemporary architects*, ed. Muriel Emanuel, 353. London: Macmillan.

M.J. Ostwald and J. Vaughan, *The Fractal Dimension of Architecture*, Mathematics and the Built Environment 1, DOI 10.1007/978-3-319-32426-5

Baltanás, José. 2005. Walking through Le Corbusier: A tour of his masterworks. London: Thames & Hudson.

Barnhart, Robert K. 2000. *Dictionary of etymology*. Edinburgh: Chambers.

Barriere, Phillipe, Sylvia Lavin, Denise Scott Brown, and Robert Venturi. 1997. Interview with Denise Scott Brown and Robert Venturi. *Perspecta* 28: 127–145.

Barros Filho, Mauro, and Fabiano Sobreira. 2005. Assessing texture pattern in slum across scales: an unsupervised approach. *CASA Working Papers* 87: 1–12.

Batty, Michael, and Paul Longley. 1994. *Fractal cities: A geometry of form and function*. San Diego: Academic Press.

Bechhoefer, William, and Carl Bovill. 1994. Fractal analysis of traditional housing in Amasya, Turkey. *Traditional Dwellings and Settlements Working Paper Series* 61: 1–21.

Bechtel, Robert B., Robert W. Marans, and William Michelson, ed. 1987. *Methods in environmental and behavioral research*. Vol. vi. New York: Van Nostrand Reinhold Co.

Beck, Haig, and Jackie Cooper. 2002. *Glenn Murcutt: a singular architectural practice*. Mulgrave, VIC: Images Pub. Group.

Benguigui, Lucien, Daniel Czamanski, Maria Marinov, and Yuval Portugali. 2000. When and where is a city fractal? *Environment and Planning B: Planning and Design* 27: 507–519. doi:10.1068/b2617.

Ben Hamouche, Mustapha. 2009. Can chaos theory explain complexity in urban fabric? Applications in traditional muslim settlements. In *Nexus Network Journal*, 11:217–242.

Bhatia, Shashank, Stephan Chalup, and Michael J. Ostwald. 2013. Wayfinding: A method for the empirical evaluation of structural saliency using 3D Isovists. *Architectural Science Review* 56: 220–231.

Blake, Peter. 1966. *Mies van der Rohe: architecture and structure*. Baltimore: Penguin Books.

Blaser, Werner. 1965. *Mies van der Rohe; the art of structure*. New York: Praeger.

Blaser, Werner. 1996. *Richard Meier: details*. Basel: Birkhäuser.

Blenkinsop, Tom G, and David J Sanderson. 1999. Are gold deposits in the crust fractals? A study of gold mines in the Zimbabwe craton. *Geological Society, London, Special Publications* 155: 141–151.

Bletter, Rosemarie Haag. 1979. Review of Five Architects. Eisenman, Graves, Gwathmey, Hejduk, Meier; Five on Five. *Journal of the Society of Architectural Historians* 38: 205–207. doi:10.2307/989447.

Blumenson, John G. 1979. *Identifying American Architecture*: *A Pictorial Guide to Styles and Terms*, 1600–1945. New York : W.W. Norton & Co.

Bognár, Botond. 2008. Beyond the bubble: the new Japanese architecture. London: Phaidon Press.

Bois, Yve-Alain, Kimball Lockhart, and Douglas Crimp. 1981. The sculptural opaque. *SubStance* 10: 23–48. doi:10.2307/3684328.

Boldt, Douglas. 2002. Fractal architecture: shaping a new century. *Friends of Kebyar* 20.1: 4–16.

Bovill, Carl. 1996. Fractal geometry in architecture and design. Boston: Birkhäuser.

Bovill, Carl. 2009. The doric order as a fractal. *Nexus Network Journal* 10: 283–289.

Box, Hal. 2007. *Think like an architect*. Austin: University of Texas Press.

Brolin, Brent C. 1976. *The failure of modern architecture*. New York: Van Nostrand Reinhold.

Brown, Clifford T., and Walter R. T. Witschey. 2003. The fractal geometry of ancient Maya settlement. *Journal of Archaeological Science* 30: 1619–1632. doi: 10.1016/S0305-4403(03)00063-3.

Brown, Clifford T., Walter R. T. Witschey, and Larry S. Liebovitch. 2005. The broken past: Fractals in archaeology. *Journal of Archaeological Method and Theory* 12: 37–78. doi:10.1007/s10816-005-2396-6.

Buczkowski, Stéphane, Soula Kyriacos, Fahima Nekka, and Louis Cartilier. 1998. The modified box-counting method: Analysis of some characteristic parameters. *Pattern Recognition* 31: 411–418. doi:10.1016/S0031-3203(97)00054-X.

Burden, Ernest. 2000. *Elements of Architectural Design*. New York: Wiley.

Burkle-Elizondo, Gerardo. 2001. Fractal geometry in Mesoamerica. *Symmetry: Culture and Science* 12: 201–214.

Burkle-Elizondo, Gerardo, and Ricardo David Valdez-Cepeda. 2001. Do the Mesoamerican artistic and architectural works have a fractal dimension? In *Emergent Nature. Patterns, Growth and Scaling in the Sciences,* ed. Miroslav Michal Novak, 431–432. Singapore: World Scientific.

Caciagli, Costantino. 2001. On precision in architecture. *Nexus Network Journal* 3: 11–15.

Çagdaş, Gulen, Gaye Gözübüyük, and Özgür Ediz. 2005. Fractal based generative design for harmony between old and new. In *Generative Art 2005, GA2005, Milan, Italy, 15–17 December,* 150–159. Milan: Domus Argenia.

Calter, Paul. A. and Kim Williams. 2015. Measuring up to Michelangelo: A methodology. In *Architecture and Mathematics from Antiquity to the Future: Volume II: The 1500s to the Future*. ed. Kim Williams and Michael J. Ostwald. 151-164. Cham: Springer.

Calloway, Stephen, and Elizabeth Cromley, eds. 1996. *The Elements of Style: An Encyclopedia of Domestic Architectural Detail*. New York: Firefly Books.

Camastra, Francesco. 2003. Data dimensionality estimation methods: A survey. *Pattern Recognition* 36: 2945–2954. doi:10.1016/S0031-3203(03)00176-6.

Capo, Daniele. 2004. The fractal nature of the architectural orders. *Nexus Network Journal* 6: 30–40. doi:10.1007/s00004-004-0004-9.

Capron, Jean-Luc. 1993. "Why?": An interview with Philippe Samyn. *Space Design* 07: 89–91.

Cardillo, Alessio, Salvatore Scellato, Vito Latora, and Sergio Porta. 2006. Structural properties of planar graphs of urban street patterns. *Physical Review. E, Statistical, Nonlinear, and Soft Matter Physics* 73: 066107.

Cassarà, Silvio. 1997. *Richard Meier*. Barcelona: Gustavo Gili.

Cassarà, Silvio. 2006. *Peter Eisenman: feints*. London: Thames & Hudson.

Chalup, Stephan, Naomi Henderson, Michael J. Ostwald, and Lukasz Wiklendt. 2008. A method for cityscape analysis by determining the fractal dimension of its skyline. In *Innovation Inspiration and Instruction: New Knowledge in the Architectural Sciences: ANZAScA 08, Newcastle, Australia, 26-28 November,* 337–344. Newcastle, NSW: University of Newcastle.

Chalup, Stephan, Naomi Henderson, Michael J. Ostwald, and Lukasz Wiklendt. 2009. Box-counting dimension analysis of edge images detected with classical edge detector. *Architectural Science Review* 52: 126–134.

Chen, Susan S., James M. Keller, and Richard M. Crownover. 1993. On the calculation of fractal features from images. *IEEE, Transactions on Pattern Analysis and Machine Intelligence* 15: 1087–1090.

Chen, Wei, Xing-xin Chen, and Li Zhou. 2010. Box-counting dimension analysis of edge images detected with classical edge detector. In *2010 International Conference on Measuring Technology and Mechatronics Automation Volume 1, ICMTMA, Changsha City, China, 13-14 March*, 756–758. IEEE. doi:10.1109/ICMTMA.2010.491.

Chermayeff, Sam, Agustín Pérez Rubio, Tomoko Sakamoto, Luis Fernández-Galiano. 2007. *Houses: Kazuyo Sejima + Ryue Nishizawa, SANAA*. Barcelona: ACTAR.

Ching, Frank. 2007. *Architecture: form, space & order*. New York: John Wiley and Sons.

Clark, Roger H., and Michael Pause. 2012. *Precedents in Architecture: Analytic Diagrams, Formative Ideas, and Partis*. New Jersey: John Wiley & Sons

Coates, Paul, Tom Appels, Corinna Simon, and Christian Derix. 2001. Current work at CECA: Dust plates and blobs. In *Proceedings of 4th International Conference on Generative Art, GA2001,Milan, Italy, 12-14 December*, unpag. Milan: Politecnico di Milano.

Colomina, Beatriz. 2009. Mies's house : exhibitionism and collectionism. In *Mies Van Der Rohe Casas Houses*, ed. Moises Puente and Anna Puyuelo. 6–7. Barcelona: Editorial Gustavo Gili.

Constant, Caroline. 2000. *Eileen Gray*. London: Phaidon.

Constant, Caroline, Wilfried Wang, Rosamund Diamond, Marietta Andreas, Brooke Hodge. 1996. *Eileen Gray: an architecture for all senses*. Tübingen: Ernst J. Wasmuth.

Cooper, Jon. 2003. Fractal assessment of street-level skylines: A possible means of assessing and comparing character. *Urban Morphology* 7: 73–82.

Cooper, Jon, and Reza Oskrochi. 2008. Fractal analysis of street vistas: A potential tool for assessing levels of visual variety in everyday street scenes. *Environment and Planning B: Planning and Design* 35: 349–363.

Cooper, Jon, David Watkinson, and Reza Oskrochi. 2010. Fractal analysis and perception of visual quality in everyday street vistas. *Environment and Planning B: Planning and Design* 37: 808–822.

Cooper, Jon, Mei-lin Su, and Reza Oskrochi. 2013. The influence of fractal dimension and vegetation on the perceptions of streetscape quality in Taipei: with comparative comments made in relation to two British case studies. *Environment and Planning B: Planning and Design* 40: 43–62. doi:10.1068/b38010.

Crompton, Andrew. 2001. The fractal nature of the everyday environment. *Environment and Planning B: Planning and Design* 28: 243–254. doi:10.1068/b2729.

Crompton, Andrew. 2002. Fractals and picturesque composition. *Environment and Planning B: Planning and Design* 29: 451–459. doi:10.1068/b12822.

Crompton, Andrew. 2012. The entropy of LEGO. *Environment and Planning B: Planning and Design* 39: 174–182. doi:10.1068/b37063.

Crompton, Andrew, and Frank Brown. 2008. A statistical examination of visual depth in building elevations. *Environment and Planning B: Planning and Design* 35: 337–348. doi:10.1068/b33084.

Crosbie, Michael J. 2011. *New York dozen: Gen X architects*. Mulgrave, VIC: Images Publishing.

Curtis, William J. R. 1986. *Le Corbusier: ideas and forms*. London: Phaidon.

Dal Co, Francesco. 1998. The world turned upside-down: The tortoise flies and the hare threatens the lion. In *Frank O. Gehry: The complete works*, ed. Francesco Dal Co and Kurt W. Forster, 39–61. New York: Monacelli Press.

Dal Co, Francesco, Hadley Arnold, and Kurt Walter Forster. 2003. *Frank O. Gehry: the complete works*. London: Phaidon.

Dantas, Jorge R. 2010. The end of euclidean geometry or its alternate uses in computer design. Proceedings of the 14th Congress of the Iberoamerican Society of Digital Graphics, SIGraDi 2010, Bogotá, Colombia, November 17-19, 161–164. Bogotá: Universidad de los Andes.

Da Silva, David, Frédéric Boudon, Christophe Godin, Olivier Puech, Colin Smith, and Hervé Sinoquet. 2006. A critical appraisal of the box counting method to assess the fractal dimension of tree crowns. *Lecture Notes in Computer Science* 4291: 751–760.

Davidson, Cynthia C., 2006. *Tracing Eisenman: Peter Eisenman complete works*. London: Thames and Hudson.

Dawes, Michael J., and Michael J. Ostwald. 2014. Prospect-Refuge theory and the textile-block houses of Frank Lloyd Wright: An analysis of spatio-visual characteristics using isovists. *Building and Environment* 80: 228–240.

Dawes, Michael J., and Michael J. Ostwald. 2015. Testing the “Wright Space”: using isovists to analyse prospect-refuge characteristics in Usonian architecture. *The Journal of Architecture* 19: 645–666.

Debailleux, Laurent. 2010. Complementary approach for vernacular wooden frame structures reconstruction. In *Digital Heritage, Third International Euro-Mediterranean Conference, Euromed 2010, Lemessos, Cyprus, November 8-13*, 441–449. Berlin: Springer. doi: 10.1007/978-3-642-16873-4

De Jonge, Krista, and Koen Van Balen 2002. Preparatory architectural investigation in the restoration of historical buildings. Leuven: Leuven University Press.

Deleuze, Gilles, and Felix Guattari. 1987. *A thousand plateaus: Capitalism and schizophrenia*. Translated by Brian Massumi. Minneapolis: University of Minnesota Press.

De Long, David G. 1994. Foreword. In *Wright in Hollywood: visions of a new architecture*, Robert L. Sweeny, xii–xvii. New York: Architectural History Foundation.

De Vico, Gionata, Vincenzo Peretti, and Gabriele A. Losa. 2009. Fractal organization of feline oocyte cytoplasm. *European Journal of Histochemistry* 49: 151–156.

Dieudonne, Jean A. 1994. *A history of algebraic and differential topology*. Boston: Birkhäuser.

Drew, Philip. 1985. *Leaves of iron: Glenn Murcutt, pioneer of an Australian architectural form*. North Ryde, NSW: Collins/Angus & Robertson.

Drew, Philip. 1994. Murcutt, Glenn Marcus. In *Contemporary architects*, ed. Muriel Emanuel, 3rd ed., 685. London: Macmillan.

Drew, Philip. 2000. Peter Stutchbury: of people and places: between the bush and the beach. Sydney: Pesaro Publishing.

Drexler, Arthur. 1975. Preface. In *Five architects: Eisenman, Graves, Gwathmey, Hejduk, Meier.*, ed. Peter Eisenman, Michael Graves, Charles Gwathmey, John Hejduk, Richard Meier, and Kenneth Frampton, 1. New York: Oxford University Press.

Eaton, Leonard. 1998. Fractal geometry in the late work of Frank Lloyd Wright. In *Nexus II: Architecture and Mathematics*, ed. Kim Williams, 23–38. Fucecchio, FI: Edizioni dell'Erba.

Edgar, Gerald. 2008. *Measure, Topology, and Fractal Geometry*. New York: Springer.

Ediz, Özgür, and Michael J. Ostwald. 2012. The Süleymaniye Mosque: a computational fractal analysis of visual complexity and layering in Sinan's masterwork. *arq: Architectural Research Quarterly* 16: 171–182. doi:10.1017/S1359135512000474.

Eglash, Ron. 1999. *African fractals: Modern computing and indigenous design*. New Brunswick: Rutgers University Press.

Eilenberger, Gert. 1986. Freedom, science, and aesthetics. In *The Beauty of Fractals*, ed. Heinz-Otto Peitgen and Peter H. Richter, 175–180. New York: Springer-Verlag.

Eisenman, Peter, Mildred S. Friedman, and Walker Art Center. 1970. *Notes on conceptual architecture: towards a definition*. Minneapolis: Walker Art Center.

Eisenman, Peter. 1975. Cardboard Architecture. In *Five architects: Eisenman, Graves, Gwathmey, Hejduk, Meier*, ed. Peter Eisenman, Michael Graves, Charles Gwathmey, John Hejduk, Richard Meier, Colin Rowe, and Kenneth Frampton, 15–17. New York: Oxford University Press.

Eisenman, Peter. 1979a. Preface. In *John Hejduk, 7 houses: January 22 to February 16, 1980*, John Hejduk. New York: Institute for Architecture and Urban Studies.

Eisenman, Peter. 1979b. In my father's house are many mansions. In *John Hejduk, 7 houses: January 22 to February 16, 1980*, John Hejduk, 9–20. New York: Institute for Architecture and Urban Studies.

Eisenman, Peter. 1984. The futility of objects: decomposition and the processes of difference. *The Harvard architecture review* 3: 64–81.

Eisenman, Peter. 1993. Folding in time. *AD profile: Folding in architecture* 102: 22–25.

Eisenman, Peter. 1999. *Diagram diaries*. New York: Universe Pub.

Eisenman, Peter. 2004. Digital scrambler: From index to codex. *Perspecta* 35: 40–53.

Eisenman, Peter, and Stephen Dobney. 1995. *Eisenman Architects: selected and current works*. Mulgrave, VIC: Images Publishing

Eisenman, Peter, Michael Graves, Charles Gwathmey, John Hejduk, Richard Meier, Colin Rowe, and Kenneth Frampton. 1975. *Five architects: Eisenman, Graves, Gwathmey, Hejduk, Meier*. New York: Oxford University Press.

Eisenman, Peter, Rosalind E. Krauss, and Manfredo Tafuri. 1987. *Houses of cards*. New York: Oxford University Press.

Eisenman, Peter, and Ariane Lourie Harrison. 2008. *Ten canonical buildings 1950-2000*. New York: Rizzoli.

Eiteljorg, Harrison. 2002. How should we measure an ancient structure? *Nexus Network Journal* 4: 14–20.

Emerson, Sheila. 1991. The authorisation of form: Ruskin and the science of chaos. In *Chaos and Order: Complex Dynamics in Literature and Science*, ed. Katherine N. Hayles, 149–166. Chicago: University of Chicago Press.

Encarnação, Sara, Marcos Gaudiano, Francisco C. Santos, José A. Tenedório, and Jorge M. Pacheco. 2012. Fractal cartography of urban areas. *Scientific Reports* 2: 1–5. doi:10.1038/srep00527.

Evans, Robin. 1995. *The projective cast: Architecture and its three geometries*. Cambridge: MIT Press.

Farmer, John, and Kenneth Richardson. 1996. *Green shift: towards a green sensibility in architecture*. Boston: Butterworth-Heinemann.

Farrelly, Elizabeth M. 1993. *Three houses: Glenn Murcutt*. London: Phaidon Press.

Feder, Jens. 1988. *Fractals*. New York: Plenum Press.

Feininger, Andreas. 1956. *The anatomy of nature: How function shapes the form and design of animate and inanimate structures throughout the universe*. Berlin: Thomas Yoseloff.

Fellows, Richard, and Anita Liu. 1997. *Research Methods for Construction*. Oxford: Blackwell.

Fernandez-Galiano, Luis. 2007. SAANA in dreams. In *Houses: Kazuyo Sejima + Ryue Nishizawa, SANAA*, ed. Chermayeff, Sam, Agustín Pérez Rubio, Tomoko Sakamoto, Luis Fernández-Galiano. 175–180. Barcelona: ACTAR.

Ferrero, Giovanni, Celestina Cotti, Michela Rossi, and Cecilia Tedeschi. 2009. Geometries of imaginary space: Architectural developments of the ideas of M. C. Escher and Buckminster Fuller. *Nexus Network Journal* 11: 305–316. doi:10.1007/s00004-008-0090-1.

Feyerabend, Paul. 1975. *Against method: Outline of an anarchistic theory of knowledge*. Atlantic Highlands: Humanities Press.

Fierro, Anette. 1997. Popular construction: Handbooks for home fronts. *Journal of Architectural Education* 51: 16–22.

Foroutan-pour, Kayhan, Pierre Dutilleul, and Donald L. Smith. 1999. Advances in the implementation of the box-counting method of fractal dimension estimation. *Applied Mathematics and Computation* 105: 195–210. doi:10.1016/S0096-3003(98)10096-6

Forster, Kurt W. 1998. Architectural choreography. In *Frank O. Gehry: The complete works*, ed. Francesco Dal Co and Kurt W. Forster, 9–38. New York: Monacelli Press.

Frampton, Kenneth. 1975. Frontality vs rotation. In *Five architects: Eisenman, Graves, Gwathmey, Hejduk, Meier*, ed. Peter Eisenman, Michael Graves, Charles Gwathmey, John Hejduk, Richard Meier, Colin Rowe, and Kenneth Frampton, 9–13. New York: Oxford University Press.

Frampton, Kenneth. 1976. Introduction. In *Richard Meier, architect: buildings and projects, 1966-1976*, Richard Meier. New York: Oxford University Press.

Frampton, Kenneth. 1992. *Modern architecture: a critical history, 3rd edition*. London: Thames & Hudson.

Frampton, Kenneth. 2005. The text-tile tectonic. In *On and by Frank Lloyd Wright: A primer of architectural principles*, ed. Robert McCarter, 170–189. London: Phaidon.

Frampton, Kenneth. 2006. *Glenn Murcutt, architect*. Sydney: 01 Editions.

Frankhauser, Pierre. 2008. Fractal geometry for measuring and modelling urban patterns. In *The Dynamics of Complex Urban Systems*, ed. Sergio Albeverio, Denise Andrey, Paolo Giordano, and Alberto Vancheri, 213–243. Heidelberg: Physica-Verlag HD.

Frascari, Marco, and Livio Volpi Ghirardini. 1998. Contra divinam proportion. In *Nexus II: Architecture and Mathematics*, ed. Kim Williams, 65–74. Fucecchio, FI: Edizioni dell'Erba.

Frederick, Matthew. 2007. *101 things I learned in architecture school*. Cambridge: MIT Press.

Friedman, Alice T. 1998. *Women and the making of the modern house: a social and architectural history*. New York: Abrams.

Friedman, Mildred S. 2009. *Frank Gehry: the houses*. New York: Rizzoli.

Fromonot, Françoise. 1995. *Glenn Murcutt: buildings and projects*. London: Thames & Hudson.

Fujimori, Terunobu. 2010. The origins of Atelier Bow-Wow's gaze. In *Behaviorology*, Yoshiharu Atelier Bow Wow, 122–129. New York: Rizzoli.

Fuller, Peter. 1987. Towards a new nature for the gothic. In *Art and Design Profile: The Post Modern Object*, ed. Andreas Papadakis, 5–10. Cambridge: Harvard University Press.

Futagawa, Yukio, and Bruce Brooks Pfeiffer. 1984. *Frank Lloyd Wright Monograph 1942-1950*. Vol. 7. Tokyo: A.D.A. Edita.

Futagawa, Yukio, and Bruce Brooks Pfeiffer. 1985a. *Frank Lloyd Wright Monograph 1914-1923*. Vol. 4. Tokyo: A.D.A. Edita.

Futagawa, Yukio, and Bruce Brooks Pfeiffer. 1985b. *Frank Lloyd Wright Monograph 1924-1936*. Vol. 5. Tokyo: A.D.A. Edita.

Futagawa, Yukio, and Bruce Brooks Pfeiffer. 1985c. *Frank Lloyd Wright Monograph 1951-1959*. Vol. 8. Tokyo: A.D.A. Edita.

Futagawa, Yukio, and Bruce Brooks Pfeiffer. 1987a. *Frank Lloyd Wright Monograph 1887-1901*. Vol. 1. Tokyo: A.D.A. Edita.

Futagawa, Yukio, and Bruce Brooks Pfeiffer. 1987b. *Frank Lloyd Wright Monograph 1907-1913*. Vol. 3. Tokyo: A.D.A. Edita.

Galofaro, Luca. 1999. *Digital Eisenman: An office of the electronic era*. Basel: Birkhäuser.

Gandelsonas, Mario. 1982. From structure to subject: the formation of an architectural language. In *House X*, Peter Eisenman, 7–30. New York: Rizzoli.

Gans, Deborah. 2000. *The Le Corbusier guide*. New York: Princeton Architectural Press.

Garner, Philippe. 1993. *Eileen Gray: design and architecture*: 1878-1976. Köln: Taschen.

Gehry, Frank O, Tim Street-Porter, Germano Celant, Mason Andrew, Ted Bickford, and Peter Arnell. 1990. *Frank Gehry, buildings and projects*. New York: Rizzoli.

Gelernter, Mark, and Virginia Dubrucq. 2004. Regionalism. In *Encyclopedia of 20th century architecture*, ed. R. Stephen Sennott, 1089–1092. New York: Fitzroy Dearborn.

Gero, John. 1998. Research in design computing: an artificial intelligence framework. In *International Conference on Artificial Intelligence for Engineering*, ed. X. Huang, S. Yang, and H. Wu, 5–12. China: HUST Press.

Ghirardo, Diane. 2003. *Architecture after modernism*. London: Thames & Hudson.

Goad, Philip. 2012. Regionalism. In *The encyclopedia of Australian architecture*, ed. Philip Goad and Julie Willis, 589. Port Melbourne, VIC: Cambridge University Press.

Goldberger, Paul. 1996. The dance of composition. In *Richard Meier: houses*, Richard Meier, 10–13. London: Thames & Hudson.

Golub, Spencer. 2004. Clockwise-counterclockwise (The vowelless revolution). *Theatre Journal* 56: 183–203.

Górski, A. Z, S. Drozdz, A. Mokrzycka, and J. Pawlik. 2012. Accuracy analysis of the box-counting algorithm. *Acta Physica Polonica* A 121: B28–B30.

Gössel, Peter, and Gabriele Leuthäuser. 2012. *Architecture in the 20th century*. London: Taschen.

Götze, Heinz. 1996. Friedrich II and the love of geometry. In *Nexus: Architecture and Mathematics*, ed. Kim Williams. 67–79. Fucecchio, FI: Edizioni dell'Erba.

Gozubuyuk, Gaye, Gulen Cagdas, and Ozgur Ediz. 2006. Fractal based design model for different architectural languages. In *The architecture co-laboratory: GameSetandMatch II : on computer games, advanced geometries, and digital technologies*, ed. Kas Oosterhuis and Lukas Feireiss, 280–286. Delft: episode publishers.

Graham, Dan. 2012. Capricorn. *Domus* 954: 144.

Grau, Juan B., Valeriano Mendeza, Ana M. Tarquis, M. Cruz Diaz, and A. Saab. 2006. Comparison of gliding box and box-counting methods in soil image analysis. *Geoderma* 134: 349–359.

Gray, Noel. 1991. Critique and a science for the sake of art: Fractals and the visual arts. *Leonardo* 24: 318.

Greenberg, Clement. 1961. *Art and culture; critical essays*. Boston: Beacon Press.

Grillo, Paul Jacques. 1975. *Form, function, and design*. New York: Dover Publications.

Groat, Linda, and David Wang. 2002. *Architectural research methods*. New York: John Wiley and Sons.

Gusheh, Maryam. 2008. *The architecture of Glenn Murcutt*. Tokyo: TOTO Shuppan.

Gutman, Robert. 1972. *People and buildings*, New York: Basic Books.

Hanson, Julienne. 1998. *Decoding homes and houses*. Cambridge: Cambridge University Press.

Hanson, Niel, and Antony Radford. 1986a. Living on the edge: a grammar for some country houses by Glenn Murcutt. *Architecture Australia* 75: 66–73.
Hanson, Neil, and Antony Radford. 1986b. On modelling the work of the architect Glenn Murcutt. *Design Computing* 1986: 189–203.
Harris, James. 2007. Integrated function systems and organic architecture from Wright to Mondrian. *Nexus Network Journal* 9: 93–102. doi:10.1007/s00004-006-0031-9.
Harvey, David. 1990. *The condition of postmodernity: an enquiry into the origins of cultural change*. Cambridge, MAS: Blackwell.
Hasegawa, Yūko. 2006. *Kazuyo Sejima + Ryue Nishizawa: SANAA*. Milan: Electa Architecture.
Hasegawa, Yūko. 2007. Radical practices in constructing relationships. In *Houses: Kazuyo Sejima + Ryue Nishizawa, SANAA*, ed. Chermayeff, Sam, Agustín Pérez Rubio, Tomoko Sakamoto, Luis Fernández-Galiano. 181–189. Barcelona: ACTAR.
Hays, K. Michael. 2002. Sanctuaries, the last works of John Hejduk: selections from the John Hedjuk archive at the Canadian Centre for Architecture, Montreal and the Menil collection, Houston. New York: Whitney Museum of American Art.
Heath, Tom, Sandy G. Smith, and Bill Lim. 2000. Tall buildings and the urban skyline: The effect of visual complexity on preferences. *Environment and Behavior* 32: 541–556.
Hebra, Alexius J. 2010. *The Physics of Metrology*. New York: Springer.
Hecker, Stefan, and Christian F. Müller. 1993. *Eileen Gray*. Barcelona: G. Gili.
Hein, Carola. 2004. Tokyo, Japan. Edited by R. Stephen Sennott. *Encyclopedia of 20th century architecture*. 1328–1332. New York: Fitzroy Dearborn.
Hejduk, John. 1974. *Fabrications*. New York: Cooper Union School of Art and Architecture.
Hejduk, John. 1979. *John Hejduk, 7 houses: January 22 to February 16, 1980*. New York: Institute for Architecture and Urban Studies.
Hejduk, John. 1985. *Mask of Medusa: works, 1947-1983*. New York: Rizzoli.
Hejduk, John, and Kim Shkapich. 1993. *Soundings: a work*. New York: Rizzoli.
Hejduk, John, and Kim Shkapich. 1995. *Adjusting foundations*. New York: Monacelli Press.
Hersey, George L. 1999. *The monumental impulse: Architecture's biological roots*. Cambridge: The MIT Press.
Hess, Alan, and Alan Weintraub. 2006. *Organic architecture: the other modernism*. Layton, UT: Gibbs Smith.
Hildebrand, Grant. 1991. *The Wright space: pattern and meaning in Frank Lloyd Wright's houses*. Seattle: University of Washington Press.
Hillier, Bill. 1996. *Space is the Machine*. Cambridge: Cambridge University Press.
Hillier, Bill, and Julienne Hanson. 1984. *The social logic of space*. New York: Cambridge University Press.
Hitchcock, Henry-Russell, and Philip Johnson. 1995. *The international style*. New York: W.W. Norton.
Hochman, Elaine S. 1990. *Architects of fortune: Mies van der Rohe and the Third Reich*. New York: Fromm International Pub. Corp.
Hoffmann, Donald. 1995. *Understanding Frank Lloyd Wright's architecture*. New York: Dover Publications.
Holl, Steven. 2010. Architecture surrounds you, in the same way as music surrounds you. *ArchIdea* 42: 4–11.
Horkheimer, Max. 1947. *Eclipse of reason*. Oxford: Oxford University Press.
Huang, Qian, Jacob R. Lorch, and Richard C. Dubes. 1994. Can the fractal dimension of images be measured? *Pattern Recognition* 27: 339–349.
Hudnut, Joseph. 1945. The post-modern house. *Architectural Record* 97: 70–75.
Isenberg, Barbara. 2009. *Conversations with Frank Gehry*. New York: Alfred A. Knopf.
Ito, Toyo. 1996. Diagram architecture. *El croquis* 77: 18–24.
Jacobs, Jane. 1961. *The death and life of great American cities*. New York: Random House.
Jahn, Graham, and Patrick Bingham-Hall. 1997. *Sydney architecture*. Sydney: Watermark Press.
James, John. 1981. *The contractors of Chartres*. London: Croom Helm.

James, John. 1982. *Chartres: the masons who built a Legend*. London: Routledge & Keegan Paul.

Jameson, Fredric. 1991. *Postmodernism, or, the cultural logic of late capitalism*. Durham: Duke University Press.

Jelinek, Herbert F., Guy N. Elston, and B. Zietsch. 2005. Fractal analysis: pitfalls and revelations in neuroscience. *Fractals in biology and medicine* IV: 85–94.

Jencks, Charles. 1973. *Modern movements in architecture*. Garden City, NY: Anchor Press.

Jencks, Charles. 1977. *The language of post-modern architecture*. New York: Rizzoli.

Jencks, Charles. 1992. *The post-modern reader*. New York: St. Martin's Press.

Jencks, Charles. 1995. *The architecture of the jumping universe: a polemic : how complexity science is changing architecture and culture*. London: Academy Editions.

Jencks, Charles. 2000. *Le Corbusier and the continual revolution in architecture*. New York: Monacelli Press.

Jencks, Charles, and Nathan Silver. 1972. *Adhocism; the case for improvisation*. New York: Doubleday.

Jodidio, Philip. 1995. *Richard Meier*. Köln: Taschen.

Joedicke, Jurgen. 1997. Organic architecture. In *The Thames & Hudson dictionary of 20th-century architecture*, Vittorio Magnago Lampugnani, tran. Barry Bergdoll, 253–255. London: Thames & Hudson.

Johnson, Anna. 2008. *The Australian house*. Balmain, NSW: Pesaro Publishing.

Johnson, Paul-Alan. 1994. *The theory of architecture: Concepts, themes and practices*. New York: Van Nostrand Reinhold.

Jones, J. Christopher. 1992. *Design methods*. New York: Van Nostrand Reinhold.

Jones, Will. 2014. *How To Read houses*. Rizzoli: New York.

Joy, Rick. 2010. A Tenacious Respect. In *Peter Stutchbury: selected projects*, Patrick Bingham-Hall, 4–5. Sydney: Pesaro Pub.

Joye, Yannick. 2006. An interdisciplinary argument for natural morphologies in architectural design. *Environment and Planning B: Planning and Design* 33: 239–252.

Joye, Yannick. 2011. A review of the presence and use of fractal geometry in architectural design. *Environment and Planning B: Planning and Design* 38: 814–828. doi:10.1068/b36032.

Joye, Yannick, and Philip Van Locke. 2007. Motivating biomorphic constructions based on complex systems science. *Systems Research and Behavioral Science* 24: 103–114. doi:10.1002/sres.742.

Karlen, Mark. 2009. *Space Planning Basics*. New Jersey: John Wiley and Sons.

Kaye, B H. 1989. *A Random Walk Through Fractal Dimensions*. Weinheim: VCH Publishers.

Kelbaugh, Doug. 2002. *Repairing the American Metropolis: Common Place Revisited*. Washington: University of Washington Press.

Kengo Kuma & Associates. 2010. *Studies in organic Architecture*. Tokyo: TOTO.

Kent, Susan. 1990. *Domestic architecture and the use of space: an interdisciplinary cross-cultural study*. Cambridge: Cambridge University Press.

Khan, Hasan-Uddin. 2009. *International style: modernist architecture from 1925 to 1965*. Köln: Taschen.

Kim, Kwang Sik, Lucien Benguigui, and Maria Marinov. 2003. The fractal structure of Seoul's public transportation system. *Cities* 20: 31–39. doi:10.1016/S0264-2751(02)00094-X.

King, Peter. 2005. *The common place: The ordinary experience of housing*. London: Ashgate.

Kipnis, Jeffrey. 1993. *Towards a new architecture*. *Architectural Design Profile: Folding in Architecture* 102: 40–49.

Kira, Moriko. 2000. *Japan, Towards totalscape: Contemporary Japanese architecture, Urban planning and landscape*. Rotterdam: NAi Publishers.

Kitchley, Jinu Louishidha. 2003. Fractals in architecture. *Architecture & design* 20: 42–48.

Kleinman, Kent, and Leslie Van Duzer. 2005. *Mies van der Rohe: the Krefeld villas*. New York: Princeton Architectural Press.

Knight, Terry Weissman. 1992. Designing with grammars. In *Computer-aided architectural design*, ed. Gerhard N. Schmitt, 33–48. Wiesbaden: Verlag Viewag.

Kobayashi, Yoshihiro, and Subhadha Battina. 2005. Housing layout design using fractals. In *Learning from the past a foundation for the future, CAAD futures 2005, Vienna, Austria, June 20-22*, 119–128. Wien: Österreichischer Kunst- und Kulturverlag.

Koch, Heinrich P. 1993. The concept of fractals in the pharmaceutical sciences. *Die Pharmazie* 48: 643–659.

Kostof, Spiro. 1977. The practice of architecture in the ancient world: Egypt and Greece. In *The architect: Chapters in the history of the profession*, ed. Spiro Kostof. New York: Oxford University Press.

Kostof, Spiro. 1985. *A history of architecture: Settings and rituals*. New York: Oxford University Press.

Kroll, Lucien. 1986. *The architecture of complexity*. London: B.T. Batsford Ltd.

Kruft, Hanno-Walter. 1994. *A history of architectural theory: from Vitruvius to the present*. New York: Princeton Architectural Press.

Kruger, Anton. 1996. Implementation of a fast box-counting algorithm. *Computer Physics Communications* 98: 224–234.

Kubala, Thomas. 1990. Finding Sullivan's Thread. *Progressive Architecture* 71: 102–104.

Kuhlmann, Dorte. 2008. Old and new organicism in architecture: The metamorphoses of an aesthetic idea. In *On growth and form: organic architecture and beyond*, ed. Sarah Bonnemaison and Philip Beesley. Halifax: TUNS Press.

Kuhn, Thomas. 1962. *The structure of scientific revolutions*. Chicago: University of Chicago Press.

Kurokawa, Kishō. 2000. *Kisho Kurokawa, architect and associates: Selected and current works*. Mulgrave, VIC: Images Publishing.

Kwinter, Sanford. 1995. The Eisenman wave. In *Eisenman Architects : Selected and current works*, ed. Stephen Dobney, 7–19. Mulgrave, VIC: Images Publishing.

Lakatos, Imre. 1976. *Proofs and refutations*. Cambridge: Cambridge University Press.

Laseau, Paul, and James Tice. 1992. *Frank Lloyd Wright: Between Principle and Form*. New York: Van Nostrand Reinhold.

Latour, Bruno. 1987. *Science in action: How to follow scientists and engineers through society*. Cambridge: Harvard University Press.

Lavin, Sylvia. 2009. Forgetting Frank. In *Frank Gehry: the Houses*, Mildred S. Friedman, 9–13. New York: Rizzoli.

Leatherbarrow, David. 2000. *Uncommon ground*. Cambridge: MIT Press.

Le Corbusier. 1923. *Vers une architecture*. Paris: G. Crès.

Levine, Neil. 2008. Introduction. In *Modern architecture being the Kahn lectures for 1930*, Frank Lloyd Wright and Neil Levine, ix–xxi. Princeton: Princeton University Press.

Lewis, Miles. ed. 2010. Architectura: Elements of Architectural Style. New York: Barron's.

Li, Jian, Qian Du, and Caixin Sun. 2009. An improved box-counting method for image fractal dimension estimation. *Pattern Recognition* 42: 2460–2469. doi:10.1016/j.patcog.2009.03.001.

Liang, Jiang, Yanqin Hu, and Hui Sun. 2013. The design evaluation of the green space layout of urban squares based on fractal theory. *Nexus Network Journal* 15: 33–49. doi:10.1007/s00004-012-0135-3.

Libeskind, Daniel. 1985. Stars at high noon. In *Mask of Medusa: works*, 1947-1983, John Hejduk, 9–14. New York: Rizzoli.

Linder, Mark. 2004. *Nothing less than literal: architecture after minimalism*. Cambridge: MIT Press.

Lorenz, Wolfgang. 2003. Fractals and fractal architecture. Masters Diss. Vienna: Vienna University of Technology.

Lorenz, Wolfgang E. 2009. Fractal geometry of architecture: Implementation of the box-counting method in a CAD-software. *Computation: The New Realm of Architectural Design. eCAADe 2009. Istanbul, Turkey, 16-19 September*, 697-704. Istanbul: eCAADe and ITU/YTU.

Lorenz, Wolfgang. 2011. Fractal geometry of architecture. In *Biomimetics - Materials, structures and processes: Examples, ideas and case studies*, ed. Petra Gruber, Dietmar Bruckner, Christian Hellmich, Heinz-Bodo Schmiedmayer, Herbert Stachelberger, and Ille C. Gebeshuber, 179–200. Berlin: Springer.

Lorenz, Wolfgang E. 2012. Estimating the fractal dimension of architecture: Using two measurement methods implemented in AutoCAD by VBA. In *Digital Physicality, eCAADe 2012, Prague, Czech Republic, 12 -14 September*, 505–513. Prague: ČVUT.

Lu, Yongmei, and Junmei Tang. 2004. Fractal dimension of a transportation network and its relationship with urban growth: a study of the Dallas - Fort Worth area. *Environment and Planning B: Planning and Design* 31: 895–911. doi:10.1068/b3163.

Lucie-Smith, Edward. 2003. *The Thames & Hudson dictionary of art terms. New edition*. World of Art. London: Thames & Hudson.

Lynn, Greg. 2004. Architecture versus sculpture. In *Barefoot on white-hot walls*, ed. Peter Noever, 160–167. Ostfildern: Hatje Cantz.

MacCormac, Richard. 2005. Form and philosophy : Froebel's kindergarten training and Wright's early work. In *On and by Frank Lloyd Wright: A primer of architectural principles*, ed. Robert McCarter, 124–143. London: Phaidon.

MacMahon, Bill. 2001. *The architecture of East Australia: an architectural history in 432 individual presentations*. Stuttgart: Edition A. Menges.

Mallgrave, Harry Francis, and David Goodman. 2011. *Introduction to architectural theory: 1968 to the present*. Malden MA: Wiley-Blackwell.

Mandelbrot, Benoit B. 1975. *Les objets fractals: forme, hasard et dimension*. Paris: Flammarion.

Mandelbrot, Benoit B. 1977. *The fractal geometry of nature*. New York: W.H. Freeman.

Mandelbrot, Benoit B. 1982. *The fractal geometry of nature*. Revised edition. San Francisco: Freeman.

Manning, Henry Parker. 1956. *Geometry of four dimensions*. New York: Dover Publications.

Marsault, Xavier. 2005. Generation of textures and geometric pseudo-urban models with the aid of IFS. In *Chaos and complexity: New Research*, ed. Franco F. Orsucci and Nicoletta Sala, 89–102. ebook: NOVA Publishers.

Martignoni, Jimena. 2008. Fractals of nature: the renovated botanical gardens of Medellín meld with the city and play with natural forms. *Landscape architecture* 98: 52.

Masi, F. 1996. *The pantheon as an astronomical instrument*. Rome: Edizioni Internazionali di Letteratura e Scienze.

Masucci, Paolo, Kiril Stanilov, and Michael Batty. 2012. Simple laws of urban growth. *arXiv*:1206.5298.

Mayer, Ralph. 1969. *A dictionary of art terms and techniques*. London: Black.

McCarter, Robert. 1999. *Frank Lloyd Wright*. London: Phaidon.

McEoin, Ewan. 2011. *Under the edge: the architecture of Peter Stutchbury*. Brooklyn, NSW: Architecture Foundation Australia.

McGillick, Paul. 2005. *Sydney architecture: a short history of Sydney architecture*. Balmain, NSW: Pesaro Pub.

Mehta, Geeta K., and Deanna MacDonald. 2011. *New Japan architecture: recent works by the world's leading architects*. Tokyo: Tuttle.

Meier, Richard. 1972. *Villa Savoye, Poissy, France, 1929-31*. Tokyo: A.D.A. EDITA.

Meier, Richard. 1984. *Richard Meier, architect, 1964/1984*. New York: Rizzoli.

Meier, Richard. 1999. *Building the Getty*. Berkeley: University of California Press.

Meier, Richard. 2003. Turning point. In *Richard Meier*, ed. Kenneth Frampton, 452–453. Milano: Electa.

Meier, Richard, Paul Goldberger, and Richard George Rogers. 1996. *Richard Meier: houses*. London: Thames & Hudson.

Meisel, Lawrence V., and Mark A. Johnson. 1997. Convergence of numerical box-counting and correlation integral multifractal analysis techniques. *Pattern Recognition* 30: 1565–1570.

Meuwissen, Joost. 2008. Aldo In wonderland: Remarks on the houses of Aldo van Eyck. *OASE* 75: 137–157.

Mical, Thomas. 2004. Hejduk, John 1929 - 2000. In *Encyclopedia of 20th century architecture, Volume 2*, ed. Stephen Sennott, 599–600. New York: Fitzroy Dearborn.

Minnite, Dianne L. 2001. Chronology. In *Out of the ordinary: Robert Venturi, Denise Scott Brown and Associates : architecture, urbanism, desig*n, David Bruce Brownlee, David G. De Long, and Kathryn B. Hiesinger. Philadelphia: Philadelphia Museum of Art in association with Yale University Press.

Mizuno, Setsuko, and Hidekazu Kakei. 1990. Fractal analysis of street forms. *Journal of architecture, planning and environmental engineering* 8: 103–108.

Moore, Charles Willard. 2004. *You have to pay for the public life: selected essays of Charles W. Moore*. Cambridge: MIT.

Moore, Charles Willard, Gerald Allen, and Donlyn Lyndon. 1974. *The place of houses*. New York: Holt, Rinehart and Winston.

Moore, R. John, and Michael J. Ostwald. 1996. The Ruskinian tradition and the science of chaos. In *Traditions and modernity in Asia and beyond*, ed. A. Kelly, K. Bieda, J. F. Zhu, and W. Dewanto, 455–466. Jakarta: Mercu Buana University.

Moore, R. John, and Michael J. Ostwald. 1997. Choral dance: Ruskin and Dædalus. *Assemblage* 32: 88–107.

Morton, David. 1975. Douglas house. *Progressive architecture* July: 38–45.

Murcutt, Glenn. 1991. Foreword. In *Leaves of iron: Glenn Murcutt, pioneer of an Australian architectural form*, ed. Philip Drew, 7–8. North Ryde, NSW: Collins/Angus & Robertson.

Murcutt, Glenn. 2007. Thoughts on the ecology of architecture. *Architecture and Urbanism* 8: 26–27.

Murcutt, Glenn. 2008. *Glenn Murcutt: Thinking drawing, working drawing*. Shohan. Tōkyō: TOTO Shuppan.

Murcutt, Glenn. 2012. Glenn Murcutt 1980: feathers of metal. *El Coquis* 163-164.

Murphy, Elizabeth. 1980. Gray, Eileen. In *Contemporary architects*, ed. Muriel Emanuel, 306. London: Macmillan.

Nango, Yoshikazu. 2010. Atelier Bow-Wow's approach to urban and architectural research. In *Behaviorology*, Yoshiharu Atelier Bow Wow, 321–340. New York: Rizzoli.

Nicolin, Pierluigi. 1998. The tao of Sejima. *Lotus international*: 6–31.

Norberg-Schulz, Christian. 1971. *Baroque architecture*. New York: Harry N. Abrams.

Norberg-Schulz, Christian. 1980. *Genius loci: towards a phenomenology of architecture*. New York: Rizzoli.

Novak, Miroslav Michal. 2006. *Complexus mundi: Emergent patterns in nature*. New Jersey: World Scientific.

Nute, Kevin. 2008. Functional versus purposive in the organic forms of Louis Sullivan and Frank Lloyd Wright. In *On growth and form: organic architecture and beyond*, ed. Sarah Bonnemaison and Philip Beesley, 44–53. Halifax: TUNS Press.

Obrist, Hans Ulrich. 2008. The father of long tails interview with Benoît Mandelbrot. In *EDGE. org*, ed. J. Brockman. http://edge.org/3rd_culture/obrist10/obrist10_index.html. Accessed 19 June 2015.

Ockman, Joan. 2011. Forward. In *IAUS, the Institute for Architecture and Urban Studies: an insider's memoir : with 27 other insider accoun*ts, Suzanne S. Frank, i–iii. Bloomington: Author House.

Oku, Toshinobu. 1990. On visual complexity on the urban skyline. *Journal of Planning, Architecture and Environmental Engineering* 6: 61–71.

Oliver, Julie. 2001. Spirit of place. *Houses*: 94–107.

Oleschko, Klaudia, Rosa Brambila, Fernando Brambila, Jean-Francois Parrot, and Pedro López. 2000. Fractal analysis of Teotihuacan, Mexico. *Journal of Archaeological Science* 27: 1007–1016. doi:10.1006/jasc.1999.0509.

Ostwald, Michael J. 1998a. Fractal traces: Geometry and the architecture of Ushida Findlay. In *Ushida Findlay*, ed. Leon van Schaik, 136–143. Barcelona: 2G.

Ostwald, Michael J. 1998b. Aperiodic tiling, Penrose tiling and the generation of architectural forms. In *Nexus II: Architecture and Mathematics*, ed. Kim Williams, 99–111. Fucecchio, FI: Edizioni dell'Erba.

Ostwald, Michael J. 2000. *Movements in twentieth century architecture: A reader in architectural theory*. Sydney: Archadia Press.

Ostwald, Michael J. 2001a. "Fractal architecture": Late twentieth century connections between architecture and fractal geometry. *Nexus Network Journal, Architecture and Mathematics* 3: 73–84.

Ostwald, Michael J. 2001b. Under siege: The Golden Mean. *Nexus Network Journal, Architecture and Mathematics* 2: 75–83.

Ostwald, Michael J. 2003. Fractal architecture: The philosophical implications of an iterative design process. *Communciation and Cognition* 36: 263–295.

Ostwald, Michael J. 2004. Freedom of form: Ethics and aesthetics in digital architecture. *The Philosophical Forum: Special issue on Ethics and Architecture* XXXV: 201–220.

Ostwald, Michael J. 2006. *The architecture of the new baroque: A comparative study of the historic and new Baroque movements in architecture*. Singapore: Global Arts.

Ostwald, Michael J. 2009. *Fractal architecture: knowledge formation within and between architecture and the sciences of complexity*. Saarbrucken: VDM Verlag.

Ostwald, Michael J. 2010a. The politics of fractal geometry in Russian paper architecture: The Intelligent Market and the Cube of Infinity. *Architectural Theory Review* 15: 125–137. doi: 10.1080/13264826.2010.495447.

Ostwald, Michael J. 2010b. Ethics and the auto-generative design process. *Building Research and Information* 38: 390–400.

Ostwald, Michael J. 2011a. Examining the relationship between topology and geometry: A configurational analysis of the rural houses (1984-2005) of Glenn Murcutt. *The Journal of Space Syntax* 2: 223–246.

Ostwald, Michael J. 2011b. A justified plan graph analysis of the early houses (1975-1982) of Glenn Murcutt. *Nexus Network Journal* 13: 737–762. doi:10.1007/s00004-011-0089-x.

Ostwald, Michael J. 2013. The fractal analysis of architecture: calibrating the box-counting method using scaling coefficient and grid disposition variables. *Environment and Planning B: Planning and Design* 40: 644–663. doi:10.1068/b38124.

Ostwald, Michael J., and Michael J. Dawes. 2013a. Miesian intersections: comparing and evaluating graph theory approaches to architectural spatial analysis. In *Graph theory: new research*, ed. Alessandra Cavalcante, 37–86. New York: Nova Scientific.

Ostwald, Michael J., and Michael J. Dawes. 2013b. Prospect-refuge patterns in Frank Lloyd Wright's Prairie houses: Using isovist fields to examine the evidence. *The Journal of Space Syntax* 4: 136–159.

Ostwald, Michael J., and Özgür Ediz. 2015. Measuring form, ornament and materiality in Sinan's Kılıç Ali Paşa Mosque: an analysis using fractal dimensions. *Nexus Network Journal* 17: 5–22. doi:10.1007/s00004-014-0219-3.

Ostwald, Michael J., and Steven Fleming. 2000. Modernism: Functionalism. In *Movements in twentieth century architecture: a reader in architectural theory*, ed. Michael J. Ostwald, 41–60. Sydney: Archadia Press.

Ostwald, Michael J., and R. John Moore. 1996a. Spreading chaos: Hayles' theory and an architecture of complexity. *Transition* 52-53: 36–53.

Ostwald, Michael J., and R. John Moore. 1996b. Fractalesque architecture: an analysis of the grounds for excluding Mies van der Rohe from the oeuvre. In *Traditions and Modernity in Asia*, ed. A. Kelly, K. Bieda, J. F. Zhu, and W. Dewanto, 437–453. Jakarta: Mercu Buana University.

Ostwald, Michael J., and R. John Moore. 1997. Unravelling the weave: An analysis of architectural metaphors in nonlinear dynamics. *Interstices* 4: 1–25.

Ostwald, Michael J., and Josephine Vaughan. 2008. Determining the fractal dimension of the architecture of Eileen Gray. In *Innovation, inspiration and instruction, ANZAScA 08, Newcastle, Australia, 26-28 November*, 9–16. Newcastle, NSW: ANZAScA and the University of Newcastle Australia.

Ostwald, Michael J., and Josephine Vaughan. 2009a. Calculating visual complexity in Peter Eisenman's architecture. In *Between man and machine integration/intuition/intelligence, CAADRIA 2009, Yunlin, Taiwan, 22-24 April*, 75–84. Yunlin: National Yunlin University of Science & Technology Department of Digital Media Design.

Ostwald, Michael J., and Josephine Vaughan. 2009b. Visual qualities in early modern and late modern architecture: A mathematical comparison of formal complexity in the houses of Gray and Sejima. In *Computing, cognition and education: Recent research in the architectural sciences*, ed. Ning Gu, Michael J. Ostwald, and Anthony Williams, 9–32. Sydney: ANZAScA.

Ostwald, Michael J., and Josephine Vaughan. 2010. The mathematics of style in the architecture of Frank Lloyd Wright: A computational, fractal analysis of formal complexity in fifteen domestic designs. In *Built environment: Design management and applications*, ed. Paul S. Geller, 63–88. Hauppauge: NOVA Science Publishers.

Ostwald, Michael J., and Josephine Vaughan. 2011. The mathematics of domestic modernism (1922-1934): An analysis of correlations between façade complexity, orientation, address and permeability. *Design Principles and Practices: An International Journal* 4: 143–162.

Ostwald, Michael J., and Josephine Vaughan. 2013a. Differentiating the Whites: Using fractal analysis to investigate the visual properties of the architecture of John Hejduk and Peter Eisenman. *Empirical Studies of the Arts* 31: 21–43.

Ostwald, Michael J., and Josephine Vaughan. 2013b. Limits and errors optimising image pre-processing standards for architectural fractal analysis. *ArS Architecture Science* 7: 1–19.

Ostwald, Michael J., Josephine Vaughan, and Stephan Chalup. 2009. A computational investigation into the fractal dimensions of the architecture of Kazuyo Sejima. *Design Principles and Practices: An International Journal* 3: 231–244.

Ostwald, Michael J., Josephine Vaughan, and Chris Tucker. 2008. Characteristic visual complexity: Fractal dimensions in the architecture of Frank Lloyd Wright and Le Corbusier. In *Nexus VII: Architecture and Mathematics*, 217–231. Turin: Kim Williams Books.

Ostwald, Michael J., and Stephen R. Wassell. 2002. Dynamic symmetries. *Nexus Network Journal, Architecture and Mathematics* 4: 123–131.

Pallasmaa, Juhani. 2012. *The eyes of the skin: architecture and the senses*. New York: John Wiley & Sons.

Park, Steven, and Le Corbusier. 2012. *Le Corbusier redrawn: the houses*. New York : Princeton Architectural Press.

Paul, James. 2004. *Introduction to the philosophies of research and criticism in education and the social sciences*. New Jersey: Prentice Hall.

Pierpont, James. 1930. Non-Euclidean geometry, a retrospect. *Bulletin of the American Mathematical Society* 36: 66–76.

Pearson, David. 2001. *The breaking wave: New organic architecture*. Stroud: Gaia.

Pears, A. R. 1975. *Dimension Theory of General Spaces*. Cambridge: Cambridge University Press.

Peitgen, Heinz-Otto, and Peter H. Richter 1986. *The beauty of fractals: Images of complex dynamical systems*. New York: Springer-Verlag.

Pogrebin, Robin. 2013. Bid for Pritzker Prize to acknowledge Denise Scott Brown. *The New York Times*, April 17.

Poppeliers, John C., and S. Allen Chambers. 2003. *What style is it: A guide to american architecture*. New Jersey: John Wiley and Sons.

Popper, Karl. 1959. *The logic of scientific discovery*. London: Hutchison and Co.

PourNejatian, Nima M., and Mohammad Mahdi Nayebi. 2010. Fractal based detection using blind box-counting method in high resolution radars. In *Proceedings of the 7th European Radar Conference, EuRAD 2010, Paris, France, 30 September–1 October*, 407–410. Piscataway, NJ: IEEE.

Prestinenza Puglisi, Luigi. 2008. *New directions in contemporary architecture: evolutions and revolutions in building design since 1988*. Hoboken, NJ : Wiley.

Puente, Moises, and Anna Puyuelo. 2009. *Mies Van Der Rohe: Houses*. Barcelona: 2G.

Radford, Antony, and Tarkko Oksala. 2007. Alvar Aalto and the expression of discontinuity. *The Journal of Architecture* 12: 257–280. doi:10.1080/13602360701469986.

Rawlings, Ashley. 2007. Political Architecture: Interview with Kisho Kurokawa. *Tokyo Art Beat*. http://www.tokyoartbeat.com/tablog/entries.en/2007/07/political-architecture-interview-with-kisho-kurokawa.html. Accessed 10 October 2012.

Rian, Iasef Md, Jin-Ho Park, Hyung Uk Ahn, and Dongkuk Chang. 2007. Fractal geometry as the synthesis of Hindu cosmology in Kandariya Mahadev temple, Khajuraho. *Building and Environment* 42: 4093–4107. doi:10.1016/j.buildenv.2007.01.028.

Richardson, Phyllis. 2004. *New sacred architecture*. London: Laurence King Publishing.

Risebero, Bill. 1982. *Modern architecture and design: an alternative history*. London: Herbert Press.

Rodin, Vladimir, and Elena Rodina. 2000. The Fractal Dimension of Tokyo's Streets. *Fractals* 8: 413–418.

Rosell, Quim. 2005. *Minimalist interiors*. New York: Harper Design.

Rowe, Colin, and Robert Slutzky. 1963. Transparency: Literal and Phenomenal. *Perspecta* 8: 45–54. doi:10.2307/1566901.

Rowe, Colin. 1996. *As I was saying: recollections and miscellaneous essays*. Volume 2. Cambridge: MIT Press.

Roy, Ankur, Edmund Perfect, William M. Dunne, and Larry McKay. 2007. Fractal Characterization of fracture networks: An improved box-counting technique. *Journal of Geophysical Research* 112: 1–9.

Ruby, Ilka, and Andreas Ruby. 2003. Trans-minimalism. In *Minimal architecture*, ed. Angeli Sachs, Philip Ursprung, Ilka Ruby, and Andreas Ruby, 120–169. Munich: Prestel Verlag.

Rybczynski, Witold. 2002. The Bilbao effect. *The Atlantic Monthly* 290: 138–142.

Rykwert, Joseph. 1971. Eileen Gray: Two houses and an interior, 1926-1933. *Perspecta* 13: 66–73.

Rykwert, Joseph. 1997. Meier. In *The Thames & Hudson encyclopaedia of 20th century architecture*, ed. Vittorio Magnago Lampugnani, 213–214. London: Thames & Hudson.

Sachs, Angeli. 2003. Essential Minimalism. In *Minimal architecture*, ed. Ilka Ruby, Andreas Ruby, Angeli Sachs, and Philip Ursprung, 28–67. Munich: Prestel Verlag.

Sakai, Satoshi, Motonori Nakamura, K. Furuya, N. Amemura, M. Onishi, I. Iizawa, J. Nakata, K. Yamaji, R. Asano, and Kuwata Tamotsu. 2012. Sierpinski's forest: New technology of cool roof with fractal shapes. *Energy and Buildings*. doi:10.1016/j.enbuild.2011.11.052.

Sala, Nicoletta. 2002. Fractals in architecture : Some examples. In *Fractals in Biology and Medicine*, ed. Gabriele A. Losa, Merlini Danilo, Theo F. Nonnenmacher, and Ewald R. Weibel, III:347–358. Basel: Birkhäuser.

Saleri, Renato. 2005. Pseudo-Urban automatic pattern generation. *Chaos in art and architecture - International Journal of Dynamical Systems Research* 1: 24.1–24.12.

Salingaros, Nikos A. 1998. A scientific basis for creating architectural forms. *Journal of Architectural and Planning Research* 15: 283–293. aft.

Salingaros, Nikos A. 2004. *Anti-architecture and Deconstruction*. Solingen: Umbau-Verlag Harald Püschel.

Salingaros, Nikos Angelos. 2005. *Principles of urban structure, Volume 4*. Amsterdam: Techne Press.

Salingaros, Nikos A. 2007. *A theory of architecture*. Solingen: Umbau-Verlag Harald Püschel.

Samper, Albert, and Blas Herrera. 2014. The fractal pattern of the French Gothic cathedrals. *Nexus Network Journal*: 16: 251–271.

Sarkar, Nirupam, and B.B. Chaudhuri. 1994. An efficient differential box-counting approach to compute fractal dimension of image. *IEEE Transactions on Systems, Man and Cybernetics*. 24: 115–120.

Schmitt, G. 1988. Expert systems and interactive fractal generators in design and evaluation. *Proceedings of the 2nd International conference on computer aided architectural design Futures, CAAD futures '87, Eindhoven, The Netherlands, 20-22 May*, 91-106. Amsterdam: Elselvier.

Schroeder, Manfred Robert. 1991. *Fractals, chaos, power laws: Minutes from an infinite paradise*. New York: W.H. Freeman.

Schulze, Franz, and Edward Windhorst. 2012. *Mies Van Der Rohe: A critical biography, new and revised edition*. Chicago: University of Chicago Press.

Scully, Vincent Joseph, and Christopher Mead. 1989. *The architecture of Robert Venturi*. Albuquerque , NM: University of New Mexico.

Sedrez, Maycon, and Alice Pereira. 2012. Fractal shape. *Nexus Network Journal* 14: 97–107. doi:10.1007/s00004-011-0099-8.

Sejima, Kazuyo. 1996. Y-House. *El Coquis* 77i: 62–69.

Sejima, Kazuyo. 1999. Explanation of the projects. *Japan Architect* 35:118–119.

Sejima, Kazuyo. 2004a. House in a Plum Grove. *El Croquis* 121/122: 278–297.

Sejima, Kazuyo. 2004b. Small House. *El Croquis* 121/122: 242–256.

Sergeant, John. 2005. Warp and woof : a spatial analysis of Wright's Usonian houses . In *On and by Frank Lloyd Wright: A Primer of Architectural Principles*, ed. Robert McCarter. 190–203. London: Phaidon.

Sherman, Daniel J. 2011. *French Primitivism and the ends of empire, 1945-1975*. Chicago: University of Chicago Press.

Sirowy, Beata. 2012. Among paradigms: Major ways of framing user-related problems in contemporary architectural discourse. *FORMakademisk* 5: 81–111.

Smith, C. Ray. 1980. Eisenman, Peter D. In *Contemporary architects*, ed. Muriel Emanuel, 231–232. London: Macmillan.

Sokal, Alan, and Jean Bricmont. 1998. *Intellectual impostures: Postmodern philosophers' abuse of science*. London: Profile.

Sommerville, Duncan M'Laren Young. 1958. *An introduction to the geometry of n dimensions*. New York: Dover.

Somol, Robert E. 1998. Still crazy after all these years. *Assemblage* 36: 84–92. doi:10.2307/3171366.

Sorkin, Michael. 1999. Frozen light. In *Gehry talks: architecture + process*, Mildred S. Friedman, 29–39. New York: Rizzoli.

Spence Roy. 1986. At Bingie Point. *The Architectural Review* 1068: 70–75.

Speyer, A. James. 1968. *Mies Van der Rohe*. Chicago: Art institute of Chicago.

Stadler, Hilar, Martino Stierli, Peter Fischli, and Museum im Bellpark. 2008. *Las Vegas studio: images from the archives of Robert Venturi and Denise Scott Brown*. Zürich: Scheidegger & Spiess.

Stamps, Arthur Earl III. 1999. Architectural detail, Van der Laan septaves and pixel counts. *Design Studies* 20: 83–97. doi:10.1016/S0142-694X(98)00025-8.

Stamps, Arthur Earl III. 2002. Fractals, skylines, nature and beauty. *Landscape and Urban Planning* 60: 163–184.

Stanley, Eugene H., and Paul Meakin. 1988. Multifractal phenomena in physics and chemistry. *Nature* 335: 405–409.

Stern, Robert. 1973. Stompin' at the Savoye. The *Architectural Forum* 138: 46–48.

Steinfeld, Edward, and Jonathan White. 2010. Inclusive housing: a pattern book : design for diversity and equality. New York: W.W. Norton & Co.

Stiny, George. 1975. *Pictorial and formal aspects of shapes and shape grammars*. Basel: Birkhäuser.

Storrer, William Allin. 1974. *The architecture of Frank Lloyd Wright, a complete catalog*. Cambridge: MIT Press.

Storrer, William Allin. 1993. *The Frank Lloyd Wright companion*. Chicago: University of Chicago Press.

Stuart, James, and Nicholas Revett. 2007. *The antiquities of Athens: measured and delineated*. New York: Princeton Architectural Press.

Stutchbury, Peter. 2010. Invisible house. *GA Houses* 115: 134–136.

Stutchbury, Peter. 2011a. Paddock house. *GA Houses* 123: 68–79.

Stutchbury, Peter. 2011b. Billabong house. *GA Houses* 120: 170–173.

Suau, Cristian. 2008. Metsapaviljonski, Form follows wood. In *The challenge of change: dealing with the legacy of the modern movement*, ed. Dirk Van Den Heuvel, Maarten Mesman, Waldo Quist, and Bert Lemmens, 27–32. Delft: IOS Press.

Swallow, Peter, David Watt, and Robert Ashton. 2004. *Measurement and recording of historic buildings*. Shaftsbury: Donhead.

Sweeney, Robert L. 1994. *Wright in Hollywood: Visions of a new architecture*. New York: Architectural History Foundation.

Taylor, Charles C., and S. James Taylor. 1991. Estimating the dimension of a fractal. *Journal of the Royal Statistical Society. Series B (Methodological)* 53: 353–364.

Taylor, Jennifer. 1990. *Australian architecture since 1960, 2nd edition*. Red Hill, ACT: National Education Division, Royal Australian Institute of Architects.

Tegethoff, Wolf. 1985. *Mies van der Rohe: the villas and country houses*. New York: Museum of Modern Art.

Terzidis, Kostas. 2006. *Algorithmic architecture*. Amsterdam: Elsevier.

Tiezzi, Enzo. 2006. *Steps towards an evolutionary physics*. Southampton: WIT Press.

Thomas, Derek. 2012. *Architecture and the urban environment*. Hoboken, NJ: Routledge.

Thomas, Isabelle, and Pierre Frankhauser. 2013. Fractal dimensions of the built-up footprint: buildings versus roads. Fractal evidence from Antwerp (Belgium). *Environment and Planning B: Planning and Design* 40: 310–329. doi:10.1068/b38218.

Thomson, Iain. 1999. *Frank Lloyd Wright: A visual encyclopedia*. London: PRC.

Trachtenberg, Marvin, Isabelle Hyman, and Marvin Trachtenberg. 1986. *Architecture: from prehistory to postmodernity*. New York: H.N. Abrams.

Trivedi, Kirti. 1989. Hindu temples: Models of a fractal universe. The *Visual Computer* 5: 243–258. doi:10.1007/BF02153753.

Tsukui, Noriko, and Venturi Scott Brown and Associates. 2009. *Venturi and Scott Brown: What turns them on*. Tokyo: A+U.

Urford, Jacqueline. 2012. The Sydney School. In *The encyclopedia of Australian architecture*, 674–676. Port Melbourne, VIC: Cambridge University Press.

Ursprung, Philip. 2003. Minimalism and minimal art. In *Minimal architecture*, Ilka Ruby, Angeli Sachs, Philip Ursprung, and Andreas Ruby, 100–102. Munich: Prestel Verlag.

Vandenberg, Maritz. 2003. *Farnsworth house: Ludwig Mies van der Rohe*. London: Phaidon.

Van Tonder, Gert.J. 2006. Changing the urban façade through image superposition. In *Proceedings of the international symposium on architecture of habitat system for sustainable development*, Fukuoka, Japan, 16-17 October, 83–89. Fukuoka: Kyushu University.

Vaughan, Josephine, and Michael J. Ostwald. 2008. Approaching Euclidean limits: a fractal analysis of the architecture of Kazuyo Sejima. In *innovation inspiration and instruction: new knowledge in the architectural sciences: ANZAScA 08, Newcastle, Australia, 26-28 November*, 285–294. Newcastle, NSW: University of Newcastle.

Vaughan, Josephine, and Michael J. Ostwald. 2009a. Nature and architecture: revisiting the fractal connection in Amasya and Sea Ranch. In *Performative ecologies in the built environment: sustainability research across disciplines, ANZAScA 09, Launceston, Tasmania, 25–27 November*. 42. Launceston:UTAS.

Vaughan, Josephine, and Michael J. Ostwald. 2009b. Refining the computational method for the evaluation of visual complexity in architectural images: Significant lines in the early architecture of Le Corbusier. In *Computation: The New Realm of Architectural Design. eCAADe 2009. Istanbul, Turkey, 16-19 September*, 689–696. Istanbul: eCAADe and ITU/YTU.

Vaughan, Josephine, and Michael J. Ostwald. 2009c. A quantitative comparison between the formal complexity of Le Corbusier's pre-modern (1905-1912) and early modern (1922-1928) architecture. *Design Principles and Practices: An international journal* 3: 359–372.

Vaughan, Josephine, and Michael J. Ostwald. 2010a. Refining a computational fractal method of analysis: testing Bovill's architectural data. In *New frontiers : proceedings of the 15th International Conference on Computer-Aided Architectural Design Research in Asia, CAADRIA 2010, Hong Kong, Hong Kong, 7-10 April*, 29–38. Hong Kong: CAADRIA.

Vaughan, Josephine, and Michael J. Ostwald. 2010b. Using fractal analysis to compare the characteristic complexity of nature and architecture: re-examining the evidence. *Architectural Science Review* 53: 323–332.

Vaughan, Josephine, and Michael J. Ostwald. 2011. The relationship between the fractal dimension of plans and elevations in the architecture of Frank Lloyd Wright: comparing the Prairie Style, Textile Block and Usonian periods. *ArS Architecture Science* 4: 21–44.

Venturi, Robert. 1966. *Complexity and contradiction in architecture*. New York: The Museum of Modern Art.

Venturi, Robert, Denise Scott Brown, and Steven Izenour. 1972. *Learning from Las Vegas: the forgotten symbolism of architectural form*. Cambridge: MIT Press.

Venturi, Robert, Denise Scott Brown, and James Steele. 1992. *Venturi Scott Brown & Associates: on houses and housing*. London: Academy Editions.

Von Moos, Stanislaus. 1979. *Le Corbusier, elements of a synthesis*. Cambridge: MIT Press.

Von Moos, Stanislaus. 1987. *Venturi, Rauch, & Scott Brown buildings and projects*. New York: Rizzoli.

Von Moos, Stanislaus. 1999. *Venturi, Scott Brown & Associates: buildings and projects, 1986-1998*. New York: Monacelli Press.

Von Seidlein, Peter C. 1997. Sullivan. In *The Thames & Hudson encyclopaedia of 20th century architecture*, ed. Vittorio Magnago Lampugnani, 323–326. London: Thames & Hudson.

Voss, Richard F. 1986. Characterization and measurement of random fractals. *Physica Scripta* T13: 27–32. doi:10.1088/0031-8949/1986/T13/004.

Voss, Richard. 1988. Fractals in nature: From characterization to simulation. In *The science of fractal images*, ed. Heinz-Otto Peitgen, Dietmar Saupe, and Michael F. Barnsley, 21–70. New York: Springer-Verlag.

Vyzantiadou, M. A., A. V. Avdelas, and S. Zafiropoulos. 2007. The application of fractal geometry to the design of grid or reticulated shell structures. *Computer Aided Design*. 39: 51–59. doi:10.1016/j.cad.2006.09.004.

Wang, Haowei, Xiaodan Su, Cuiping Wang, and Rencai Dong. 2011. Fractal analysis of urban form as a tool for improving environmental quality. *International Journal of Sustainable Development & World Ecology* 18: 548–552. doi:10.1080/13504509.2011.603760.

Wang, Wen, Ma Xueqiang, and Liu Hong. 2008. A computer-aided harmonious architecture design method based on fractals. In *Natural Computation, 2008. ICNC '08. Fourth International Conference on*, 7:323–326. doi:10.1109/ICNC.2008.2.

Watt, D, and P Swallow. 1996. *Surveying historic buildings*. Shaftsbury: Donhead.

Wen, Kuo-Chung, and Yu-Neng Kao. 2005. An analytic study of architectural design style by fractal dimension method. In *2005 Proceedings of the 22nd ISARC*, 1–6. Ferrara, Italy: ISARC.

Westheimer, Gerald. 1991. Visual discrimination of fractal borders. *Proceedings: Biological Sciences* 243: 215–219. doi:10.2307/76616.

Whiting, Sarah. 2004. Eisenman, Peter D. 1932 -. In *Encyclopedia of 20th century architecture, Volume 1*, ed. Stephen Sennott, 1:394–396. New York: Fitzroy Dearborn.

Willis, Julie, and Philip Goad. 2012. Search for Australian Style. In *The encyclopedia of Australian architecture*, 55–56. Port Melbourne, Vic.: Cambridge University Press.

Willmert, Todd. 2006. The "ancient fire, the hearth of tradition": combustion and creation in Le Corbusier's studio residences. *arq: Architectural Research Quarterly* 10: 57–78. doi:10.1017/S1359135506000121.

Wright, Frank Lloyd. 1908. In the cause of architecture. *The Architectural Record* xxiii: 155–220.

Wright, Frank Lloyd. 1954. In the nature of materials: Integrity. In *On and by Frank Lloyd Wright: A Primer of Architectural Principles*, ed. Robert McCarter. London: New York.

Wright, Frank Lloyd. 1960. *Frank Lloyd Wright: Writings and buildings*. Cleveland, Ohio: World Publishing Co.

Xie, W, and W Xie. 1997. Fractal-based analysis of time series data and features extraction. *Journal of Signal Processing*. 13: 98–104.

Yamagishi, Ryoichi, Shigeru Uchida, and Shin-Ichi Kuga. 1988. An experimental study of complexity and order of street-vista. *Journal of architecture, planning and environmental engineering* 2: 27–35.

Yessios, Chris I. 1987. A fractal studio. In *Integrating computers into the architectural curriculum*. 169–182. Raleigh: ACADIA.

Yu, L, D Zhang, K Wang, and W Yang. 2005. Coarse iris classification using box-counting to estimate fractal dimensions. *Pattern Recognition* 38: 1791–1798.

Zarnowiecka, Jadwiga C. 1998. Chaos, databases and fractal dimension of regional architecture. In *Computerised craftsmanship in architectural education*, 267–270. Paris: eCAADe.

Zarnowiecka, Jadwiga C. 2002. In search of new computer tools: what does Bovill really measure in architecture? In *Connecting the real and the virtual - design e-ducation 20th eCAADe Conference Proceedings*, 342–345. Warsaw: eCAADe.

Zimmerman, Claire. 2006. *Mies van der Rohe: 1886-1969 : the structure of space*. Köln: Taschen.

Zukowsky, John, ed. 1998. *Japan 2000: architecture and design for the Japanese public*. Munich: Prestel.

Index

M.J. Ostwald and J. Vaughan, *The Fractal Dimension of Architecture*, Mathematics and the Built Environment 1, DOI 10.1007/978-3-319-32426-5